Studies in Computational Intelligence

Volume 488

Series Editor

J. Kacprzyk, Warsaw, Poland

For further volumes:
http://www.springer.com/series/7092

Abdelmalek Amine · Otmane Ait Mohamed
Ladjel Bellatreche
Editors

Modeling Approaches and Algorithms for Advanced Computer Applications

Editors
Abdelmalek Amine
Département d'informatique
Saida University
Centre Universitaire Taher Moulay de Saida
Algeria

Ladjel Bellatreche
LIAS/ISAE-ENSMA
Futuroscope Chasseneuil Cedex
France

Otmane Ait Mohamed
Department of Electrical and Computer Engineering
Concordia University
Montreal
Quebéc
Canada

ISSN 1860-949X ISSN 1860-9503 (electronic)
ISBN 978-3-319-03304-4 ISBN 978-3-319-00560-7 (eBook)
DOI 10.1007/978-3-319-00560-7
Springer Cham Heidelberg New York Dordrecht London

© Springer International Publishing Switzerland 2013
Softcover re-print of the Hardcover 1st edition 2013
This work is subject to copyright. All rights are reserved by the Publisher, whether the whole or part of the material is concerned, specifically the rights of translation, reprinting, reuse of illustrations, recitation, broadcasting, reproduction on microfilms or in any other physical way, and transmission or information storage and retrieval, electronic adaptation, computer software, or by similar or dissimilar methodology now known or hereafter developed. Exempted from this legal reservation are brief excerpts in connection with reviews or scholarly analysis or material supplied specifically for the purpose of being entered and executed on a computer system, for exclusive use by the purchaser of the work. Duplication of this publication or parts thereof is permitted only under the provisions of the Copyright Law of the Publisher's location, in its current version, and permission for use must always be obtained from Springer. Permissions for use may be obtained through RightsLink at the Copyright Clearance Center. Violations are liable to prosecution under the respective Copyright Law.
The use of general descriptive names, registered names, trademarks, service marks, etc. in this publication does not imply, even in the absence of a specific statement, that such names are exempt from the relevant protective laws and regulations and therefore free for general use.
While the advice and information in this book are believed to be true and accurate at the date of publication, neither the authors nor the editors nor the publisher can accept any legal responsibility for any errors or omissions that may be made. The publisher makes no warranty, express or implied, with respect to the material contained herein.

Printed on acid-free paper

Springer is part of Springer Science+Business Media (www.springer.com)

Preface

This volume constitues the proceedings of the Fourth International Conference on Computer Science and Its Applications (CIIA 2013), which was held during May 4–6, 2013 in Saida, Algeria. CIIA is focused on the various aspects of advances in computer science and its applications and features world renowned keynote speakers and excellent technical program addressing the technology landscape and current challenges in different topics distributed into four tracks: Computational Intelligence, Security & Network Technologies, Information Technology, and Computer Systems and Applications. This event is a great opportunity for local and foreign attendees to exchanges knowledge and experience in various fields of IT.

This year the program committee received 390 papers, each of which was refereed by two or three reviewers selected by the program committee. Of these submissions, 42 papers were accepted for presentation at the conference and publication in this volume.

The organizers are grateful to Yamine Ait Ameur, Frédéric Boniol, Mohand Boughanem and Abdelwahab Hamou-Lhadj for agreeing to give invited talks at CIIA 2013. An abstract of each invited talk in included in this proceeding volume.

We sincerely thank all authors who submitted papers to this conference. We were pleased indeed by the number and the quality of the submissions. We congratulate those whose proposal was accepted, and we hope that the comments of the reviewers have been constructive and encouraging for the other authors.

Given the large number of proofreading and the heavy responsibility, we would like to thank all the track-chairs and their respective Program Committee members for their excellent job during the selection process. Their help in finding and assigning reviewers is much appreciated. Clearly, all this would not have been possible without the valuable report provided by sub-reviewers.

This conference could not have been possible without the moral and financial support from the Taher Moulay University of Saida, la DG-RSDT, and lATRST.

We would like to thank our local organizers at Faculty of Science and Technology especially the Department of Mathematics and Computer Science for their help in many aspects of planning and running CIIA 2013.

Finally, we thank EasyChair Team for creating and supporting the outstanding EasyChair Conference Management system.

April 2013

Abdelmalek Amine
Otmane Ait Mohamed
Ladjel Bellatreche

Organization

Honorary Chairs

Berrezoug Belgoumene
Ghouti Djellouli

General Chair

Abdelmalek Amine

Program Committee Chairs

Otmane Ait Mohamed
Ladjel Bellatreche

Program Committee

Chouarfia Abdallah
El Hassan Abdelwahed
Abdelkader Adla
Sultan Ahmed
Mohamed Ahmed-Nacer
Yamine Ait Ameur
Idir Ait-Saadoune
Reza Akbarinia
Samir Aknine
Eisa Alanazi
Rashid Ali
Mahklouf Alliouat
Ilham Alloui
Abbes Amira
Abdelkrim Amirat
José Enrique Armendáriz-Iñigo
Yacine Atif
Maurizio Atzori
Madjid Ayache
Latifa Baba-Hamed
Atmani Baghdad
Amar Balla
Fatiha Barigou
Mahamed Batouche
Henri Bauer
Ghalem Belalem
Hafida Belbachir
Boualem Benatallah
Karima Benatchba
Reda Bendraoun
Djamal Benslimane
Sidi Mohamed Benslimane
Abdelkader Benyettou
Kamel Berkaoui
Elisa Bertino
Nik Bessis

Mohamed Bettaz
Azeddine Bilami
Ismaïl Biskri
Frederic Boniol
Thouraya Bouabana Tebibel
Karim Bouamrane
Lydia Boudjeloud
Nacer Boudjlida
Mahmoud Boufaida
Zizette Boufaida
Boubakeur Boufama
Mohand Boughanem
Aoued Boukelif
Abdellah Boukerram
Kamel Boukhalfa
Jalil Boukhobza
Azedine Boulmakoul
Omar Boussaid
Belattar Brahim
Laurence Capus
Hubert Cardot
Jesus Carretero
Eugenio Cesario
Brahim Chaib-Draa
Yllias Chali
Allaoua Chaoui
Maryline Chetto
Mohamed Amine Chikh
Salim Chikhi
Dickson Chiu
Annie Choquet-Geniet
Alain Crolotte
Alfredo Cuzzocrea
Pasqua D'Ambra
Mourad Debbabi
Amar Bensaber Djamel
Yassine Djouadi
Mahieddine Djoudi
Habiba Drias
Bourennane El-Bay
Abderrahim El-Qadi
Zakaria Elberrichi
Mohamed Erradi
Mostafa Ezziyyani
Nader Fahima
Sébastien Faucou
Eric Feron
Cherif Fodil
Benjamin C. M. Fung
Faiez Gargouri
Ahmad Ghazal
Abdelghani Ghomari
Nacira Ghoualmi-Zine
Claude Godart
Anastasios Gounaris
Emmanuel Grolleau
Zahia Guessoum
Fabrice Guillet
Mohand-Said Hacid
Abdenour Hadid
Allel Hadj-Ali
Haffaf Hafid
Djamila Hamdadou
Abdelwahab Hamou-Lhadj
Chihab Hanachi
Saad Harous
Sachio Hirokawa
Jérôme Hugues
Abdessamad Imine
Ali Jaoua
Stéphane Jean
Imed Kacem
Faraoun Kamel Mohamed
Rezaul Karim
Saroj Kaushik
Okba Kazar
Samir Kechid
Samee-Ullah Khan
Hamamache Kheddouci
Chabane Khentout
Mohamed-Khireddine Kholladi
Mohamed Tahar Kimour
Mouloud Koudil
Sanjay Kumar Madria
Yasine Lafifi
Mustapha Lalam
Farouk Lalaoui
Chaker Larabi
Sekhri Larbi
Ahmed Lehireche
Pascal Lienhardt
Wenming Liu
Giovanni Livraga
Moussa Lo
Haibing Lu
Sofian Maabout
Zidani Madjid
Jamal Malki
Mimoun Malki
Tiziana Margaria
Farhi Marir
Carlos Maziero
Mohamed El Bachir Menai
Takao Miura
Senouci Mohamed
Babahenini Mohamed Chaouki
Bandar Mohammad
Benmohammed Mohammed
Anirban Mondal
Mattia Monga
Malek Mouhoub
Nassereddine Mouhoub
Hassan Mountassir
Abdelouahab Moussaoui
Abdi Mustapha Kamel
Ahlem Nabli
Kazumi Nakamatsu
Ahmad Nesir-Zghoul
Lhouari Nourine
Mohammed Ouali
Mourad Chabane Oussalah
Claire Pagetti
Marc Pantel
Fernando-Silva Parreiras
Makan Pourzandi
Binod Kumar Prasad
Chalal Rachid

Abdellatif Rahmoun
Sivaram Rajeyyagari
Adjoudj Reda
Hamou Reda Mohamed
Pascal Richard
Paolo Rosso
Mohammed Ali Roula
Samira Sadaoui
Lakhdar Sais
Sherif Sakr
Abdel-Badeeh Salem
Loé Sanou
Sid-Ahmed Selouani
Benaoumeur Senouci
Hamid Seridi
Hossain Shahriar
Ana Simonet
Michel Simonet
Hala Skaf-Molli
Yahya Slimani
Anna Squicciarini
Amrudee Sukpan
Andrea Tagarelli
Noria Taghezout
Abdelmalik Taleb-Ahmed
Said Talhi
Domenico Talia
David Taniar
Rafael Tolosana
Paolo Trunfio
Trihn Tuan
Satya Valluri
Gilles Venturini
Lingyu Wang
Komminist Weldemariam
Virginie Wiels
Leandro-Krug Wives
Marek Wojciechowski
Robert Wrembel
Fatos Xhafa
Yinglong Xia
Belabbas Yagoubi
Benmimoun Youcef
Jessie Zhang
Bing Zhou
Samir Zidat
Abdelhafid Zitouni

Additional Reviewers

Abouenour, Lahsen
Al-Kateb, Mohammed
Alnafie, Emdjad
Amblard, Frédéric
Amirat, Abdelkrim
Belhoul, Yacine
Belkhodja, Leila
Ben Yaghlane, Boutheina
Benmoussa, Yahia
Bounhas, Ibrahim
Chouiref, Zahira
Daoui, Mehammed
Diaf, Naser
Douar, Brahim
Draa, Amer
Elmansouri, Raida
Franco Salvador, Marc
Huo, Wei
Jabbour, Said
Khalid, Osman
Khan, Atta Ur Rehman
Khundker, Nafisa
Kougka, Georgia
Lagares, Angel
Layeb, Abdesslem
Maya, Benabdelhafid Souilah
Memar, Sara
Miura, Takao
Motta, Jesus Antonio
Mourtada, Aref
Oliveira, Júlia
Olivier, Pierre
Oufella, Sarah
Ouhammou, Yassine
Ouzzif, Mohammed
Saidouni, Djamel Eddine
Salhi, Yakoub
Serhani, Mohamed
Shirai, Masato
Sonia, Gueraich
Souhila, Arib
Soumia, Bendekkoum
Tlili, Raja
Wakabayashi, Kei
Yahiaoui, Saïd

Contents

Invited Talks

Track 1: Computational Intelligence (CI)

Track 2: Security Network Technologies (SNT)

Track 4: Computer Systems and Applications (CSA)

New Challenges for Future Avionic Architectures

Frédéric Boniol

2 avenue Edaourd Belin
Toulouse, 31055
France
frederic.boniol@onera.fr

Abstract. Electronic sets operated on aircraft are usually summarized as "avionic architectures" (for "aviation electronic architecture"). Since the 70s, avionic architectures, composed of digital processing modules and communication buses, are supporting more and more avionic applications such as flight control, flight management, etc. Hence, avionic architectures have become a central component of an aircraft. They have to ensure a large variety of important requirements: safety, robustness to equipment failures, determinism, and real-time. In response to these requirements, aircraft manufacturers have proposed several solutions. This article has a twin objectives: firstly to survey the state of the art of existing avionic architectures, including the IMA (for Integrated Modular Avionic) architecture of the most recent aircraft; and secondly to discuss two challenges for the next generation of avionic architectures: reconfiguration capabilities, and integrating COTS processing equipment such as multi-core processors. We believe that these two challenges will be central to the next generation of IMA architectures (called IMA-2G for IMA-2d generation).

A. Amine et al. (Eds.): *Modeling Approaches and Algorithms*, SCI 488, p. 1.
DOI: 10.1007/978-3-319-00560-7_1 © Springer International Publishing Switzerland 2013

Checking System Substitutability: An Application to Interactive Systems

Yamine Ait Ameur[1] and Abdelkrim Chebieb[2]

[1] IRIT/ENSEEIHT
2 Rue Charles Camichel
31071 Toulouse Cedex 7
yamine@enseeiht.fr
[2] ESI
BP 68M OUED SMAR,
16270, EL HARRACH ALGER
k_chebieb@esi.dz

Abstract. The capability to substitute a given system by another one is a property useful for dealing with adaptation, maintenance, interoperability, reliability, etc. This talk proposes a formally based approach for checking the substitutability of a system by another one. It exploits the weak bi-simulation relationship.

In this talk a system is seen as a state-transition system. Two systems are observed to check if one may be substituted by the other preserving their behaviour. The weak bi-simulation relationship is revisited to handle systems that have different sets of labels by defining a relation on labels. A transformation of the systems to be compared is defined according to the relation defined on labels. Classical weak bi-simulation is then used to model check the substitutability property.

The approach is illustrated on the case of plastic interactive systems. We show how an interactive system supporting a set of interactive tasks can be replaced by another interactive system that performs the same tasks with different interaction devices. Relations on labels are borrowed from an ontology of interaction and of interaction devices. A case study will be used along the talk to illustrate how the proposed approach practically works.

A. Amine et al. (Eds.): *Modeling Approaches and Algorithms*, SCI 488, p. 3.
DOI: 10.1007/978-3-319-00560-7_2 © Springer International Publishing Switzerland 2013

The Role of Software Tracing in Software Maintenance

Abdelwahab Hamou-Lhadj

Software Behaviour Analysis (SBA) Research Lab
Concordia University, Montreal, QC, Canada
abdelw@ece.concordia.ca

Abstract. Our society depends greatly on software systems for many critical activities including finance, health, education, telecommunications, aerospace, and more. Maintaining these systems to keep up with new user needs and ever changing technologies is an important task, but also a challenging and costly one. Research shows that software maintenance activities can take up to 80% of the time and effort spent during the lifecycle of software. The constant challenge for software maintainers is to understand what a system does before making any changes to it. In an ideal situation, this understanding should come from system documentation, but, for a variety of reasons, maintaining sufficiently good documentation has been found to be impractical in many organizations.

In this talk, I will discuss techniques to aid software engineers in understanding software systems. I will focus on the techniques that permit the understanding of system behaviour through tracing. Tracing encompasses three main steps: run the system, observe what it does, and make sense of how and why it does it in a certain way. The major challenge is that typical traces can be overwhelmingly large, often millions of lines long. I will present a set of techniques that we have developed to simplify the analysis of large traces. I will report on lessons learned from industrial projects in which my research lab has played a leading role. I will conclude my talk with an outlook on future challenges and research directions.

A. Amine et al. (Eds.): *Modeling Approaches and Algorithms*, SCI 488, p. 5.
DOI: 10.1007/978-3-319-00560-7_3 © Springer International Publishing Switzerland 2013

Information Retrieval and Social Media

Mohand Boughanem

University Paul Sabatier of Toulouse - IRIT lab
Mohand.boughanem@irit.fr

The social Web (Web 2.0) changed the way people communicate, now a large number of online tools and platforms, such as participative encyclopedias (e.g., wikipedia.org), social bookmarking platforms (e.g., connotea.org from the Nature Publishing Group), public debate platforms (e.g., agoravox.fr), photo sharing platforms (e.g., flickr.com), micro blogging platforms (e.g., blogger.com, twitter.com), allow people to interact and to share contents. These tools provide to users the ability to express their opinions, to share content (photos, blog posts, videos, bookmarks, etc.); to connect with other users, either directly or via common interests often reflected by shared content; to add free-text tags or keywords to content; users comment on content items. All these user-generated contents need not only to be indexed and searched in effective and scalable ways, but they also provide a huge number of meaningful data, metadata that can be used as clues of evidences in a number of tasks related particularly to information retrieval. Indeed, these user-generated contents have several interesting properties, such as diversity, coverage and popularity that can be used as wisdom of crowds in search process. This talk will provide an overview of this research field. We particularly describe some properties and specificities of these data, some tasks that handle these data, we especially focus on two tasks namely searching in social media (ranking models for social IR, (micro)blog search, forum search, real time social search) and exploiting social data to improve a search.

A. Amine et al. (Eds.): *Modeling Approaches and Algorithms*, SCI 488, p. 7.
DOI: 10.1007/978-3-319-00560-7_4 © Springer International Publishing Switzerland 2013

Machine Learning Tool for Automatic ASA Detection

Mohammed El Amine Lazouni[1], Mostafa El Habib Daho[1], Nesma Settouti[1], Mohammed Amine Chikh[1], and Saïd Mahmoudi[2]

[1] Biomedical Engineering Laboratory. Tlemcen Algeria
[2] Computer Science Department - Faculty of Engineering - University of Mons - Belgium
mine_lazouni@yahoo.fr, mostafa.elhabibdaho@gmail.com, {nesma.settouti,mea_chikh}@mail.univ-tlemcen.dz, Said.Mahmoudi@umons.ac.be

Abstract. The application of machine learning tools has shown its advantages in medical aided decision. This paper presents the implementation of three supervised learning algorithms: the C4.5 decision tree classifier, the Support Vector Machines (SVM) and the Multilayer Perceptron MLP's in MATLAB environment, on the preoperative assessment database. The classification models were trained using a new database collected from 898 patients, each of whom being represented by 17 features and included in one among 4 classes. The patients in this database were selected from different private clinics and hospitals of western Algeria.

In this paper, the proposed system is devoted to the automatic detection of some typical features corresponding to the American Society of Anesthesiolo-gists sores (ASA scores). These characteristics are widely used by all Doctors Specialized in Anesthesia (DSA's) in pre-anesthesia examinations. Moreover, the robustness of our system was evaluated using a 10-fold cross-validation method and the results of the three proposed classifiers were compared.

Keywords: ASA score, DSA, Pre-anesthesia consultation, SVM, MLPs, C4.5.

1 Introduction

Risk is ubiquitous in medicine but anesthesia is an unusual specialty as it routinely involves deliberately placing the patient in a situation that is intrinsically full of risks. The Patient's safety depends on the management of those risks.The anesthetic risk classification is of prime importance not only in carrying out the day-to-day anesthetic practice but it coincides with surgical risks and morbidity condition as well [3].

In Algeria there are about 7000 DSA's, but this number is insufficient to insure all the tasks that have to be performed for the patient's safety [5]. The main problem is that despite their small number, their presence is indispensable in each hospital or clinic. Indeed, they have to insure the pre-anesthesia examinations of all patients who need general or local anesthesia. Moreover, they have to

A. Amine et al. (Eds.): *Modeling Approaches and Algorithms*, SCI 488, pp. 9–16.
DOI: 10.1007/978-3-319-00560-7_5 © Springer International Publishing Switzerland 2013

be present in the operating room during surgery and after that during the post-operative period. The realization of these different tasks is really hard to perform. That is why, we propose in this work an artificial intelligence based approach allowing to bring assistance to DSA's.

The related works in preoperative patient classification were carried out by Peter et al. in [6]. The authors of this work have developed an automatic instrument used for grading the level of the anesthetic patient risk, with a modified version presented by Hussman and Russell [6]. So far, the risk prediction has been carried out using statistical analysis tools, which lacks the desired precision [2].

The aim of the study prsented in this work is to apply and analyze three different machine learning methods: the C4.5 decision tree classifier, the Support vector Machines (SVM), and the Multilayer Perceptron techniques for the classification of the ASA physical status.

In this paper, we target two distinct objectives: the database construction and data classification. To this aim, we divide this work as follows. In section II, we describe the database used and we discuss its different parameters. After that in section III, we review some basic theory concepts. Section IV presents the experimental results and the discussion. Finally, we shall summarize the main points of our prototype and conclude the paper.

2 Data Collection

In this section we present the creation of the database. The database has been obtained with the help of the DSA's. The patients in this database were selected from

Table 1. Database parameters

Sex	488 males and 410 females
Age	between 2 months and 105 years
Medical backgrounds	Hypertension Diabetes Respiratory failure Heart failure (HF)
ECG	Heart rate 1 (bpm) Heart rate 2 (bpm) Heart rate3 (bpm) Steadiness of heart rate Pace maker Atrioventricular block Left ventricular hypertrophy
Oxygen saturation	Take a measure (%)
Blood sugar or blood glucose level	Take a measure (g/l)
Blood pressure (BP)	Systole (mmHg) Diastole (mmHg)
Classes	Physical Status according to the examination by the DSA

different private clinics and hospitals of western Algeria: TLEMCEN hospital, ORAN CANASTEL hospital, ORAN HAMMOU BOUTELILIS clinic, ORAN NOUR clinic, and TLEMCEN LAZOUNI clinic.

We have to note that the unavailability of a standardized database in this field prompted us to create this personal database. In all, 898 subjects were introduced in the data collection among whom 488 males and 410 females.

This database is devoted to the detection of the ASA physical status. It is characterized by 17 parameters presented in table1.

In table 2, we present the details of the database used according to ASA physical status.

Table 2. The clinical data related to the 898 patients of the database and their distribution according to ASA classes

ASA Physical Status	1	2	3	4
Number of patients	219	395	232	52
Mean age (year)	57,62	67,49	65,35	79,07
Mean heart rate 1 (bpm)	79,18	79,05	97,29	109
Mean heart rate 2 (bpm)	78,45	79,67	99,35	110
Mean heart rate 3 (bpm)	79,32	80,32	100,57	109
Mean oxygen saturation	98,58	98,75	93,58	89
Mean blood glucose level	1,25	1,48	2,89	3,42
Mean blood pressure (systole)	123	135	155	169
Mean blood pressure (diastole)	81	95	102	110

The ASA physical status allows to evaluate the anesthetic risks and to obtain a predictive parameter of surgical mortality. We have selected patients with ASA Physical Status 1, 2, 3 and 4. The number of patients where the ASA physical status is 4 is really small compared to those of ASA physical status 1, 2, and 3 because ASA physical status 4 are subject to a seversystemic disease that is a constant threat to life.

The output or classes for the database take the values '1', '2', '3', or '4'.

* '1' means that a patient is in ASA physical status 1.
* '2' means that a patient is in ASA physical status 2.
* '3' means that a patient is in ASA physical status 3.
* '4' means that a patient is in ASA physical status 4.

In claas '1', there are 219 patients that is 24.38% of the cases, in class '2', 395 patients that is 43.98% of the cases, in class '3', 232 patients that is 25.84% of the cases, and in class '4' 52 patients only that is 05.80% of the cases.

In 1939, a committee of the American Society of Anesthetists, Inc, composed of Drs Sakland, Rovenstine and Taylor, was empowered "to study, examine, experiment, and devise a system for the collection and tabulation of statistical data" in anesthesia [7]. With foresight, the committee anticipated future problems regarding definitions of and uses for the proposed classifications. Consequently, they proposed a system ready to modification that is similar in its essential elements to the one in use today. Their recommendations were published by the American Society of Anesthesiologists (ASA) in 1941 [7]. The physical status evaluation of a patient in agreement with the ASA classification is the first and very important step to predict the perioperative risks to which the patient will be submitted. The outcomes of surgical treatment and anesthetic methods are commonly stratified as ASA classes as given in Table 3.

Table 3. ASA Physical Status [3]

ASA Physical Status	Patient Status
1	Normal healthy patient
2	A patient with mild systemic disease (that does not limit activity)
3	A patient with severe systemic disease (limits activity, but not incapacitating)
4	A patient with severe systemic disease that is a constant threat to life
5	A moribund patient who is not expected to survive with or without the operation
6	A declared brain-dead patient whose organs are being removed for donor purposes

In this database, we could not select patients with ASA Physical Status 5 and 6 because they were dying.

3 Methods

In this work we have used three supervised learning algorithms: the C4.5 decision tree classifier, the Support Vector Machines (SVM) and the Multilayer Perceptron MLP's classification in order to detect the ASA physical status. The proposed system was evaluated using a 10-fold cross-validation method and the results of the three proposed classifiers were compared.

The preoperative assessment data analysis and the anesthetic risk classification (detection for the ASA physical status) were carried out using a MATLAB software. It is designed in a flexible manner to try out existing methods on new databases.

- The Support vector machines or support vector networks [1] are supervised learning models with associated learning algorithms that analyze data and recognize patterns, used for classification and regression analysis. A Support Vector Machine performs classification by constructing an N-dimensional

hyperplane that optimally separates the data into two categories. SVM models are closely related to the neural networks. In fact, a SVM model using a sigmoid kernel function is equivalent to a two-layer perceptron neural network. The support vector machines are known for their good performance in a multi-class system.
- The Multilayer Perceptron network is the most widely used neural network classifier. The MLPs are universal approximators. They are valuable tools in problems when one has little or no knowledge about the form of the relationship between input vectors and their corresponding outputs.
- The C4.5 decision tree learner produces decision tree models. The algorithm uses the greedy technique to induce decision trees for classification. A decision-tree model is built by analyzing training data and the model is used to classify unseen data. The C4.5 generates decision trees, where the nodes evaluate the existence or significance of individual features.

The Cross-validation, or rotation estimation, is a technique for assessing how the results of a statistical analysis will generalize to an independent data set. It is mainly used in settings where the goal is prediction, and one wants to estimate how accurately a predictive model will perform in practice. [4]

The database D is randomly splited into k mutually exclusive subsets (the folds) $D1, D2, \ldots Dk$ of approximately equal size. The inducer is trained and tested k times, each time $t \in 1, 2, ...k$. It is trained on D/Dt and tested on Dt.

The cross-validation estimate of accuracy=(The overall number of correct classification)/(The number of instances in the database).

Table 4. Predictive performance of the classifiers

	C4.5									
Model	#1	#2	#3	#4	#5	#6	#7	#8	#9	#10
Accuracy (%)	88.18	84.36	91.55	83.62	93.74	83.67	87.72	93.61	89.54	84.44
Accuracy average	88.01%									
Correctly classified instances	765									
incorrectly classified instances	133									
	MLPs									
Model	#1	#2	#3	#4	#5	#6	#7	#8	#9	#10
Accuracy (%)	86.31	94.25	91.77	95.71	85.49	94.99	93,45	90,71	92,41	89,22
Accuracy average	91.43%									
Correctly classified instances	791									
incorrectly classified instances	107									
	SVM									
Model	#1	#2	#3	#4	#5	#6	#7	#8	#9	#10
Accuracy (%)	90.36	93.61	89.75	96.81	96.59	92.77	95,47	92,11	90,68	94,31
Accuracy average	93.25%									
Correctly classified instances	808									
incorrectly classified instances	90									

4 Performance Evaluation

The results of the experiments are summarized in table 2. The performances of the three models were evaluated on the basis of two criteria: the prediction accuracy, and the error rate. These values and are illustrated in Figures 1, and 2.

As shown in Fig 1, and 2, the Support Vector Machines predict better than other algorithms. Among the three classifiers used for the experiment, the Multi-Layer Perceptron algorithm and the Support Vector Machines algorithm provides more or less the same prediction accuracy. The accuracy rate of the C4.5 decision tree algorithm is the lowest among the three machine learning techniques.

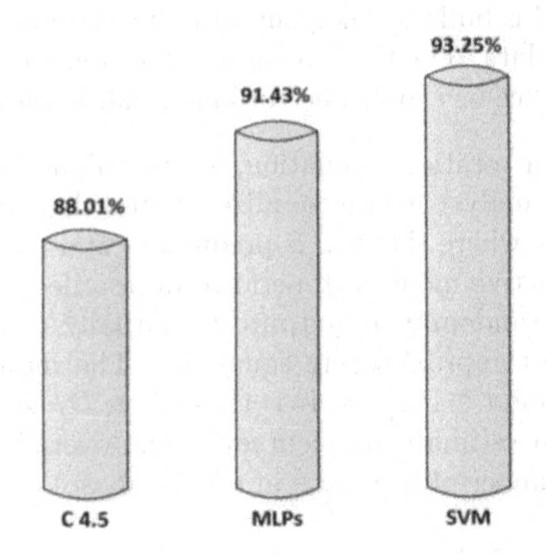

Fig. 1. Prediction accuracy histogram

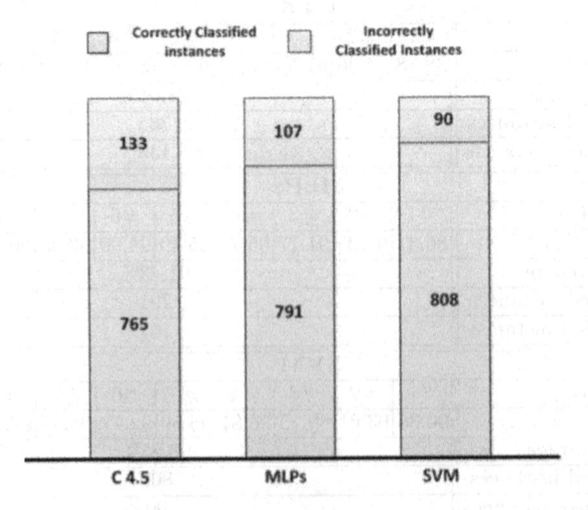

Fig. 2. Error rate histogram

Fig 2, presents the number of incorrectly classified instances, The SVM machine learning is the best system for this classification.

The performance of the learning techniques is highly dependent on the nature of the training data. The confusion matrix (table 5) was used to evaluate the classification error rate.

Table 5. Confusion matrix of the classifiers

		ASA1	ASA2	ASA3	ASA4
C4.5	ASA1	190	22	7	0
	ASA2	22	330	43	0
	ASA3	3	12	207	10
	ASA4	0	4	10	38
MLPs	ASA1	203	14	2	0
	ASA2	23	334	37	1
	ASA3	2	13	209	8
	ASA4	0	0	7	45
SVM	ASA1	210	8	1	0
	ASA2	20	339	34	2
	ASA3	0	8	214	10
	ASA4	0	1	6	45

From Table 5 it is shown that:

- 190 patients with ASA physical status 1 among 219 ones, 330 patients with ASA physical status 2 among 395, 207 patients with ASA physical status 3 among 232 and 38 patients with ASA physical status 4 among 52 were recognized correctly by the C4.5 classifier.
- 203 patients with ASA physical status 1 among 219, 334 patients with ASA physical status 2 among 395, 209 patients with ASA physical status 3 among 232 patients and 45 patients with ASA physical status 4 among 52 patients were recognized correctly by the MLP's.
- 210 patients with ASA physical status 1 among 219 patients, 339 patients with ASA physical status 2 among 395, 214 patients with ASA physical status 3 among 232 patients and 45 patients with ASA physical status 4 among 52 patients were recognized correctly by the SVM.

In the confusion matrix good results correspond to large numbers down the main diagonal and small numbers are found off the diagonal. From the confusion matrix given in Table 5, it is observed that SVM classifier gives better results than the MLP's classifier which its self gives better results than the C4.5 classifier. The results strongly suggest that machine learning can aid DSA's in the prediction of preoperative anesthetic risks. It is hoped that more interesting results will follow in further exploration of data.

5 Conclusion

In this paper, we have presented a medical decision support system based on three supervised learning algorithms: (1) C4.5 Decision tree classifier, (2) Support vector Machines (SVM) and (3) Multilayer Perceptron for helping Doctors specialized in Anesthesia to detect the ASA physical status. In particular, the system has been developed with 898 medical cases collected from patients in western Algeria.

Furthermore, the robustness of the proposed system was examined using a 10-fold cross-validation method and the results showed that the SVM-based decision support system is better compared to MLPs and C4.5 classifiers. The results could achieve average classification accuracy at 93.25%.

In this study, the parameter values related to the number of patients in the dataset, the distribution of samples and the anthropometric operation errors by different anesthetists and so on may fluctuate the results. A better solution for this issue is to get much more medical records to optimize the performance and stability of the system.

As future work, we also plan to improve the proposed model using some additional classifiers. We plan also to include more details on the proposed diagnosis such as accepted or refused patient for surgery, to choose the best anesthetic technique for surgery either general or local anesthesia, the patient's tracheal intubation either easy or hard, and finally the development of a fuzzy-based computer aided diagnosis for obtaining more detailed results.

References

1. Cristianini, N., Shawe-Taylor, J.: An Introduction to Support Vector Machines and other kernel-based learning methods. Cambridge University Press (2000)
2. Hussman, J., Russell, R.: Memorix: Surgery. Tech. rep., p. 66. Chapman & Hall Medical (1997)
3. Karpagavalli, S., Jamuna, K., Vijaya, M.: Machine learning approach for preoperative anaesthetic risk prediction. International Journal of Recent Trends in Engineering 1(2) (May 2009)
4. Kohavi, R.: A study of cross-validation and bootstrap for accuracy estimation and model selection. In: The Fourteenth International Joint Conference on Artificial Intelligence IJCAI (1995) ISSN 1137-1143
5. Lazouni, M.A., El Habib Daho, M., Chikh, M.A.: Un système multi-agent pour l aide au diagnostic en anesthésie. In: Biomedical Engineering International Conference (BIOMEIC 2012), Tlemcen, Algeria, October 10-11, p. 82 (2012) ISSN 2253-0886
6. Lutz, P.K.: The medical algorithms project, ch31. anaesthesiology, section: Preoperative patient classification and preparation. Online Excel 681-687 (2008)
7. Sakland, M.: Grading of patients for surgical procedures. Anesthesiology 2, 281–284 (1941)

Stator Faults Detection and Diagnosis in Reactor Coolant Pump Using Kohonen Self-organizing Map

Smail Haroun[1], Amirouche Nait Seghir[1], and Said Touati[2]

[1] Laboratoire des Systèmes Electriques et Industriels (LSEI),
Faculté d'électronique et Informatique, USTHB, Bp32 El alia Bab Ezzouar, Alger, Algérie
haroun.smail@yahoo.com, naitseghir_a@yahoo.fr
[2] Département de Génie électrique (DGE), Centre de Recherche Nucleaire de Birine (CRNB),
Bp 180, Ain Oussera, Algerie
saidtouati@yahoo.fr

Abstract. Nuclear power industries have increasing interest in using fault detection and diagnosis (FDD) methods to improve availability, reliability, and safety of nuclear power plants (NPP). In this paper, a procedure for stator fault detection and severity evaluation on reactor coolant pump (RCP) driven by induction motor is presented. Fault detection system is performed using unsupervised artificial neural networks: the so-called Self-Organizing Maps (SOM). Induction motor stator currents are measured, recorded, and used for feature extraction using Park transform, Zero crossing times signal, and the envelope, then statistical features are calculated from each signal which serves for feeding the neural network, in order to perform the fault diagnosis. This network is trained and validated on experimental data gathered from a three-phase squirrel-cage induction motor. It is demonstrated that the strategy is able to correctly identify the stator fault and safe cases. The system is also able to estimate the extent of the stator faults.

Keywords: Self-Organizing Map, Reactor coolant Pump, Fault Detection and diagnosis.

1 Introduction

The reactor coolant pump (RCP) is one of the kernel components of NPP, which is in charge of conveying the heat generated in the nuclear reactor core to the steam generator [1], the coolant pump is a centrifugal pump driven by a three phase induction motor. In spite of their robustness and high efficiency, the induction motors can be the seat of an important variety of failures [2]. One of the most widespread faults in the induction motor is the inter-turn short circuits in the stator winding. These incipient faults, if it left undetected, can lead to motor failure, and its performance will directly influence the safe and reliable operation of the RCP. Furthermore, if the pump failed, a hazard would be caused consequently. Therefore, predicting the developing faults in the pump quite necessary for achieving high reliability and safety [1].

A. Amine et al. (Eds.): *Modeling Approaches and Algorithms*, SCI 488, pp. 17–26.
DOI: 10.1007/978-3-319-00560-7_6 © Springer International Publishing Switzerland 2013

Many researchers have focused their attention on fault detection and diagnosis (FDD) in nuclear power plants [3].

Recently, Artificial Intelligence (AI) techniques have been proposed for the noninvasive fault detection in NPP and its equipment. These AI-based techniques include neural network [4-5], genetic algorithms [6] and pattern recognition by support vector machine [1], or the combination of different techniques like neural networks - Genetic Algorithms [7] and neural nets - Fuzzy logic [8].

The Kohonen Self-Organizing map combined with the current frequency spectrum analysis is used for identification of mechanical unbalances in [10], and for broken rotor bar fault of induction motor in [11].

In this paper, we present an automatic fault classification methodology for fault detection and diagnosis based on Self-Organizing Maps combined with signal processing methods applied to the three phase currents for motor stator faults detection and severity evaluation in a reactor coolant pump. The current preprocessing techniques used are extended Park Vector, zero crossing times, and the envelope of the three phase currents. Then statistical features are calculated from each signal which serves as input of the SOM. The network is trained and tested using experimental results on a real induction machine.

2 Self-organizing Maps

The Self-Organizing Map (also known as Kohonen map) is an unsupervised artificial neural network which is a powerful method for clustering and visualization of high dimensional data [9]. The SOM algorithm implements a nonlinear topology preserving mapping from a high-dimensional input data space onto a low dimension discrete space (usually 2D), called the topological map.

2.1 SOM Architecture

A SOM model is composed by two layers of neurons. One of them, called input layer (composed by N neurons, one for each input variable), is responsible for receiving and transmitting information from outside to the output layer. The output layer (formed by M neurons) is in charge of information processing and the construction of map features. Usually, neurons in the output layer are organized in two dimensional map [10], as shown in Fig. 1. Note that each neuron of the network is completely connected to all the nodes of the input layer. The number of neurons determines the accuracy and generalization capability of the SOM and it is determined by the heuristic equation (1).

$$M = 5\sqrt{N} \tag{1}$$

Where M is the number of neurons and N is the number of samples of the training data. According to Eq. (2), the ratio between side-lengths of the map (n_1 and n_2) is the square root of the ratio between the two biggest eigenvalues of the training data (e_1 and e_2) [11].

$$n_1/n_2 = \sqrt{e_1/e_2} \tag{2}$$

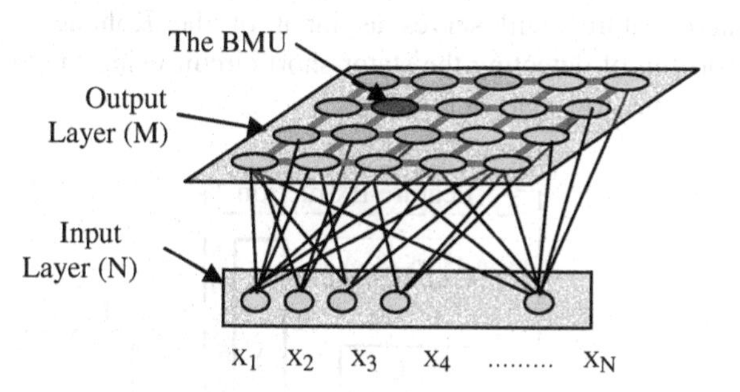

Fig. 1. SOM Architecture

2.2 Training Algorithm

During training procedure, the weight vectors are adapted in such a way that close observations in the input space would activate two close neurons of the SOM [12]. The SOM is trained iteratively. At each training step, a sample input data vectors X is randomly presented from the training data sets, and the distance between the data and all the weight vectors of the SOM is calculated. The neuron whose weight vector is closest to the input vector is called the best-matching unit, often denoted *bmu*:

$$\left\| X - W_{bmu} \right\| = \min i \left\{ \left\| X - W_i \right\| \right\} \quad , i \in \{1, \ldots, m\} \tag{3}$$

Where *Wbmu* is the best-matching unit weight vector. After finding the *bmu*, the weight vectors of the SOM are updated. The weight vectors of the *bmu* and its topological neighbors are moved closer to the input data vector. The weight-updating rule of the unit *i* is:

$$W_i(\tau+1) = W_i(\tau) + \varepsilon(\tau) h_{bmu}(i,\tau) \left[X(\tau) - W_i(\tau) \right] \tag{4}$$

where τ is time, $\varepsilon(\tau)$ is a learning rate and $h_{bmu}(i,\tau)$ is defined as the neighborhood kernel function around the *bmu*. Usually, $\varepsilon(\tau)$ is a decreasing function of time and should be between 0 and 1. The Gaussian neighborhood function is chosen.

3 Fault Detection Strategy

In this work, the analysis of the three stator currents is used for stator fault detection in induction motor. The adopted procedure of diagnosis calls three tools for preprocessing of the stator currents, these tools are the magnitude of the Park's currents, extraction of the envelope, and the signal of the zero crossing times (ZCT).

To reduce the dimension of data, and to improve the classification performances, statistical indicators were calculated as a fault features from these three signals.

The calculated features will serves as input of the Kohonen neural network (SOM). The algorithm of detecting the stator short circuit using the SOM neural network is depicted by the flowchart in Fig. 2.

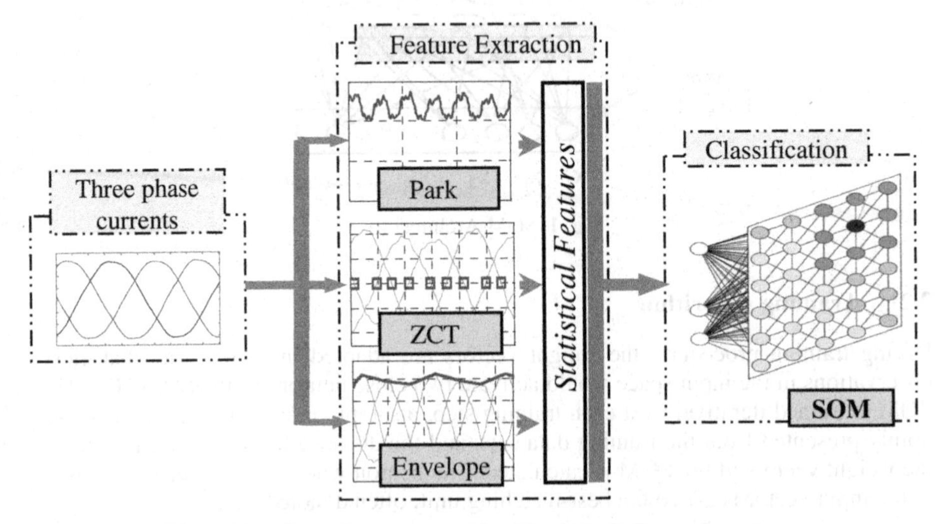

Fig. 2. Flowchart of the proposed intelligent fault diagnosis system

3.1 Park's Vector Magnitude

This technique is based on the Park's Vector Approach; it provides greater insight into the severity of the faults. The instantaneous line currents of the stator are transformed into the Park's vector using (5-6) [13].

$$i_{sd} = (\sqrt{2}/\sqrt{3})i_{sa} - (1/\sqrt{6})i_{sb} - (1/\sqrt{6})i_{sc} \tag{5}$$

$$i_{sq} = (1/\sqrt{2})i_{sb} - (1/\sqrt{2})i_{sc} \tag{6}$$

Under ideal conditions, for a healthy motor, Lissajou's curve isq= f (isd) shows a perfect circle where the instantaneous magnitude is constant. An unbalance due to turn faults results in an elliptic representation of the Park's Vector. The magnitude of the Park's Vector given by (7) will contain a frequency that is twice the fundamental frequency. The amplitude of this frequency is proportional to the degree of unbalance.

For better feature extraction, this resulting signal is normalized by subtracting his mean value.

$$i_{sdq}(t) = \sqrt{i_{sd}(t)^2 + i_{sq}(t)^2} \tag{7}$$

3.2 The Three-Phase Stator Current Envelope

An envelope is the geometric "line shape" of a modulation in the amplitude of the three-phase stator currents due to motor faulty conditions. Inter-turn short-circuits causes a profile modification of the three phase stator current leading to an envelope cyclically repeated at a rate equal to the power frequency (f) [14].

The procedures used here to obtain the three phase stator current envelope can be summarized in the following steps [15]:

- In the first step, the ripple of the three-phase stator current is isolated as shown in Fig. 3.a.
- The second step, which is the process of "envelope identification", consists of extracting from the three phase currents only the positive peak of each period in each phase, as shown in Fig. 3.b.
- In the third step, these few points are interpolated to smoothly represent the dynamic behavior of the three phase stator current envelope, as shown in Fig. 3.c.
- The fourth and last step is the normalization that centers the signal at zero mean. (Fig. 3.d).

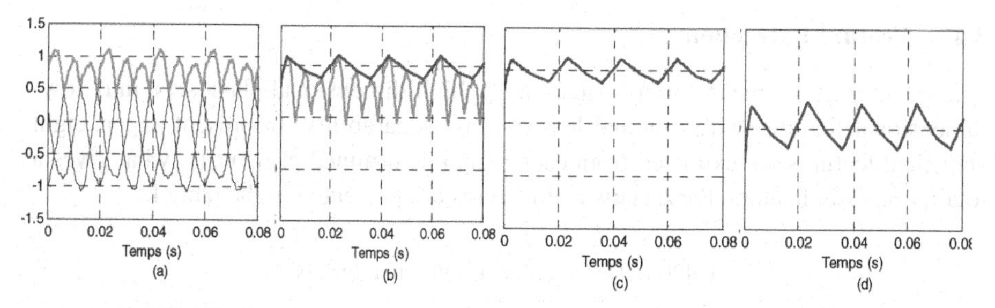

Fig. 3. The process used for obtaining the 3-phase current envelope

3.3 Zero Crossing Times

The monitoring of successive zero crossing times method has been successfully applied in speed measurement and fault detection. The ZCT signal consists of a series of data values, obtained at each zero crossing time of the 3-phase current. The values of data are defined as the time difference between two adjacent zero-crossing times (Tn - Tn-1) minus the natural reference time of the ZCT signal [16]:

$$T_{ZC}(n) = T(n) - T(n-1) - T_{ref} \tag{8}$$

In a real application, it is impossible to find the time that the current is exactly equal to zero due to the discrete sampling. Thus, Fig. 4 illustrate the algorithm for calculating approximate zero points [16]. Assume that the current is linear in a small time interval (i.e. sampling time). At the sampling time *(n – 1)* and *(n)*, $I(n-1) \times I(n) < 0$. Based on this assumption, the following equation is given by:

$$\frac{I(n)-0}{I(n)-I(n-1)} = \frac{T(n)-T(k)}{T(n)-T(n-1)} \tag{9}$$

Then, the approximate time $T(k)$ can be calculated by:

$$T(k) = T(n) - \frac{I(n)[T(n) - T(n-1)]}{I(n) - I(n-1)} \tag{10}$$

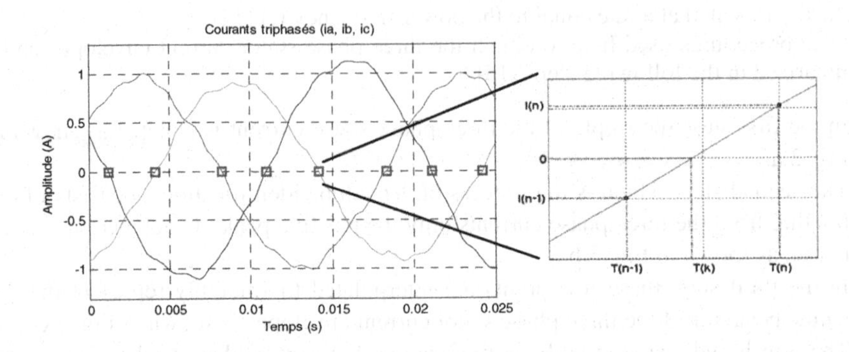

Fig. 4. Calculation of approximate ZC times from a stator current

3.4 Feature Extraction

Feature extraction means transformation of high-dimensional data sets into low-dimensional features with minimal loss of class separability. In present paper, eight statistical features are extracted from each signal determined previously, what gives a totality of 24 indicators; these statistical features are presented in the table I.

Table 1. The extracted Feature parameters

Indicator	*Expression*	*Indicator*	*Expression*
Rms	$X_{RMS} = \sqrt{\frac{1}{N}\sum_{i=1}^{n}(x_i^2)}$	Kurtosis	$X_{KURT} = \frac{1}{N}\sum_{i=1}^{n}\left(\frac{x_i - \bar{x}}{\sigma}\right)^4$
Standard Deviation	$\sigma = \sqrt{\frac{1}{N}\sum_{i=1}^{N}(x_i - \bar{x})^2}$	Crest Indicator	$X_{CI} = \frac{\max\lvert x_i\rvert}{\sqrt{\frac{1}{N}\sum_{i=1}^{n}(x_i^2)}}$
Variance	$\sigma^2 = \frac{1}{N-1}\sum_{i=1}^{N}(x_i - \bar{x})^2$	Clearance Indicator	$X_{CLI} = \frac{\max\lvert x_i\rvert}{\left(\frac{1}{N}\sum_{i=1}^{n}\sqrt{\lvert x_i\rvert}\right)^2}$
Skewness	$X_{SKEW} = \frac{1}{N}\sum_{i=1}^{n}\left(\frac{x_i - \bar{x}}{\sigma}\right)^3$	Shape Indicator	$X_{SI} = \frac{\sqrt{\frac{1}{N}\sum_{i=1}^{n}(x_i^2)}}{\frac{1}{N}\sum_{i=1}^{n}\sqrt{\lvert x_i\rvert}}$

4 Implementation and Results

4.1 Test Rig and Measurement

In order to verifying the proposed stator short circuit diagnostics technique, the experimental set up shown in Fig.5 is used. It consists of a 1.5 kW (LS 90 Leroy Somer), 2 poles pairs squirrel cage induction machine. To practically introduce a shorted-turns fault in the motor, the induction machine has been re-wound and two tapes at 16 and 32 turns in the stator phase winding were soldered, corresponding to 6% and 12% fault of the global number of turns in one phase as shown in Fig.6. For safety and for connection purpose, the tapes were connected to the motor terminal box. The reactor coolant pump torque is simulated by means of an electromagnetic brake. Current Hall Effect sensors were placed the line current cables for measuring of the currents in line. These currents are sampled and recorded on a personal computer at 5 kHz sampling frequency using the National Instruments USB-6008 data acquisition (DAQ) device.

Fig. 5. The laboratory setup for induction motor stator faults experiments

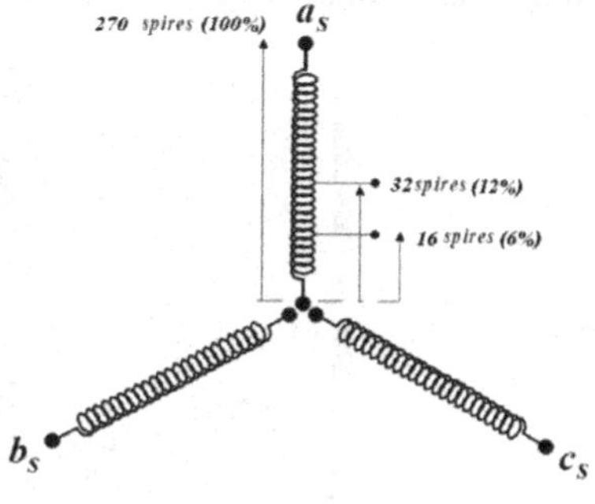

Fig. 6. Stator winding with the fault taps

In our experimentations, the motor was tested in the healthy case and two faulty cases with 6% and 12% stator inter-turns fault for three different load conditions (respectively 0 N.m, 3 N.m and 4 N.m).

4.2 Application of the Diagnostic Strategy

The collected data from experiment consist of 360 samples (120 representatives from each fault case and from the healthy case). After the acquisition of the stator currents, the Park's vector magnitude, ZCT signal and the envelope were extracted from the currents samples using MATLAB, then, the statistical features were calculated from each signal to construct a data base (size is 360 × 24). Half of this data base serves to train the SOM, and the rest were kept for testing of the ANN. The self-organizing map network was implemented by using the SOM toolbox, for Matlab [11] developed in the Helsinki University of Technology. The data array was normalized before being admitted to the neural network, by normalization of the variance of vector components to unity, and its mean to zero. Then, a label and a color is associated to each case: ***(heal) green*** for the healthy, ***(cc6) yellow*** for the 6% stator fault, and ***(cc12) red*** for the 12% stator fault case .

4.3 Results

The result of the SOM can be to interpret either by a topological two-dimensional grid containing the labels associated to the classes or by the Unified distance matrix (U-matrix), after the training phase. The first representation offers a topological knowledge of the whole data projection, whereas the second provides visualization of the distance between closest units in the map.

The Figures (7, 8, and 9) show the obtained maps for the studied cases, by using each signal features separately, whereas figure 10 show the obtained map by using all the extracted features from the three signals. The U-matrix shows no clear separation between different cases, but from the labels it seems that they correspond to three clusters.

By visual inspection of the obtained maps, we notice that classification by the ZCT signal features is better than classification by the envelope features and those of the Park's currents magnitude in the quality of the number of activated neurons and the distribution of the classes on the map. But the use of all the included indicators gives the best result where we notice a very clear separation between the different classes, with a minimal number of activated neurons.

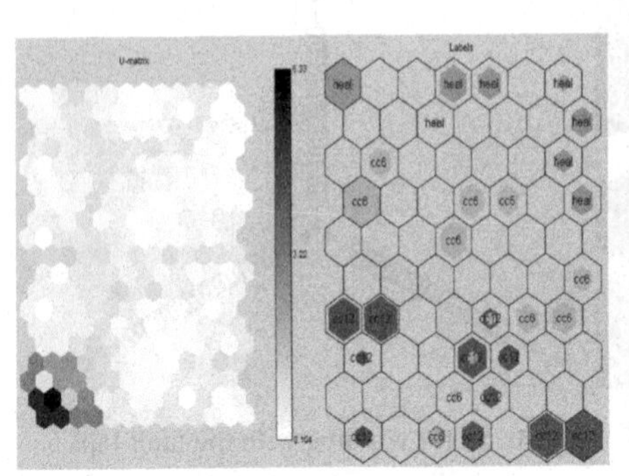

Fig. 7. Training using Park's features

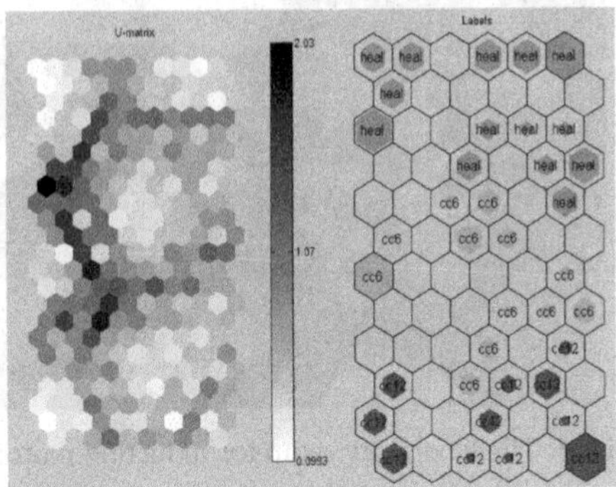

Fig. 8. Training using envelope features

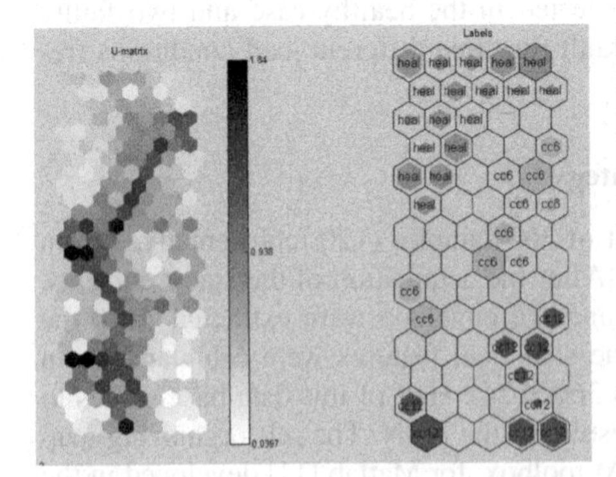

Fig. 9. Training using ZCT signal features

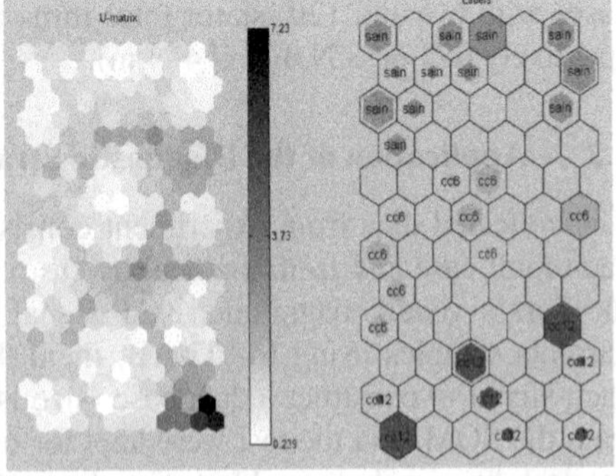

Fig. 10. Training using the included features

Classification performance can be analyzed by projection of the testing data sets on the trained maps then find, for each data sample, the best matching unit from the map. Then, the class label of that unit is given to the sample [11]. Classification accuracy can be evaluated as fraction of correctly classified input samples. The classification performances of the SOMs are given in Table 2. This results show that the ZCT signal has the best features for stator fault s detection and severity evaluation.

Table 2. Classification results

Training Features	Input Class \ Output Results	Healthy (%)	6% SC (%)	12% SC (%)	Not Classified (%)	Total Correct Classification (%)
Park	Healthy	**93.33**	0	0	6.66	96.11
	6% SC	0	**98.33**	0	1.66	
	12% SC	0	0	**96.66**	3.33	
Envelope	Healthy	**96.66**	0	0	3.33	90.55
	6% SC	0	**93.33**	3.33	3.33	
	12% SC	0	11.66	**81.66**	6.66	
ZCT	Healthy	**100**	0	0	0	97.77
	6% SC	0	**95**	0	5	
	12% SC	0	0	**98,33**	1.66	
ALL	Healthy	**100**	0	0	0	97.22
	6% SC	0	**91.66**	0	8.33	
	12% SC	0	0	**100**	0	

5 Conclusion

This study presents a methodology for detection of stator faults in induction motor of reactor coolant pump by classifying them using self-organizing map (SOM), this methodology incorporates most appropriate features, which are extracted from the three phase currents. Three pre-processing techniques, Park's vector magnitude, envelope extraction, and the zero crossing times signal are investigated to facilitate fault classification, and to reduce the complexity of the neural network.

The results obtained with this approach show that it possible to detect the stator fault in an early stage. The system is also able to estimate the extent of the stator faults. In a future work, the clustering of the SOM is proposed to avoid the not classified cases. Further investigation will also be conducted to the implementation of this strategy in a real time application, in order to satisfy the safety requirements of the nuclear power plants.

References

1. Yan, G., Zhu, Y.: Application research of local support vector machines in condition trend prediction of reactor coolant pump. In: Yu, W., Sanchez, E.N. (eds.) Advances in Computational Intelligence. AISC, vol. 61, pp. 35–43. Springer, Heidelberg (2009)
2. Aboubou, A., Sahraoui, M., Ghougal, A., Zouzou, S.E.: Analyse du contenu spectral de la tension de neutre de la machine asynchrone en vue de son diagnostic. Courrier du Savoir – N°06, pp.95–102 (June 2005)
3. Maa, J., Jiang, J.: Applications of fault detection and diagnosis methods in nuclear power plants: A review. Progress in Nuclear Energy 53, 255–266 (2011)
4. Weerasinghe, M., Barry Gomm, J., Williams, D.: Neural networks for fault diagnosis of a nuclear fuel processing plant at different operating points. Control Engineering Practice 6, 281–289 (1998)
5. Bae, H., Chun, S.-P., Kim, S.: Predictive Fault Detection and Diagnosis of Nuclear Power Plant Using the Two-Step Neural Network Models. In: Wang, J., Yi, Z., Żurada, J.M., Lu, B.-L., Yin, H. (eds.) ISNN 2006. LNCS, vol. 3973, pp. 420–425. Springer, Heidelberg (2006)
6. Yangping, Z., Bingquan, Z., DongXin, W.: Application of genetic algorithms to fault diagnosis in nuclear power plants. Reliability Engineering and System Safety 67, 153–160 (2000)
7. Xiao-cheng, S., Chun-ling, X., Yuan-hui, W.: Nuclear power plant fault diagnosis based on genetic-RBF neural network. Journal of Marine Science and Application 5(3), 57–62 (2006)
8. Ming-Yu, F., Xin-Qian, B., Ji, S.: Fault Diagnosing System of Steam Generator for Nuclear Power Plant Based on Fuzzy Neural Networks. Journal of Marine Science and Application 1(1), 41–46 (2002)
9. Kohonen, T.: Self-Organizing Maps. Springer, Berlin (2001)
10. Bossio, J.M., De Angelo, C.H., Bossio, G.R., García, G.O.: Fault Diagnosis on Induction Motors Using Self-Organizing Maps. In: 9th IEEE/IAS International Conference on Industry Applications, INDUSCON 2010 (2010)
11. Aroui, T., Koubaa, Y., Toumi, A.: Clustering of the Self-Organizing Map based Approach in Induction Machine Rotor Faults Diagnostics. Leonardo Journal of Sciences, 1–14 (2009)
12. Vesanto, J., Himberg, J., Alhoniemi, E., Parhankangas, J.: Self-organizing map in Matlab: the SOM Toolbox. In: Proceedings of the Matlab DSP Conference 1999, Espoo, Finland, November 16-17, pp. 35–40 (1999), `http://www.cis.hut.fi/projects/somtoolbox`
13. Cardoso, A.J.M., Cruz, S.M.A., Fonseca, D.S.B.: Inter-Turn Stator Winding Fault Diagnosis in Three-phase Induction Motors, by Park's Vector Approach. IEEE Transactions on Energy Conversion 14(3) (September 1999)
14. Arabacı, H., Bilgin, O.: Detection of Rotor Bar Faults by Using Stator Current Envelope. In: Proceedings of the World Congress on Engineering, WCE 2011, London, U.K., July 6 - 8, vol. II (2011)
15. da Silva, A.M.: Induction motor fault diagnostic and monitoring methods. A Master Thesis, Marquette University, Milwaukee, Wisconsin (May 2006)
16. Ukil, A., Chen, S., Andenna, A.: Detection of stator short circuit faults in three-phase induction motors using motor current zero crossing instants. Electric Power Systems Research 81, 1036–1044 (2011)

A New Approach for the Extraction of Moving Objects

Farou Brahim[1,2], Seridi Hamid[2], and Akdag Herman[3]

[1] Computer Science Department, Badji Mokhtar University, Annaba, Algeria
[2] LabSTIC, Guelma University, POB 401, 24000 Guelma, Algeria
{farou,seridi}@labstic.net
[3] LIASD, Paris 8 University, 93526 Saint-Denis, France
Herman.akdag@ai.univ-paris8.fr

Abstract. We propose in this paper a background subtraction system for image sequences extracted from fixed camera using Gaussian Mixture Models (GMM) and the analysis of color histograms. This system can achieve best accuracy than a simple GMM while maintaining the same computational resources. Images extracted from the video will first be divided into several areas of equal size where the behavior of each area is monitored by the analysis of color histograms. For each new frame the color histograms of the zones will be calculated and parts reported to have significant variation in histogram will be updated at the background model. Test carried out show that this approach present best results than a simple GMM. This improvement is important for processing in real time environment.

Keywords: Machine Vision, Video surveillance, Gaussian mixtures, modeling the background, image processing.

1 Introduction

Video surveillance is an active area in computer vision that allows analyzing video sequences and gives vast amount of data storage and display. Surveillance tasks have been executed by human beings who continually observe computer monitors to detect unauthorized activity over many cameras [1]. A computer vision system, can take the place of the human operator and monitor immediate unauthorized behavior and long-term suspicious behavior. The used methods to solve the problem of the video surveillance in its different levels [2] [3], are namely, extraction, identification and tracking of moving objects. This process begins first with the extraction of objects that do not belong in the background [4]. The method most commonly used is the modeling of the environment [5].

In the literature these approaches are divided into three parts: extraction methods of the background based on the comparison between the current image and the image of reference [6], the temporary difference that permits to cutting the video into a set of fragments separated in a lapse of constant time and the use of the optical flow estimating movements using a set of points similar to an object through several fragments [6] [7]; the second step is to classify objects extracted from the first step using techniques such as contours detection and specific descriptors of an object [6] [7]; the last step is

A. Amine et al. (Eds.): *Modeling Approaches and Algorithms*, SCI 488, pp. 27–36.
DOI: 10.1007/978-3-319-00560-7_7 © Springer International Publishing Switzerland 2013

the tracking of objects across the scène. In this paper we propose a novel approach for extracting moving objects based on GMM and Color histograms.

The remaining of this paper is organized as follows. In section 2 we give a brief description of some background subtraction system. Section 3 summarizes the proposed system. Finally, section 4 reports obtained results on a real challenging dataset.

2 State of the Art

The contributions made in recent years in this area revolve around two main axes: the features used for image representation and the methods used for modeling the background. Some others works based on hybrid methods (features / characteristics, features/models and models/styles) have also contributed to the improvement of video surveillance.

[8] have used the pixels probabilities to measure the characteristics of low texture and modeling of local states with invariant scales and an adaptive learning of the background for a multi-modal representation. [9] proposed an improved modeling of the background using an approach based on the region and sparse as model characteristics of the background. An adjustment of the algorithm is performed depending on the dense regions and smooth textured scenes. Another region-based approach was proposed in [10] where two regions of motion are constructed. The first is a subtraction between the current frame and it's adjacent and the second is a subtraction between adjacent of the next frame. The difference between the regions of movement and the intersection of neighboring regions is calculated to select the candidate regions. The update is performed for regions having stable probabilities among the selected regions. [11] proposed a block modeling combining colors and textures to represent the background. [12] proposed a similar work with integrating a local binary pattern, RGB color space, Sobel filter as characteristics and the Choquet's fuzzy integral to reduce uncertainty effects. [13] used an orthogonal transformation of non-separable wavelets information frames for modeling the background with a set of post processing to eliminate shadows and noise. Another use of wavelets is proposed [14], where each region of the model-based region is represented by multi resolution wavelets with K Gaussian mixtures. For a representation and a direct adaptation of the background from a compressed video [15] have used the discrete cosine transform and AC coefficients. A non-parametric approach proposed by [16] where they applied a condensation algorithm to minimize the number of samples used for modeling the probability density of the background. The detection of moving objects is provided by Markov random fields. [17] proposed an extension of nonparametric models based on spatio-temporal modeling of the background, they used 3x3 blocks centered on the pixel in question and calculated the density of the core relative to 9 neighbors of that pixel. Other approaches have been proposed for modeling the background such as the sequential density approximation [18], the median filter on several levels with a codebook [19] and Bayesian decision [20] [21] . In [20] the posterior probability of observing a pixel as background or foreground is estimated from the frequency of application of the pixel, whereas in [21] propose a Bayesian framework using spectral and spatial-temporal for modeling the background. The background is represented by the most significant characteristics for each pixel. The Bayesian classification, in both

approaches, will classify each new fragment in the two classes (foreground and background) based on the main feature. Gaussian mixtures are among the most widely used methods for modeling the environment [22]. They are efficient if the video does not contain any local variation where changes in light [22]. However, learning a cluttered background and with multiple moving objects remains the major problem of this method [22] [23]. The work of [24] showed that GMM offers a compromise between extraction quality and execution time compared to other methods for extracting the background. In [25] the authors proposed a model of the background based on mixtures of Gaussian and used a real parallel to the implementation of this method on images of 1024x1280 pixels resolution at a rate greater than 30 frames per second (fps). To eliminate the problem of sudden changes in brightness, [26] proposed two different models of the background, the first when the environment is bright and the other when it is dark by using an analyzer for brightness to choose the most appropriate model for the extraction of moving objects. Another model is proposed in [27] using as features of tensor fields instead of the color space that are used in the GMM. For modeling the background [28] proposed hybridization between Gaussian mixtures and k-means approach. This model is based on the collection of intensities of each pixel, where each intensity is analyzed by a mixture of Gaussians which is then divided into two clusters by the k-means method. Another hybridization was proposed in [29] which aims to eliminate problems caused by camera motion by using fuzzy logic type 2 for treating uncertainty generated by the GMM during the modeling of the background. The combination of background models and adaptive Gaussian mixtures is proposed in [30]. The first model detects changes in the fine textures in the background using a motion detection algorithm, while the second models the background using a selective algorithm for estimating the movement hierarchy. Work as those proposed in [31] [32] apply an alpha rate adaptive learning GMM allowing to solve the problem of coefficient updating GMM which greatly influences the performance of the method.

In this paper, we will focus on the detection of moving objects in video surveillance through a fixed camera using GMM. To overcome the problems of local variation and variation in brightness, first, proposed system allows segmentation of the original image, extracted from the video into several areas. Each area will be assigned to one thread that will monitor the behavior of the area based on the analysis of histograms. Only segments that have undergone a change will be updated by the GMM.

3 System Description

The proposed system transforms the captured video from RGB to HSV frames because the RGB representation is not adequate due to the influence of light on the description of objects [33] where HSV model is recognized to be the closest model of human perception and ensures a better result since the brightness affects only one component [33].

The first captured frame is divided into N segments of equal size to minimize local variations. The choice of the number of segments depends on the initial size of the images, the environment, and the field of use. On segmentation imposes a decrease in performance and sub segmentation a decrease in quality which implies that the choice of the number of segments must be a compromise between performance and quality.

3.1 Background Modeling

This step creates a very stable model of the background due to hybridization between the individual capabilities of threads lookouts and the power of mixtures of Gaussians.

We use multithread approach to accelerate the operations and to simplify actions. In fact, each thread will have the role of monitoring the behavior of the area to which he was assigned. For each area, we create and save the histogram of the latter in his memory. Thus each new treatment, the agent can detect changes in the area by comparing the current state of the histogram with the state memorized.

First, we will create and save the histogram of each area. Then, in each new treatment, we calculate and compare the current state of the histogram with the memorized state. If the comparison is greater than a certain threshold T, a signal will be triggered indicating the detection of a change in the area.

In the literature, there are several methods to measure the similarity (dis-similarity) between two probability distributions P and Q [34] [35], the most used are the Bhattacharyya distance (Equation 1, 2)

$$D_B(p,q) = -ln\big(BC(p,q)\big) \tag{1}$$

where

$$BC(p,q) = \textstyle\sum_{x \in X} \sqrt{p(x)q(x)} \tag{2}$$

Gaussian Mixture Models.

This section briefly describes a mixture of the Gaussians-based background subtraction method proposed in [36]. This method describes the probability of observing a pixel value, Xt , at time t as (6).

$$P(X_t) = \textstyle\sum_{i=1}^{k} w_{i,t}\, \eta\, (P_t\, , \mu_{i,t}\, , \Sigma_{i,t}) \tag{3}$$

where K is the number of Gaussians, which is usually set to be between 3 and 7; $W_{i,t}$, $\mu_{i,t}$ and $\Sigma_{i,t}$ are weight, mean and the covariance matrix of the i-th Gaussian in the mixture at time t, respectively. For computational efficiency, red, green and blue pixel values are assumed to be independent and have the same variances. To update this model, the following on-line K-means approximation is used [36]. Every new pixel value is checked against the K Gaussian distributions to determine whether this value is within 2.5 standard deviation of one of them. If none of the distributions includes this pixel value, the least probable distribution is replaced with a distribution that's mean; variance and weight are set to the current pixel value, pre-determined high variance and low weight, respectively. The weights of the K distributions at time t are updated as (4).

$$w_{k,t} = (1-\alpha)\cdot w_{k,t-1} + \alpha\, M_{k,t} \tag{4}$$

where α is a learning rate, and $M_{k,t}$is 1 for the distribution which includes the current pixel value within its 2.5 standard deviation and 0 for the other distributions. After updating the weights, they are renormalized to make their summation become one.

The parameters of the distribution which includes the current pixel value within its 2.5 standard deviation are updated as (5) (6).

$$\mu_{k,t} = (1 - \varphi_k) \cdot \mu_{k,t-1} + \varphi_k \cdot P_t \tag{5}$$

$$\sigma_{k,t}^2 = (1 - \varphi_k) \cdot \sigma_{k,t-1}^2 + \varphi_k (P_t - \mu_{k,t})^T (P_t - \mu_{k,t}) \tag{6}$$

$$\varphi_t = \alpha\, \eta(P_t / \mu_k \sigma_k) \tag{7}$$

where φ_t is a learning factor for adapting distributions. The parameters of the other distributions remain the same. To decide whether $X_t$is included in the background distributions, the distributions are ordered by the value of $w_{k,t} / \sigma_{k,t}$ and the first B distributions which satisfy (8) are chosen as the background distributions.

$$\beta = \arg\min(\textstyle\sum_{k=1}^{b} w_{k,t} > B) \tag{8}$$

where B is a measure of the minimum portion of the data that should be accounted for by the background.

If $X_t$is within 2.5 standard deviation of one of these B distributions, it is decided as a background pixel.

3.2 Detection of Moving Objects

We performed a localization of objects by analyzing histograms of horizontal and vertical projection. Using an iterative process, we try at each step to segmenting the image in parts according to the number of peaks between two minima in the histogram. We apply the same process to each part until attain a non-segmental part. Each peak between two minimums represents a moving object.

4 Experimental Results

The system presented in this paper is implemented in Java on a computer with an Intel Core i5 2.67 GHz power and 4GB of memory capacity.

This section seeks to highlight the results obtained from tests carried out on the basis of videos. The videos used are different from the context, the light intensity, fluctuations in the abundance of objects, number and type of people, movements caused by the active elements of nature such as wind, clouds, dust, the noise and jostling movements of the camera.

We use eight videos to evaluate the performance of our system. Database is divided into two categories. We took from the first category three videos, V1 is taken by the authors in the parking of the computer science department at Guelma University, V2 is taken in a public park and V3 is taken in rooms of computer science department. The second category contains public videos (D1, D2, D3, D4 and D5) [37] representing respectively a campus, highway I, highway II, intelligent room and a laboratory.

Table 1. description of used database

Video	Frame Number	Resolution	Environment
V1	1077	352 X 288	Parking of department
V2	555	352 X 288	Public park
V3	616	352 X 288	Office department
D1	1178	320 X 240	campus
D2	439	320 X 240	Highway I
D3	499	320 X 240	Highway II
D4	299	320 X 240	Intelligent room
D5	886	320 X 240	Laboratory

Table 1 shows some details about the videos used in system testing. The background modeling parameters used in our system is set as the flowing, the number of Gaussians (K), the learning rate (α) and the measure of the minimum portion (B) were set respectively to 5, 0.001 and 0.3. These parameters were empirically chosen to have slight and stable system.

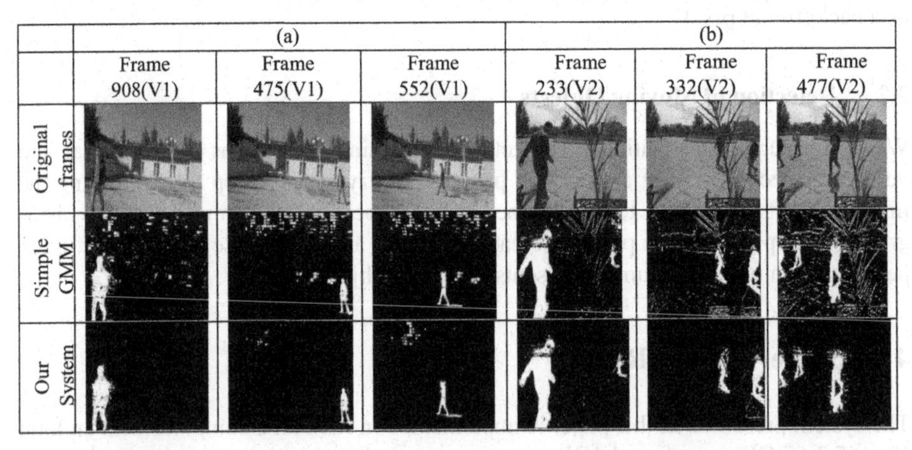

Fig. 1. A simple comparison between our approach and a simple GMM. (a) is taken in the parking of the computer science department and (b) is taken in in a public park.

V1 and V2 are taken in an environment with wind and a frequent agitation of trees as well as the appearance of flying objects and animals.

Fig.1 (a) shows that the system gives very good performance compared to a simple GMM when operating in an environment with one moving object. Fig.1 (b) shows that our system has also a good performance when operating in an environment with multiple moving objects. Fig.2 shows the effectiveness of our system in partially dark environments with very large variations in brightness caused by the reflection of light in a narrow environment. Parts of the body not affected by the light remain undetected, but the quality of the contours allows us to recover the moving object. Fig.2 clearly shows the contribution of our approach compared to the use of a simple GMM.

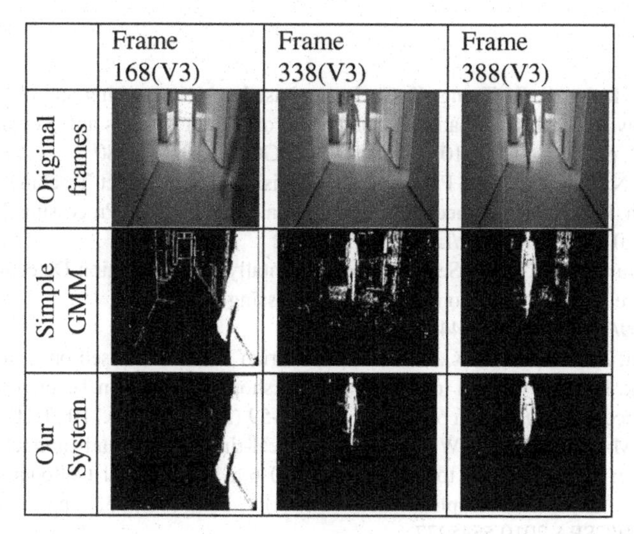

Fig. 2. Comparison between a simple GMM and our system in an office administration

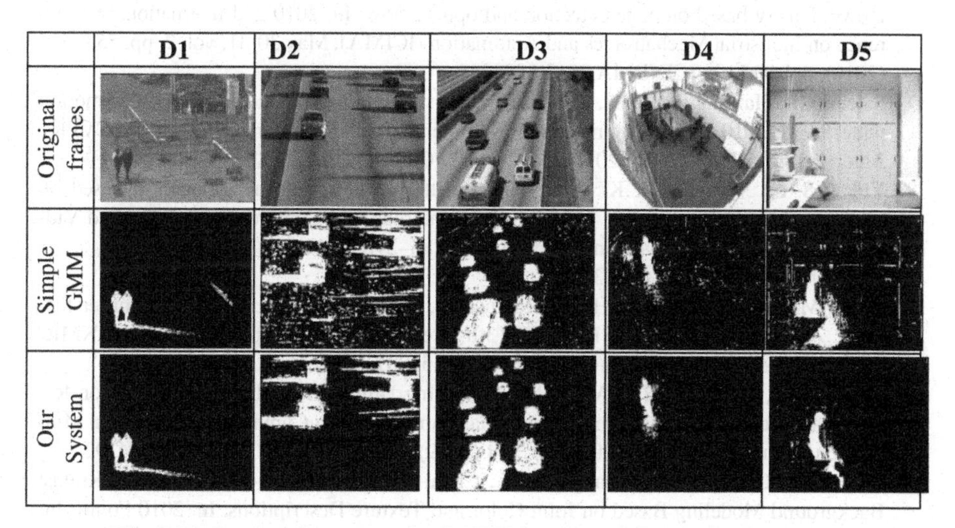

Fig. 3. Comparison between a simple GMM and our system on public videos

5 Conclusion

We presented in this paper a background subtraction system for image sequences extracted from fixed camera using Gaussian Mixture Models (GMM) and the analysis of color histograms to overcome the major problem in Gaussian mixtures such as: brightness variations and local variations. Tests conducted on different environments with complex background show that this approach can achieve best accuracy than a simple GMM while maintaining the same computational resources.

References

1. Lai, J.-C., Huang, S.-S., Tseng, C.-C.: Image-based vehicle tracking and classification on the highway. In: 2010 International Conference on Green Circuits and Systems (ICGCS), June 21-23, pp. 666–670 (2010), doi:10.1109/ICGCS.2010.5542980
2. Bishop, A.N., Savkin, A.V., Pathirana, P.N.: Vision-Based Target Tracking and Surveillance With Robust Set-Valued State Estimation. IEEE Signal Processing Letters 17(3), 289–292 (2010), doi:10.1109/LSP.2009.2038772
3. Woo, H., Jung, Y., Kim, J., Seo, J.: Environmentally Robust Motion Detection for Video Surveillance. IEEE Transactions on Image Processing PP(99), 1, doi:10.1109/TIP.2010.2050644
4. Li, Z., Wei, Z., Yin, B., Ji, X., Shan, R.: Pedestrian Detection Based on a New Two-Step Framework. In: 2010 Second International Workshop on Education Technology and Computer Science (ETCS), March 6-7, vol. 3, pp. 56–59 (2010), doi:10.1109/ETCS.2010.160
5. Hedayati, M., Zaki, W.M.D.W., Hussain, A.: Real-time background subtraction for video surveillance: From research to reality. In: 2010 6th International Colloquium on Signal Processing and Its Applications (CSPA), May 21-23, pp. 1–6 (2010), doi:10.1109/CSPA.2010.5545277
6. Kim, G.-J., Eom, K.-Y., Kim, M.-H., Jung, J.-Y., Ahn, T.-K.: Automated measurement of crowd density based on edge detection and optical flow. In: 2010 2nd International Conference on Industrial Mechatronics and Automation (ICIMA), May 30-31, vol. 2, pp. 553–556 (2010), doi:10.1109/ICINDMA.2010.5538248
7. Saleemi, I., Hartung, L., Shah, M.: Scene understanding by statistical modeling of motion patterns. In: 2010 IEEE Conference on Computer Vision and Pattern Recognition (CVPR), June 13-18, pp. 2069–2076 (2010), doi:10.1109/CVPR.2010.5539884
8. Yuk, J.S.-C., Wong, K.-Y.K.: An efficient pattern-less background modeling based on scale invariant local states. In: 2011 8th IEEE International Conference on Advanced Video and Signal-Based Surveillance (AVSS), pp. 285–290 (2011), doi:10.1109/AVSS.2011.6027338
9. Uzair, M., Khan, W., Ullah, H., Ur Rehman, F.: Background modeling using corner features: An effective approach. In: IEEE 13th International Multitopic Conference, INMIC 2009, pp. 1–5 (2009), doi:10.1109/INMIC.2009.5383113
10. Lei, T., Fan, Y., Li, L.: The Algorithm of Moving Human Body Detection Based on Region Background Modeling. In: International Symposium on Computer Network and Multimedia Technology, CNMT 2009, pp. 1–4 (2009), doi:10.1109/CNMT.2009.5374792
11. Lin, C.-Y., Chang, C.-C., Chang, W.-W., Chen, M.-H., Kang, L.-W.: Real-Time Robust Background Modeling Based on Joint Color and Texture Descriptions. In: 2010 Fourth International Conference on Genetic and Evolutionary Computing (ICGEC), pp. 622–625 (2010), doi:10.1109/ICGEC.2010.159
12. Azab, M.M., Shedeed, H.A., Hussein, A.S.: A new technique for background modeling and subtraction for motion detection in real-time videos. In: 2010 17th IEEE International Conference on Image Processing (ICIP), pp. 3453–3456 (2010), doi:10.1109/ICIP.2010.5653748
13. Gao, D., Jiang, Z., Ye, M.: A New Approach of Dynamic Background Modeling for Surveillance Information. In: 2008 International Conference on Computer Science and Software Engineering, vol. 1, pp. 850–855 (2008), doi:10.1109/CSSE.2008.601
14. Mendizabal, A., Salgado, L.: A region based approach to background modeling in a wavelet multi-resolution framework. In: 2011 IEEE International Conference on Acoustics, Speech and Signal Processing (ICASSP), pp. 929–932 (2011), doi:10.1109/ICASSP.2011

15. Wang, W., Yang, J., Gao, W.: Modeling Background and Segmenting Moving Objects from Compressed Video. IEEE Transactions on Circuits and Systems for Video Technology 18(5), 670–681 (2008), doi:10.1109/TCSVT.2008.918800
16. Luo, X., Bhandarkar, S.M., Hua, W., Gu, H.: Nonparametric Background Modeling Using the CONDENSATION Algorithm. In: IEEE International Conference on Video and Signal Based Surveillance, AVSS 2006, p. 3 (2006), doi:10.1109/AVSS.2006.81
17. Vemulapalli, R., Aravind, R.: Spatio-temporal nonparametric background modeling and subtraction. In: 2009 IEEE 12th International Conference on Computer Vision Workshops (ICCV Workshops), pp. 1145–1152 (2009), doi:10.1109/ICCVW.2009.5457574
18. Wang, H., Ren, M.-W., Yang, J.-Y.: Background Modeling Method Based on Sequential Kernel Density Approximation. In: Chinese Conference on Pattern Recognition, CCPR 2008, pp. 1–6 (2008), doi:10.1109/CCPR.2008.44
19. Ma, J., Li, S.: Moving Target Detection Based on Background Modeling by Multi-level Median Filter. In: The Sixth World Congress on Intelligent Control and Automation, WCICA 2006, vol. 2, pp. 9974–9978 (2006), doi:10.1109/WCICA.2006.1713948
20. Xu, S.: Dynamic Background Modeling for Foreground Segmentation. In: Eighth IEEE/ACIS International Conference on Computer and Information Science, ICIS 2009, pp. 599–604 (2009), doi:10.1109/ICIS.2009.102
21. Li, L., Huang, W., Gu, I.Y.-H., Tian, Q.: Statistical modeling of complex backgrounds for foreground object detection. IEEE Transactions on Image Processing 13(11), 1459–1472 (2004), doi:10.1109/TIP.2004.836169
22. Yu, J., Zhou, X., Qian, F.: Object kinematic model: A novel approach of adaptive background mixture models for video segmentation. In: 2010 8th World Congress Intelligent Control and Automation (WCICA), pp. 6225–6228 (2010), doi:10.1109/WCICA.2010.5554402
23. Zhang, L., Liang, Y.: Motion Human Detection Based on Background Subtraction. In: 2010 Second International Workshop Education Technology and Computer Science (ETCS), vol. 1, pp. 284–287 (2010), doi:10.1109/ETCS.2010.440
24. Hedayati, M., Zaki, W.M.D.W., Hussain, A.: Real-time background subtraction for video surveillance: From research to reality. In: 2010 6th International Colloquium Signal Processing and Its Applications (CSPA), pp. 1–6 (2010), doi:10.1109/CSPA.2010.5545277
25. Li, X., Jing, X.: FPGA based mixture Gaussian background modeling and motion detection. In: 2011 Seventh International Conference on Natural Computation (ICNC), vol. 4, pp. 2078–2081 (2011), doi:10.1109/ICNC.2011.6022578
26. Cheng, F.-C., Huang, S.-C., Ruan, S.-J.: Illumination-Sensitive Background Modeling Approach for Accurate Moving Object Detection. IEEE Transactions on Broadcasting 57(4), 794–801 (2011), doi:10.1109/TBC.2011.2160106
27. Caseiro, R., Henriques, J.F., Batista, J.: Foreground Segmentation via Background Modeling on Riemannian Manifolds. In: 2010 20th International Conference on Pattern Recognition (ICPR), pp. 3570–3574 (2010), doi:10.1109/ICPR.2010.871
28. Charoenpong, T., Supasuteekul, A., Nuthong, C.: Adaptive background modeling from an image sequence by using K-Means clustering. In: 2010 International Conference on Electrical Engineering/Electronics Computer Telecommunications and Information Technology (ECTI-CON), pp. 880–883 (2010)
29. El Baf, F., Bouwmans, T., Vachon, B.: Fuzzy statistical modeling of dynamic backgrounds for moving object detection in infrared videos. In: IEEE Computer Society Conference on Computer Vision and Pattern Recognition Workshops, CVPR Workshops 2009, pp. 60–65 (2009), doi:10.1109/CVPRW.2009.5204109

30. Doulamis, A., Kalisperakis, I., Stentoumis, C., Matsatsinis, N.: Self Adaptive background modeling for identifying persons' falls. In: 5th International Workshop on Semantic Media Adaptation and Personalization (SMAP), pp. 57–63 (2010), doi:10.1109/SMAP.2010.5706861
31. Kan, J., Li, K., Tang, J., Du, X.: Background modeling method based on improved multi-Gaussian distribution. In: International Conference on Computer Application and System Modeling (ICCASM), vol. 2, pp. 214–218 (2010), doi:10.1109/ICCASM.2010.5619314
32. Suo, P., Wang, Y.: An improved adaptive background modeling algorithm based on Gaussian Mixture Model. In: 9th International Conference on Signal Processing (ICSP), pp. 1436–1439 (2008), doi:10.1109/ICOSP.2008.4697402
33. Haq, A.-U., Gondal, I., Murshed, M.: Automated multi-sensor color video fusion for nighttime video surveillance. In: IEEE Symposium on Computers and Communications (ISCC), pp. 529–534 (2010), doi:10.1109/ISCC.2010.5546791
34. Djouadi, A., Snorrason, O., Garber, F.: The quality of training-sample estimates of Bhattacharyya coefficient. IEEE Transactions on Pattern Analysis and Machine Intelligence 12(1), 92–97 (1990)
35. Lin, J.: Divergence Measures Based on the Shannon Entropy. IEEE Transactions on Information Theory 37(i) (1991)
36. Charoenpong, T., Supasuteekul, A., Nuthong, C.: Adaptive background modeling from an image sequence by using K-Means clustering. In: International Conference on Electrical Engineering/Electronics Computer Telecommunications and Information Technology (ECTI-CON), pp. 880–883 (2010)
37. Prati, A., Mikic, I., Trivedi, M., Cucchiara, R.: Detecting Moving Shadows: Formulation, Algorithms and Evaluation. IEEE Transactions on Pattern Analysis and Machine Intelligence 25(7), 918–923 (2003)

Seeking for High Level Lexical Association in Texts

Ismaïl Biskri[1], Louis Rompré[1], Christophe Jouis[2], Abdelghani Achouri[1], Steve Descoteaux[1], and Boucif Amar Bensaber[1]

[1] LAMIA – Université du Québec à Trois-Rivières (Qc), Canada
{Biskri,Rompre,Achouri,Descoteaux,Bensaber}@uqtr.ca
[2] LIP6- Université Pierre et Marie Curie, Paris, France
Christophe.jouis@lip6.fr

Abstract. Searching information in a huge amount of data can be a difficult task. To support this task several strategies are used. Classification of data and labeling are two of these strategies. Used separately each of these strategies have certain limitations. Algorithms used to support the process of automated classification influence the result. In addition, many noisy classes can be generated. On the other hand, labeling of document can help recall but it can be time consuming to find metadata. This paper presents a method that exploits the notion of association rules and maximal association rules, in order to assist textual data processing, these two strategies are combined.

Keywords: Classification, Maximal Association Rules, Extraction of Knowledge.

1 Introduction

It's difficult to determine the exact size of the Web. Some statistics can give an estimate. For example, Google announced in 2008 that they have indexed 1,000 billion Web pages. Today the number of Web pages indexed by Google is estimated over 50, 000 billion! The Web contains different kinds of textual documents such as books, technical papers or user reviews. This huge amount of data makes it possible to find information on any topic. For instance a short navigation on the Web can provide enough information to properly plan a trip; it is possible to find available hotel rooms, attractions, transportation etc. Although the size of the Web is an advantage, it adds some disadvantage during selection of information to consider. In fact, it's impossible to read the content of all documents. Some documents ignored may contain relevant information. That is why several strategies are employed to facilitate the exploration of textual documents.

Labeling is one of the most common strategies even for multimedia documents like audio documents [1]. This strategy is to add descriptive metadata to documents. The goal is to find those who promote the recall when searching. The easiest method is manual labeling. Although this method is relatively efficient, the quality of the results depends on a priori knowledge of the person responsible for labeling documents. However, in a context of abundance, labeling cannot be done manually. One of the solutions proposed to solve this problem is to automate the process. We believe that association rules and maximal association rules can be used to assist the selection of relevant metadata.

A. Amine et al. (Eds.): *Modeling Approaches and Algorithms*, SCI 488, pp. 37–46.
DOI: 10.1007/978-3-319-00560-7_8 © Springer International Publishing Switzerland 2013

2 Classification of Data

The main goal of classification is to group into classes objects that share similar property. Thus, classification can make finding information easier when the classes are consistent. When a classification is automatically generated, documents are organized in such a way that documents whose raw data shared similarities are grouped together. The creation of classes of similarities is strongly influenced by the presence or absence of significant properties within the data. To perform this task several algorithms are used to extract characteristics from documents. These characteristics are used as unit of information to represent the properties of documents. Units of information need to be enough "discriminant" to support the classification process. Tools such as neural networks are used to produce classes of similarities. These tools take as inputs the vector form of the documents and they give as output the classes of similarities. Documents are judged similar when they contain same units of information at a frequency that is almost identical. Therefore the unit of information selected influences the results obtained. Unfortunately there is no theory to guide the choice of the optimal unit of information [2]. The consequence is when classifications are created automatically many classes of similarities may be produced and many of them can be noisy. Such classifications may seem insignificant. Despite this, the automated classification is able to generate interesting pre-arrangements.

When data is grouped into subsets some regular patterns may appear. Studying these patterns can show unexpected properties of the data. These unexpected properties are hidden knowledge. Hidden knowledge is a source of information that can be used to bring out some tags properly describing the content of the data. Discovering hidden knowledge into classes of similarities is possible when an analysis is done by an expert. However, the number of classes and the noise contained in these classes are barriers when analysis is performed on classification automatically generated. We suggest using the association rules and the maximal association rules to assist extraction of hidden knowledge. The following section reviewed the definition of association rules and maximal association rules as defined respectively in [3].

3 Association Rules and Maximal Association Rules

Association rules are frequently used in data mining. They allow finding regularities into transactions. The interest for association rules mainly came from the work of Agarwal on transactional databases analysis [3]. Agrawal has shown that it is possible to define rules to illustrate the relationship between items that co-occur in commercial transactions. He showed that association rules can be used to identify which products are frequently bought together. For instance clients who buy items x and y often buy item z. In that context the transactions are a list of products. However, the concept of transactions can be redefined in a more generalized form than the one suggested by Agrawal. In fact, a transaction can be simply defined as a finite subset of data. This generalization allows applying association rules to many domains. The only challenge is to adapt the concept of transaction to targeted domain. When association rules are applied to classes of similarities then classes themselves are considered as transactions.

To explain what an association rule is, consider a set of data consisting of multiple transactions $T_i = \{t_1, t_2, t_3, \dots, t_n\}$. An association rule is denoted $t_i \rightarrow t_k$ where t_i is called the antecedent t_k is called the consequent. It expresses relationship tween t_i and t_k. The quality of an association is calculated using a measure m and a predefined threshold σ_m. Thus, a rule is considered as a good quality rule if $m(t_i \rightarrow t_k) \geq \sigma_m$. Several measures are proposed in the literature and many studies are dedicated to their evaluation [4, 5]. Among existing measures support and confidence are the most common. When association rules are applied to classes of similarities support and confidence are defined as follow:

Support (X) : Let X be a subset of elements spread in different classes of similarities then the support of this subset is denoted $S(X)$. The support of X is equal to the number of classes that contain X. The calculation of support is given by equation 3.1 where n equals the total number of classes.

$$S(X) = \sum_{i=1}^{n} \delta up(T_i) \text{ where } \delta up(T_i) = \begin{cases} 0, & X \not\subseteq T_i \\ 1, & X \subseteq T_i \end{cases} \tag{3.1}$$

A subset X is considered frequent when $S(X)$ is greater than a predefined old σ_s.

Support. (X → Y) : If X and Y are two subsets of elements such as $X \neq \emptyset$ and $Y \neq \emptyset$ then the support of the association rule $X \rightarrow Y$ is denoted $S(X \rightarrow Y)$. The support of an association rule is equal to the number of classes that contain both X and Y. Thus, $S(X \rightarrow Y) \Leftrightarrow S(X \cup Y)$. The calculation of $S(X \rightarrow Y)$ is given by equation 3.2 where n equal the total number of classes.

$$S(X \rightarrow Y) = \sum_{i=1}^{n} \delta up(T_i) \text{ where } \delta up(T_i) = \begin{cases} 0, & X \not\subseteq T_i \vee Y \not\subseteq T_i \\ 1, & X \subseteq T_i \wedge Y \subseteq T_i \end{cases} \tag{3.2}$$

Confidence. $C(X \rightarrow Y)$: The confidence of an association rule $X \rightarrow Y$ is denoted $C(X \rightarrow Y)$. The confidence of an association rule is equal to the number of classes that contain both the antecedent and consequent among the classes that contain the antecedent. The calculation of $C(X \rightarrow Y)$ is given by equation 3.3.

$$C\left(X \rightarrow Y\right) = \frac{S\left(X \rightarrow Y\right)}{S(X)} \tag{3.3}$$

Typically association rules are extracted using APRIORI algorithm [6]. This algorithm used the support and confidence measures to restrict the number of associations considered. Association rules provide a significant advantage: they are able to extract hidden knowledge in classes of similarities even if they are numerous and noisy. Despite their potential, association rules may ignore relevant associations with low frequency of occurrence. For example, if an item X often appears with an item Y and less often with another item Z then it's probable that the association between X and Y was retained and not the association between X, Y and Z. The reason is that the confidence of the relationship between X, Y and Z would be too low compare with the relationship with X and Y.

A maximal association rules is denoted $X \xrightarrow{max} Y$. They are proposed to overcome the constraint related to the frequency of occurrence which applies to association rules (Feldman et al., 1997). Maximal association rules are used to obtain exclusion associations. The form of this kind of association is given by equation 3.4 where S_1, S_2 , S_3 and S_4 are subset of data.

$$S_1 \cup \overline{S_2} \rightarrow S_3 \cup \overline{S_4} \quad (3.4)$$

Special measures are defined to establish the quality of maximal association rules [7]. These measures are the M-Support and the M-Confidence.

M-Support. $S_{max}(X)$: The maximal support or m-support of a subset of items X is denoted $S_{max}(X)$. If E_i and X be two subsets of elements such as $X \subseteq E_i$ then $S_{max}(X)$ is equal to the number of classes that contain X and no other item from Y.

$$E_i \cap T_i = X \quad (3.5)$$

Consider the following elements to illustrate what is the m-support:

$T_1 = \{t_1, t_2\}$
$T_2 = \{t_3, t_5\}$
$T_3 = \{t_3, t_4, t_6\}$
$E_1 = \{t_1, t_2, t_3, t_4\}$
$X = \{t_3\}$ where $X \subseteq E_1$.

Given these elements, $S_{max}(X) = 1$ because:

$E_1 \cap T_1 = \{t_1, t_2, t_3, t_4\} \cap \{t_1, t_2\} = \{t_1, t_2\} \neq X$,
$E_1 \cap T_2 = \{t_1, t_2, t_3, t_4\} \cap \{t_3, t_5\} = \{t_3\} = X$
$E_1 \cap T_3 = \{t_1, t_2, t_3, t_4\} \cap \{t_3, t_4, t_6\} = \{t_3, t_4\} \neq X$.

M-Support. $S_{max}\left(X \xrightarrow{max} Y\right)$: $S_{max}(X \xrightarrow{max} Y)$ is the number of classes that $S_{max}(X)$ and S(Y).

M-Confidence. $C_{max}\left(X \xrightarrow{max} Y\right)$: The maximal confidence or m-confidence is given by equation 3.6 where $|D(X, g(Y))|$ is a subset consisting of the classes that $S_{max}(X)$ and that contain at least one item of Y.

$$C_{max}(X \xrightarrow{max} Y) = \frac{S_{max}(X \xrightarrow{max} Y)}{|D(X, g(Y))|} \quad (3.6)$$

The algorithm used to extract maximal association rules is similar to the one used to extract regular association rules. It extracts rules whose m-support and m-confidence are above predetermined thresholds. Maximal association rules are complementary to the regular association rules. They highlight associations that regular rules tend to ignore. However, if only the maximal association rule is used it may result in loss of interesting associations (Amir et al., 2005). The combined use of association rules and maximal association rules is very interesting during analysis of classes of similarities because they are able to bring out various hidden knowledge.

4 Methodology

Association rules and maximal association rules employ measures that are generic enough and consistent to allow extraction of relevant associations hidden in noisy classes regardless of the classification method used. An association that frequently appears in classes generated with different classification methods (different classifiers or same classifier with different parameters) is called a strong association [8]. Such associations are useful to consolidate results obtained using different classification strategies. In sum, strong associations allow to highlight constant relations that can well describe the content.

The proposed methodology is to use the association rules and maximal association rules to extract recurring patterns within classes of similarities produced using various methods of classification. Captured associations are used like high level descriptors of the content. The process that leads to the extraction of strong associations is composed of the following four major steps: (1) Preparation of the data; (2) Application of a classification method; (3) Extraction of strong associations; (4) Evaluation of results. Steps 1 and 2 can be repeated several times to create a large dataset while steps 3 and 4 are more related to the analysis and data mining.

Preparation of the data: The first step is to create vector representations of data. To do this, the text data is manually imported into the system. In a future work, a Web robot would be used to automatically adding data into the system. Imported data are segmented into sentences, paragraphs or freely depending of the configuration of the system. One vector representation is created per segment. The unit of information considered for build these vector representations is n-gram of characters. A n-gram of characters is defined here as a sequence of *n* successive characters. For instance, in the word *computer* the n-grams are: *com*, *omp*, *mpu*, *put*, *ute* and *ter* when *n* is fixed to 3. The choice of n-gram of characters as a unit of information is justified by the fact that cutting into sequences of *n* consecutive characters is possible in most languages. It's necessary that the unit of information is supported by several languages due of the multiplicity of languages used on the Web. Also n-grams of characters tolerate a certain ratio of distortion [10, 11]. To restrict the size of the lexicon, some processing like deletion of hapax (n-gram that appears only once) can be applied.

Application of a classification method: The second step is to apply a classification method among those available. Three classification methods are available: Fuzzy-ART [12], K-Means [13] and SOM [14]). The choice of these methods is not dictated by specific technical reasons. It is motivated by the interest of these methods in the literature [15, 16]. In fact, the modular architecture of our system encourages the integration of new methods of classification. After this step, classifications are produced.

Extraction of strong associations: Extraction of strong associations is carried out using the method outlined in [9] where the similarity classes are treated as transactions. The vocabulary is used to build the subset E_1 in which X is selected. In other words, E_1 is a subset of the vocabulary and X $\subseteq E_1$. In the simplest scenarios E_1 coincide with X. Otherwise, the value of X can be chosen by a user related his objective or randomly selected by the system. Depending on X, the system generates the subsetE_2. This subset is used to define the possible consequents. E_2 is built by subtracting X of the union of all transactions (classes of similarities) where X appears. For illustration consider the following elements:

$T_1 = \{t_1, t_2, t_3,\}$ $T_2 = \{t_2, t_4, t_5\}$ $T_3 = \{t_1, t_5\}$

Let say X = $\{t_3\}$ then E_2' equal the union of all transactions who contain t_3 more exactly the union of T_1 and T_3 or $\{t_1, t_2, t_3, t_5\}$. E_2 is created by subtracting X to E_2' : $E_2 = E_2' - \text{X} = \{t_2, t_3, t_5\}$. Each subset contained in E_2 is considered as a candidate. Thus, the subsets $\{t_2\}$, $\{t_3\}$, $\{t_5\}$, $\{t_2, t_3\}$, $\{t_2, t_5\}$, $\{t_3, t_5\}$ and $\{t_2, t_3, t_5\}$ may be consequents. Measures of support, confidence, m-support and m-confidence are finally calculated for each subset. To avoid a computational cost too significant, the cardinality of the consequents can be fixed.

Evaluation of results: Evaluation is the final step. To support this step, the system provides an interface for easily navigating between rules, classes of similarities and their origin.

5 Experimentations

To realize our experimentations we used a system previously developed to assist interpretation of class of similarities [8, 9]. This system was implemented in C#. The results of the analyses are stored in XML databases.

The experiments we will show (We limit ourselves to show just maximal associations and scores of each association (M-support and M-Confidence). We assume that the reader is sufficiently familiar with the methods of classification and we do not need to show classes of similarities) were applied to a corpus extracted from Tripadvisor.fr. It consists of 45 guest reviews on a Parisian hotel. Reviews are written in French. The choice of this type of corpus is dictated by our desire to have a diverse vocabulary without being too large. The aim of these experiments is not to demonstrate the relevance of some classifiers or to demonstrate the ability of our approach to handle large documents but rather to assess the ability of our approach to extract strong associations between lexical units. To do this, several classifications have been produced using K-Means, Fuzzy-ART and SOM. The same vector representations were used as input for the three methods. To support a certain deformation data associated with the presence of errors in spelling we used tri-grams of characters as the unit of information. To reduce the number of tri-grams considered, we removed the tri-grams that contain spaces, numbers, special characters and those whose frequency of occurrence was less than 3. The size of the vector equals the number of distinct trigram enumerated in the corpus after cleaning. A vector was created for each corpus review. Moreover, to get the same number of classes (even if the classes, because of their vocabulary, are not similar) and have as much as possible the same basis of comparison, we set the classifiers in order to obtain 24 classes of similarities. For a sample of selected words (*Hôtel, Chambre, Déjeuner*), we tried to find in the classes of similarity obtained with all three classifiers, the words that co-occur with them. We find that despite the differences we observe in the results of the three classifications, the process of extracting the maximal association rules allows identifying the strongest associations which are spread over all classes obtained for each classification. In the case of our experiment, these associations appear to be, approximately, the same for all three classifications. We give in the tables below associations (whose m-support is greater than 1) extracted from classes obtained with the three classifiers. Higher are the values of M-Support and M-Confidence, stronger are the associations.

We note that in the case where X = *Hôtel*, the first 3 extracted associations are the same regardless of the classifier used (tables 1, 2 and 3). In the case where X = *Chambre*, 4 of the first 5 extracted associations are the same (tables 4, 5 and 6). In the case where X = *Déjeuner*, the first 2 extracted associations are the same regardless of the classifier used (tables 7, 8 and 9). However, if we just consider results of Fuzzy-Art and SOM, we note that 5 of the first 6 extracted associations are the same (tables 8 and 9).

Table 1. First maximal association rules (classes obtained with K-Means)

X	Y	M-Support	M-Confidence
Hôtel	*Gare*	3	75%
	Chambre	3	75%
	Petit	3	75%
	Nuit	2	50%
	Dan	2	50%
	Nord	2	50%

Table 2. First maximal association rules (classes obtained with Fuzzy-Art)

X	Y	M-Support	M-Confidence
Hôtel	*Petit*	6	100%
	Chambre	5	83,33%
	Gare	4	66,66%
	Personnel	3	50%
	Déjeuner	3	50%
	Salle	3	50%
	Nord	3	50%

Table 3. First maximal association rules (classes obtained with SOM)

X	Y	M-Support	M-Confidence
Hôtel	*Petit*	4	80%
	Chambre	4	80%
	Gare	4	80%

Table 4. First maximal association rules (classes obtained with K-Means)

X	Y	M-Support	M-Confidence
Chambre	*Gare*	8	57,14%
	Petit	6	42,86%
	Nord	6	42,86%
	Salle	4	28,57%
	Hôtel	3	21,43%
	Dan	3	21,43%
	Déjeuner	3	21,43%

Table 5. First maximal association rules (classes obtained with Fuzzy-Art)

X	Y	M-Support	M-Confidence
Chambre	*Petit*	8	58,82%
	Hôtel	6	29,41%
	Nord	6	35,29%
	Gare	4	41,18%
	Déjeuner	3	23,53%

Table 6. First maximal association rules (classes obtained with SOM)

X	Y	M-Support	M-Confidence
Chambre	*Gare*	10	62,5%
	Petit	9	56,5%
	Nord	7	43,75%
	Accueil	5	31,25%
	Hôtel	4	25%
	Déjeuner	4	25%
	Situer	4	25%

Table 7. First maximal association rules (classes obtained with K-Means)

X	Y	M-Support	M-Confidence
Déjeuner	*Petit*	3	100%
	Chambre	3	100%

Table 8. First maximal association rules (classes obtained with Fuzzy-Art)

X	Y	M-Support	M-Confidence
Déjeuner	*Chambre*	4	100%
	Petit	4	100%
	Gare	3	75%
	Nord	3	75%
	Hôtel	3	75%
	Salle	2	50%
	Bain	2	50%

Table 9. First maximal association rules (classes obtained with SOM)

X	Y	M-Support	M-Confidence
Déjeuner	*Chambre*	4	100%
	Petit	4	100%
	Gare	3	75%
	Hôtel	2	50%
	Salle	2	50%
	Situer	2	50%

Associations extracted from the classes of similarity are used to identify in the corpus relevant information that may represent the general labels of the object on which reviews were issued. To do this our system can identify in the corpus all segments that contain the associations and from which the labels will be extracted semi-automatically. In the case of our examples: *Hôtel* (hotel) is strongly associated with *Gare*, regardless of the classifier used, because the hotel is located next to a railway station (*gare*) and this association was often mentioned. It would become a label. It is important to mention that we show here that association rules approach can extract cooccurrences common to all three classifications. Therefore, these cooccurrences are called: strong associations. We completed our experiment with applying the process of association rules extraction over all classes obtained without distinguishing classifiers that allowed get them. Our initial goal was to verify the persistence of associations derived from classes in each classification. Our second goal was to test a process of meta-classification. We obtained the strong associations given in tables 10.

Table 10. First maximal association rules

X	Y	M-Support	M-Confidence
Hôtel	*Petit*	11	84,62%
	Chambre	10	76,92%
	Gare	9	69,23%
	Nord	6	46,15%
	Déjeuner	5	38,46%
	Accueil	4	30,76%
Chambre	*Petit*	22	53,66%
	Gare	21	51,22%
	Nord	16	39,02%
	Hôtel	10	24,39%
	Salle	8	19,51%
	Déjeuner	8	19,51%
Déjeuner	*Chambre*	8	100%
	Petit	8	100%
	Gare	6	75%
	Hôtel	5	62,5%
	Nord	5	62,5%
	Salle	4	50%

We note that the strongest associations are persistent. The "meta-classification" highlights common associations extracted from classes obtained by the three classifiers.

With these experiments, we have demonstrated that it is possible to extract strong associations in classes of similarity, regardless of the classifier used. These associations are relevant clues due to the regularity of their co-occurrence. The user (which is not necessarily an expert of the domain or a language engineer) can, according to the associations rules and their strength (given by the M-Support and the M-Confidence), select lexical descriptors that he/she thinks appropriate. Thus, in the examples discussed above, beside to *hôtel* as descriptor, the user can decide that *petit* and *gare* are important descriptors of the hotel. These descriptors remain clues that allow the user direct access to the part (or parts) of the text to which extracted associations refer.

6 Conclusion

Several strategies are employed to facilitate the exploration of textual documents. The labeling is one of the most common. The amount of textual documents available on networks required some mechanism to assist the selection of tags used. In this paper we had shown that association rules and maximal association rules can be applied to extract strong association in a set of classes of similarities. Strong associations can be considerate like stable descriptors of content. We believe that when the antecedent of a valuable rule is used as descriptor of a document it can be useful to add the consequent. Following that assumption can assist the selection of metadata. Because the proposed method is based on co-occurrence relations, it is not limited to providing assistance when labeling a document. The proposed method can be used for several other applications like lexical disambiguation, information retrieval, knowledge extraction and computer assisted reading.

References

1. Bertin-Mahieux, T., Eck, D., Mandel, M.: Automatic Tagging of Audio: The State-of-the-Art. In: Wang, W. (ed.) Machine Audition: Principles, Algorithms and Systems. IGI Publishing (2010)
2. Estes, W.K.: Classification and Cognition. Oxford University Press (1994) ISBN 0-19-510974-0
3. Agrawal, R., Imielinski, T., Swami, A.: Minning association rules between sets of items in large databases. In: Proceedings of the ACM SIGMOD Conference on Management of Data, pp. 207–216 (1993)
4. Le Bras, Y., Meyer, P., Lenca, P., Lallich, S.: Mesure de la robustesse de règles d'association. In: Proceedings of the QDC 2010, Hammamet, Tunisie (2010)
5. Vaillant, B.: Mesurer la qualité des règles d'association : études formelles et expérimentales, Thesis École Nationale Supérieure des Télécommunications of Bretagne (2006)
6. Agrawal, A., Srikant, R.: Fast Algorithms for Mining Association Rules. In: Poceedings of the 20th International Conference on Very Large Database, pp. 487–499 (1994)
7. Amir, A., Aumann, Y., Feldman, R., Fresko, M.: Maximal Association Rules: A Tool for Mining Associations in Text. Journal of Intelligent Information System 25(3), 333–345 (2005)
8. Biskri, I., Rompré, L., Descoteaux, S., Achouri, A., Amar Bensaber, B.: Extraction of Strong Associations in Classes of Similarities. In: Proceedings of IEEE/ICMLA, Boca Raton, Florida, USA (2012)
9. Biskri, I., Hilali, H., Rompré, L.: Extraction de relations d'association maximales dans les textes. Actes du JADT, 173–182 (2010)
10. Miller, E., Shen, D., Liu, J., Nicholas, C., Chen, T.: Techniques for Gigabyte-Scale N-gram Based Information Retrieval on Personal Computers. In: Proceedings of the PDPTA 1999, Las Vegas, U.S.A. (1999)
11. Damashek, M.: Gauging Similarity with n-Grams: Language-Independent Categorization of Text. Science 267, 843–848 (1995)
12. Carpenter, G., Grossberg, S.: Fuzzy ART: Fast Stable Learning and Categorisation of Analog Patterns by an Adaptative Resonance System. Neural Network 4, 759–771 (1991)
13. MacQueen, J.: Some Methods for classification and Analysis of Multivariate Observations. In: Proceedings of the 5th Berkeley Symposium on Mathematical Statistics and Probability. University of California Press (1967)
14. Kohonen, T.: The Self-Organisation Map. Proceedings of the IEEE 78(9), 1464–1480 (1990)
15. Anderson, J.: An Introduction to Neural Network. MIT Press (1995) ISBN 0-262-01144-1
16. Haykin, S.: Neural Networks: A Comprehensive Foundation. Macmillan College Publishing Company (1994) ISBN 0-02-352761-7

Statistical and Constraint Programming Approaches for Parameter Elicitation in Lexicographic Ordering*

Noureddine Aribi[1,2] and Yahia Lebbah[1,2]

[1] Laboratoire LITIO, Université d'Oran, BP 1524, El-M'Naouer, 31000 Oran, Algérie
{aribi.noureddine,ylebbah}@gmail.com
[2] Laboratoire I3S/CNRS, Université de Nice - Sophia Antipolis, Nice, France

Abstract. In this paper, we propose statistical and constraint programming approaches in order to tackle the parameter elicitation problem for the lexicographic ordering (LO) method. Like all multicriteria optimization methods, the LO method have a parameter that should be fixed carefully, either to determine the optimal solution (best tradeoff), or to rank the set of feasible solutions (alternatives). Unfortunately, the criteria usually conflict with each other, and thus, it is unlikely to find a convenient parameter for which the obtained solution will perform best for all criteria. This is why elicitation methods have been populated in order to assist the Decision Maker (DM) in the hard task of fixing the parameters. Our proposed approaches require some prior knowledge that the DM can give straightforwardly. These informations are used in order to get automatically the appropriate parameters. We also present a relevant numerical experimentations, showing the effectiveness of our approaches in solving the elicitation problem.

Keywords: Parameter Elicitation, Multicriteria Optimisation, Constraint Programming, Statistics, Lexicographic ordering.

1 Introduction

Many methods for solving multicriteria optimization problems exist, and it is not so simple to choose a method well adapted to a given problem. Moreover, even after a multicriteria method has been selected, different parameters (e.g., some weights, some utility functions,...) need to be determined. This may be even more difficult, and elicitation methods are sometimes used to assist the decision maker in this task.

In this paper, we focus on the lexicographic ordering method [8], which is appropriate when a strict dominance relation between the criteria can be established. The parameter of this method is a total ordering of the criteria. We assume that we have some sampling data (alternatives or observations) along

* This work is supported by TASSILI research program 11MDU839 (France, Algeria). This work is also supported by PNR research project 70/TIC/2011 (Algeria).

A. Amine et al. (Eds.): *Modeling Approaches and Algorithms*, SCI 488, pp. 47–56.
DOI: 10.1007/978-3-319-00560-7_9 © Springer International Publishing Switzerland 2013

with their outcome values (cf. [1]). For instance, consider a situation in which a seller wishes to explain/justify the price of his different products, say digital cameras, to his customers. A digital camera is characterized by several criteria: its number of mega-pixels, its weight, the maximal zoom range,... The need of the seller is to have a justification for his customers, which is both simple and consistent with its products. The customers of this seller do not understand subtle tradeoffs between criteria. Indeed, they can only understand that one criterion (e.g., zoom range) is more important than another one (e.g., weight). So, the challenge here consists to find a permutation between the criteria that best justifies the product's prices. In this paper, we discuss how the Spearman's coefficients can be used to heuristically elicit a robust permutation of criteria for a lexicographic ordering method. We have compared this approach with a multiple linear regression approach, and with an exact method based on Constraint Programming [14]. This comparison was performed by measuring the discrepancy between the ordering obtained and the ordering given by the outcome.

The rest of this paper is organized as follows. Section 2 gives a formulation of the multicriteria elicitation problem. Section 3 illustrates the statistical and exact approaches on a motivating example. Section 4 describes with details the three proposed approaches. Our experiments are described and discussed in Sections 5 & 6. Section 8 concludes the paper with some perspectives.

2 Problem Formulation

The problem of parameter elicitation related to the lexicographic ordering method can be described with the following main elements:

- A set of criteria: $X = \{X_1, ..., X_n\}$, where $n \geq 2$;
- A set of alternatives $A = \{A_1, ..., A_m\}$;
- A set of preference relations, given either by:
  - An outcome vector (Y), so that one outcome value is associated to each solution. Values in Y are either gathered experimentally, or collected from responses of a questionnaire survey of preference (cf. [1]).
  - or a set of symbolic preferences: $\{(A_i^k, A_j^k) \mid A_i^k \prec_{DM} A_j^k, i, j \in [1..m]\}_{k=1}^{L}$, where $A_i^k \prec_{DM} A_j^k$ means that the DM finds A_i^k better than A_j^k.

In our approaches we make the assumption that we have prior information about the preferences of decision maker, and we focus on how to use these information, rather than how to get them.

Definition 1 (Lexicographic order). *[10] The criteria are ordered by the decision maker according to their perceived importance. Hence, the solution (alternative) with the best value for the most important criterion is selected. Tied solutions are evaluated using their performances in the second most important criterion, and so on. Formally, let $x, y \in \mathbb{R}^n$,*

$$x <_{lex} y \Longleftrightarrow (x_0 < y_0) \vee [(x_0 = y_0) \wedge \langle x_1, ..., x_{n-1}\rangle <_{lex} \langle y_1, ..., y_{n-1}\rangle] \quad (1)$$

In order to evaluate and compare fairly between our elicitation approaches, we have defined an optimality criterion described below.

Definition 2 (Discrepancy measure).
Let $A_Y = (A_1, ..., A_n)$ *be the optimal ordering of the alternatives according to their outcome values,* Y*. More precisely,* $rank(A_i, Y^{\uparrow}) < rank(A_{i+1}, Y^{\uparrow}), i = 1..n-1$*, where* $rank(A_i, Y^{\uparrow})$*, denotes the rank of the outcome value of* A_i *in the sorted outcomes vector,* $Y^{\uparrow}$*. Let now* $A_\theta = (A_{\sigma(1)}, ..., A_{\sigma(n)})$ *be the lexicographic ordering of the alternatives according to a permutation between criteria* θ*. Hence, the proposed discrepancy measure between* A_θ *and* A_Y*, is given by the sum of truth values:*

$$D(A_\theta, A_Y) = \sum_{i=1}^{n-1} v_i, \; where \; v_i = \begin{cases} 1 \; if \; \sigma(i) > \sigma(i+1), & \text{\% violation case} \\ 0 \; otherwise & \text{\% non-violation case} \end{cases} \tag{2}$$

Definition 3 (Optimal permutation). *A permutation* θ^* *is said to be optimal, if and only if for any permutation* θ*,*

$$D(A_\theta, A_Y) \geq D(A_{\theta^*}, A_Y) \tag{3}$$

Note that for a multicriteria problem with n criteria, we have $n!$ possible permutations. So, the space of possible permutations has a size being exponential in the number of criteria. Therefore, the challenge consists to find the best permutation without going into all possibilities.

3 Motivating Example

Let us reconsider the introductory example, in which we have one outcome variable Y (the *price*), that the seller wishes to explain/justify to his customers. A sampling data of eighteen apparatus are shown in Table (1), where X_1 indicates the *weight*, X_2 the *resolution* of image sensor, and X_3 the *zoom* of the apparatus.

Table 1. Sample data for the Seller Problem

#	1	2	3	4	5	6	7	8	9	10	11	12	13	14	15	16	17	18
X_1	440	510	470	400	340	390	250	250	360	220	320	280	310	300	280	200	270	210
X_2	2	2	4	4	6	8	8	12	14	14	18	20	20	22	24	26	28	30
X_3	5	2	2	4	12	1	5	6	8	7	7	9	10	9	3	12	14	13
Y	389	416	421	425	434	449	461	465	468	473	478	484	485	488	527	529	532	566

Our approaches aim to find a permutation between criteria that minimizes the optimality criterion given by Equation (2). Following the Spearman-based approach [5], we compute the correlation coefficient, r_s, for each criterion. So, we obtain a vector of coefficients, $\langle -\underline{.763}, \underline{.997}, \underline{.689} \rangle$ for X_1, X_2 and X_3 respectively. Underlined values (e.g., $\underline{.997}$) means that the result is statistically significant using a significance level of 5%. Next, the sorting permutation of this vector (in decreasing order), i.e., $\langle X_2, X_3, X_1 \rangle$ heuristically designates the most interesting permutation, θ_{rs}, according to this approach. Similarly, the Multiple Linear Regression approach [12] seeks to find the most prominent permutation by examining the vector of the standardized regression coefficients (so called *beta weights*). For the studied example, the vector of beta weights is equals to $\langle \overline{.11}, \underline{.98}, \overline{.19} \rangle$. Here, the overlined values indicates that the result is statistically insignificant according to MLR approach. The permutation derived from this vector is then $\theta_{mlr} = \langle X_2, X_3, X_1 \rangle$, which is similar to the permutation, θ_{rs}, given by the Spearman-based approach. Unlike the MLR approach, the Spearman approach has reported significant coefficients, which makes this approach much robust compared to MLR approach. Finally, the result of the exact method based on Constraint Programming has revealed the same permutation, i.e., $\theta_{cp} = \theta_{rs} = \theta_{mlr}$, meaning that, for our case study, the results of the statistical approaches was optimal. In this paper, we would like to highlight the relevance of the proposed statistical approaches, by comparing them with the CP approach.

4 Approaches for Parameter Elicitation

In the present work, we propose three approaches to address the problem of parameter elicitation regarding the lexicographic ordering (LO) method.

4.1 Spearman's Correlation Coefficient Based Approach

In what follows we focus on how the monotonicity relationship can be exploited to heuristically determine a robust permutation between criteria.

Definition 4 (Spearman's Rank Correlation Coefficient). *Spearman's rank correlation coefficient (r_s) [5] is a nonparametric measure of correlation between two variables X and Y. It assesses how well a monotonic function could describe the relationship between two ranked variables.*

Spearman's correlation coefficient varies from -1 to $+1$ and the absolute value of r_s describes the strength of the monotonic relationship. Prior to using the Spearman based approach, we should check the *Null hypothesis* (H_0). This hypothesis states that there is no association between the variables in *the underlying population.* Usually, a *Significance test* (p-value) [17], is used to investigate whether to accept or reject H_0, according to a significance level[1], α, most commonly equals to .05 or 5%. So, a low enough p-value (i.e., p-value $< \alpha$) is considered to be sufficient to reject H_0, and thus, the result is said to be statistically significant.[2]

[1] Significance levels indicate how likely a result is due to chance.

[2] The p-value does not measure the significance of the correlations, but it might be useful in deciding whether a correlation exists at all.

The Proposed Algorithm. Algorithm (1) implements the statistical approach based on Spearman's rank coefficient (Spearman's *rho*). It is described as follows: First, we start by converting the outcome variable Y to ranks (*inst.* 1/RANK function). Then, we iterate over all criteria in X (*loop* 2−5). So, for each criterion X_i, we compute its ranked vector x_i (*inst.* 3). After that, the Spearman's *rho* is computed for the two vectors of ranks (*inst.* 4/COMPUTERHO function). The resulting correlation coefficient is added to the list of correlation coefficients (*inst.* 5). Instruction (6) sorts the list of correlation coefficients, *rhoList*, and stores the result in p. Finally, Algorithm (1) returns p as a robust permutation between criteria (*inst.* 8).

Algorithm 1. STATELICIT finds a Spearman-based permutation between criteria

```
   Input: A finite set of criteria: X = {X_1, ..., X_n}; a vector of outcomes Y
   Output: The optimal permutation between criteria, p
1  y ← RANK(Y)
2  for i ← 1 to n do
3  |   x_i ← RANK(X_i)
4  |   rho ← COMPUTERHO(x_i, y)
5  |_  rhoList ← rhoList ∪ {rho}
6  p ← SORT(rhoList)
7  return p
```

As we will see in the experimental section, this algorithm usually succeeds to find the optimal permutation. However, we have found some particular cases where this heuristic approach does not capture the optimal solution.

4.2 Multiple Linear Regression-Based Approach

This statistical approach is based on Multiple Linear Regression (MLR) [12], and aims to infer a robust parameter (permutation) from the standardized regression coefficients, most commonly known as *beta weights*.

Background. Multiple linear regression is a type of regression for predicting the value of one dependent variable from the values of two or more independent variables. The matrix representation of the MLR model is: $Y_{m1} = X_{m(n+1)}B_{(n+1)1} + U_{m1}$ where B is the degree of explanation of each predictor, while U is the residual vector. The optimal solution of an MLR problem, is computed by minimizing the sum of squared residuals, $Min(\hat{U}^T\hat{U})$. Therefore, the optimal solution is given by the following formula:

$$\hat{B} = (X^T X)^{-1} X^T Y \tag{4}$$

where the hat symbol is used to denote the estimated regression coefficients. When investigating the MLR approach, several important assumptions should be fulfilled, namely, sample size[3], linearity, normality, homoscedasticity, autocorrelation, multicollinearity, outliers (cf. [2]). Moreover, the quality of the multiple regression is measured using the R^2 coefficient [16]. It is interpreted as the proportion of the variance in the dependent variable Y that is predictable (or explained) from the independent variables. Besides, beta weight coefficients are heavily relied on to assess variable importance [11]. A beta weight for an independent variable indicates the expected increase or decrease in the dependent variable, in standard deviation units, given a one standard-deviation increase in independent variable with all other independent variables held constant.

Elicitation. The idea behind this approach consists to predict the robust permutation between criteria, by looking at the relative importance of each criterion. In the context of this paper, the independent variables consist of the set of criteria, while the dependent variable refers to the outcome variable. After the elicitation problem has been transformed to an MLR model, this approach proceeds through three steps, (i) see whether the MLR assumptions are fulfilled; (ii) solve the regression model using Equation (4), and check the goodness of the model fit using R^2 coefficient; (iii) sort the resulting beta weights in decreasing order, to heuristically retrieve a robust permutation between criteria. Thus, the criterion with the highest beta weight corresponds to the most important criterion, and so on. Obviously, the major drawback of this approach is the linearity assumption, since in practice most relationships are nonlinear. Moreover, even if all the underlying assumptions are met, this approach may mess the optimal solutions.

4.3 Constraint Programming Approach

Modern Constraint Programming languages aim at making easy problems formulation and solving [14]. One of the key success of CP is global constraints design, such as ALL-DIFFERENT [13] and ELEMENT constraints [9] exploited in our CP optimization model. This model is described in Algorithm (2), where:

- D_p (*inst.* 1), specifies the domain of variables in p.
- Variables in p, must be pairwise different (*inst.* 2).
- The preferences are derived from the outcome vector, and are handled using *Reified Constraints*[4] (*loop.* 3−4). So, for each preference relation $A_i \prec_{DM} A_j$ we post the reified constraint $(A[i] <_{lex} A[j]) \iff b$, where $b \in \{0,1\}$. When $A_i \simeq_{DM} A_j$ nothing is done, since each alternative (A_i and A_j) outranks the other, and hence this case does not affect the discrepancy degree.
- The objective is modelled using a *cost function* (*inst.* 5), with which the number satisfied preferences is maximized.

---

[3] A rule of thumb suggested in [16] that the sample size should satisfy the equation $m \geq 50 + 8n$, where n is the number of criteria (predictors).

[4] Reified constraints reflect the validity of a constraint C into a 0/1 finite domain.

Algorithm 2. LEXMAXCSP finds the optimal ordering between criteria

Input: l preference relations; n criteria; m alternatives: $A_l, ..., A_m$
Output: p: Optimal permutation between criteria
1 $D_p \leftarrow \{1, ..., n\}$
2 ALL-DIFFERENT(p)
3 **for** $k \leftarrow 1$ ***to*** l **do**
4 $\quad$ $preference[k] \leftarrow$ LEXLE($A[i], A[j], p$)
5 MAXIMIZE($\sum_{k=1}^{l} preference[k]$)
6 **return** p

Algorithm 3. LEXLE finds a permutation between criteria, so that $x <_{lex} y$

Input: x and y: two vectors of n integer values
Output: A permutation p
1 $lex_{n-1} \leftarrow (x[p[n-1]] < y[p[n-1]]) \vee ((x[p[n-1]] = y[p[n-1]]) \wedge (x[p[n]] < y[p[n]]))$
2 **for** $k \leftarrow n-2$ ***downto*** 1 **do**
3 $\quad$ $lex_k \leftarrow (x[p[i]] < y[p[i]]) \vee ((x[p[i]] = y[p[i]]) \wedge (lex_{k-1}))$
4 **return** lex_1

Algorithm (3) is an iterative version of the recursive Formula (1). Thus, given two vectors X an Y, this algorithm looks for a permutation between criteria, that allows to X to be less than Y in the lexicographical sense of the term. Unlike the statistical approaches presented earlier, the proposed CP approach computes the optimal permutation by construction.

5 Experimental Results

We evaluated experimentally the results of our elicitation approaches on a set of realistic instances [16,4]. Experiments were run using the following frameworks: (i) IBM ILOG Concert C++ and CP Optimizer (version 12.5) to implement and solve our CP model. (ii) IBM SPSS Statistics (version 20.0) to test our statistical approaches. These experiments have been done on x86_64 machine with Intel Xeon (x5460) 4xCores @3.16 Ghz and 8 Gb of free memory. Additionally, we have used the discrepancy measure using Equation (3) to compare between the proposed approaches.

Now, let us reconsider our motivating example (see Section 3), for which we have attempted to apply the MLR approach. This approach proceeds through three main steps. In the first step, we should verify that the MLR assumptions are met by interpreting the SPSS output sheet. Because of lack of space, we do not present details about this step. As a second step, we evaluate the usefulness of our regression using the R^2 coefficient, which is given at 94.8%. That is a good regression. Notice that we cannot proceed to the next step unless this condition is fulfilled. The final step consists to access the contribution of each criterion in predicting the variance of the outcome variable, in an attempt to determine a permutation between criteria that optimizes the discrepancy measure. H_0 hypothesis was accepted for both X_1 and X_3 criteria, since their p-values were above the significance level at $\alpha = 5\%$. This means that the computed permutation cannot be generalized to the underlying population. However, the *Null* hypothesis was rejected for X_2 criterion, because its coefficient is significant (i.e., p-value $< 5\%$). Now, to heuristically predict a robust permutation between criteria, we need to sort the obtained vector of beta weights (i.e., $\langle .11, .89, .19 \rangle$) in decreasing order. Lastly, we compute the sorting permutation, i.e., $\langle X_2, X_3, X_1 \rangle$, as the robust permutation between criteria. The major limitation of this approach is the linearity assumption. Knowing that even if all assumptions are met, MLR does not give the optimal solution. Therefore, we will allow to use it even if certain assumptions are not verified, and see the relevance of the solution compared to optimal solution provided by the exact method.

Table (2) shows the results of our experiments conducted on a data set of 11 instances. The Column CP reports the solving time (in seconds) of the exact method, with a timeout (TO) of 3,600 *sec* in all experiments. Whereas, columns nbC, nbA, provide the number of criteria, and the number of alternatives respectively. In columns MLR and SP, the right-hand side of the slash symbol refers to the optimal discrepancy measure computed by the CP approach, while the left-hand side designates the discrepancy measure of the statistical approaches. Since both statistical approaches are based on heuristics, their results are not directly comparable with the results of the exact method. However, we decided to report the degree of optimality instead of running time. As we can see from the results, the MLR approach was outperformed by both CP and Spearman-based approaches in almost all benchmarks. Additionally, compared to the results of the CP approach, it is worth observing that the Spearman based approach obtains the optimal value in most of the studied instances.

Table 2. Experimental results for the proposed approaches. *nbC, nbA* are number of criteria and alternatives respectively, *MLR, SP* indicate the discrepancy measure of both MLR and Spearman-based approaches, compared with the discrepancy measure of the CP approach, # is the benchmark number.

#	nbC	nbA	MLR	SP	CP
1	3	465	**226**/226	**226**/226	.10
2	5	465	214/213	**213**/213	1.22
3	6	465	242/217	219/217	6.34
4	8	465	228/222	**222**/222	5.77
5	8	369	186/162	**162**/162	10.97
6	9	434	210/203	206/203	1184.00
7	11	175	84/76	**76**/76	2.73
8	11	465	231/212	214/212	386.55
9	18	312	148/–	139/–	TO
10	19	445	75/72	**72**/72	57.55
11	44	369	176/160	161/160	2,793.69

6 Discussion

Our constraint programming (CP) approach gives the optimal permutation by construction. However, the computed permutation is proved to be optimal only for the sampling data. Instead, the statistical approaches, in particular, the Spearman's based approach, are more convenient, due to their inferential power in drawing a conclusion on the whole decision space. To illustrate this phenomena, let us suppose that for a given multicriteria problem, the optimal permutation, with respect to the discrepancy criteria (see Equation 2), is given by $\langle X_1, X_2, \{X_3, ..., X_k\}\rangle$, with a discrepancy measure of v. Meaning that, whatever permutation of $\langle X_3, ..., X_k\rangle$ we choose, we will find v violations. Whereas the statistical approaches access the strength of each criterion, and thus return a single and a robust permutation, according to a specific significance level, in time linear in the number of criteria and alternatives. As it has been mentioned before, the MLR approach has a severe drawback, which is in fact, due to the linearity assumption. Nevertheless, if a nonlinearity appears, one may be able to incorporate into the model an appropriate linearizing techniques (e.g., *log*, or the family of powers and roots transformations).

7 Related Works

The problem of elicitation in multicriteria decision making has been also tackled by many works in decision making community. The main focus of these works is on preference elicitation [6,3,7,15]. These works encompass both (pro)active and iterative learning procedures, and are advocated as means to restrict the domain of admissible preferences which enables to make a good decision. Here, we suppose that the multicriteria decision making is done in the context of a multicriteria method. For instance, the paper [1] tackles parameter elicitation in the context of OWA methods. The main contribution of our work is on the use of the Spearman correlation coefficient, and CP techniques to handle efficiently the elicitation problem in the Lexicographic method.

8 Conclusion and Perspectives

In this paper we have proposed to use the Spearman based approach toward solving the problem of parameter elicitation regarding the Lexicographic ordering method. We have used a discrepancy measure to compare this approach with a multiple linear regression approach, and with an exact method. We have shown that the MLR approach was very restrictive, since it involves several hypotheses difficult to meet in practice. We have also showed a relevant empirical results, carried on some problems provided in the literature. These results are encouraging enough to merit further investigation of others multicriteria methods, especially when fairness becomes a concern.

Due to the simplicity, speed, and robust results provided by the Spearman based approach, we believe that combining this approach along with an exact optimization method is very interesting to investigate.

Acknowledgements. The authors would like to thank Olivier Lhomme for his valuable support.

References

1. Beliakov, G.: How to build aggregation operators from data. International Journal of Intelligent Systems 18, 903–923 (2003)
2. Berry, W.D.: Understanding Regression Assumptions. Quantitative Applications in the Social Sciences, vol. 92. SAGE Publications (1993)
3. Boutilier, C., Regan, K., Viappiani, P.: Simultaneous elicitation of preference features and utility. In: Proceedings of the Twenty-fourth AAAI Conference on Artificial Intelligence (AAAI 2010), pp. 1160–1167. AAAI Press, Atlanta (2010)
4. Brase, C.H., Brase, C.P.: Understandable Statistics: Concepts and Methods. Cengage Learning (2011)
5. Corder, G.W., Foreman, D.I.: Nonparametric Statistics for Non-Statisticians: A Step-By-Step Approach. Wiley (2009)
6. Delecroix, F., Morge, M., Delecroix, F.: An algorithm for active learning of lexicographic preferences. In: Pirlot, M., Mousseau, V. (eds.) Proc. of the Workshop from Multiple Criteria Decision Aiding to Preference Learning, pp. 115–122 (November 2012)
7. Escoffier, B., Lang, J., Öztürk, M.: Single-peaked consistency and its complexity. In: Proceedings of the 2008 Conference on ECAI 2008: 18th European Conference on Artificial Intelligence, pp. 366–370. IOS Press, Amsterdam (2008)
8. Figueira, J., Greco, S., Ehrgott, M.: Multiple Criteria Decision Analysis: State of the Art Surveys. Springer, Boston (2005)
9. Van Hentenryck, P.: Constraint satisfaction in logic programming. Logic Programming. MIT Press (1989)
10. Marler, R.T., Arora, J.S.: Survey of multi-objective optimization methods for engineering. Structural and Multidisciplinary Optimization 26, 369–395 (2004), doi:10.1007/s00158-003-0368-6
11. Nathans, L.L., Oswald, F.L., Nimon, K.: Interpreting multiple linear regression: A guidebook of variable importance. Practical Assessment, Research & Evaluation 17(9), 21–62 (2012)
12. Pedhazur, E.J.: Multiple regression in behavioral research: explanation and prediction. Harcourt Brace College Publishers (1997)
13. Régin, J.-C.: A filtering algorithm for constraints of difference in csps. In: Hayes-Roth, B., Korf, R.E. (eds.) AAAI, pp. 362–367. AAAI Press, The MIT Press (1994)
14. Rossi, F., Van Beek, P., Walsh, T.: Handbook of Constraint Programming. Foundations of Artificial Intelligence, vol. 35. Elsevier (2006)
15. Roy, B., Bouyssou, D.: Aiding Decisions with Multiple Criteria: Essays in Honor of Bernard Roy. International Series in Operations Research & Management Science. Springer (2002)
16. Tabachnick, B.G., Fidell, L.S.: Using Multivariate Statistics, 5th edn. Allyn & Bacon, Inc., Needham Heights (2006)
17. Zar, J.H.: Significance testing of the spearman rank correlation coefficient. Journal of the American Statistical Association 67(339), 578–580 (1972)

Some Global Measures for Shape Retrieval

Saliha Bouagar and Slimane Larabi

Computer Science Department
University of Sciences and Technology HOUARI BOUMEDIENE
{sbouagar,slarabi}@usthb.dz

Abstract. In this paper, we propose an efficient shape retrieval method. The idea is very simplistic, it is based on two global measures, the ellipse fitting and the minimum area rectangle. In this approach we don't need any information about the shape structure or its boundary form, as in most shape matching methods, we have only to compute the relativity between the surface of the shape and both of the minimum area rectangle encompassing it and its ellipse fitting. The proposed method is invariant to similarity transformations (translation, isotropic scaling and rotation). In addition, the matching gives satisfying results with minimal cost. The retrieval performance is illustrated using the MPEG-7 shape database.

Keywords: ellipse fitting, minimum area rectangle, shape retrieval.

1 Introduction

In shape-based retrieval methods, the major goal is to search all shapes in the database that are similar to a query according to a predefined representation. The challenge is to retrieve in the first few images the most similar shapes to the query with a minimal response time. In the literature, the most popular shape representation methods can be classified into two categories: structural-based methods and boundary-based methods.

The first ones use topological information as the medial axis transform (MAT) to extract the skeleton which represents the shape using a graph. Based on skeletons, many approaches have been developed for shape matching. Liu and Geiger [1] proposed to match shape axis tree defined by the locus of midpoints of optimally corresponding boundary points. However, their algorithm does not preserve the coherence of the shape. A variant of the medial axis, shock graph, was considered by Sharvit et al. [2] for shape recognition. Siddiqi et al. [3] converted the shock graph to rooted trees and compared shapes by matching the graph trees based on sub graph isomorphism. But the choice of the oldest shock as the root of the tree is arbitrary and may lead to erroneous matches. A hierarchical representation using axial shape description was proposed in [4,5]. In this approach, hierarchy by scale was used to construct a series of skeletons.

For the boundary-based methods, a set of rich features present along the contour is exploited. In the literature various approaches have been proposed to represent the

A. Amine et al. (Eds.): *Modeling Approaches and Algorithms*, SCI 488, pp. 57–64.
DOI: 10.1007/978-3-319-00560-7_10 © Springer International Publishing Switzerland 2013

boundary shape. A large number of these methods [6,7] use local geometric features, such as curvature, angle and arc length, to identify mathematical landmarks on the shape boundary. The curvature scale space (CSS) representation [8] has been shown to be a robust shape representation. Based on the scale space filtering technique applied to the curvature of a closed boundary curve, the representation behaves well under perspective transformations of the curve. However, the method involves a complicated and time-consuming search.

Davies et al. [9] proposed to apply the minimum description length (MDL). They considered the point correspondence problem as one of finding the parameter-rization for each shape. However, like other global shape correspondence methods [10], the MDL-based approach uses a very complicated and highly nonlinear cost function.

In our contribution, we represent a shape by global parameters characterizing the whole shape. The method exploits the characteristics of the minimum area rectangle [11] and the ellipse fitting [12,13] to achieve the retrieval. Thanks to the uniqueness of these attributes, we can represent a shape by some relative measures involving the shape surface and its proportional dimension with respect to its ellipse fitting and its minimum rectangle. Obviously these attributes don't provide us any knowledge about the form as in common shape representation methods, but empirical experiments show its efficiency in shape retrieval even in translation, rotation, scale change, reflection and moderate deformation transformation. Another advantage of the method is its simplicity and its cheap cost.

The paper is organized as follow: in the next paragraph we explain the principle of the method, in section 3 we gives the matching scheme and we show experimental results in section 4, discussion and conclusion are presented at the end.

2 Global Measures

2.1 The Minimum Area Rectangle

The minimum area rectangle (MAR) associated to the silhouette is defined as the smallest rectangle minimizing the area between it and the convex hull enclosing the silhouette [11]. The convex hull is the smallest convex polygon for which every point of the silhouette touches its edges or it is inside it [14, 15], for each silhouette, a unique convex hull is associated. In the same way, the MAR is unique for each silhouette.

Fig. 1. The MAR and the convex hull (dashed polygon correspond to the convex hull, the rectangle enclosing it is the MAR)

The MAR depends only on the global form of the shape, so it is invariant to translation, rotation, scale change and mirroring. These elements are detailed below:

- Translation : the translation transformation doesn't affect the MAR, because the convex hull doesn't depend on the position of the shape but only on the disposition of the boundary points.
- Rotation : changing the orientation of a shape will change also the orientation of the MAR, but its dimensions remain unchanged.
- Scale change : a scale change affects the area occupied by the shape and so the surface of the MAR. however the shape is not distorted; the proportion of the global form is preserved. As a result, we can infer a relation between different MAR corresponding to different scales of a given shape, it corresponds to the ratio of the width and the length of the MAR.
- Mirroring : when a vertical or a horizontal reflection is applied to the shape, its surface is conserved and so for the MAR.

2.2 The Ellipse Fitting

The ellipse fitting associated to a set of points (see figure 2) is defined as the ellipse for which the sum of the squares of the distances to these points is minimal[16]. Accurate and optimized methods for ellipse fitting localization was the scope of several works [18,19,20].

Fitting ellipse to given points in the plane is a problem that arises in many application areas, as computer graphics, observational astronomy (orbit estimation), structural geology (strain in natural rocks), etc. In our work, the ellipse can serve as a geometric primitive which allows reduction and simplification of initial image.

Thanks to the uniqueness of the ellipse fitting associated to a shape, we can obtain the invariance to translation, rotation, scale change and mirroring as detailed for the minimum rectangle.

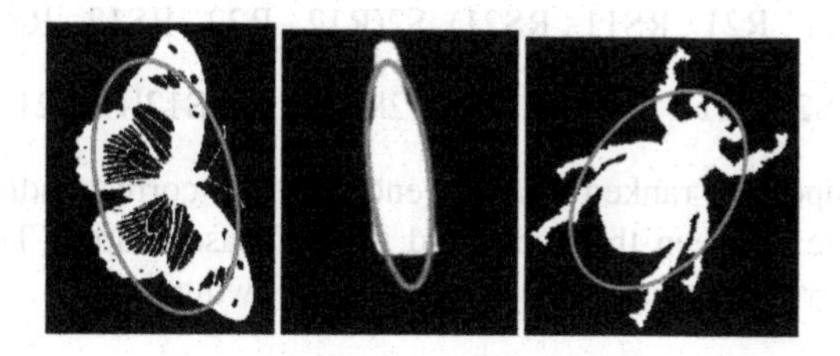

Fig. 2. The ellipse fitting corresponding to some shapes

2.3 Shape Representation Using Global Measures

In our representation we exploit the uniqueness of the MAR and the ellipse fitting to obtain relative measures. Therefore, we associate to the shape four attributes:

R1: the ratio of the width and the length of the MAR enclosing it,
RS1: the ratio of the shape surface and the MAR surface,
R2: the ratio of the major axis and the minor axis of the ellipse fitting,

RS2: the ratio of the shape surface and the ellipse fitting surface,

We can formulate this representation as follow: let be S a given shape, S is represented by the 4_uplet (R1 , R2 , RS1 , RS2):

$$S \rightarrow (R1 , R2 , RS1 , RS2) \tag{1}$$

In this manner, when one of the transformations listed in the previous paragraph is applied to S, R1, R2, RS1, RS2 remain unchanged thanks to the MAR and the ellipse fitting properties (see figure 3).

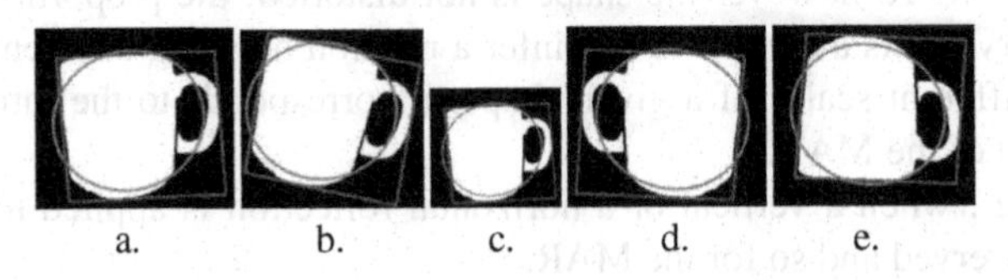

Fig. 3. An example of invariance of the representation under several transformations. In: a. the initial shape S(0.90, 0.87, 0.74, 1.02), b. planar rotation S(0.90, 0.86, 0.73, 1.13), c. scale change S(0.91, 0.85, 0.71, 1.09) d. horizontal flip S(0.90, 0.8, 0.76, 1.05), e. vertical flip S(0.89, 0.87, 0.74, 1.07).

3 Shape Matching Scheme

In our approach we don't achieve a point to point matching or structural matching. We just do a global comparison using intrinsic properties relative to the coarse form of the shapes.

The matching process is achieved in two passes: an offline pass in which we calculate the (R1 , R2 , RS1 , RS2) for every entry in the database. In the second pass we calculate the (R1 , R2 , RS1 , RS2) of the query shape and retrieve similar shapes from the database with respect to a similarity measure SM, where SM is defined for two shapes S1(R11 , R21 , RS11 , RS21), S2(R12 , R22 , RS12 , RS22) as :

$$SM= |S1-S2| =|R11- R12|+|R21- R22|+|RS11- RS12|+|RS21-RS22| \tag{2}$$

The retrieved shapes are ranked in an ascending order corresponding to the minimal similarity measure between the query and the database shapes. Therefore the algorithm which retrieves the shapes could have the following form:

```
Initialize shape  query to S_q(R1_q,R2_q,RS1_q,RS2_q)
Initialize shape S_i(S_iR1_i,R2_i,RS1_i,RS2_i)∈DataBase ∀ i:=1 to
n / n: number of the database entries
For i:=1 to n do
SM_i:= |S_q-S_i| ;
For i:=1 to n do
Sort (SM_i);
```

In the SMi vector, the first entries correspond to the most similar shapes to the query.

Throw the matching process, it is so clear that the method is very simplistic and has low cost. In fact, the comparison of the query to any shape of the database consists to compare four values: R1 , R2 , RS1 , RS2 contrarily to other methods in which the complexity is higher and the matching algorithm involves optimization methods.

4 Experimentation

The validation of our approach was achieved using shapes from the MPEG-7 shape database [17], organized in 70 classes of objects, each class contain 20 images. The classes represent different kinds of objects: animals, tools, foods, geometric forms, engines,...

Two kinds of experiments were conducted over the data set: the matching of a query to shapes from the same class and from the whole database.

The first test consists to take a query shape from a given class and calculate the distance between it and the remaining shapes of the same class (according to formula (2)). In figure 4, we show the obtained results for the class "bat". At the top left column we have the query shape and the other columns illustrate the retrieved responses from the same class (without considering the query). The retrieved results are ranked according to their similarity measure, for example, the first response (image number 1) is the most similar shape to the query among all instances of the class, the corresponding similarity measure value is 0.075472, it corresponds to the smallest value of the obtained results.

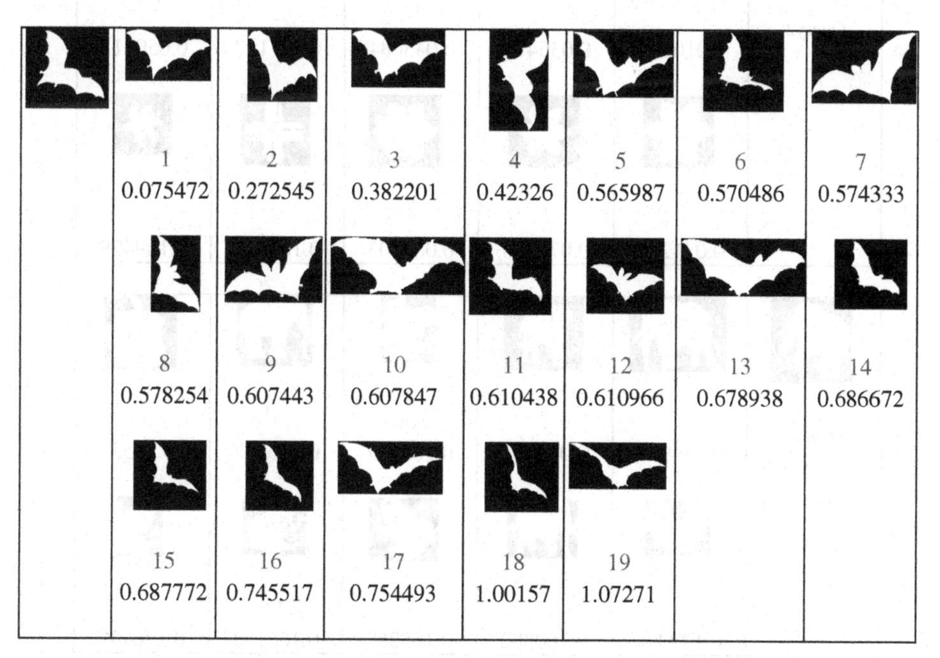

Fig. 4. An example of the obtained results for shape retrieval of the same class

Intuitively, this result is correct, it is clear that visually the retrieved shape is the most similar shape to the query one, because it is the same shape on which a planar rotation is applied. This experiment performs perfectly when the considered class contains shapes corresponding to translation, rotation, scale change or mirroring transformations of the query shape. However, in case of deformation, the first

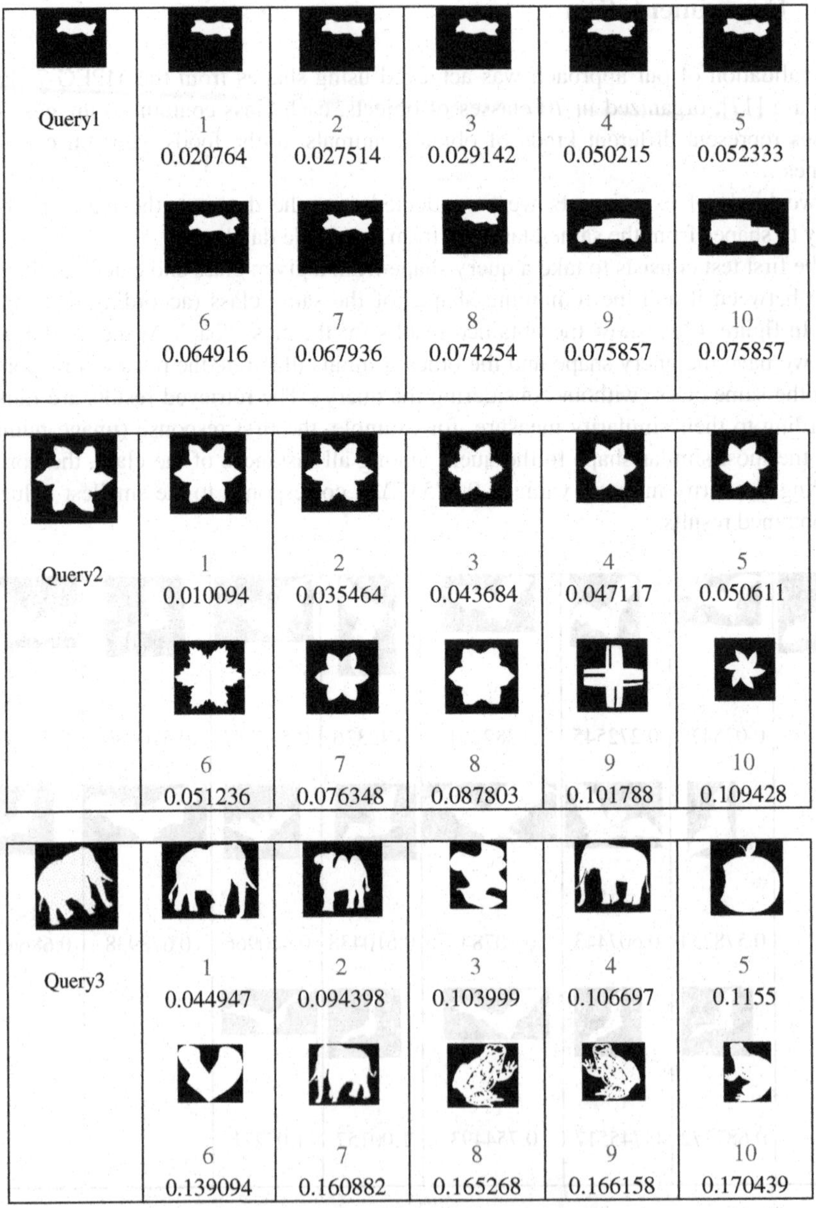

Query1	1 0.020764	2 0.027514	3 0.029142	4 0.050215	5 0.052333
	6 0.064916	7 0.067936	8 0.074254	9 0.075857	10 0.075857

Query2	1 0.010094	2 0.035464	3 0.043684	4 0.047117	5 0.050611
	6 0.051236	7 0.076348	8 0.087803	9 0.101788	10 0.109428

Query3	1 0.044947	2 0.094398	3 0.103999	4 0.106697	5 0.1155
	6 0.139094	7 0.160882	8 0.165268	9 0.166158	10 0.170439

Fig. 5. Retrieval of shapes from different classes

retrievals don't correspond systematically to the most similar shapes, because the similarity measure doesn't take into account the coarse form or the details of the shape.

Furthermore, the method gives accurate responses, the top four retrieved images corresponds exactly to the same shape as the query (after a geometric transformation: rotation, scale change, perspective). This experiment illustrates the high accuracy of our method in the shape matching problem when we deal with objects of the same class under geometric transformations.

The second experiment consists to retrieve similar shapes to the query from the whole database 70 classes x 20 images. Several tests are conducted to evaluate the behavior of the shape retrieval when we increase considerably the research space. In each test we submit a given query (from the database) and we search the most similar shapes from the whole dataset (according to formula 2), we consider the top 10 most similar shapes. Three examples are given in figure 5, we have in the first column the query shape and in the other columns the top 10 similar shapes from different classes. The corresponding values of the similarity measure are reported below the responses.

This experiment shows that even we enlarge considerably the search space, we still retrieve correct responses in the top results. In fact, for all tested queries (taken from several classes) we have obtained at least 1 shape from the same class in the top 3 retrieved shapes. This result is very interesting since we don't use any information about the shape topology or the boundary, only global measures are used.

5 Conclusion

In this work we have proposed a fast and simplistic shape retrieval method. The method is based only on two global attributes : the ellipse fitting and the minimum area rectangle, which can give us crucial information about the shape without involving complicated representations as graphs or multi-scale representation of voluminous attributes. The representation is invariant to several kinds of transformations. It allows a very fast retrieval process with an appreciable accuracy matching responses.

The results conducted over a well known dataset (MPEG-7 shape database) are very convincing.

In future works, we can apply this method in the field of indexing large shape database, or object classification.

References

1. Liu, T.-L., Geiger, D.: Approximate tree matching and shape similarity. In: Proceedings of the IEEE International Conference on Computer Vision, Corfu, Greece, pp. 456–462 (1999)
2. Sharvit, D., Chan, J., Tek, H., Kimia, B.B.: Symmetry-based indexing of image databases, J. Visual Commun. Image Representation 9(4), 366–380 (1998)

3. Siddiqi, K., Bouix, S., Tannenbaum, A., Zucker, S.W.: The Hamilton–Jacobi skeleton. In: Proceedings of the IEEE International Conference on Computer Vision, Corfu, Greece, pp. 828–834 (1999)
4. Pizer, S., Olivier, W., Bloomberg, S.: Hierarchical shape description via the multiresolution symmetric axis transform. IEEE Transactions on PAMI 9, 505–511 (1987)
5. Rom, H., Medioni, G.: Hierarchical decomposition of axial shape description. IEEE Transactions on PAMI 15, 973–981 (1993)
6. Gdalyahu, Y., Weinshall, D.: Flexible syntactic matching of curves and its application to automatic hierarchical classification of silhouettes. IEEE Trans. Pattern Anal. Mach. Intell. 21(12), 1312–1328 (1999)
7. Petrakis, E.G.M., Diplaros, A., Milios, E.: Matching and retrieval of distorted and occluded shapes using dynamic programming. IEEE Trans. Pattern Anal. Mach. Intell. 24(11), 1501–1516 (2002)
8. Mokhtarian, F.: Silhouette-based isolated object recognition through curvature scale space. IEEE Trans. Patt. Anal. Mach. Intell. 17, 539–544 (1995)
9. Davies, R., Twining, C., Cootes, T., Waterton, J., Taylor, C.: A minimum description length approach to statistical shape modeling. IEEE Trans. Med. Imag. 21(5), 525–537 (2002)
10. Hill, A., Taylor, C., Brett, A.: A framework for automatic landmark identification using a new method of nonrigid correspondence. IEEE Trans. Pattern Anal. Mach. Intell. 22(3), 241–251 (2000)
11. Schneider, P.J., Eberly, D.H.: Geometric Tools for Computer Graphics. Textbook Binding (2002)
12. Coope, I.D.: Circle fitting by linear and nonlinear least squares. Journal of Optimization Theory and Applications 76(2), 381 (1993)
13. Kanatani, K., Rangarajan, P.: Hyperaccurate ellipse fitting without iterations. In: Proc. Int. Conf. Computer Vision Theory and Applications (VISAPP 2010), Angers, France, vol. 2, pp. 5–12 (May 2010)
14. Sedgewick, R.: Algorithms. Addison-Wesley- the classic in the field (1983)
15. Graham, R.L.: An efficient algorithm for determining the convex hull of a finite planar set. Information Processing Letters 1, 132–133 (1972)
16. Gander, W., Golub, G.H., Strebel, R.: Least-Squares Fitting of Circles and Ellipses. BIT 34, 558–578 (1994)
17. Latecki, L.J., Lakämper, R., Eckhardt, U.: Shape descriptors for non-rigid shapes with a single closed contour. In: Computer Vision and Patt. Recog.: CVPR 2000, pp. 1424–1429 (2000)
18. Kanatani, K.: Hyperaccurate Ellipse Fitting Without Iterations. In: Proc. Int. Conf. Computer Vision Theory and Applications (VISAPP 2010), Angers, France, vol. 2, pp. 5–12 (May 2010)
19. Stojmenovic, M., Nayak, A.: Direct Ellipse Fitting and Measuring Based on Shape Boundaries. In: Mery, D., Rueda, L. (eds.) PSIVT 2007. LNCS, vol. 4872, pp. 221–235. Springer, Heidelberg (2007)
20. Yu, J., Kulkarni, S.R., Vincent Poor, H.: Robust Fitting of Ellipses and Spheroids, arXiv:0912.1647v1 (December 2009)

Clustering with Probabilistic Topic Models on Arabic Texts

Abdessalem Kelaiaia[1] and Hayet Farida Merouani[2]

[1] Computer sciences Department, University of may 08, 1945, Guelma, Algeria
Sam_kelaiaia@Yahoo.fr
[2] LRI Laboratory, Computer sciences Department, University of Badji Mokhtar, Annaba, Algeria
Hayet_Merouani@Yahoo.fr

Abstract. Recently, probabilistic topic models such as LDA (Latent Dirichlet Allocation) have been widely used for applications in many text mining tasks such as retrieval, summarization, and clustering on different languages. In this paper we present a first comparative study between LDA and K-means, two well-known methods respectively in topics identification and clustering applied on Arabic texts. Our aim is to compare the influence of morpho-syntactic characteristics of Arabic language on performance of first method compared to the second one. In order to study different aspects of those methods the study is conducted on benchmark document collection in which the quality of clustering was measured by the use of two well-known evaluation measure, F-measure and Entropy. The results consistently show that LDA perform best results more than K-means in most cases.

Keywords: Clustering, topics identification, Arabic text, LDA, K-means, pre-processing, stemming.

1 Introduction

Document clustering is a fundamental and enabling tool for efficient document organization. Recently, probabilistic topic models methods such as LDA (Latent Dirichlet Allocation) have been used in clustering (recently integrated in Mallet framework [16], Gensim framework [17], hierarchical clustering [19]) and take good results. Arabic language is greatly inflectional and derivational language which makes text difficult task. To our knowledge, and until this writing, there is no study that highlights the influence of the morpho-syntactic characteristics of this language on performance of such methods in document clustering. For this reason, this paper will compare the influence of morpho-syntactic characteristics of Arabic language on performance of LDA and K-means which are well-known methods in topics identification and clustering.

The rest of paper is organized as follows. The next section presents document clustering. In the two following subsections, we will present the two methods we are using in the present study. Section 3 describes the clustering evaluation; section 4

A. Amine et al. (Eds.): *Modeling Approaches and Algorithms*, SCI 488, pp. 65–74.
DOI: 10.1007/978-3-319-00560-7_11 © Springer International Publishing Switzerland 2013

presents Arabic language and related works; section 5 presents in details our experimentation procedure and evaluation, section 6 describes and discusses results, finally section 7 concludes.

2 Document Clustering

Clustering is a process of grouping objects represented in the same form in uniform groups (clusters). In document clustering objects become documents (texts). The need for such grouping is explained by the large number of texts that are often contained in a document collection. As two follows sections we will describes the two methods we will use in present study.

2.1 K-means Presentation

The K-means method belongs to the family of partitioning algorithms. This type of algorithm and its variants are best known in the community of data classification. In this type, each cluster is represented by an average (mean) or a weighted average called the "centroid" [23] which is the closest to all other elements in the cluster. This centroid is calculated using “Equation 1”.

$$C_j = \frac{1}{n_j} \sum_{X_i \in J} X_i \tag{1}$$

where C_j is the centroïd of cluster j, Xi is an element of this cluster, and n_j is the number of these elements.

2.2 LDA Presentation

Since its first introduction by [4], the Latent Dirichlet Allocation (LDA) continues to attract a considerable interest from the statistical machine learning and natural language processing communities. The idea behind LDA is that each document in the collection is modeled as a mixture over an underlying set of topics, and each topic is modeled as a probability distribution over the terms in the vocabulary. According to this, the process of generating a collection is as follows (here we describe the smoothed LDA with symmetric Dirichlet priors [10]):

1. for each topic *z*, a multinomial distribution ϕ_z is sampled from a Dirichlet distribution (β).
2. for each document *d*, a multinomial distribution θ_d over topics is sampled from a Dirichlet distribution (α) .
3. for each word *w* in the document *d*, a single topic is chosen according to distribution θ_d.
4. each word is sampled from a multinomial distribution ϕ_z over words specific to the sampled topic *z*.

Thus, the likelihood of generating a whole collection is:

$P(D_{1..N}|\alpha, \beta) =$

$$\iint \prod_{z=1}^{K} P(\phi_z/\beta) \prod_{d=1}^{N} P(\theta_d/\alpha) \left(\prod_{i=1}^{N_d} \sum_{z_i=1}^{K} P\left(\frac{z_i}{\theta}\right) P(w_i/z, \phi) \right) d\theta d\phi \quad (2)$$

where K is the number of topics and *N* is the total number of documents in collection.

2.3 LDA and Clustering

According to [14] generally, there are two ways of using topic models for document clustering. The first approach uses a topic model to reduce the dimension of representation of documents (from word representation to topic representation) and then applies a standard clustering algorithm like K-means in the new representation whereas, the other approach uses topic models more directly. The idea is that each topic z becomes, after estimating the parameters ϕ and θ, a new cluster and the documents assigned to this cluster are the documents with the highest probability "Equation 3" of assigning of the topic z to these documents.

$$arg\, max_{Z=1..K}\, \theta_d \quad (3)$$

In the present study, we will use the second approach which allows us to measure the performance of LDA compared to traditional methods of clustering like K-means.

3 Clustering Evaluation

To evaluate the quality or goodness of produced clusters, two types of measures are usually used, internal and external measures [23, 15]. When we do not have an external knowledge about (predefined sets of classes) that allows us to compare different produced clusters, we use the first type, and otherwise, we use the second one.

Many external measures are presented in the literature. To provide further evidence for the results in this study, we use two well-known evaluation measures, F-measure, and Entropy.

3.1 F-measure

The F-measure [13] is a harmonic combination of two measures, precision and recall [18], which has a long history in information retrieval domain for clustering evaluation. The cluster is viewed as the result of a query for a specific class and class is viewed as a relevant set of documents relevant for a query. Precision is the fraction of correctly retrieved documents (attribution of document from cluster to correct class) ("Equation 4"), and recall is the fraction of correctly retrieved (classed) documents out of all matching documents ("Equation 5").

$$Precision\,(i,j) = \frac{n_{ij}}{n_j} \quad (4)$$

$$Recall(i,j) = \frac{n_{ij}}{n_i} \quad (5)$$

where n_{ij} is the number of documents of class i in cluster j, n_j is the number of documents of cluster j and n_i is the number of documents of class i.

The F-measure of cluster j and class i is then given by:

$$F(i,j) = \frac{(2 * Precision\ (i,j) * Recall\ (i,j))}{(Precision\ (i,j) + Recall\ (i,j))} \quad (6)$$

Note that [15] present another definition of precision and recall calculated on entire collection;

$$Precision = \frac{A}{A+C} \quad (7)$$

$$Recall = \frac{A}{A+B} \quad (8)$$

3.2 Entropy

The entropy is a function of the distribution of classes in the resulting clusters [24]; it indicates the quantity of disorder in these clusters [22]. The lower value of entropy indicates a better clustering (higher homogeneity). The greater entropy means that the clustering is not good.

Given a class i and a cluster j, the entropy of a cluster j is defined as:

$$E_j = -\frac{1}{\log q}\sum_{i=1}^{q}\frac{n_{ij}}{n_j}\log\frac{n_{ij}}{n_j} \quad (9)$$

where n_{ij} is the number of documents of the class i that were assigned to the cluster j, q is the total number of classes in the document collection, and n_j is of documents in cluster j.

Thus, the overall entropy is weight average of all cluster entropies:

$$Entropy = \sum_{j=1}^{k}\frac{n_j}{n}E_j \quad (10)$$

where n_j is the size of cluster j, k is the number of clusters, and n is the total number of documents in whole document collection.

4 Arabic Language

4.1 Particularities

The Arabic language has an alphabet containing 28 consonants that change their layout according to their position. Unlike English, Arabic is an agglutinative language; articles, prepositions and pronouns stick to adjectives, nouns, verbs and particles which they relate, which creates ambiguities during morphological analysis.

An Arabic word can represent a phrase in English; it may be composed of a stem (base), proclitics such as prepositions or conjunctions, prefixes and suffixes, which express grammatical features and indicate the functions of cases, verb mode and modalities (number, gender, etc.) and enclitics, which are personal pronouns [8]. For example the word أتأكلونها, which mean: "do you eat it?" is decomposed as follows:

Table 1. Arabic word decomposition

Enclitic	Suffix	Stem	prefix	Proclitic
هَا	ونَ	أكُل	تَ	أ

4.2 Arabic Language and Clustering

The nature of the Arabic language, the writing system, writing orientation, omission of vowels and morphological structure has slowed research into this language, especially in automatic classification (categorization or clustering). In the literature, most of the research is focused on the morphological aspect of this language [12] via developing preprocessing tools such as stemming and their influence on information retrieval or on supervised classification (categorization), but only a small number of research projects focus on document clustering, we identified two major works, on a morphological analysis based on the language and using the n-gram. The authors [21] used a statistical approach (based on the technique of entropy maximization) for the clustering of an Arab-based articles covering several areas such as politics, economics, etc. [11] developed an algorithm (integrated into the standard software clusters TEMIS Insight Discoverer) that, from descriptors in Arabic, contains similar documents in classes according to their semantic similarity and proximity topic.

4.3 Arabic Language and Probabilistic Topic Models

Regarding the use of probabilistic topics models in Arabic language, we mentioned a single major work [5] which investigates the influence of root-based stemming approach on LDA supervised classification.

5 Experimentation Procedure

5.1 Document Collections

To evaluate the performance of the two methods, LDA and K-means, we used OSAc (Open Source Arabic Corpus) a document collection member of OSAC (Open Source Arabic Corpora), collected from multiple sites by Motaz K. Saad [20] and describes in "Table 2".

Table 2. Document collections

Category	Number of documents	Number of words (Million)	Size (Mbytes)
Astronomy	557		
Law	944		
Economy	3102		
Education	2626		
Entertainment	982		
History	3233		
Recipe	2373		
Story	726		
Religion	3171		
Health	2296		
Sport	2419		
Total (11)	22429	18,18	182

5.2 Text Preprocessing

Preprocessing aims to standardize the representation of texts to be classified. There are commonly four steps:

First, we convert text files to UTF-8 encoding and remove non letters, punctuation marks and diacritics.

Second, we need to give a transliterated form to each word in the document collection to be used in the clustering process.

Third, we need, after tokenization, remove stop words such as 'أين', 'كان', 'حيث', etc, since they are frequent and carry no information. During our study we collected about 875 stop words.

Fourth, we need to stem the word to its origin, which means we only consider the root form of words. Stemming aims to obtain the lexical root or stem for words in natural language, by removing affixes attached to them, i.e. it's regrouping under a single identification words whose root is common. For example, the words يحمل, محمل, حملة are flexions of stem حمل. For this, stemmers are developed; they are generally designed for a specific language on which a certain expertise should be developed. [12] considers that the use of a dictionary for stems and morphological analysis are other forms of stemming. Several stemming algorithms have been studied for different languages. For Arabic, there are several stemmers, the most famous are Al-Stem [6] and StemmerLight10 [12]. In our case we choose to employ Al-Stem due to its performance [7] and we re-implemented it to work with entire text.

Note that after the third and fourth steps we save each document in cleaned and stemmed form in the goal to test the influence of removing of the stop words and stemming on clustering process. According to this, we will have three forms for each document collection, raw (transliterated) form, cleaned form, and stemmed form.

5.3 Clustering Process

In this phase we submitted the preprocessing results of the document collection (raw form, cleaned form, and stemmed form) on the clustering process with LDA and K-means methods.

5.4 Evaluation

After the clustering processes, we compute the evaluation measures, F-measure and Entropy for all the clustering results compared to predefined structure showing in Table 2.

6 Results and Discussion

We present the results of our experiments in Figure 1 summarized in Table 3 which clearly demonstrate at first sight that LDA perform best results more than K-means. As mentioned above the results showing in Figure 1 are computed according to the predefined structure (Table 2).

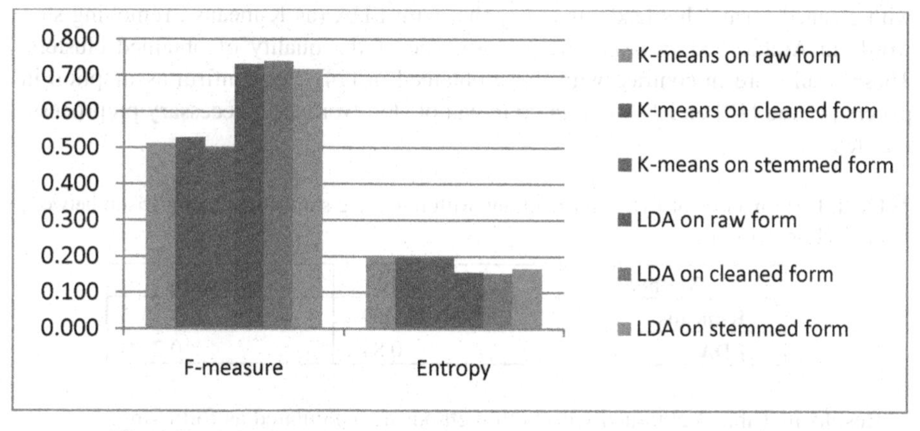

Fig. 1. Performances of LDA and K-means under F-measure and Entropy

6.1 Performance of LDA over K-means

From Table 3, we observe that LDA provides a substantial performance improvement over K-means with two metrics on the three collection forms. This performance is in contrast to research of [14] which states that the use of probabilistic topic models in clustering is not as accurate as traditional clustering methods such as K-means in respect to the functioning of this topic models. [14] research has been conducted on Reuters 21578 and TDT2 document collections. Both collections are in English, this may be the cause of decreasing of performance of probabilistic topic models with respect to K-means.

Table 3. Performances of LDA over K-means on the OSAc document collection in raw, cleaned, and stemmed forms

	F-measure	Entropy
Raw form	21,96 %	4,44 %
Cleaned form	21,12 %	4,53 %
Stemmed form	22,00 %	3,27 %

Note that for the two metrics (F-measure and Entropy), the performance of LDA over K-means on each form in Table 3 is calculated according to "Equation 11" and "Equation 12".

$$Form\ performance_{\ F-measure} = F - measure_{LDA} - F - measure_{K-means} \quad (11)$$

$$Form\ performance_{\ Entropy} = Entropy_{LDA} - Entropy_{K-means} \quad (12)$$

6.2 Influence of Elimination of the Stop Words on Clustering Quality

Table 4 shows the comparison of the results of clustering performed on raw and cleaned forms with both methods where both methods appear to perform better results with cleaned form. This leads us to say that with LDA (as K-means), removing stop words in Arabic text increase the performance of the quality of obtained clusters. These results are in contrast with those obtained in [14] and confirm assumption in papers [4] and [10] on LDA that the removal of stop words is a necessary preprocessing step.

Table 4. Performances of LDA and K-means with removing stop words (comparison between raw and cleaned forms)

Methods	F-Measure	Entropy
K-means	1,69%	0,16 %
LDA	0,84%	0,24 %

Results in Table 4 obtained with both methods are calculated as following:

$$K - means\ performance_{metric} = metric_{cleaned\ form} - metric_{raw\ form} \quad (13)$$

$$LDA\ performance_{metric} = metric_{cleaned\ form} - metric_{raw\ form} \quad (14)$$

6.3 Influence of Stemming on Clustering Quality

Table 5 shows the comparison of the results of clustering performed on raw and stemmed forms with both methods. In this table, we notice that the stemming has decreased quality of the obtained clusters. This is mostly due to the effect of stemming, which remove the flexions of words that have not the same root, so the documents relating to the same topic will have a greater chance of being in the different cluster.

Table 5. Performances of LDA and K-means with stemming (comparison between raw and stemmed forms).

Methods	F-Measure	Entropy
K-means	-1,47%	-0,08%
LDA	-1,42%	-1,09%

7 Conclusion

The present work compares between LDA and K-means in order to examine the reaction of LDA in clustering of Arabic texts which is a very flexional language. The experiment is conducted on benchmark Arabic document collection (OSAc, Open Source Arabic Corpus). Two metrics are used, F-measure and Entropy. We start by doing the comparison between the results obtained by the two methods on the document collection and the predefined structure. The results consistently show a clear improvement of LDA over K-means on raw, cleaned, and stemmed forms of document collection. Afterwards, we investigate the influence of the preprocessing tasks, elimination of stop words and stemming on performance of the studied methods. We found that both methods show a good improvement in the first case (elimination of stop words). In the second one, ambiguity caused by stemming was decreased the performance of both methods.

Bearing in mind the obtained results, our future work will extend the present study to other important parameters such as lemmatization, a very important preprocessing operation in Arabic language, and investigate other variant of LDA such as Dynamic Topic Model (DTM) [2], Correlated Topic Model (CTM) [3].

References

1. Abbas, M., Smaili, K., Berkani, D.: Multi-Category Support Vector Machines for Identifying Arabic Topics. Advances in Computational Linguistics, Special issue of Journal of Research in computing Science 41, 217–226 (2009)
2. Blei, D., Lafferty, J.: Dynamic topic models. In: Proceedings of the 23rd International Conference on Machine Learning (2006)
3. Blei, D., Lafferty, J.: A correlated topic model of science. Annals of Applied Statistics 1(1), 17–35 (2007)
4. Blei, D.M., Ng, A.Y., Jordan, M.I.: Latent dirichlet allocation. Journal of Machine Learning Research 3, 993–1022 (2003)
5. Brahmi, A., Ech-cherif, E., Benyettou, A.: Arabic texts analysis for topic modeling evaluation. Information Retrieval 14 (2011)
6. Darwish, K., Oard, D.W.: Evidence combination for Arabic-English retrieval. In: TREC, pp. 703–710. NIST, Gaithersburg (2002)
7. Darwish, K., Hassan, H., Emam, O.: Examining the Effect of Improved Context Sensitive Morphology on Arabic Information Retrieval. In: Proceedings of the ACL Workshop on Computational Approaches to Semitic Languages, Ann Arbor, USA, pp. 25–30 (2005)

8. Diab, M., Hacioglu, K., Jurafsky, D.: Automatic Tagging of Arabic Text: From Raw Text to Base Phrase Chunks. In: Proceedings of the 5th Meeting of the North American Chapter of the Association for Computational Linguistics/Human Language Technologies Conference (HLT-NAACL 2004), USA, pp. 149–152 (2004)
9. El Sulaiti, L.: L'arabe contemporain. Radio Qatar, Qatar (2003)
10. Griffiths, T.L., Steyvers, M.: Finding scientific topics. Proceedings of the National Academy of Science 101, 5228–5235 (2004)
11. Huot, CH., Coupet, P.: Le Text Mining sur la langue Arabe : application au traitement des sources ouvertes. TEMIS SA, Paris, France (2005)
12. Larkey, L.S., Ballesteros, L., Connell, M.E.: Arabic Computational Morphology. In: Light Stemming for Arabic Information Retrieval. Springer (2007)
13. Larsen, B., Aone, C.: Fast and effective text mining using linear time document clustering. In: Proceedings of the Conference on Knowledge Discovery and Data Mining, pp. 16–22 (1999)
14. Lu, Y., Mei, Q., Zhai, C.: Investigating task performance of probabilistic topic models: an empirical study of PLSA and LDA. Inf. Retrieval 14(2001), 178–203 (2011)
15. Manning, C.D., Raghavan, P., Schütze, H.: Introduction to Information Retrieval, pp. 327–331. Cambridge University Press, Cambridge (2008)
16. Mccallum, A.K.: MALLET: A Machine Learning for Language Toolkit (2002), `http://mallet.cs.umass.edu`
17. Řehůřek, R., Sojka, P.: Gensim – Python Framework for Vector Space Modelling, NLP Centre, Faculty of Informatics, Masaryk University, Brno, Czech Republic (2011), `http://radimrehurek.com/gensim/`
18. Van Rijsbergen, C.J.: Information Retrieval, 2nd edn. Buttersworth, London (1979)
19. Rosenzvi, M., Griffiths, T., Steyvers, M., Smyth, P.: The author-topic model for authors and documents. In: Proceedings of the 20th Conference on Uncertainty in Artificial Intelligence, Banff, Alberta, Canada (2004)
20. Saad, M.K., Achour, W.: OSAC: Open Source Arabic Corpora, 6th ArchEng International Symposiums. In: The 6th International Symposium on Electrical and Electronics Engineering and Computer Science, pp. 118–123. European University of Lefke, Cyprus (2010)
21. Sawaf, H., Zaplo, J., Ney, H.: Statistical Classification Methods for Arabic News Articles. In: Proceedings of the ACL/EACL Workshop on ARABIC Language Processing: Status and Prospects, Toulouse, France (2001)
22. Shannon, C.E.: A mathematical theory of communication. Bell System Technical Journal 27, 379–423, 623–656 (1948)
23. Steinbach, M., Karypis, G., Kumar, V.: A Comparison of Document Clustering Techniques. In: KDD Workshop, Text Mining, Minnesota, USA (2000)
24. Zhao, Y., Karypis, G.: Criterion functions for document clustering: Experiments and analysis, Technical Report #01-40, University of Minnesota (2001)

Generating GCIs Axioms from Objects Descriptions in $\mathcal{EL}$-Description Logics

Zina Ait-Yakoub[1] and Yassine Djouadi[2]

[1] Department of Computer Science
University of Tizi-Ouzou, BP 17, RP, Tizi-Ouzou, Algeria
ait.yakoub@hotmail.fr
[2] IRIT Toulouse
118 Route de Narbonne, 31062 Toulouse, Cedex 9, France
djouadi@irit.fr

Abstract. Description Logic are well appropriate for knowledge representation. In such a case, intensional knowledge of a given domain is represented in the form of a terminology (TBox) which declares general properties of concepts relevant to the domain. The terminological axioms which are used to describe the objects of the considered domain are usually manually entered. Such an operation being tiresome, Formal Concept Analysis (FCA) has been already used for the automatic learning of terminological axioms from object descriptions (i.e. from concept instances). However, in all existing approaches, induced terminological axioms are exclusively restricted to the conjunctive form, that is, the existential constructor ($\exists r.C$) is not allowed. In this paper, we propose a more general approach that allows to learn existentially quantified general concept inclusion (GCIs) axioms from object descriptions given as assertions in the $\mathcal{EL}$ language.

1 Introduction

Description Logics (DLs) [8] are a well-investigated family of logic-based knowledge representation formalisms, which are employed in various application domains, such as natural language processing, configuration, databases, and bio-medical ontologies, but their most notable success so far is the adoption of the DL-based language OWL [12] as standard ontology language for the semantic web [14]. From the Description Logic point of view, an ontology is a finite set of general concept inclusion axioms (GCIs) of the form $C \sqsubseteq D$, where C,D are concept descriptions using an appropriate concept description language.

Actually, the construction of ontologies is usually performed manually by knowledge engineers. Such a construction is a tedious and tricky task. However, the most important arising problem concerns the computing of a minimal Terminological Base (TBox) of general concept inclusion axioms.

Based on the lattice theory, Formal Concept Analysis (FCA) [9] can be used to analyze data given in the form of a formal context. Particularly, FCA provides efficient algorithms for computing a minimal basis of all implications holding in a given formal context. In this spirit, FCA has been already used for the automatic generation of terminological axioms from object descriptions (i.e. from concept instances). In such a case, implications are assimilated to terminological axioms.

A. Amine et al. (Eds.): *Modeling Approaches and Algorithms*, SCI 488, pp. 75–84.
DOI: 10.1007/978-3-319-00560-7_12 © Springer International Publishing Switzerland 2013

However, it may be remarked that in all existing approaches, generated terminological axioms are restricted to the conjunctive form of the $\mathcal{EL}$-language. That is, the existential constructor ($\exists r.C$) is not allowed for such generated terminological axioms.

In an original way, we propose in this paper a more general approach that allows to induce the minimal set of all general concept inclusion (GCIs) axioms from object descriptions given as assertions. That is, unlike existing approaches, our method allows to generate GCIs existentially quantified.

This work, which constitutes a first attempt in this direction, will make use of a simple DL language, namely the $\mathcal{EL}$ one. Note that, our proposed approach is not restricted to $\mathcal{EL}$. It will certainly be possible to generalize it to other DLs, which will be investigated in further researches.

This paper is organized as follows. Section 2 gives a background on formal concept analysis. Section 3 presents description logics, whereas section 4 relates previous works on the common utilization of FCA and DL. Our proposition is presented in section 5. An illustrative example is also given in the following section. We conclude and present our future researches.

2 Formal Concept Analysis

Formal concept analysis (FCA for short) [9] consists of inducing granules of knowledge called formal concepts from an $Objects \times Attributes$ binary relation. FCA relies on the notion of a formal context which consists of a triple $K = (G, M, I)$ where G is the set of objects, M the set of attributes and I the binary relation s.t. $I \subseteq G \times M$. $(g, m) \in I$ means that the object g satisfies the attribute m. Relation I can be viewed as a table where, for instance, rows correspond to objects, columns to properties, and a table entry contains "×" or nothing, depending on whether the object satisfies or not the corresponding attribute.

Given a formal context $K = (G, M, I)$ and given two sets $A \in 2^G$ and $B \in 2^M$, a powerset operator $(.)'$ (called also Galois derivation operator) is dually defined among the sets 2^G and 2^M as follows:

$$A' = \{m \in M | (g, m) \in I \text{ for all } g \in A\}$$
$$B' = \{g \in G | (g, m) \in I \text{ for all } m \in B\}$$

A formal concept of K is a pair $\langle A, B \rangle$ with $A \subseteq G, B \subseteq M$ such that $A' = B$ and $B' = A$. A is called the extent and B the intent of the formal concept $\langle A, B \rangle$. The set of all formal concepts is denoted by $\mathcal{L}(G, M, I)$. A formal concept corresponds to a maximal rectangle full of crosses in the table representing a formal context. For brevity, we write g' and m' instead of $\{g\}'$ and $\{m\}'$ respectively.

From a formal context, one may also induce the so-called *attribute implications* [9]. Let $K = (G, M, I)$ be a formal context and $P, C \in 2^M$, an attribute implication of the form $P \to C$ is defined as:

$$P \to C \Leftrightarrow P' \subseteq C' \Leftrightarrow C \subseteq P'')$$

That is, for every g from G: if every attribute from the premise P applies to the object g, then every attribute from the conclusion C also applies to g. The set P being called the premise and C being called the conclusion of this implication.

Since the set of all attribute implications may contain many redundancies, it is more appropriate to use a condensed (i.e. minimal) representation. Among such representations, the Duquenne-Guigues [7] basis is a minimal set of implication from which we can find every other implications that hold through inference.We give hereafter its definition.

Definition 1. *The set of implications* $\{ X \rightarrow X'' \mid X$ *is a Pseudo-Intent* $\}$ *is called a Duquenne-Guigues Basis.*

Where a pseudo-intent is defined as:

Definition 2. *For a formal context* (G, M, I)*, a set* $P \subseteq M$ *will be called Pseudo-Intent if* $P'' \neq P$ *and* $Q'' \subseteq P$ *holds for every Pseudo-Intent* $Q \subset P$.

3 Description Logics

Description logics [8] (DL for short) are decidable fragments of first-order logic used to represent and reason on knowledge. In order to define concepts in a DL knowledge base, one starts with a set of concept names (unary predicates) N_C, a set of role names (binary predicates) N_R and a set of objects names N_O. The DL paradigm aims to build concept descriptions or, in short, concepts using constructors. The set of constructors determine the expressive power of the DL. In this paper, we restrict ourselves to the DL $\mathcal{EL}$, whose every concept name is a concept description and, for any concept description C and D and any role r, top-concept ($\top$), conjunction ($C \sqcap D$), and existential restriction ($\exists r.C$) are also concept descriptions. The semantics of $\mathcal{EL}$-concept descriptions is defined in terms of an interpretation $\mathcal{I} =< \Delta^{\mathcal{I}}, .^{\mathcal{I}} >$. The domain $\Delta^{\mathcal{I}}$ of $\mathcal{I}$ is a non-empty set of individuals (objects) and the interpretation function $.^{\mathcal{I}}$ maps each concept name $A \in N_C$ to a subset $A^{\mathcal{I}}$ of $\Delta^{\mathcal{I}}$ and each role $r \in N_R$ to a binary relation $r^{\mathcal{I}} \subseteq \Delta^{\mathcal{I}} \times \Delta^{\mathcal{I}}$.

A knowledge base consists of an ABox and a TBox. An ABox (Assertions Box) is a finite set of assertions of the form $A(a)$ or $r(a, b)$, where A is a concept name, r is a role name, and a, b are individual names from a set N_O. Interpretations of ABoxes must additionally map each individual name $a \in N_O$ to an element $a^{\mathcal{I}}$ of $\Delta^{\mathcal{I}}$. An interpretation $\mathcal{I}$ is a model of the ABox A iff it satisfies all its assertions, i.e., $a^{\mathcal{I}} \in A^{\mathcal{I}}$ for all concept assertions $A(a)$ in A and $(a^{\mathcal{I}}, b^{\mathcal{I}})$ for all role assertions $r(a, b)$ in A.

A TBox (Terminology Box) represents intensional knowledge of a problem domain which declares general properties of concepts relevant to the domain. One may distinguish two types of Tboxes. In the most basic type, a TBox contains *concept definitions* of the form $A = C$ which define a concept name A by a concept description C. Concept descriptions are terms built from primitive concepts by means of language constructors provided by the DL.

In the more general type, TBoxes contain universally true implications, so-called *general concept inclusion* (GCI) axioms of the form $C \sqsubseteq D$, where both C and D are arbitrary concept descriptions. A GCI $C \sqsubseteq D$ holds iff the extension of C is a subset of the extension of D. Hence, D is implied whenever C holds. From an application point of view, the utility of general TBoxes for DL knowledge bases has long been observed. If $C \sqsubseteq D$ and $D \sqsubseteq C$, we say that D is *equivalent* to C ($C = D$). The TBox $\mathcal{T}$ is called acyclic if it contains only equivalence statements where the left-hand side is not used in the concept description on the right-hand side implicitly or explicitly and the TBox is called cyclic if it contains equivalence statements where the left-hand side is used in the concept description on the right-hand side implicitly or explicitly.

Our proposed approach is dedicated to the second type of Tboxes, namely the ones containing general concept inclusion (GCI) axioms. We give hereafter some useful definitions for the rest of the paper.

Definition 3. *The concept description D subsumes the concept description C w.r.t the TBox $\mathcal{T}$ ($C \sqsubseteq_{\mathcal{T}} D$) iff $C^{\mathcal{I}} \subseteq D^{\mathcal{I}}$ for all models $\mathcal{I}$ of $\mathcal{T}$. We write $C \sqsubseteq D$ iff C is subsumed by D. Two concept descriptions C, D are called equivalent w.r.t. $\mathcal{T}$ iff they subsume each other, i.e., $C \equiv_{\mathcal{T}} D$ iff $C \sqsubseteq_{\mathcal{T}} D$ and $D \sqsubseteq_{\mathcal{T}} C$*

Definition 4. *Given a collection $C_1, ..., Cn$ of $\mathcal{L}$ concept description, the least common subsume (lcs) of $C_1, ..., C_n$ in $\mathcal{L}$ is the most specific $\mathcal{L}$-concept description that subsumes $C_1, ..., C_n$, i.e, it is an $\mathcal{L}$-concept description D such that 1) $C_i \sqsubseteq D$ for $i = 1, ..., n$ (D is a common subsumer) 2) If E is an $\mathcal{L}$-concept description satisfying $C_i \sqsubseteq E$ for $i = 1, ..., n$, then $D \sqsubseteq E$ (D is least)*

4 Cross-Fertilizing FCA and DL: A Survey

There are two main tendencies for cross-fertilizing FCA and DL between themselves. The first category aims to enrich the language of FCA by borrowing constructors from DL languages [13] whereas, the second category aims to employ FCA methods for solving problems encountered in knowledge representation with DLs [1, 2, 3, 4, 5, 6, 10]. Since this paper, is concerned by the second category, we survey in the following related existing approaches.

In [1], Baader et al. restrict themselves to the case where objects have only a partial description. In the sense that, for some attributes, it is not known whether they are satisfied by the object or not. This extension is necessary in order to deal with the open-world semantics of the description logic knowledge bases, and explore them using attribute exploration.

Baader [2] has used FCA for an efficient computation of an extended subsumption hierarchy of a set of DL concepts. More precisely, he used attribute exploration for computing the subsumption hierarchy of all conjunctions of a set of DL concepts. The main motivation for this work was to determine the interaction between defined concepts, which might not easily be seen by just looking at the subsumption hierarchy of defined concepts.

In [3], Baader and Sertkaya are interested in computing the subsumption hierarchy of all least common subsumers (lcs) of subsets of set of description logic concepts (S), without having to compute the least common subsumer for all subsets of S, using methods from formal concept analysis.

There are very few works which are concerned by learning (inducing) the TBox from object descriptions. In order to obtain complete knowledge about the subsumption relationships in the given model between arbitrary $\mathcal{FLE}$ concepts, Rudolph [6] gives a multi-step exploration algorithm. For each step, he generates implication base from a formal context by applying the attribute exploration method and generates the attribute set for the next exploration step. Rudolph points out that, at an exploration step, there can be some concept descriptions in the attribute set that are equivalent, i.e., attributes that can be reduced. To this aim, he introduces a method that he calls empiric attribute reduction. In principle, it is possible to carry out infinitely many exploration steps, which means that the algorithm will not terminate. In order to guarantee termination, the same author restricts the number of exploration steps.

Another approach which consists of completing the TBox with terminological axioms learned from assertions contained in an ABox is proposed in [10]. This approach translates data from DL formalism, that is instances of the ABox, to a form compliant with FCA (i.e. lattices). More precisely, authors adapt classical FCA algorithms in order to build sets of concept definitions from object descriptions. However, these approaches deal exclusively with the conjunctive form of concept descriptions. For this purpose, the proposition described in the next section aims to learn GCIs containing conjunction of concepts as well as conjunction of roles existentially restricted.

5 Proposition

5.1 Theoretical Aspects

Let us consider the following notations and abbreviations. Given an object $O_i \in N_O$, we consider a mapping τ which associates to each object O_i its corresponding $\mathcal{EL}$-concept description C_i ($C_i = \tau(O_i)$). Let also $concepts(C_i)$ denotes the set of all concept names occurring in C_i, $roles(C_i)$ denotes the set of all role names occurring in an existential restriction of C_i, and $restrict_r(C_i)$ denotes the concept description occurring in an existential restriction on the role r of C_i. For a nonempty subset $\{a_1, \ldots, a_k\}$ of concept names, $\bigsqcap A$ denotes the conjunction $a_1 \sqcap \ldots \sqcap a_k$ of concepts. For a nonempty subset $\{r_1, \ldots, r_s\}$ of role names, we denote the conjunction $r_1 \sqcap \ldots \sqcap r_s$ by $\bigsqcap R$. An object description based on concepts and roles may now be abbreviated as follows:

$$C_i = \tau(O_i) = \bigsqcap_{A \in concepts(C_i)} A \sqcap \bigsqcap_{r \in roles(C_i)} \bigsqcap_{E \in restrict_r(C_i)} \exists r.(E)$$

Example 1. Let C :Man$\sqcap$ Father$\sqcap$hasChild(Man$\sqcap$ Father) be the description of the object Bob (i.e. $C = \tau$(Bob)). Thus, $concepts(C)$=$\{Man, Father\}$, $roles(C)$=$\{hasChild\}$, and $restrict_{hasChild}(C)$ =$\{Man \sqcap Father\}$.

The obvious analogy between an object description C_i related to DL paradigm and a tuple (row) of a formal context related to FCA theory leads us to generate the context formal $K_C = (N_O, N_C \cup N_R, \mathcal{I})$ where the Cartesian product $\mathcal{I}$ (*i.e.* $\mathcal{I} \subseteq N_O \times (N_C \cup N_R)$) is now obtained using the following algorithm. The formal context $K_C = (N_O, N_C \cup N_R, \mathcal{I})$ is defined from in objects descriptions, where each object O_i is described by $\mathcal{EL}$-concept description $C_i = \tau(O_i)$. We propose to determine all entries of $\mathcal{I}$ using the following algorithm:

Algorithm. Gen_Formal_Context

Require: N_O, N_C, N_R
Begin
1: $\mathcal{I} := \{\emptyset\}$;
2: **For** each $o \in N_O$ **Do**
3:    **For** each $A \in concept(\tau(o))$ **Do**
4:       $\mathcal{I} := \{(o, A)\} \cup \mathcal{I}$;
5:    **End For**;
6:    **For** each $r \in roles(\tau(o))$ **Do**
7:       $\mathcal{I} := \{(o, r)\} \cup \mathcal{I}$;
8:    **End For**;
9: **End For**
End

According to the method proposed in [7], we may define the Duquenne-Guigues base or stem base DG_{K_C} of all attribute implications of the formal context K_C as follows. Let us first recall that such a base is not only irredundant, but also has minimal cardinality among all other bases induced from K_C.

Definition 5. *The Duquenne-Guigues base DG_{K_C} of all implications is given as:*
$DG_{K_C} = \{ \{A_1, ..., A_m, r_1, ..., r_n\} \rightarrow \{A_{m+1}, ..., A_p, r_{n+1}, ..., r_k\} \mid \{A_1, ..., A_m, r_1, ..., r_n\}$ *is a Pseudo-Intent* $\wedge \{A_{m+1}, ..., A_p, r_{n+1}, ..., r_k\} = \{A_1, ..., A_m, r_1, ..., r_n\}''\}$

Where $A_1, ..., A_m, A_{m+1}, ..., A_p \in N_C$ and $r_1, ..., r_n, r_{n+1}, ..., r_k \in N_R$. $\{\{A_1, ..., A_m, r_1, ..., r_n\} \rightarrow \{A_{m+1}, ..., A_p, r_{n+1}, ..., r_k\}$ is the general form of each implications $\in DG_{K_C}$ but, two subsets $(\phi(DG_{K_C}), \psi(DG_{K_C}))$ can be distinguished as formalized in the two following definitions.

Definition 6. *We call $\phi(DG_{K_C}) = \{ (\{m_1, ..., m_k\} \rightarrow \{m_{k+1}, ..., m_p\}) \in DG_{K_C} \mid m \in \{m_1, ..., m_k, m_{k+1}, ..., m_p\} \Rightarrow m \in N_C\}$ the set of implications of DG_{K_C} without the implications which have the role names r belonging to the premise or the conclusion, i.e., the implications of $\phi(DG_{K_C})$ are of the form $\{A_1, ..., A_m\} \rightarrow \{A_1, ..., A_m\}$.*

Definition 7. *We call $\psi(DG_{K_C}) = \{ (\{m_1, ..., m_k\} \rightarrow \{m_{k+1}, ..., m_p\}) \in DG_{K_C} \mid \exists m \in \{m_1, ..., m_k, m_{k+1}, ..., m_p\} \wedge m \in N_R\}$ the set of implications of DG_{K_C} with the implications which have at least one role names r belonging to the premise or the conclusion.*

Unlike existing approaches which are restricted to attribute implications *imp* without roles (i.e. $imp \in \phi(DG_{K_C})$), our proposed approach allows to induce attribute implications *imp* containing roles (i.e. $imp \in \phi(DG_{K_C}) \cup \psi(DG_{K_C})$). For this purpose, we need to have the type of each role. Given $imp_i \in \psi(DG_{K_C})$, let $\psi(DG^*_{K_C})$ denotes the set of all implications of $\psi(DG_{K_C})$ s.t. a type is assigned for each role of each implication (example: if $\exists r.C$ is a concept description with existential restriction, then r is a role name and C is a type of r). It comes that an implication of $\psi(DG^*_{K_C})$ is of the form:

$Imp_i = \{\{A_1, ..., A_m, \exists r_1.(restrict^*_{r_1}(Imp_i)), ..., \exists r_n.(restrict^*_{r_n}(Imp_i))\} \longrightarrow \{A_{m+1}, ..., A_p, \exists r_{n+1}.(restrict^*_{r_{n+1}}(Imp_i)), ..., \exists r_k.(restrict^*_{r_k}(Imp_i))\}$.

Where $restrict^*_r(imp_i)$ denotes the type of the role r $r \in \{m_1, ..., m_k, m_{k+1}, ..., m_p\}$). The two following propositions establish the theoretical framework which allows to obtain the set $\psi(DG^*_{K_C})$.

Proposition 1. *Assume that* $Imp_i = \{\{m_1, ..., m_k\} \rightarrow \{m_{k+1}, ..., m_p\}\}$ *is an implication of* $\psi(DG_{K_C})$*, that r is the role names of* $\{m_1, ..., m_k, m_{k+1}, ..., m_p\}$*, and that* $\{O_1, ..., O_s\}$ *is the set of objects corresponding to* $\{m_1, ..., m_k, m_{k+1}, ..., m_p\}'$*, then for each r:*
$restrict^*_r(Imp_i) = lcs(\sqcap resctict_r(\tau(O_1)), ..., \sqcap resctict_r(\tau(O_l)))$

Proof. Due to lack of space, proof is left for a long version of this paper.

In [11] it is shown that the lcs of two or more $\mathcal{EL}$-concept description always exists and it can be computed in polynomial time.

Proposition 2. $B_1 \rightarrow B_2$ *is an implication holds in* $\psi(DG^*_{K_C})$*, then* $\sqcap B_1 \sqsubseteq \sqcap B_2$.

Proof. Assume that the subsumption relationship $\sqcap B_1 \sqsubseteq \sqcap B_2$ does not hold, this is the case iff there exists an objet $g \in (\sqcap B_1)^I$ and $g \notin (\sqcap B_2)^I$, this is equivalent to $g \in m^I$ for all $m \in B_1$,and there exists $p \in B_2$ such that $g \notin p^I$ (using the semantics of the conjunction operator). By definition, $g \in B'_1$, and $g \notin B'_2$ then $B'_1 \not\subseteq B'_2$, this shows that the implication $B_1 \rightarrow B_2$ does not hold. Obeviously, all of the conclusions we have made are reversible.

5.2 Learning Algorithm

The proposed algorithm generates *GCIs* (General Concept Inclusion) in order to obtain a TBox with a minimal number of *GCIs* from which we can find every other that hold through inference.It takes as input $\psi(DG_{K_C})$ and $\phi(DG_{K_C})$, It takes also K_C. The algorithm "*Gen_GCIs*" is given as follows:

Algorithm. Gen_GCIs

Require: $\psi(DG_{K_C})$,$\phi(DG_{K_C})$, K_C
Begin
1: $\psi(DG^*_{K_C}) := \{\emptyset\}$;
2: **For** each $Imp = \{\{m_1, ..., m_k\} \rightarrow \{m_{k+1}, ..., m_p\}\} \in \psi(DG_{K_C})$ **Do**
3: $Imp^* := Imp$;
4: $\{O_1, ..., O_s\}$ is the set of objects, corresponding to $\{m_1, ..., m_k, m_{k+1}, ..., m_p\}'$;
5: **For** each $r \in \{m_1, ..., m_k, m_{k+1}, ..., m_p\}$ **Do**
6: $restrict^*_r(Imp) := lcs(\sqcap resctict_r(\tau(O_1)), ..., \sqcap resctict_r(\tau(O_s)))$;
7: replace r by $\exists r.restrict^*_r(Imp)$ in Imp^*;
8: **End For**;
9: $\psi(DG^*_{K_C}) := \psi(DG^*_{K_C}) \cup \{Imp^*\}$;
10: **End For**
11: **For** each $Imp^* = (A \rightarrow B) \in (\psi(DG^*_{K_C}) \cup \phi(DG_{K_C}))$ **Do**
12: Add the GCI ($\sqcap A \sqsubseteq \sqcap B$) to the TBox ;
13: **End For**
End

---

6 Illustrative Example

Let us illustrate our proposition through the following example:

N_C ={Man, Woman, Father, Mother, Parent, Aunt, Uncle}, N_r ={hasBrother, hasSister} and N_O ={Hocine, Amine, Idir, Lili, Nora, Ali, Aziz, Sonia, Lilia, Lyes, Samir, Ania, Lina}.

We consider the following set of objects described by concept descriptions:

Hocine: Man ⊓ Father ⊓ Parent
Amine: Man
Idir: Man ⊓ Father ⊓ Parent
Lila: Woman ⊓ Mother ⊓ Parent
Nora: Woman
Ali: Man ⊓ ∃ hasSister.(Aunt ⊓ Woman ⊓ Mother ⊓ Parent) ⊓ Uncle
Aziz: Man ⊓ ∃ hasSister.(Man ⊓ Father ⊓ Parent) ⊓ Uncle
Sonia: Woman ⊓ ∃ hasSister.(Mother ⊓ Parent ⊓ Woman) ⊓ Aunt
Lilia: Woman ⊓ ∃ hasBrother.(Man ⊓ Father ⊓ Parent) ⊓ Aunt
Lyes: Man ⊓ ∃ hasSister.(Woman ⊓ Mother ⊓ Parent) ⊓ Uncle
Samir: Man ⊓ ∃ hasBrother.(Uncle ⊓ Father ⊓ Mother ⊓ Parent) ⊓ Uncle
Ania: Woman ⊓ ∃ hasSiste.(Mother ⊓ Aunt ⊓ Parent ⊓ Woman) ⊓ Aunt
Lina: Woman ⊓ ∃ hasBrother.(Man ⊓ Parent ⊓ Uncle) ⊓ Aunt

Initially, we generate the context formal $K_C = (N_O, N_C \cup N_R, \mathcal{I})$ using the algorithm "*Gen_Formal_Context*":

N_O:={ Hocine, Amine, Idir, Lila, Nora, Ali, Aziz, Sonia, Lilia, Lyes, Samir, Ania, Lina }

$N_C \cup N_R$:={Man, Woman, Father, Mother, Parent, Aunt, Uncle, hasSister, hasBrother}

$\mathcal{I}$:={(Hocine, Man), (Hocine, Father), (Hocine, Parent), (Amine, Man), (Idir, Man), (Idir, Father), (Idir, Parent), (Lila, Woman), (Lila, Mother), (Lila, Parent), (Nora, Moman), (Ali, Man), (Ali, hasSister), (Ali,Uncle), (Aziz, Man), (Aziz, hasBrother), (Aziz, Uncle), (Sonia, Woman), (Sonia, hasSister), (Sonia, Aunt), (Lilia, Woman), (Lilia, hasBrother), (Lilia, Aunt), (Lyes, Man), (Lyes, hasSister), (Lyes, Uncle), (Samir, Man), (Samir, hasBrother), (Samir, Uncle), (Ania, Woman),(Ania, hasSister),(Ania, Aunt), (Lina, Woman), (Lina, hasBrother), (Lina, Aunt)}

Implications of Duquenne-Guigues resulting from the context formal $K_C = (N_O, N_C \cup N_R, \mathcal{I})$ are as follow:

$\phi(DG_{K_C})$ ={{Father} → {Man, Parent}, {Mother} → {Woman, Parent}, {Man, Parent} → {Father}, {Woman, Parent} → {Mother}, {Aunt} → {Woman}, {Uncle} → {Man}}

$\psi(DG_{K_C})$ ={{Woman, hasSister} → {Aunt}, {Woman, hasBrother} → {Aunt}, {Man, hasSister} → {Uncle}, {Man, hasBrother} → {Uncle}} Now, we compute $restrict_r^*(Imp_i)$ for each role of each implication $\in \psi(DG_{K_C})$ using the algorithm "*Gen_GCIs*"

- Imp_1: {Woman, hasSister} → {Aunt}
{Sonia, Ania} = {Woman, hasSister, Aunt}′ (to use the formal context K_C)

$$resctict^*_{hasSister}(imp_1) = lcs(\sqcap resctict_{hasSister}(\tau(Sonia)), \sqcap resctict_{hasSister}(\tau(Ania)))$$

$$resctict^*_{hasSister}(imp_1) = lcs((Mother \sqcap Parent \sqcap Woman), (Mother \sqcap Aunt \sqcap Parent \sqcap Woman))$$

$$resctict^*_{hasSister}(imp_1) = (Parent)$$

the implication {woman, ∃ hasSister.(Parent)} → {Aunt} is added to $\psi(DG^*_{K_C})$

All other implications ∈ $\psi(DG_{K_C})$ are computed of same manner that we have computed $resctict^*_{hasSister}(imp_1)$ then obtain:

$\psi(DG^*_{K_C})$ ={ {woman, ∃ hasSister.(Parent)} → {Aunt}, {woman, ∃ hasBrother.(Parent)} → {Aunt}, {man, ∃ hasSister.(Parent)} → {Uncle},{man, ∃ hasBrother.(Parent)} → {Uncle}} At the end of the algorithm "*Gen_GCIs*", the following GCI axioms have been found: father ⊑ man ⊓ parent
mother ⊑ woman ⊓ parent
man ⊓ parent ⊑ father
woman ⊓ parent ⊑ mother
Aunt ⊑ woman
Uncle ⊑ man
woman ⊓ ∃ hasSister.(Parent) ⊑ Aunt
woman ⊓ ∃ hasBrother.(Parent) ⊑ Aunt
man ⊓ ∃ hasSister.(Parent) ⊑ Uncle
man ⊓ ∃ hasBrother.(Parent) ⊑ Uncle

7 Conclusion

Cross fertilizing both description logics and formal concept analysis seems an appealing domain of research. There are many tendencies in this direction. Learning concept definitions from object descriptions is mainly addressed in this spirit. However, in all existing approaches such concept definitions are restricted to concepts names. In this paper, we enlarge the learning process and propose an approach which allows to learn General Concept Inclusion (GCI) containing concept names as well as existentially quantified roles. For this purpose, we have established the appropriate propositions. As direct future work, we intend first to generalize our proposition to more expressive DL families. We intend also to address incomplete (missing, uncertain, imprecise) objects descriptions.

Acknowledgments. This work was supported by the National Research Program (PNR) of the Directorate General for Scientific Research and Technological Development (DGRSDT) - Ministry of Higher Education and Scientific Research under the grant 14/IF/2011.

References

[1] Baader, F., Ganter, F., Sattler, U., Sertkaya, B.: Completing description logic knowledge bases using formal concept analysis. LTCS-Report LTCS-06-02, Chair for Automata Theory, Inst. for Theoretical Computer Science, TU Dresden, Germany, (2006)

[2] Baader, F.: Computing a Minimal Representation of the Subsumption Lattice of all conjunctions of Concepts Defined in a Terminology. In: Ellis, G., Levinson, R.A., Fall, A., Dahl, V. (eds.) Proceedings of the 1st International KRUSE Symposium on Knowledge Retrieval, Use and Storage for Efficiency, pp. 168–178 (1995)

[3] Baader, F., Sertkaya, B.: Applying formal concept analysis to description logics. In: Eklund, P. (ed.) ICFCA 2004. LNCS (LNAI), vol. 2961, pp. 261–286. Springer, Heidelberg (2004)

[4] Stumme, G.: The Concept Classification of a Terminology Extended by Conjunction and Disjunction. In: Foo, N.Y., Göbel, R. (eds.) PRICAI 1996. LNCS, vol. 1114, pp. 121–131. Springer, Heidelberg (1996)

[5] Baader, F., Molitor, R.: Building and Structuring Description Logic Knowledge Bases Using Least Common Subsumers Concept Analysis. In: Ganter, B., Mineau, G.W. (eds.) ICCS 2000. LNCS, vol. 1867, pp. 290–303. Springer, Heidelberg (2000)

[6] Rudolph, S.: Relational exploration: Combining Description Logics and Formal Concept Analysis for knowledge specification. Ph.D. dissertation, Fakultat Mathematik und Naturwissenschaften, TU Dresden, Germany (2006)

[7] Guigues, J.L., Duquenne, V.: Familles minimales d'implications informatives resultant d'un tableau de donnees binaires. Math. Sci. Humaines 95, 5–18 (1986)

[8] Baader, F., Calvanese, D., McGuinness, D., Nardi, D., Patel-Schneider, P.F. (eds.): The Description Logic Handbook: Theory, Implementation, and Application. Cambridge University Press (2003)

[9] Ganter, B., Wille, R.: Formal Concept Analysis: Mathematical Foundations. Springer, Berlin (1999)

[10] Bazin, A., Ganacia, J.: Completing terminological Axioms with Formal Concept Analysis. In: ICFCA 2012, International Conference on Formal Concept Analysis, pp. 30–39 (2012)

[11] Baader, F., Küsters, R., Molitor, R.: Computing least common subsumers in description logics with existential restrictions. In: Proc. of the 16th Int. Joint Conf. on Artificial Intelligence (IJCAI 1999), pp. 96–101 (1999)

[12] Horrocks, I., Patel-Schneider, P.F., van Harmelen, F.: From SHIQ and RDF to OWL: the making of a web ontology language. Journal of Web Semantics 1(1), 7–26 (2003)

[13] Prediger, S.: Terminologische Merkmalslogik in der Formalen Begriffsanalyse. In: Stumme, G., Wille, R. (eds.) Begriffliche Wissensverarbeitung – Methoden und Anwendungen, pp. 99–124. Springer, Heidelberg (2000)

[14] Baader, F., Horrocks, I., Sattler, U.: Description logics as ontology languages for the semantic web. In: Hutter, D., Stephan, W. (eds.) Mechanizing Mathematical Reasoning. LNCS (LNAI), vol. 2605, pp. 228–248. Springer, Heidelberg (2005)

Automatic Phonetization of Arabic Text

Fayçal Imedjdouben and Amrane Houacine

Faculty of Electronics and Computer Science, U.S.T.H.B, Algiers, Algeria
{fimedjdouben,ahouacine}@usthb.dz

Abstract. We present in this paper a system for automatic phonetization of text dedicated to the standard Arabic language. The general methodologies, as well as technical details are given. The phonetic transcription constitutes a fundamental step for the development of any text to speech system. The system developed here consists of two stages. In the first stage are processed the exceptions based on an Arabic language exception database. The second step consists in processing the remaining text content by using phonetic transcription rules we have elaborated. Validation is performed through phonetic transcription of a well established corpus of standard Arabic text.

Keywords: Arabic language, phonetic transcription, text to speech.

1 Introduction

The automatic phonetization of the text is a prerequisite element for the realization of a text to speech synthesizer. This one is intimately related to the characteristics of the language considered and makes it possible to establish a correspondence between the graphemes constituting the text and a sequence of phonemes which will be used to produce the pronunciation of the text [1-3].

The research work which treats specifically the phonetic transcription of the Arabic language is very little. We particularly mention the works of M. Elshafei Ahmed [4], Mansour M. Al-ghamdi [5], Tahar Saidane [6], and Yousif A. El-imam [7]. The latter have presented a phonetic transcription system which depends entirely on the succession of sounds in the words, as well as he considered the exception words that do not obey to established rules of phonetic transcription. These exception words are gathered in a lexicon of the exceptions.

In what follows, we will describe the various stages followed for the realization of our system of automatic phonetization, by specifying the rules of phonetic transcription elaborated for the automatic phonetization of the standard Arabic text.

2 Arabic Language

In order to realize the phonetic transcription of the Arabic text, a study of the language characteristics is necessary over various aspects (grammatical, phonological,

A. Amine et al. (Eds.): *Modeling Approaches and Algorithms*, SCI 488, pp. 85–94.
DOI: 10.1007/978-3-319-00560-7_13 © Springer International Publishing Switzerland 2013

phonetics, and coding of the language). Thereafter we included all the problems associated to the transcription in order to generate our lexicon of the exceptions, as well as description of our base of rules for phonetic transcription. This last treats the whole of the following forms:

- The consonants which are about 28 consonants.
- Short vowels symbolized by (َ ُ ِ).
- Long vowels symbolized by (اَ وُ يِ).
- The doubly of vowel which is called **tanwin** symbolized by (ً ٍ ٌ).
- **Sukun** symbolized by « ْ ».
- **Ya maqsurah** symbolized by « ى ».
- The gemination called **chadda** symbolized by « ّ ».
- The elision of the consonant **alif** symbolized by « ا ».
- The lunar letters (ا ب ج ح خ ع غ ف ق ك م ه و ي), which do not assimilate the consonant **lam** symbolized by « ل » of the article « ال ».
- The solar letters (ت ث د ذ ر ز س ش ص ض ط ظ ل ن), which assimilate the consonant **lam** « ل » of the article « ال ».
- Suppression of the short vowels, as well as **tanwin** at the end of a sentence.
- Replacement of the long vowels by short vowels when the long vowels are followed by the article « ال ».
- The different graphic representation of the consonant **alif** which are symbolized by (آ ئ ؤ إ أ).
- Suppression of the consonant **ta marbota** symbolized by « ة » at the end of a sentence and to replace it by consonant **ha** symbolized by « ه ». Apart from that it is replaced by the consonant **ta** symbolized by « ت ».

3 Automatic Phonetization System

The architecture of our system of automatic phonetic transcription is composed of two phases of sequential processing: The first stage consists in a pre-processing of the text. This last is followed by the second phase which comprises the phonetic transcription in accordance with the characteristics of the language.

3.1 Text Pre-processing

The characters of the Arabic language do not belong to the **ASCII** code, hence the need to use another code which takes into account the Arabic language. This one is the **Unicode** standard (Table 1). This last allows to code all the characters used by the Arabic language and to exchange data of text between different platforms and systems.

Table 1. Unicode standard for the Arabic characters

Unicode (Hex)	0	1	2	3	4	5	6	7	8	9	A	B	C	D	E	F
062		ء	آ	أ	ؤ	إ	ئ	ا	ب	ة	ت	ث	ج	ح	خ	د
063	ذ	ر	ز	س	ش	ص	ض	ط	ظ	ع	غ					
064		ف	ق	ك	ل	م	ن	ه	و	ى	ي	ـً	ـٌ	ـٍ	ـَ	ـُ
065	ـِ	ـّ	ـْ													

The objective of the pre-processing is the disambiguation of the text at entry of the automatic system of phonetization, so that the resulting text does not comprise any ambiguity for the later linguistic processing. The stages consist of:

- Reading of the text to the two byte format because of the Unicode standard.
- Treatment of the punctuations, spaces and line jumps.
- Suppression of the diacritic signs that are added to the article « ال ».
- Suppression of the consonant **alif** symbolized by « ا » at the beginning of a sentence and replacing it by the consonant **hamza** symbolized by « ء ».
- Organization of the text entry in the form of cells, where each cell contains a word. This operation is made to facilitate the phase of the phonetic transcription based on a lexicon of the exceptions.

3.2 Phonetic Transcription

The phonetic transcription consists in producing the pronunciation corresponding to the text in entry under the form of a list of phonemes. The phonetization of a text is not an easy task because there is not a direct correspondence between the graphemes and the phonemes [8-10]. We divided the phonetic transcription module in two phases of processing:

- Phonetization based on a lexicon of the exceptions.
- Phonetization based on rules.

Phonetization Based on a Lexicon of the Exceptions. This phase consists of a simple comparison between the elements of the starting text and a list of words established (Table 2) as admitting exceptions (28 special words and 9 abbreviations). Note that the comparison itself is not done on the characters which constitute the word but according to their codes in **Unicode** standard (Table 1). This method consists in directly providing the phonetization corresponding to the word of exception, without passing by the base of rules of the phonetic transcription. But the disadvantage of this method is that it does not take into account the position of the word treated in the sentence (the word which precedes it or which follows it).

From This Fact We Established about Ten Rules to Solve This Problem. These Rules Treat Following Cases:

- The lunar letters.
- The solar letters.
- Suppression of the diacritics signs at the end of the word when situated at the end of a sentence.
- Suppression of the long vowels and replacing them by short vowels when the long vowels are followed by the article « ال ».

Table 2. A sample of our lexicon of the exceptions

Special words	Phonetic transcription	Abbreviations	Phonetic transcription
أولَئِكَ	?ula:?ika	أنا	?an-ba?ana:
هَكَذَا	ha:kaDa:	اه	?in-taha:
الرَّحْمَنُ	?a-rraX-ma:nu	ثنا	XadaTana:
ذَلِكَ	Da:lika	الخ	?ila:_?a:xirihi
هَذَا	ha:Da:	رحه	raXimahu_lla:hu

The generation of our lexicon of the exceptions (Fig. 1) is done automatically, this gives us a facility to handle and integrate any other exception, and this task is done simply by using two text documents (.txt). The first document contains the words of exceptions and the second document contains the same words of exceptions but corrected so that the phonetic transcription will be generated correctly. The lexicon of the exceptions generated contains at the same time the words of exceptions coded in **Unicode** standard used at the time of the comparison, as well as their appropriate phonetic sequences according to **SAMPA** notation [11].

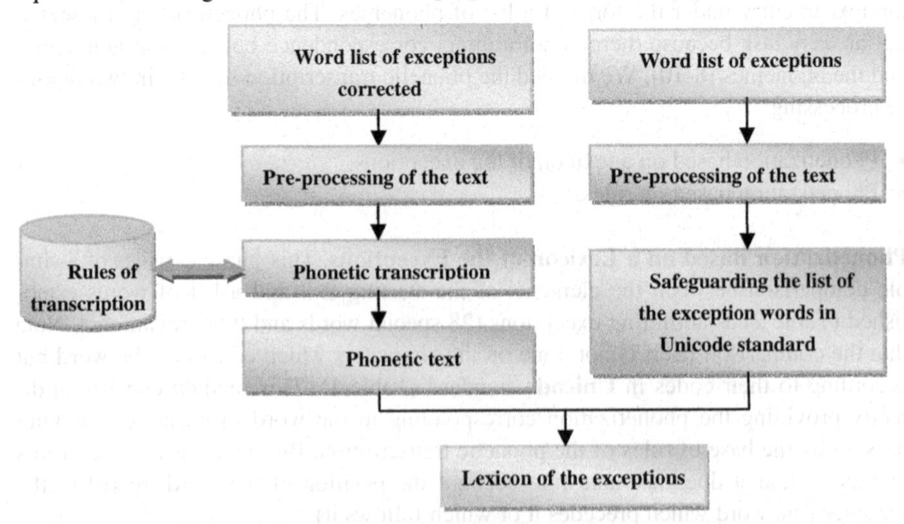

Fig. 1. Diagram of generation of the lexicon of the exceptions

Phonetization Based on Rules. The base of rules of our system of phonetization uses the Unicode standard (the test of the characters is performed using the Unicode standard representation). This latter gives to our system portability and functionality even if the machine or the platform is not configured for Arabic language.

The base of rules which we have elaborated comprises 100 rules. These rules take into account the whole of the graphic realizations of the Arabic language which are 44 graphemes to generate a phonetic sequence which is 34 phonemes (28 consonants, 3 short vowels, 3 long vowels). This base of rules contains at the same time the rules of phonetic transcription and the phonemes resulting from each rule without passing by a table of grapheme-phoneme conversion (phonetic is incorporated in the rules).

The methodology followed to realize the rules of phonetic transcription is of following form: each grapheme is replaced by one or more phonemes according to its left contexts, its right contexts, or both at the same time. The advantage of this approach is to permit variability in the form of the rules in our base of rules. Each grapheme treated may have its own rules, so the rules do not obey a fixed structure.

This grapheme specific approach assures a better precision of the phonetic result and a rapidity of transcription since for some graphemes, it is not useful to systematically test the right and left contexts.

We present here some established rules. Note that the order of application of these rules is very important. An inadequate order can generate transcription errors. The phonetic result is represented according to **SAMPA** notation (Table 3).

Rule1: **[Phe] = LC + C**

Where **Phe** is the phonetic result, **LC** corresponds to the left context of the character tested, and C is the character tested.

This rule indicates that a character C, preceded by a character **LC**, will have as a phonetic transcription **Phe.**

As example: [ ] = 'ـُ' + 'و' This indicates that the consonant waw 'و' (represented in Unicode by **'648'**), preceded by the short vowel damma 'ـُ' (represented in Unicode by **'64F'**), will be dumb.

Rule2: **[Phe] = C + RC**

Where **RC** corresponds to the right context of the character tested.

This rule indicates that a character C, followed by a character **RC**, will have as a phonetic transcription **Phe.**

As example: **[u:]** = 'ـُ' + 'و' This indicates that the short vowel damma 'ـُ' (represented in Unicode by **'64F'**), followed by the consonant waw 'و' (represented in Unicode by **'648'**), will have as a phonetic transcription **[u:]**.

Rule3: **[Phe] = LC2 + LC1 + C + RC**

Where **LC1** is the first left context of the character tested, and **LC2** is the second left context of the character tested.

This rule indicates that a character C, preceded by characters **LC1, LC2** and followed by a character **RC**, will have as a phonetic transcription **Phe**.

As example: [ ] = **'00D'** + ' ا ' + ' ل ' + **'SL'** This indicates that the consonant lam ' ل ' (represented in Unicode by **'644'**), preceded by the consonant alif ' ا ' (represented in Unicode by **'627'**), and of the character **'00D'** (beginning of sentence), and followed of solar letters **'SL'**, will have no phonetic transcription, so the letter lam ' ل ' is dumb.

Rule4: **[Phe] = LC2 + LC1 + C**

This rule indicates that a character C, preceded by characters **LC1, LC2**, will have as a phonetic transcription **Phe**.

As example: **[a:]** = ' ل ' + ' ا ' + ' َ ' This indicates that the short vowel fatha ' َ ' (represented in Unicode by **'64E'**), preceded by the consonant alif ' ا ' (represented in Unicode by **'627'**), and of the consonant lam **'ل'** (represented in Unicode by **'644'**), will have as a phonetic transcription **[a:]**.

Rule5: **[Phe] = LC + C + RC1 + RC2**

Where **RC1** is the first right context of the character tested, and **RC2** is the second right context of the character tested.

This rule indicates that a character C, preceded by a character **LC**, and followed by characters **RC1, RC2**, will have as a phonetic transcription **Phe**.

As example: **[?] + [a] = '00D'** + ' ا ' + ' ل ' + **'SL'** This indicates that the consonant alif ' ا ' (represented in Unicode by **'627'**), preceded by the character **'00D'** (beginning of sentence), and followed by the consonant lam ' ل ' (represented in Unicode by **'644'**), and of solar letters **'SL'**, will have as a phonetic transcription the succession of two phonemes **[?]** and **[a]**.

Rule6: **[Phe] = C + RC1 + RC2**

This rule indicates that a character C, followed by characters **RC1, RC2**, will have as a phonetic transcription **Phe**.

As example: **[u] + [w]** = ' ُ ' + **' و '** + **'SV'** This indicates that the short vowel damma ' ُ ' (represented in Unicode by **'64F'**), followed by the consonant waw **' و '** (represented in Unicode by **'648'**), and of short vowels **'SV'**, will have as a phonetic transcription the succession of two phonemes **[u]** and **[w]**.

The other rules elaborated consist in associating to each grapheme, one or more phonemes directly without passing by the left and right contexts. These kinds of rules are called **direct rules** and are invoked only after having verified that all above mentioned six rules are not applicable for current character. The structure is of following form: **[Phe]** = C. This rule indicates that a character C, will have as a phonetic transcription **Phe**.

As example: **[i] + [n]** = ' ٍ '. This indicates that the diacritical sign kasratan ' ٍ ' (represented in Unicode by **'64D'**), will have as a phonetic transcription the succession of two phonemes **[i]** and **[n]**.

Here is an example of application of these rules on the word 'الطعام' (food). We begin by transcribing the word letter by letter:

- The consonant **alif** 'ا' is transcribed by the phonemes **[?]** + **[a]** according to **rule5**.
- The consonant **lam** 'ل' is transcribed by non-presence of phoneme according to **rule3**.
- The consonant **ta** 'ط' is transcribed by the phonemes **[t`]** + **[t`]** according to **rule2**.
- The gemination **chadda** 'ّ' is transcribed by non-presence of the phoneme according to a **direct rule**.
- The vowel **fatha** 'َ' is transcribed by the phoneme **[a]** according to a **direct rule**.
- The consonant ayn 'ع' is transcribed by the phoneme **[H]** according to a **direct rule**.
- The vowel **fatha** 'َ' is transcribed by the phoneme **[a:]** according to **rule2.**
- The consonant **alif** 'ا' is transcribed by non-presence of the phoneme according to **rule1.**
- The consonant **mim** 'م' is transcribed by the phoneme **[m]** according to a **direct rule**.

Finally obtain the following phonetic transcription: **?at`t`aHa:m**

Table 3. Grapheme-phoneme correspondence of the Arabic language according to SAMPA notation

N°	Grapheme	Name	Phoneme	N°	Grapheme	Name	Phoneme
0	ء	Hamza	?	20	ف	Fa	f
1	ئ ؤ إ أ ا	Alif	?	21	ق	Qaf	q
2	ب	Ba	b	22	ك	Kaf	k
3	ت ة	Ta	t	23	ل	Lam	l
4	ث	Tha	T	24	م	Mim	m
5	ج	Jim	Z	25	ن	Nun	n
6	ح	Hha	X	26	ه	Ha	h
7	خ	Kha	x	27	و	Waw	w
8	د	Dal	d	28	ي	Ya	j
9	ذ	Thal	D	29	َ	Fatha	a
10	ر	Ra	r	30	ُ	Damma	u
11	ز	Zay	z	31	ِ	Kasra	i
12	س	Sin	s	32	ىَ اَ آ	Long fatha	a:
13	ش	Shin	S	33	وُ	Long damma	u:
14	ص	Sad	s`	34	يِ	Long kasra	i:
15	ض	Dad	d`	35	ْ	Sukun	-
16	ط	Ta	t`	36	ً	Fathatan	an
17	ظ	Zha	D`	37	ٌ	Dammatan	un
18	ع	Ayn	H	38	ٍ	Kasratan	in
19	غ	Ghayn	G	39		The silence	_

4 Results

In order to test our system of automatic phonetization we considered an Arabic corpus of sentences realized by Mansour M. Al-ghamdi [12]. This corpus contains a list of 367 sentences, composed of 1835 words, and 12940 graphemes. The number of words per sentence is 2 to 9 words per sentence.

For the verification of the phonetic transcription generated automatically by our system, we proceeded to a manual transcription of the 367 sentences. And, using the automatic transcription system, we obtained a transcription of all these sentences correctly with a full rate of success. Results on the 20 first phrases are illustrated in (Table 4). Note that, according to Mansour M. Al-ghamdi, this Arabic corpus is characterized by the richness, balance and phonetic diversity, and if a text to speech system pronounces this corpus of Arabic sentence correctly, so it will be able to pronounce any other Arabic text also correctly. From this fact it may be attended that our system of automatic phonetization will be able to transcribe any other Arabic text correctly.

Table 4. Results of the phonetic transcription for the first 20 sentences of the corpus

N°	Arabic sentence	Manual transcription using **SAMPA** notation	Automatic transcription using **SAMPA** notation
1	مِنْ بَخْس نِعْمَةِ اللّهِ دَفْنُهَا	min_bax-si_niH-mati_lla:hi_ daf-nuha:	min_bax-si_niH-mati_lla:hi_ daf-nuha:
2	أبْجَلَنِي هَذَا الطّعَامُ	?ab-Zalani:_ha:Da_ t`t`aHa:m	?ab-Zalani:_ha:Da_t`t`aHa:m
3	الأعْشَى الشّاعِرُ مِنْ أذهَى الشّعَرَاءِ	?al-?aH-Sa_SSa:Hiru_ min_ ?ad-ha_SSuHara:?	?al-?aH-Sa_SSa:Hiru_ min_?ad-ha_SSuHara:?
4	مِنْ أضْأَلِ الطّعَامِ رَغِيفُ القمْحِ	min_?ad`-?ali_t`t`aHa:mi_ raGi:fu_ l-qam-X	min_?ad`-?ali_t`t`aHa:mi_ raGi:fu_ l-qam-X
5	رَأى النّائِمُ أضْغَاثَ أحْلامِ	ra?a_nna:?imu_?ad`-Ga:Ta_ ?aX-la:m	ra?a_nna:?imu_?ad`-Ga:Ta_ ?aX-la:m
6	أضْهَى الرّجُلُ إبِلَهُ	?ad`-ha_rraZulu_?ibilah	?ad`-ha_rraZulu_?ibilah
7	كَانَ حَاتِمُ الطّائِيّ يُكْرِمُ الأضْيَافَ	ka:na_Xa:timu_t`t`a:?ijju_ juk-rimu_ l-?ad`-ja:f	ka:na_Xa:timu_t`t`a:?ijju_ juk-rimu_ l-?ad`-ja:f
8	الأفْغَانُ مُقَاتِلُونَ أفذَاذٌ	?al-?af-Ga:nu_ muqa:tilu:na _?af-Da:D	?al-?af-Ga:nu_ muqa:tilu:na_ ?afDa:D
9	إنْطَلَقَتْ أفْوَاجُ الحُجّاجِ فِي جَوٍّ أغْبَرَ	?in-t`alaqat_?af-wa:Zu_ l-XuZZa:Zi_ fi:_Zawwin_ ?aG-bar	?in-t`alaqat_?af-wa:Zu_ l-XuZZa:Zi_ fi:_Zawwin_ ?aG-bar
10	اللّبَنُ مِنْ أغْنَى الأغْذِيَةِ	?a-llabanu_min_?aG-na_ l-?aG-Dijah	?a-llabanu_min_?aG-na_ l-?aG-Dijah
11	أحَطْتُ بِأضْدَادِ هَذِهِ الكَلِمَةِ	?aXat`-tu_bi?ad`-da:di_ ha:Dihi_l-kalimah	?aXat`-tu_bi?ad`-da:di_ ha:Dihi_l-kalimah
12	يَعِيشُ كَثِيرٌ مِنَ الأشْخَاصِ فِي أحْزَانٍ	jaHi:Su_kaTi:run_mina_ l-?aS-xa:s`i_ fi:_?aX-za:n	jaHi:Su_kaTi:run_mina_ l-?aS-xa:s`i_fi:_?aX-za:n

Table 4. (*continued*)

13	حِينَمَا تَثُورُ الأَشْجَانُ تَتَغَيَّرُ الأَمْزِجَةُ	Xi:nama:_taTu:ru_ l-?aS-Za:nu_ tataGajjaru_ l-?am-ziZah	Xi:nama:_taTu:ru_l-?aS-Za:nu _tataGajjaru_l-?am-ziZah
14	أَحْضَرْتُ مِنْ الشَّارِقَةِ أَشْيَاءَ لِلْبَيْعِ	?aX-d`ar-tu_min_SSa:riqati_ ?aS-ja:?a_lil-baj-H	?aX-d`ar-tu_min_ SSa:riqati_ ?aS-ja:?a_ lil-baj-H
15	أَكَلْتُ بَيْضَ الدَّجَاجِ	?akal-tu_baj-d`a_ddaZa:Z	?akal-tu_baj-d`a_ddaZa:Z
16	فِي هَذِهِ الأَكْوَابِ أَمْشَاجٌ مِنَ الشَّرَابِ	fi:_ha:Dihi_l-?ak-wa:bi_ ?am-Sa:Zun_ mina_SSara:b	fi:_ha:Dihi_l-?ak-wa:bi_ ?am-Sa:Zun_mina_SSara:b
17	أَمَّا فِعْلُ الْخَيْرِ فَحَسَنٌ	?amma:_fiH-lu_l-xaj-ri_ faXasan	?amma:_fiH-lu_l-xaj-ri_ faXa-san
18	فِي الْمُسْتَشْفَى أَمْصَالٌ ضِدَّ الأَمْرَاضِ	fi_l-mus-taS-fa:_?am-s`a:lun _d`idda_l-?am-ra:d`	fi_l-mus-taS-fa:_?am-s`a:lun_ d`idda_l-?am-ra:d`
19	نَسْمَعُ الأَنْغَامَ عِنْدَ عَزْفِ الْمَعَازِف	nas-maHu_l-?an-Ga:ma_ Hin-da_Haz-fi_l-maHa:zif	nas-maHu_l-?an-Ga:ma_ Hin-da_Haz-fi_l-maHa:zif
20	أَنْذَرَ الرَّجُلُ صَدِيقَهُ بِوَقْفِ مُسَاعَدَتِهِ	?an-Dara_rraZulu_ s`adi:qahu_biwaq-fi_ musa:Hadatih	?an-Dara_rraZulu_s`adi:qahu_ biwaq-fi_ musa:Hadatih

5 Conclusion

In this paper we presented the various steps which we followed to build a system for automatic phonetization for the standard Arabic language. Our rule base of the phonetic transcription allowed a satisfactory transcription on a well established standard Arabic corpus.

We adopted a strategy of insertion of the transcription of the exceptions from a pre-established base. And we also established that it is not always necessary to consider the left and right contexts in the transcription process. Exploiting these two aspects allows a saving of time in the text processing step.

Finally, due to the fact that our rules are not developed in a generic approach, any specificity in the language phonetics may easily be added. Also, considering more corpus sentences in standard Arabic language, covering different fields of interest, will be important to assure a full rate of confidence to our system.

References

1. Zeki, M., Khalifa, O.O., Naji, A.W.: Development of an Arabic Text-To-Speech System. In: IEEE International Conference on Computer and Communication Engineering, Kuala Lumpur, Malaysia, pp. 1–5 (2010)
2. Elshafei, M., Al-muhtaseb, H., Al-ghamdi, M.: Techniques for high quality Arabic speech synthesis. Information Sciences 140, 255–267 (2002)
3. Hamad, M., Hussain, M.: Arabic Text-To-Speech Synthesizer. In: IEEE Student Conference on Research and Development, pp. 409–414 (2011)

4. Elshafei, M.A.: Toward an Arabic text-to-speech system. Arabian Journal of Science and Engineering 16, 339–373 (1991)
5. Al-ghamdi, M., Al-muhtaseb, H., Elshafei, M.: Phonetic Rules in Arabic Script. King Saud University Journal: Computer Sciences and Information 16, 1–25 (2004)
6. Saidane, T., Zrigui, M., Ben Ahmed, M.: La Transcription Orthographique-Phonétique de la Langue Arabe. In: RÉCITAL, Fès (2004)
7. El-imam, Y.A.: Phonetization of Arabic: rules and algorithms. Computer Speech and Language 18, 339–373 (2004)
8. Saidane, T., Zrigui, M., Ben Ahmed, M.: Un système de synthèse de la parole arabe par concaténation de polyphèmes: Les résultats de l'utilisation d'un lissage linéaire. In: 3rd International Conference: Sciences of Electronic, Technologies of Information and Telecommunications, Tunis (2005)
9. Zemirli, Z.: ARAB_TTS: An Arabic Text To Speech Synthesis. In: IEEE International Conference on Computer Systems and Applications, pp. 976–979 (2006)
10. Rashad, M.Z., El-bakry, H.M., Isma'il, I.R., Mastorakis, N.: An Overview of Text-To-Speech Synthesis Techniques. In: 4th International Conference on Communications and Information Technology, Corfu Island, Greece, pp. 84–89 (2010)
11. SAMPA, Speech Assessment Methods Phonetic Alphabet, `http://www.phon.ucl.ac.uk/home/sampa/arabic.htm`
12. Al-ghamdi, M., Al-hamid, A.H., Al-dasuqi, M.M.: Database of Arabic Sounds: Sentences. Technical Report, King Abdulaziz City of Science and Technology, Saudi Arabia (2004)

A Novel Region Growing Segmentation Algorithm for Mass Extraction in Mammograms

Ahlem Melouah

Badji-Mokhtar Annaba University P.O. Box 12, 23000, Annaba, Algeria
melouahlem@yahoo.fr

Abstract. This article presents an automatic mass extraction approach by application of a novel region growing algorithm. The region-growing process is guided by regional features analysis consequently; the result will be a robust algorithm able of respecting various image characteristics. The evaluation of the proposed approach was carried out on all MiniMIAS database mammograms containing circumscribed lesions. All masses from various characters of background tissues are well detected.

Keywords: region growing algorithm, features, mass detection, mammogram.

1 Introduction

Mammography is by far the only effective screening method for breast cancer detection in early stage [1]. Mass extraction in mammograms is a difficult and ambiguous task. Tumors can be missed because they are obscured by glandular tissues and it is therefore difficult to observe their boundaries. Many researchers try to propose systems to perform efficiently this task, called segmentation. The region growing algorithm appears as the natural choice in the masses segmentation [2].

Though conventional region-growing is an excellent pixel-based segmentation method, unfortunately, it presents serious drawbacks. The performance of segmentation algorithms often depends on two parameters: initial seed and threshold selection. A correct seed point choice is the basic requirement for region growing and a good growing stop criteria based on a well threshold selection is the guarantee for promising segmentation results. It is difficult to achieve an appropriate setting for these parameters because it requires an experienced user in application fields both algorithm and knowledge. In order to overcome the second difficulty, we suggest a novel region growing algorithm which simultaneously segments a mass and develops a stopping criterion.

This paper is organized as follows: Section 2 describes the region growing algorithm and discusses previous works, Section 3 explains the proposed approach in detail: preprocessing step is described in Section 3.1, training base elaboration and masses extractions by the new region growing algorithm are detailed in Sections 3.2 and 3.3 respectively, Section 4 shows experiment results and Section 5 draws conclusions.

A. Amine et al. (Eds.): *Modeling Approaches and Algorithms*, SCI 488, pp. 95–104.
DOI: 10.1007/978-3-319-00560-7_14 © Springer International Publishing Switzerland 2013

2 Background

Region growing is based on the following principles [3], [4]: suppose that we start with a seed pixel s located somewhere inside the suspected lesion and wish to expand from that pixel to fill a coherent region. Local pixels around the seed are examined to determine the most similar ones. So the next 4- or 8-neighboring pixels are checked for similarity so that the region can grow. Let's define a similarity measure M such as if pixels in the 4- or 8-neighboring region are similar, they are added to the region. First, consider a pixel q adjacent to pixel s. We can add pixel q to pixel s's region if $M(s,q)<T$ for some threshold T. We can then proceed to the other neighbors of s and do likewise. If pixels s and q are similar $M(s, q)$ produces a low result if not, the result is high. Suppose that $M(s,q)<T$ and we add pixel q to pixel s's region. We can now consider the neighbors of q and add them in the same way, if they are similar enough. So, the region continues to grow until there are no remaining similar pixels among the 4- or 8-neighbors.

Now a few unanswered questions must be raised:

1. How do we choose the seed point s?
2. How do we define similarity measure M and which threshold T do we use?

Region growing algorithm has successfully been used as a segmentation technique in medical images [5], [6], [7], [8], [9]. In mammography and for masses segmentation, some works propose region growing methods which discuss how to select a seed point automatically [10], [11]. But, since the region growing algorithm can produce an accurate mass contour if a suitable threshold is chosen [12], most works focus on the region growing process by defining a similarity measure M based on an automatic selection of the threshold T. Thus, setting an optimal threshold automatically becomes the most important task in improving this algorithm [12].

Kupinski et al. [13] presented an approach where the masses were segmented using a technique that combines region growing algorithm with likelihood analysis. The approaches consist of using the same seed point with several threshold values, thereafter, applying a measurement of probability in order to choose the most probable area. This approach was used again by Kinnard et al. [14]. CAO et al. [12], [15] proposed the same principal with adding edge gradients information to ameliorate segmentation results.

Pohlman et al. [5] proposed an adaptive region growing method: given an initial seed point and cutoff factor (k), the mean (μ) and standard deviation (σ) were calculated for a 5x5 window. Neighboring pixel was included in the segmented region if its gray level was within $\pm k\ \sigma$ gray level values of the mean gray level value of all pixels belonging in the region. As region grew, μ and σ were updated, therefore, the inclusion criteria $\pm k\ \sigma$ was updated. k=2.0 for each initial segmentation attempt, and changed by intervals of 0.1. k was increased if the tumor was enclosed along with the background. k was decreased if the tumor was not enclosed. A similar principal was used by Mencattini et al. [2] and Rabottino et al. [16].

3 Methodology

Conventional region growing is an effective pixel-based segmentation method and is robust to tissue interference [12]. Since features masses descriptors can have a strong effect upon masses segmentation; this work proposes a novel algorithm to perform an adaptive growing method based on regional feature analysis. Figure 1 presents the flowchart of the proposed method for breast masses extraction in mammograms. The method includes two big steps. Initially, a training base is elaborated. At this stage, expert assistance is essential. The expert surrounds masses by delimiting their borders. Secondly, a region growing segmentation step with an automatic threshold choice is applied in order to determine masses borders. To improve the studied image quality, a preprocessing stage is added to the preceding ones.

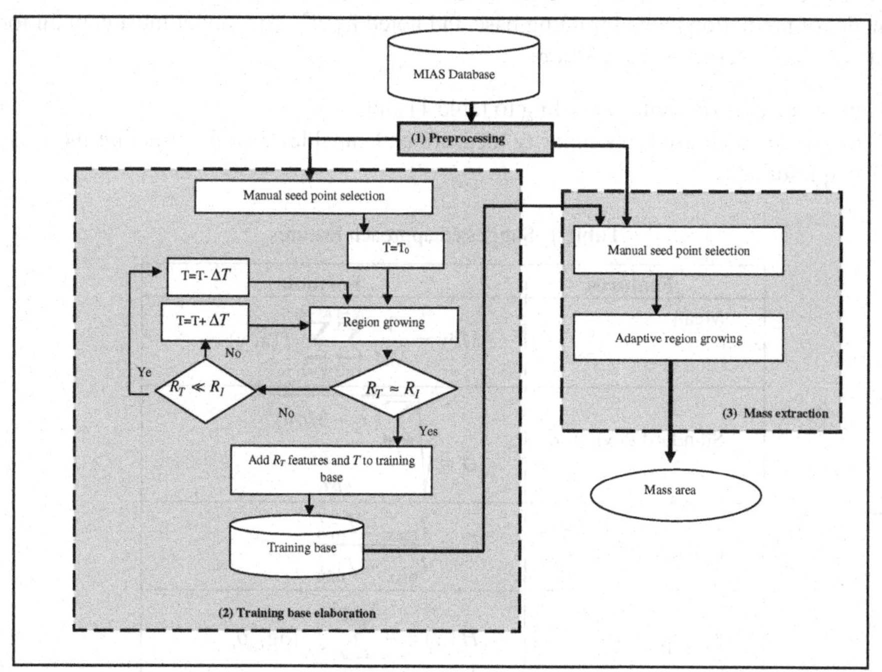

Fig. 1. Flowchart of the proposed approach

3.1 Preprocessing

Most mammograms and ultrasound images suffer from a poor image quality caused by unwanted speckle noise and low contrast. So that, mass detection becomes effective, sequences of preprocessing steps are designed to enhance the region of interest intensity and remove the noise effects.

Two-stages preprocess consisting of gradient enhancement and median filter-based smoothing is used.

The first stage eliminates the noise in the image background by using a median filter. Then, the second stage enhances gradient values of pixels with high intensities which are assumed to be potential mass pixels. To increase the gradient values of mass pixels the procedure of Yang et al. is used [17].

$$I_1(x,y) = Gradient\ enhancement\ \{I(x,y)\} = g(x,y) + I(x,y) \qquad (1)$$

Where $I(x,y)$ is the gray level value of the pixel located at the position (x,y) and $g(x,y)$ is the gray level value of the gradient of the pixel (x,y).

3.2 Training Base Elaboration

This stage objective is to extract features from the cancerous zones in order to build a model of prediction named training base and noted by B^R. The model must contain for each mass two crucial information:

1. mass features (definite according to table 1) and,
2. threshold value used in similarity measure and capable of well extracting the mass in question.

Table 1. Suggested approach features

Features	Formula
Mean	$Moy = \frac{1}{NM}\sum_{x=0}^{N-1}\sum_{y=0}^{M-1} f(x,y)$
Standard deviation	$\sigma = \sqrt{\frac{\sum_i (x_i - Moy)^2}{NM}}$
Contrast	$C = \frac{I_{max} - I_{min}}{I_{max} + I_{min}}$
Entropy	$H(x) = -\sum_{i=0}^{NM-1} p_i \log_i p_i$
Regularity	$R = 1 - \frac{1}{(1+\sigma)}$
Uniformity	$U = \sum_{i=0}^{NM-1} p_i^2$

The entries in this phase are mammograms where the masses are identified beforehand by an expert.

Let us note by I an entry image and by R_I the mass area in I encircled by the expert. A seed s is then placed, manually, inside R_I and a criterion of similarity noted M_T based contrast is adopted

$$M_T = Contraste^R < T \quad (2)$$

with T the used threshold firstly initialized with a given value T_0.

With using M_T and s, an iterative region growing process is launched. Let Suppose R_T the result area of this process.

$$R_T = Region\ growing\ (M_T,\ s) \quad (3)$$

The visual comparison of R_I and R_T leads to one of the following situations:

1. $R_T \approx R_I$: R_T and R_I are equivalent; this situation is illustrated by figure 2(a,b). In this case, R_T features are calculated (according to table 1) then inserted with T value in the training base B^R;
2. $R_T << R_I$: R_T area is definitely smaller than R_I area as shown in the figure 2(a,c). This situation implies that the value of the threshold T authorized aggregation but only in limited number of pixels. In order to increase this number, it is necessary to increase the value of the threshold T by adding a step ΔT to this last.

$$T = T + \Delta T \quad (4)$$

3. $R_T >> R_I$: This case is the opposite of the preceding ones; R_T area is more important than R_I area. Figure 2(a,d) illustrates this situation. The threshold T value allows the aggregation of a high number of pixels, which implies that it is necessary to reduce the value of the threshold T as follows:

$$T = T - \Delta T \quad (5)$$

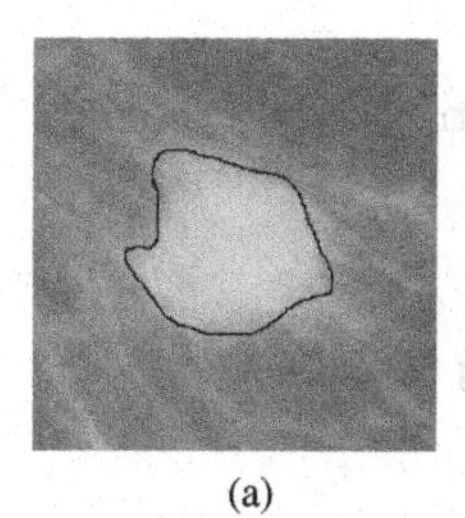
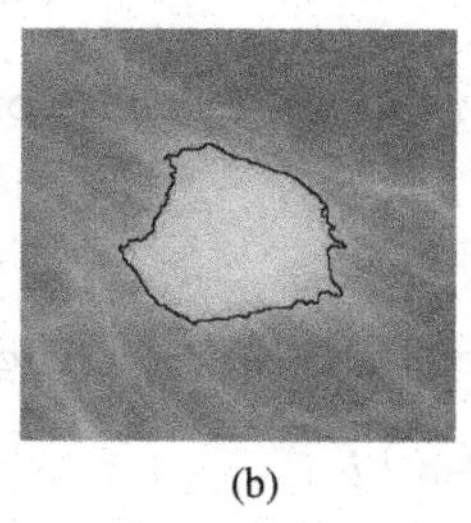
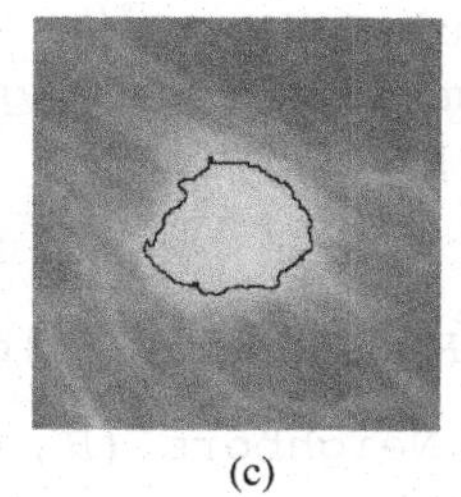
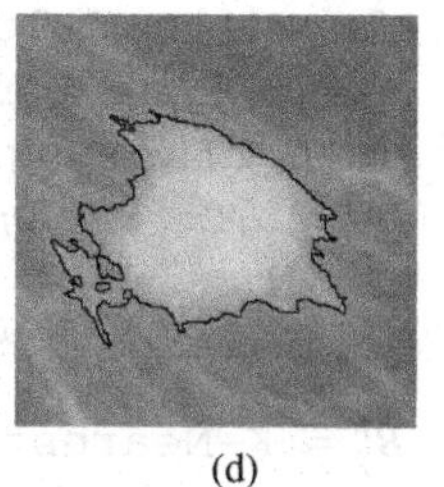

(a) (b) (c) (d)

Fig. 2. Region growing process results with different thresholds values (a) image I with masse R_I surrounded by the expert; (b) image I with region R_T extracted by region growing process T=0.10, R_T *and* R_I are equivalent; (c) image I with region R_T extracted by region growing process T=0.05, R_T is smaller than R_I; (d) image I with region R_T extracted by region growing process T=0.18, R_T is bigger than R_I

For the situation 2 and the situation 3, the region growing process must start again but, with using the new value of T. R_T area resulting from this process is compared to R_I. This scenario is repeated until equivalence of R_T and R_I.

3.3 Masses Detection by a Novel Region Growing Algorithm

A successful specification of a structure in a medical image requires information on a seed location and specification of region homogeneity characteristics. The first step of region growing consists of selecting the seed point which is inside the breast lesion. Due to the low quality of mammograms, most of region growing methods require the seed point be selected manually in advance. Let us note by *s* the seed point.

This stage objective is to extract the mass area by using the seed point *s* and the training base B^R as entries. The similarity criterion based contrast M_T (1) is used in this stage. Notice that, the threshold value *T* becomes adaptive: the threshold value *T* is modified at each region growing process iteration according to the feature of the area in the course of formation. However and at the beginning, the *T* value is maintained fixed due to the small size of the formed area. The first threshold value, noted by T_0, is the maximum value of all the thresholds values appearing in the training base B^R. T_0 is calculated as follows:

$$T_0 = \underset{i=0}{\overset{n-1}{Max}}\{T_i^{B^R}\} \tag{6}$$

with T_i threshold values in training base B^R and *n* the size of the training base B^R.

Since formation of an acceptable size area (ten pixels), the threshold value becomes adaptive and will be specified by taking into account the feature of the area in the course of formation. The algorithm above (Calcul-T) describes the threshold *T* calculation steps: suppose R_i the area extracted at region growing iteration *i*; after R_i feature calculation, the k-Nearest-Neighbors algorithm (kNN) is applied in order to determine among all training base B^R entries, the entry *j* which shows feature nearest to those of R_i. The threshold value of B^R at entry *j* will be selected as the threshold value *T* which will be used in the iteration *i+1* of region growing process.

```
Algorithm Calcul-T (R_i, B^R)
/*R_i : area formed at region growing iteration i */
/*B^R: training base*/
VAR   T: Real;
BEGIN
Determine C_{R_i} /*R_i features calculate according table 1*/;
B_j^R = K-Nearest-Neighbors (B^R, C_{R_i});
T= B_j^R .Threshold /* threshold value of B^R at entry j */;
Return T /*threshold value*/;
END.
```

So the novel adaptive region growing algorithm will became:

```
Algorithm: RegionGrowing (s, B^R)
/*s: seed*/
/*B^R :training base*/
```

```
CONST  N =10;
VAR    List, R_M: list;
       T,T_0: integer;
BEGIN
List={s}
R_M={s}
T_0 = Max_{i=0}^{n-1}{T_i^{B^R}}
FOR i=0 to N DO
  e=List.Element
  remove e from List
  determine Vs /*e neighbors*/
  FOR each element v of Vs DO
    IF  M_{T_0}(R_M + v) =True DO
      R_M=R_M+v
      add v to List
    ENDIF
  ENDFOR
ENDFOR
WHILE List.empty=False DO
  T= Calcul-T(R_M, B^R)
  e=List.Element
  remove e from List
  determine Vs /* e neighbors */
  FOR each element v of Vs DO
    IF  M_T (R_M + v) =True DO
      R_M=R_M+v
      add v to List
    ENDIF
  ENDFOR
ENDWHILE
Return R_M /*mass area*/;
END.
```

4 Experimental Results

For this work, the MiniMIAS database provided by the Mammographic Image Analysis Society (MIAS) is used. This database contains left and right breast images for a total of 161 patients. MiniMIAS consists of a variety of normal mammograms as well as mammograms with different characteristics and several abnormalities.

The evaluation of the proposed approach was carried out on all mammograms containing circumscribed lesions. The smallest lesion extends to 18 pixels in radius, while the largest one to 197 pixels. For the training procedure, fifty percent from all studied mammograms which had been randomly selected were used.

Three results of the proposed adaptive region growing approach are showed in Figure 3. The value of ΔT was fixed at 0.01. The following remarks can be deduced from the obtained results:

- The suggested approach allows the detection of masses from various characters of background tissues (Fatty, Dense-glandular and Fatty-glandular); for all tested images, the novel region growing algorithm succeeded in determining the masses as shown in figure 3.
- The position of the seed affects the quality of the results: a bad placement of the seed distorts the results of the approach.

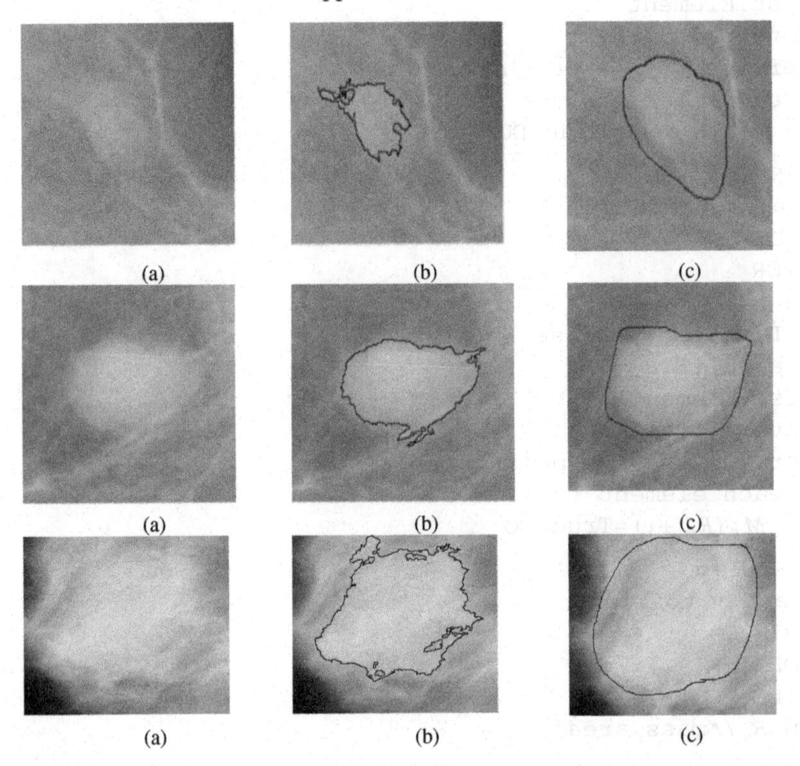

Fig. 3. Mass extraction (a) original image (b) mass extracted by the proposed approach (c) masse surrounded by the expert

5 Conclusion

A masses extraction approach by a novel region growing algorithm was detailed in this work. The approach comprises three stages: a preprocessing stage, a training base elaboration stage and a mass extraction stage. The preprocessing stage consists of applying two operators one for contrast enhancement and another for noises elimination. The training base is a base which contains masses features and thresholds values which allow their extractions; at this stage, masses were identified beforehand by the expert. The extraction operation consists of applying the novel region growing algorithm which starts from a manually placed seed. In the suggested algorithm, threshold value is determinate, automatically, according to the features of the area in the course of formation. The importance of this adaptive region growing mass extraction approach lies in its power to extract masses for all tested breast densities.

It would be worthful to validate these results on other databases as DDSM database. In addition, it is necessary to study other homogeneity criteria in order to improve the approach results. Furthermore, it is possible to apply other classification methods, like the neural networks, fuzzy kNN.

Acknowledgement. The authors are thankful to the radiologist Beledjhem Nedjhem (radiology service of Ibn Rochd CHU Annaba hospital) and Dr.Aouras Hayette (genecology service of Ibn Rochd CHU Annaba hospital) for accepting the analysis of mammograms used. The author also wishes to thank Dr.Djeddi and Dr.Amari (oncology service of Pont Blanc Annaba hospital) for their help.

References

1. Department of Health and Human Services, Centers for disease control and prevention: The National Breast and Cervical Cancer Early Detection Program. At-A-Glance. U.S. Resource ID: 4776 (1998)
2. Mencattini, A., Rabottino, G., Salmeri, M., Lojacono, R., Colini, E.: Breast mass segmentation in mammographic images by an effective region growing algorithm. In: Blanc-Talon, J., Bourennane, S., Philips, W., Popescu, D., Scheunders, P. (eds.) ACIVS 2008. LNCS, vol. 5259, pp. 948–957. Springer, Heidelberg (2008)
3. Adams, R., Bischof, L.: Seeded Region Growing. IEEE Transactions on Pattern Analysis and Machine Intelligence 16, 641–647 (1994)
4. Hojjatoleslami, S.A., Kittler, J.: Region growing: a new approach. IEEE Transactions on Image Processing 7, 1079–1084 (1998)
5. Pohlman, S., Powell, K.A., Obuchowski, N.A., Chilcote, W.A., Grundfest-Broniatowski, S.: Quantitative Classification of Breast Tumors in Digitized Mammograms. Medical Physics 23, 1337–1346 (1996)
6. Sahiner, B., Chan, H.P., Wei, D., Petrick, N., Helvie, M.A., Adler, D.D., Goodsit, M.M.: Image Feature Selection by a Genetic Algorithm: Application to Classification of Mass and Normal Breast Tissue. Medical Physics 23, 1671–1684 (1996)
7. Petrick, N., Chan, H.P., Sahiner, B., Helvie, M.A.: Combined adaptive enhancement and region-growing segmentation of breast masses on digitized mammograms. Medical Physics 26, 1642–1654 (1999)
8. Pohle, R., Toennies, K.D.: Segmentation of medical images using adaptive region growing. In: Proceedings of SPIE Medical Imaging, vol. 4322, pp. 1337–1346 (2001)
9. Shan, J., Cheng, H.D., Wang, Y.: A novel automatic seed point selection algorithm for breast ultrasound images. In: Proceedings of International Conference on Pattern Recognition, Tampa, Finland, pp. 1–4 (2008)
10. Hejazi, M.R., Ho, Y.-S.: Automated Detection of Tumors in Mammograms Using Two Segments for Classification. PCM. In: Proceedings of the 6th Pacific-Rim Conference on Advances in Multimedia Information Processing, vol. 1, pp. 910–921 (2005)

11. Senthilkumar, B., Umamaheswari, G., Karthik, J.: A novel region growing segmentation algorithm for the detection of breast cancer. crossref, pp. 1–4 (December 28, 2010)
12. Cao, Y., Hao, X., Zhu, X., Xia, S.: An adaptive region growing algorithm for breast masses in mammograms. Front. Electr. Electron. Eng. 5, 128–136 (2010)
13. Kupinski, M.A., Giger, M.L.: Automated Seeded Lesion Segmentation on Digital Mammograms. IEEE Transactions on Medical Imaging 17, 510–517 (1998)
14. Kinnard, L., Lo, S.-C.B., Wang, P., Freedman, M.T., Chouikha, M.A.: Maximum-Likelihood Automated Approach to Breast Mass Segmentation. In: Proceedings of the IEEE Symposium on Biomedical Imaging, pp. 241–244 (July 2002)
15. Cao, Y., Hao, X., Xia, S.: An improved region-growing algorithm for mammographic mass segmentation. In: Proceedings of SPIE 2009, vol. 7497 (2009), doi:10.1117/12.833044
16. Rabottino, G., Mencattini, A., Salmeri, M., Caselli, F., Lojacono, R.: Performance Evaluation of a Region Growing Procedure for Mammographic Breast Lesion Identification. Computer Standard and Interfaces Elsevier 33, 128–135 (2011)
17. Yang, S.-C., Wang, C.-M., Chung, Y.-N., Hsu, G.-C., Lee, S.-K., Chung, P.-C., Chang, C.-I.: A computer-aided system for mass detection and classification in digitized mammograms. Biomedical Engineering-Applications Basis & Communication 17, 215–228 (2005)

Coverage Enhancement in Wireless Video-Based Sensor Networks with Rotating Capabilities

Nawel Bendimerad and Bouabdellah Kechar

Oran University, Faculty of Sciences, Department of Computer Science
Research Laboratory in Industrial Computing and Networks (*RIIR*),
P.O. Box 1524 El M'Naouar, ORAN, Algeria
bendimerad.nawel@yahoo.fr, kechar.bouabdellah@univ-oran.dz

Abstract. Video monitoring for infrastructure surveillance and control is a special case of wireless sensor network applications in which large amounts of data are sensed and processed in real-time, and then communicated over a wireless network called a Wireless Video Sensor Network (WVSN). Effectively managing of such as networks is a major challenge. Considering that, a critical issue is the quality of the deployment from the sensing coverage viewpoint. Area coverage is still an essential issue in a WVSN. In this paper, we study the area coverage problem in WVSN with rotatable sensors, which can switch to the best direction to get high coverage. After the video sensors are randomly deployed, each sensor calculates its next new direction to switch in order to obtain a better coverage than previous one. In order to extend the network lifetime, our approach is to define a subset of the deployed nodes called a cover set to be active while the other nodes can sleep. Our proposed scheme is implemented and tested in OMNeT++ and shows significant enhancement in terms of average percentage of coverage.

Keywords: Wireless Video Sensor Network, coverage, cover set, scheduling, surveillance, OMNeT++.

1 Introduction

A Wireless Video Sensor Network (WVSN) is composed of a set of sensor nodes equipped with miniaturized cameras. In this paper, we are more particularly interested in WVSN for surveillance applications such as monitoring country borders. In this kind of networks, hundreds or thousands of video nodes with minimal memory, processing and resolution are deployed in an area of interest. These nodes collaborate between each other to ensure a surveillance task according to a given application and to transmit, via an ad-hoc network, useful video data to a sink station. Desired features of such a surveillance infrastructure are high reliability, longest lifetime, and largest coverage. In this paper we are interested by the coverage problem. Our main objective is to ensure and maintain high coverage of a region of interest with rotatable video sensors.

A. Amine et al. (Eds.): *Modeling Approaches and Algorithms*, SCI 488, pp. 105–114.
DOI: 10.1007/978-3-319-00560-7_15 © Springer International Publishing Switzerland 2013

The problem of coverage in Wireless Sensor Networks (WSN) was largely studied. The existing works on the connected coverage problem in scalar sensor networks [1], [2], [3], [4] used omnidirectional sensors with disk-like sensing coverage. Thus, two scalar nodes are likely to be redundant if they are close to each other. Different from WSN, in WVSN the sensing range of a video sensor node is represented by its Field of View (*FoV*) (described in Section 2.1). In addition, two video sensor nodes can be redundant and still be far from each other. Consequently, the coverage region of a video node is determined by its location and orientation.

In order to ensure a successful completion of the issued sensing tasks, sensor nodes must be deployed appropriately to reach an adequate coverage level. Therefore, in our work, we consider that a sensor can face to several directions. There are different ways to extend the sensing ability of WVSN. One way is to put several video sensors of the same kind on one sensor node, each of which faces to a different direction. Another way is to equip the sensor node with a mobile device that enables the node to move around [5]. The third way that we adopt is to equip the sensor node with a device that enables the sensor on the node to switch (or rotate) to different directions.

Like traditional WSN, power consumption is a fundamental concern in WVSNs. Sensing video data usually consumes even more energy than sensing scalar data. To assure a better coverage with minimum energy consumption in a randomly deployment area, we have to use redundant nodes (nodes that cover the same region) leading to overlaps among the monitored areas. Therefore, a common approach is to define a subset of the deployed nodes called a cover set to be in active mode while the other nodes can turn in sleep mode. The result is a schedule of the activity of sensor nodes in such a way that guarantees the deployment area coverage as well as the network connectivity [6].

In this paper we used the cover set construction approach and Video Node's Scheduling proposed in [6]. Our contribution is to provide a coverage enhancement using rotatable sensors. After randomly deployed video sensor nodes, before the cover set construction phase, each video sensor calculates its second and third direction using its first direction to which the sensor faces when it is deployed. Finally, the video sensor switches to the best direction in order to assure high area coverage. The obtained results show that our approach outperforms the proposed model in [6] in terms of coverage and prolong the network lifetime since we found more number of cover sets.

The remainder of the paper is organized as follows. In Section 2, the coverage and sensor nodes scheduling already introduced by [6] are presented. In Section 3, we explain our Rotatable Video Sensing Model. Then we will present some preliminary results on how to further increase network coverage when video sensor nodes have rotation feature.

2 Coverage and Sensor Nodes Scheduling

We can find three different kinds of coverage in WVSN:

- **Known Targets Coverage Problem.** Where we have to determine a subset of connected video nodes that covers a given set of target-locations scattered in a 2D plane.

- **Barrier coverage Problem.** Where the aim is to achieve a static arrangement of nodes that minimizes the probability of undetected penetration through the barrier.
- **Area Coverage Problem.** Which seeks to find a subset of connected video nodes that ensures the coverage of the entire region of deployment in a 2D plane.

Most of the previous works have considered the known-targets coverage problem [5], [7], [8], [9], [10]. The objective is to ensure at all-time the coverage of some targets with known locations. For example, the authors in [10] organize sensor nodes into mutually exclusive subsets that are activated successively, where the size of each subset is restricted and not all of the targets need to be covered by the sensors in one subset. In [7], a directional sensor model is proposed, where a sensor is allowed to work in several directions. The idea behind this is to find a minimal set of directions that can cover the maximum number of targets. It is different from the approach described in [5] that aims to find a group of non-disjoint cover sets, each set covering all the targets to maximize the network lifetime.

The goal of the barrier coverage is to deploy a chain of wireless sensors in the barrier, which usually is a long belt region, to prevent mobile objects from crossing the barrier undetected [11], [12]. In [13] the authors study the line-based sensor deployment strategies and provide both analytical results and interesting observations about how line-based sensor deployment can improve the barrier coverage performance than two-dimensional uniform sensor deployment.

Concerning the area coverage problem, the existing works focus on finding an efficient deployment pattern so that the average overlapping area of each sensor is bounded [14], [15], [16]. The authors in [17] analyze new deployment strategies for satisfying some given coverage probability requirements with directional sensing models. A model of directed communications is introduced to ensure and repair the network connectivity.

Different from the above works, our paper mainly focuses on the area coverage problem and more specifically on scheduling of randomly deployed rotatable video sensor nodes. The local coverage objective of each sensor node is to ensure, at all time, the coverage of this area either by itself (if it is active) or by a subset of its neighbors, called the cover set.

2.1 Video Sensor Nodes and Cover Sets

In WVSN, the sensing range of sensor nodes is replaced with the camera's *FoV*. The *FoV* as illustrated in figure 1 is defined as the maximum volume visible from the camera. The camera therefore is able to capture images of distant areas and objects that appear within the camera's depth of field, which is the distance between the nearest and the farthest object that the camera can capture sharply.

Definition 1 - The *FoV* of a video sensor node is defined as a sector denoted by a 4-tuple $v(P,Rs,\vec{V},\alpha)$ where:

- P refers to the position of v,
- Rs is its sensing range,

- $\vec{V}$ refers to the vector representing the line of sight of the camera's *FoV* which determines the sensing direction, and
- α is the offset angle of the *FoV* on both sides of $\vec{V}$ (an Angle Of View denoted (*AoV*) is represented by 2 α).

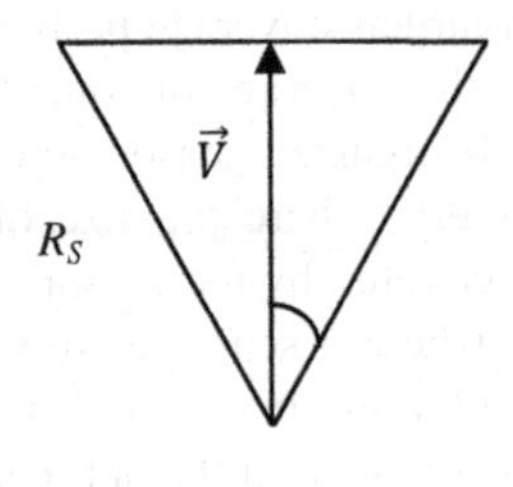

Fig. 1. Video sensing model

Fig. 2. Example of *FoV* Coverage

Definition 2 **-** We define a cover set $Co_i(v)$ of a video node v as a subset of video nodes such that: $\cup_{v' \in Co_i(v)}$ (v''s *FoV* area) covers v's *FoV* area.

Definition 3 **-** *Co*(v) is defined as the set of all the cover sets of node v.

In order to determine whether a sensor's *FoV* is completely covered or not by a subset of neighbor sensors we have used the approach defined by [6] that the aim is to use significant points of a video sensor's *FoV* to determine cover sets. These cover sets may not completely cover sensor v's *FoV* but a high percentage of it. The existing works [1], [2], [3], [4] that use omni-directional coverage seek to construct only disjoints sets of video nodes. In our case, we have the possibility that two or more covers have some video nodes in common. Therefore, selecting one cover set can also reduces the life time of the sensor it has in common with another cover set.

In the example illustrated by Figure 2, v's *FoV* is represented by (*abcg*). To find the set of covers, node v has to find the sets: *AG = {v2, v3}*, *BG = {v1, v4}* and *CG = {v1, v5}*. Then, it can construct the set of cover sets as follows: *Co(v) = {{v}, {v2, v1}, {v3, v1}, {v2, v4, v5}, {v3, v4, v5}}*.

2.2 Video Node's Scheduling

The activity of video sensor nodes operates in rounds as follows [6]:

- Every node orders its sets of covers according to their cardinality, and gives priority to the cover sets with minimum cardinality.
- If two cover sets or more in *Co*(v) have the same cardinality, priority is given to the cover set with the highest level of energy.
- For each round, after receiving the activity message of its neighbors, the sensor node tests if the active nodes belongs to a cover sets. If yes, it goes in sleep mode, else, it decides to be active and diffuses its decision.

- A video node v starts with the first cover set $Co_1(v) \epsilon Co(v)$ (which has the lowest cardinality) and tests if it is satisfied. A cover set is comprised by a set of video nodes and if one node is or become inactive this cover set cannot be satisfied.
- This process is repeated for each cover set and at every round.

3 Rotatable Video Sensing Model

In order to enhance the coverage model presented above, we have added to each sensor node in the area the capability to change its *FoV*. This function is added at the first step just after deployment before the cover sets construction phase.

To address our proposition, we need to make the following assumptions:

- All video sensors have the same sensing range (*Rs*), communication range (*Rc*) and sensing angle α, where $0 < \alpha < 2\pi/3$.
- Video sensors within *Rc* of a sensor are called the sensor's neighboring nodes.
- The sensing direction of each video sensor is rotatable.
- Each sensor knows its location information and determines the location of its neighboring sensors by communication.
- The target region is on a two-dimensional plane with non obstacle.

Figure 3 shows a Rotatable Video Sensing Model. The point *P* is the position of the video sensor that can switch to three possible directions $\vec{V}_1$, $\vec{V}_2$, and $\vec{V}_3$. $\vec{V}_1$ is the direction to which the sensor faces when it is deployed and the shadowed sector above $\vec{V}_1$ is the sensing region of the video sensor when it works in $\vec{V}_1$.

After the first deployment, since we have the line of sight of the first FoV_1, the video sensor can calculate the second and the third direction with them respective *FoV* as follows:

$$\vec{V}_2=\vec{V}_1+(2*PI)/3 \quad (1)$$

$$\vec{V}_3=\vec{V}_1+(4*PI)/3 \quad (2)$$

And, by using the extra graphical support, we can calculate the second FoV_2 and the third FoV_3 of the video sensor.

After a neighborhood discovery, the video sensor calculates for each direction the number of neighbors that can cover his *FoV*.

Finally, the video sensor switches to the direction that has the minimum number of neighbors in order to ensure a maximum coverage.

The procedure to select a *FoV* is summarized in Algorithm1.

The approach we described in this paper is completely distributed: every video node calculates its local solution for coverage. Each video node v can cover a sector area thanks to its *FoV*. Then, its local coverage objective is to ensure, at all time, the coverage of this area either by itself (if it is active) or by a subset of its neighbors.

In our work, we added to the original model the rotation possibility to the second $\vec{V}_2$ or the third $\vec{V}_3$ direction in order to ensure a best coverage of the area.

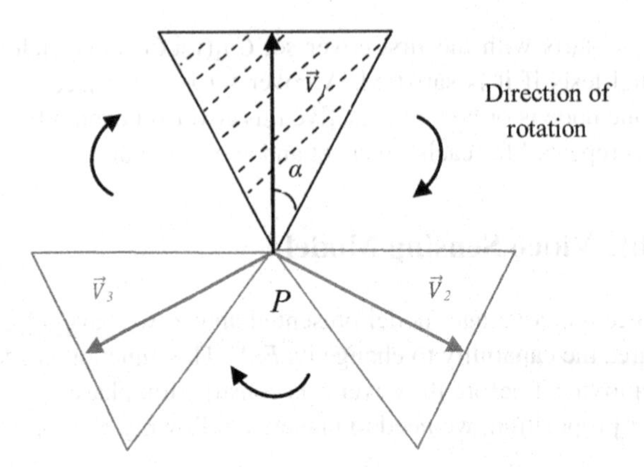

Fig. 3. Rotatable Video Sensing Model

```
Algorithm.1:  FoV selection
  1:   Input: N(v)  : represents the set of neighbors of node
v; V⃗1: first direction of v.
  2:  Output: FoV
      //Calculate the second direction
  3:  V⃗2=V⃗1+(2*PI)/3
      //Calculate the third direction
  4:  V⃗3=V⃗1+(4*PI)/3
  5:  Calculate FoV2 using the extra graphical support
  6:  Calculate FoV3 using the extra graphical support
  7:  For all neighbors v'∈ N(v) do
  8:  Compute  the  number  of  neighbors  that  can  cover  at
least 2 points of each FoVi , i=1,3
  9:  End for
 10: If FoV2 is the less covered direction Then
 11:               FoV=FoV2
 12: Else
 13:               FoV=FoV3
 14: End if
```

4 Experimental Results

In this section, we simulate and analyze the performance of our scheme and by comparing it with the proposed model by [6].

4.1 Simulation Environment

The effectiveness of our proposal is validated through simulations using the discrete event simulator OMNet++ [18]. Video sensor nodes have the same communication and sensing ranges. The position P and the direction $\vec{V}$ of a sensor node are chosen randomly. A simulation starts by a neighborhood discovery. Each node gathers positions and directions of its neighbors, selects an optimal direction and finds the cover sets. Then, round by round each node decides to be active or not. The rest of parameters are summarized in table1.

Table 1. Simulation parameters

Parameters	Value
Communication range	30m
Sensing range	20m
AoV	36°
Depth of view	25m
Battery lifetime	100 units
Area	75m * 75m

In our simulations we focus on the coverage results. We varied the deployed nodes density from 75 to 250 nodes, and we noted at each round the percentage of area coverage. At the end of a round an active node decreases its battery lifetime by one unit. Simulation ends as soon as the subset of nodes with power left is disconnected (when all active nodes have no neighbors anymore). In order to reduce the impact of randomness, each simulation is executed 5 times then gets the average value.

4.2 Performance Metrics

We have used the following three metrics to compare the performance of our approach:

Initial percentage of coverage. Is the percentage of coverage of the whole area when all sensors are active. It is the maximum coverage we could achieve as the nodes are randomly deployed.

Average percentage of coverage. Is calculated when all sensor nodes have determined their cover sets and have computed for each cover set the percentage of coverage. Note that we used the is_inside( ) feature of the extra graphical library to know whether a given point is covered by a sensor's *FoV* which is modeled with a triangle object.

Average number of cover sets. After the cover sets construction phase, each node calculate the number of cover sets found. Then at the end of the simulation (when the number of active nodes is below a given threshold) the simulation displays the statistics collected for average number of cover sets for all the sensors.

4.3 Performance Results

Initial percentage of coverage by varying the network density. We start our simulations by computing the initial percentage of coverage.

Figure 4 shows the initial percentage of area coverage using the existing approach denoted by WVSN and our rotatable video sensing model denoted by $WVSN_R$. We notice that increasing the network density increases the initial percentage of coverage using the two approaches. The results obtained by our model are better because we choose always at first the direction to cover the less covered region by other sensors.

Average percentage of coverage by varying the network density. We now compare the average percentage of coverage of the two propositions obtained from the beginning of the simulation until the end of the network lifetime.

Figure 5 indicates that the average percentage of coverage found using our model is much higher than the results obtained by the second model. However, when the network density increases (more than 150 sensor nodes) we can guarantee a good sensing coverage of the deployment area using the model without rotation.

Average number of cover sets by varying the network density. Figure 6 shows the relationship between the average number of cover sets and the network density. The results presented in this figure indicate that the average number of cover sets increases almost linearly when the number of video sensors increases.

The number of cover set a sensor node can have depends on the network density and on the randomly set line of sight at each simulation run. It should be noted that the increase in number of video sensors can lead to finding more cover sets in the case of our scheme.

Average number of cover sets by varying the communication range. We now present the results obtained by varying the communication range in order to study the capability of our approach to perform under different communication range. In figure 7 the number of nodes is fixed at 100 and we vary the communication range from 30 to 150 m.

When the communication range increases, we can see that the average number of cover sets also increases. This is because the larger the communication range is, the more sensing neighbours a sensor node will have, which will lead to more number of cover sets.

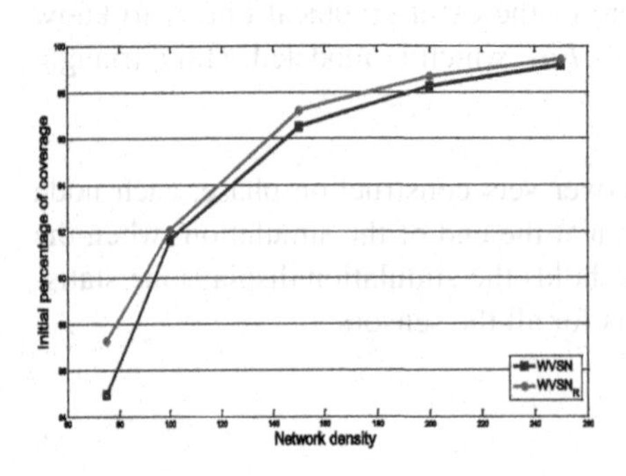

Fig. 4. Initial percentage of coverage by varying the network density

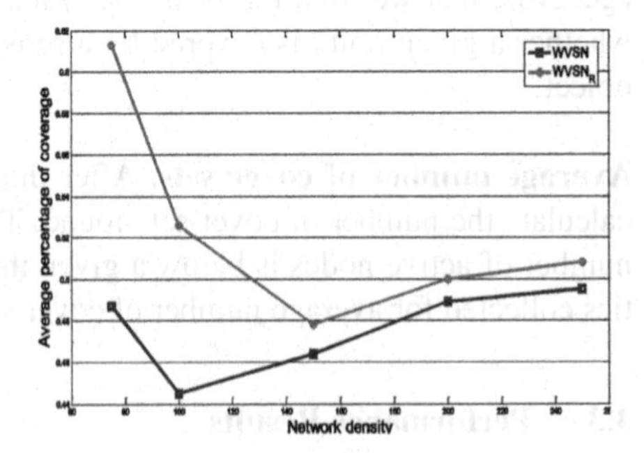

Fig. 5. Average percentage of coverage by varying the network density

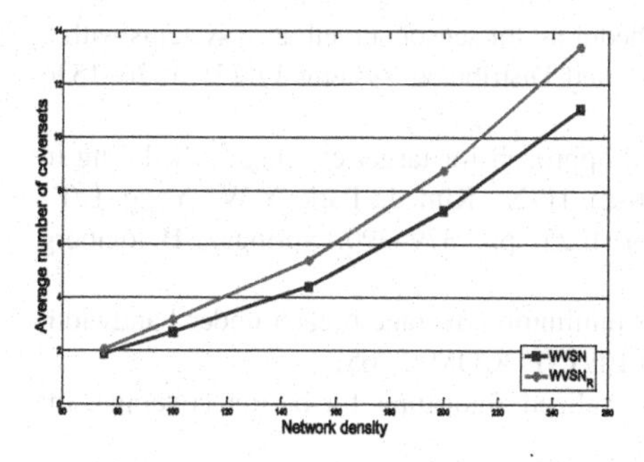

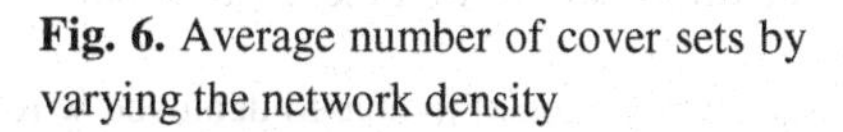

Fig. 6. Average number of cover sets by varying the network density

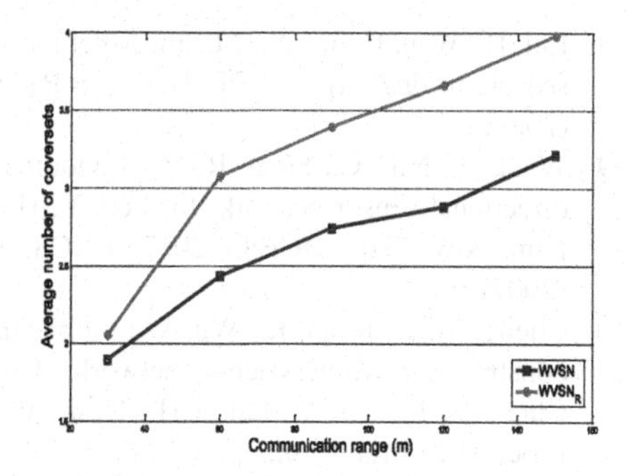

Fig. 7. Average number of cover sets by varying the communication range

5 Conclusion

In this paper, we proposed an extension of an existing model by adding the function of rotation to each sensor node in the network in order to enhance the coverage of a deployment area. Coverage analyses in WVSN were presented. A deployment coverage quality parameter was proposed for coverage estimation.

The simulation results indicate that our proposed approach performs better in area coverage with a higher percentage of coverage and number of cover sets.

As part of our future work, we plan to evaluate the energy consumption by each sensor in the WVSN using a better scheduling algorithm.

References

1. Wang, C., Thai, M.T., Li, Y., Wang, F., Wu, W.: Coverage based information retrieval for lifetime maximization in sensor networks. In: IEEE GlobeCom (2007)
2. Bahi, J., Makhoul, A., Mostefaoui, A.: Hilbert mobile beacon for localization and coverage in sensor networks. International Journal of Systems Science 39(11), 1081–1094 (2008)
3. Gallais, A., Carle, J., Simplot-Ryl, D., Stojmenovic, I.: Localized sensor area coverage with low communication overhead. IEEE Trans. on Mobile Computing (TMC) 7(5), 661–672 (2008)
4. Bahi, J., Makhoul, A., Mostefaoui, A.: Improving lifetime and coverage through a mobile beacon for high density sensor networks. In: The Second IEEE International Conference on Sensor Technologies and Applications, SENSORCOMM 2008, pp. 335–341 (2008)
5. Cai, Y., Lou, W., Li, M., Li, X.Y.: Target-oriented scheduling in directional sensor networks. In: 26th IEEE International Conference on Computer Communications, INFOCOM 2007, pp. 1550–1558 (2007)
6. Pham, C., Makhoul, A.: Performance study of multiple cover-set strategies for mission-critical video surveillance with wireless video sensors. In: IEEE WiMOB (2010)
7. Ai, J., Abouzeid, A.A.: Coverage by directional sensors in randomly deployed wireless sensor networks. Journal of Combinatorial Optimization 11(1), 21–41 (2006)

8. Liu, H., Wan, P., Jia, X.: Maximal lifetime scheduling for sensor surveillance systems with k sensors to one target. IEEE Trans. on Parallel and Distributed Systems 17(12), 1526–1536 (2006)
9. Wang, J., Niu, C., Shen, R.-M.: Randomized approach for target coverage scheduling in directional sensor network. In: Lee, Y.-H., Kim, H.-N., Kim, J., Park, Y.W., Yang, L.T., Kim, S.W. (eds.) ICESS 2007. LNCS, vol. 4523, pp. 379–390. Springer, Heidelberg (2007)
10. Cheng, M.X., Ruan, L., Wu, W.: Achieving minimum coverage breach under bandwidth constraints in wireless sensor networks. In: IEEE INFOCOM (2005)
11. Chen, A., Kumar, S., Lai, T.H.: Designing localized algorithms for barrier coverage. In: Proc. ACM MobiCom, pp. 63–74 (2007)
12. Wang, Y., Cao, G.: Barrier coverage in camera sensor networks. In: Proc. ACM MobiHoc, pp. 119–128 (2011)
13. Saipulla, A., Westphal, C., Liu, B., Wang, J.: Barrier coverage of linebased deployed wireless sensor networks. In: Proceeding of IEEE Infocom (2009)
14. Yu, Z., Teng, J., Bai, X., Xuan, D., Jia, W.: Connected coverage in wireless networks with directional antennas. In: Proceeding IEEE INFOCOM, pp. 2264–2272 (2011)
15. Kasbekar, G., Bejerano, Y., Sarkar, S.: Lifetime and coverage guarantees through distributed coordinate-free sensor activation. In: Proceeding ACM MobiCom, pp. 169–180 (2009)
16. Ma, H., Zhang, X., Ming, A.: A coverage-enhancing method for 3d directional sensor networks. In: Proc. IEEE INFOCOM, pp. 2791–2795 (2009)
17. Ma, H., Liu, Y.: Some problems of directional sensor networks. International Journal of Sensor Networks 2(1-2), 44–52 (2007)
18. OMNeT++, `http://www.omnetpp.org/` (last accessed February 20th, 2013)

Improving Wireless Sensor Networks Robustness through Multi-level Fault Tolerant Routing Protocol

Zibouda Aliouat and Makhlouf Aliouat

Computer science Department Ferhat Abbas University Sétif, Algeria
{aliouat_zi,aliouat_m}@yahoo.fr

Abstract. Wireless Sensor Networks (WSN) are now commonly admitted as a promising networking paradigm bringing huge benefits in industrial and socio-economical domains. However, although the applications dedicated to WSNs are in constant increase and research activities are intensely carried on, WSNs remain hindered by some weaknesses and lacks preventing them to reach their full maturity. These lacks may be seen even in basic WSN services like data routing in which energy optimization, data security, fault tolerant nodes, etc, have to be improved. So, we are dealing with the fault tolerance issue of routing protocols in order to dependably ensure data forwarding from nodes to sink. This is indispensable for a WSN to successfully accomplish its mission.

In this paper, we propose an energy economical and Fault Tolerant Multilevel Routing Protocol, borrowing idea from the well known TEEN and LEACH protocols. Simulation results via NS2 simulator showed convincing and interesting protocol performances.

Keywords: WSN, Multi-level clustering, Fault tolerance, Energy Saving, Routing protocols.

1 Introduction

A WSN is composed by a set of sensor nodes, communicating each other via wireless links, capable of capturing physical phenomena (Temperature, humidity, radioactivity, etc…) from a given environment, and transform them into digital values which are routed to one or several base stations (Sink) [1]. These sensors can be randomly deployed in hostile environments and even inaccessible to the human being in order to monitor and/or control a particular phenomenon.

Being often scattered randomly in harmful and even unreachable areas and in a huge number, sensor nodes of a WSN are vulnerable to various types of failures such as: energy lack, sensor component dysfunction like radio or sensing unit breakdown, physical damages etc. Since repairing or replacing a failed sensor node is often infeasible because of surroundings dangerousness, it is thus crucial for a WSN to overcome failure occurrences corrupting its nodes in order to survive until successful mission accomplishment. However, this desirable objective should not be reached without fault tolerance assistance. Therefore, all WSN services need to be resilient to failures whatever their causes. Because routing data from any source node to a base station is

A. Amine et al. (Eds.): *Modeling Approaches and Algorithms*, SCI 488, pp. 115–124.
DOI: 10.1007/978-3-319-00560-7_16 © Springer International Publishing Switzerland 2013

a crucial task, making routing protocols more robust, facing so possible node failures is a valuable aim contributing to the success of the application hosted by a WSN [2, 3]. It is then of primary importance to include a fault-tolerance mechanism within all the protocols to be implemented at different levels of the various WSN architecture layers.

In this paper, we propose a routing protocol named IIFTMRP (Inter-Intra Fault Tolerant sensitive energy aware Multi-level Routing Protocol), which is a hierarchical routing protocol based on nodes organization in clusters with an inter clusters multi-hop communication. IIFTMRP is then provided with fault tolerance capabilities making data routing more reliable and consequently ensuring a WSN more robustness capable of successfully carrying out its mission.

The rest of the paper is organized as follows: in section 2, we discuss some related works in multilevel routing protocols while highlighting their weaknesses. Section 3 is dedicated to present characteristics of multi-level routing protocols, while fault tolerance attribute is integrated in these protocols in section 4. Sections 5 and 6 present our proposals, and their performance evaluations are discussed in section 7. Section 8 concludes our paper.

2 Related Works

One of the main goals of clustering algorithms for WSNs is to cope with scalability and to reduce the energy consumption of nodes [4-6]. In a typically large WSN, nodes may be located far away from the base station so, it is very difficult for these sensor nodes to communicate with base station and minimize the energy consumption at the same time, unless if they are organized as multilevel clustering schemes. That organization provides better connectivity and reduces the energy consumption as well. As the node energy budget is the most important resource, and data routing is a vital service in WSN, routing and clustering are intensely studied in WSNs current research [1, 7, 8, 9]. In this context, Low Energy Adaptive Clustering Hierarchy (LEACH) [7] is the first and well known hierarchical (clustering) routing protocol which was proposed to raise this challenge. Although LEACH can carry through the clustering process, in order to prolong WSN lifetime, it does not guarantee good clusterheads distribution and remains lacking efficiency in large-scale WSNs.

For the same raisons, Manjeshwar and agrawal [8] have proposed a clustering protocol named TEEN (Threshold sensitive Energy Efficient sensor Network protocol) for critical applications where the change of certain environment parameters may be sudden and must be taken in consideration immediately in the WSN. So, the architecture of the network is based on a hierarchical nodes grouping at several levels where the closest nodes from each other, form clusters. Then the process of clustering goes to the second level until the base station is reached.

Xinfang et al. [10] proposed a multilevel clustering algorithm based on mathematical model of clustering virtual back-bone. This algorithm rotates the CH role to all nodes to evenly distribute the energy consumption. It reduces the number of relays for data transmission by reducing the number of nodes in the tree configuration.

Another multilevel clustering algorithm proposed by Rasid et al. [11] to reduce the network energy consumption by determining the optimal number of CH nodes in multilevel network. The proposed algorithm chooses a set of nodes to become level 1 CHs from all the nodes. Level 2 CHs are selected from the set of level 1 CHs. This process is repeated until highest level of CHs is reached.

As a result, none of the previous protocols (to quote only those among others) has taken into account the fault tolerance node capabilities, enabling WSNs to operate normally until successful mission end.

3 Multilevel Routing Protocol (MRP)

The idea to provide MRP with fault tolerance ability is borrowed from the hierarchical routing protocol TEEN [8]. Like TEEN, MRP is appropriate for critical applications, because it is conceived to be sensitive to sudden changes of the environmental attributes such as temperature, pressure, humidity etc... The network architecture of sensors is based on clusters at several levels where the closest nodes start by forming clusters and this process will be repeated until the base station is reached (Fig. 1).

After the clusters have been formed as in [7], each clusterhead transmits to its member nodes two threshold values: a Hardware Threshold HT, which is the minimal value of the controlled parameter (supervised) and Software Threshold ST representing the maximum value of the controlled parameter [8]. The node which detects a sensed value out of the interval]HT, ST[has to announce it to the base station by transmitting a message to its clusterhead which in turn transmits it to its neighbor clusterhead and so on until that message arrives at the base station.

In order to avoid the problem of collisions between the simultaneous emissions, we used a TDMA scheduling between the member nodes of the same cluster and a CDMA code between clusterheads.

The MRP algorithm is held in rounds. Each round is made up of an initialization phase and a transmission phase as for protocol LEACH. In the initialization phase, clusterheads are elected, clusters are formed and also the closest neighbors to each clusterhead are selected. The phase of data transmission consists of data gathering for each member of every cluster as well as the inter clusterhead data communication.

Announce Phase: In this phase, the clusterheads are elected and thus the clusters are formed in the following way: each node calculates the probability Pi(t) enabling it to know if it is CH or not during the current round. It generates a random number I ranging between 0 and 1. If this number I is lower than Pi(t) then the node will become CH [7].

Organization Phase of clusters: after the CHs nodes are elected, they multicast a warning or advertising message ADV containing CH identifier and the values of the thresholds using the protocol CSMA/CA to avoid collisions between CHs. As a reply to ADV reception, each node sends a JOIN message to the nearest CH to be its member. The distance between node and CH is based on signal strength of received ADV message; stronger is the signal nearer is the sender node.

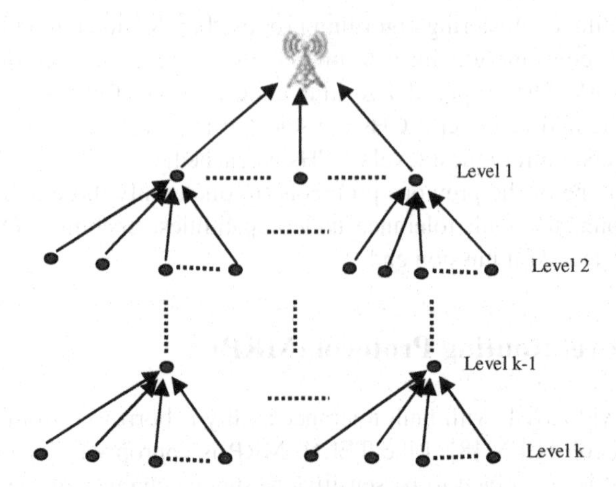

Fig. 1. Multi-level cluster-heads organization

Scheduling Phase and seeking for CH neighbors: This phase consists of two sub phases. The first one allows confirming the cluster member nodes, so once JOIN messages are received by CHs, they send back an acknowledgment message with TDMA scheduling containing a list of sensor nodes that have chosen this CH. The node is then able to surely determine if it is a member of cluster, just checking if the list contains its node identifier. The second phase allows, for each CH of level I, to determine a nearest CH neighbor pertaining to level J in a network divided into k levels. After having calculated the rays of the various levels using the formula $r = n/k$ where r is a ray and n is the side length of the nodes deployment surface (as a square, see simulation parameters), each CH can know the level to which it belongs, and consequently it seeks for a CH nearest to it in the level immediately higher (if it is not empty of course). The neighbor of CHs of the first level is the base station.

Data Transmission Phase: This phase is the longest one, in which data are collected by sensor nodes from their surroundings, sent them out to CHs where they are aggregated and forwarded to CHs higher level until reaching the base station.

The MRP requires that only data with values greater than the threshold ST and less than the threshold HT will be sent out to the CH, this minimizes the number of data emissions and then optimizes energy consumption.

4 Fault Tolerance in MRP

A sensor node is composed of four major units namely: sensing unit, processing unit, transmission unit and finally energy supplying unit. According to node activity energy supplying unit is more likely to break down than the others. Such breakdown can appear at any time and thus at whatever phase. The possible breakdowns in each unit may be:

— **Sensing Unit:** either it does not function, i.e. it does not sense any more, therefore there will be no data to transmit, or the sensed values are erroneous with regards to verifications established according to likelihood tests.
— **Processing unit:** the difficulty can arise from the CPU which may provide erroneous computation results.
— **Data Transmission Unit:** in this unit the breakdown can be either in emission or in reception or both. In all cases, there will be anomalous behavior.
— **Energy Supplying Unit:** the depletion of the battery can lead first to a performance degradation that will converge to the death of the sensor node..

In the hierarchical routing protocols, two types of nodes exist: standard nodes which are used for sensing events, and clusterheads which are in charge of clusters management. So, a failed node may be either a standard node or a clusterhead. Due to the important role played by CHs in WSN, we then focus our work on their robustness. Indeed, when a CH fails, all its member nodes will be prevented to send out their sensed data. Thus, a part of a WSN area (the one covered by the cluster) will be ignored by the network end user. This information loss may be of prime importance and may lead to cause mission failure. Since the energy supplying unit fails progressively and the processing unit is more resilient to failures [12], so in the sequel, we will deal with sending and receiving failures (omission failures) in CH nodes.

4.1 Failures in Emission

The messages sent by a clusterhead are ADV (HT and ST) and TDMA towards standard nodes, and DATA messages to neighbor CHs. If a CH does not send ADV, it will not receive JOINs, and standard nodes having received ADVs from other CHs will join the nearest of these CHs. In the same way, a failed CH in emission will be unable to forward data received from its low level neighbor CH, so these data may never attain the base station. In such a situation, the lost data may be crucial for the end user to take decisions, and its WSN application should be merely aborted. So, the incurred damages may be unacceptable.

4.2 Failures in Reception

Normally a CH must receive JOIN and data messages from standard nodes as well as data from CHs of lower levels. Any JOIN message not received by a CH makes the sender node not able to send its sensed data because of lack TDMA slot. Since a CH cannot receive data from its member nodes, it is also unable to receive data from its low level neighbor CHs. Therefore, data sent from a certain number of CHs linked to the failed one do not reach the base station.

Given that the sensed data are the main goal or the imperative raison of a WSN, and the crucial role to be assumed by clusterheads is to successfully accomplish this goal via the data forwarding process to the base station, robustness of CHs is of primary importance. We are then interested by CH failure in data reception/emission and how to overcome this drawback. To this end, we will focus, in a first time, on providing CHs with an Inter clusters Fault Tolerance Routing Protocol (IFTMRP). In this case, each CH node has to choose two neighbor CHs instead of one. The aim is to get back (in case of failure) CH data of the lower levels but the drawback is that, data of cluster members will be lost. So, to overcome this problem, we use an alternative solution in which data of member nodes are retrieved from a vice CH which has to replace the CH in the event of failure. This is the objective devoted to the Inter Intra cluster Fault Tolerant Multilevel Routing Protocol (IIFTMRP). The inter clusters failure tolerance is implicit, since all data sent to CH are received by the associated vice CH. It has, therefore, not need to seek for a second neighbor.

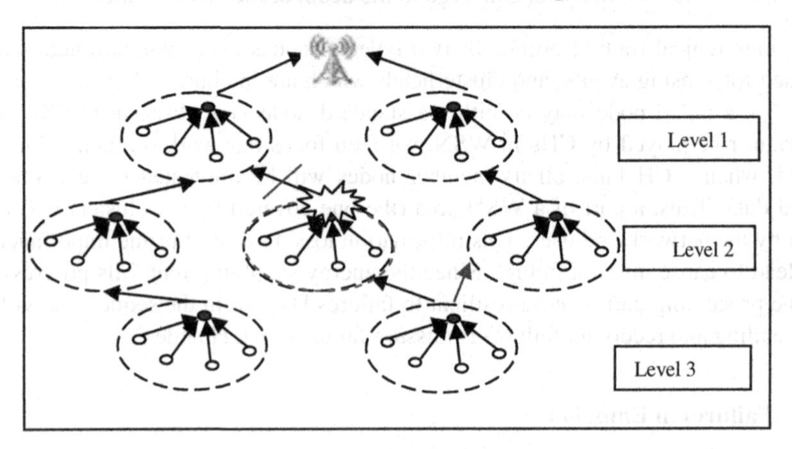

Fig. 2. Inter cluster recovery

5 Inter cluster Fault Tolerant Multi-level Routing Protocol (IFTMRP)

The proposed FTMR Protocol tolerating inter cluster faults is given as follows:

- Each CH chooses two neighbor CHs instead of one.
- If the first CH fails (in receiving data), the data will be relayed to the second.
- We assume that a single neighbor fails at a time (tolerating a single failure during a TDMA round).
- When a CH sends its data to the neighbor CH, it waits a short amount of time to receive the associated acknowledgment.
- When data are received, the CH must send an acknowledgment to the transmitter CH.
- If the sender does not receive an acknowledgment, it transmits the data to the second neighbor CH.

In the Inter-Cluster Recovery, nodes in the cluster with their CH failed are left transmitting to no one until the next round begins with a whole new selection process. The CH of lower level transmits data to another CH at the level of failed CH as in Fig. 2.

6 Inter-Intra cluster Fault Tolerant Multi-level Routing Protocol (IIFTMRP)

The proposed Protocol IIFTMRP acts as follows:

- Each CH selects a node among its cluster members with the highest energy level as a vice CH.
- All nodes in the cluster know the vice CH chosen.
- The vice CH sends data only when its CH is failed.
- During the slot reserved for the vice CH, if the CH has not received data, it sends a message to the vice CH noticing that he has failed; it sends nothing, otherwise.
- If the vice CH receives a warning message from the CH, it takes over, it will then be responsible for transmitting the received data to higher level CH.
- Otherwise, the CH sends its data itself to higher level CH.
- When a CH sends its data to a chosen neighbor CH, it does not need to wait for an acknowledgment, even to choose two CHs. Because if its neighbor is failed the vice CH retrieves the data.

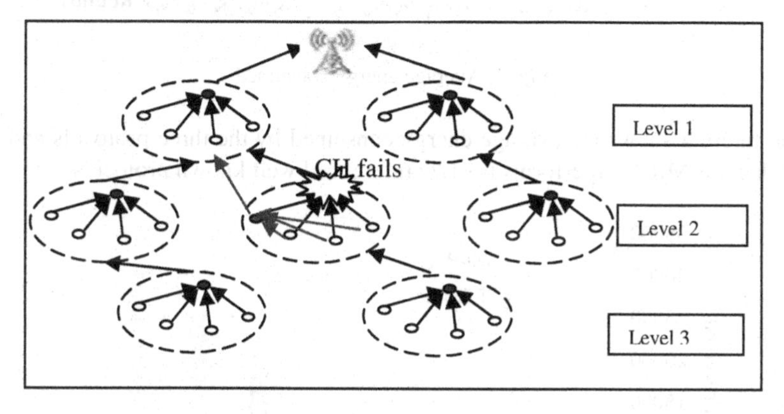

Fig. 3. Inter-intra cluster recovery

All data sent from lower CHs to the failed CH are received by the associated vice CH. Therefore, not need to seek for a second neighbor CH (Fig.3).

7 Simulation Results

Our experimentation model is established on 100 nodes with the node 101 set as a base station, dispersed on a surface of 1000 * 1000 m. At the beginning of the simulation, the initial energy is equal to 2 joules for every node except the base station which supposed with enormous energy budget of 5000 Joules. We assume that all nodes have a fixed position during the entire simulation period. The parameters of our simulation are summarized in the following table:

Table 1. Simulation Parameters

Paramters	Values
Deployment Surface	(1000*1000)
Base station Position	(50 ,175)
Sensor nodes Number	100
Initial energy budget for each sensor node	2j
Round time duration	20s

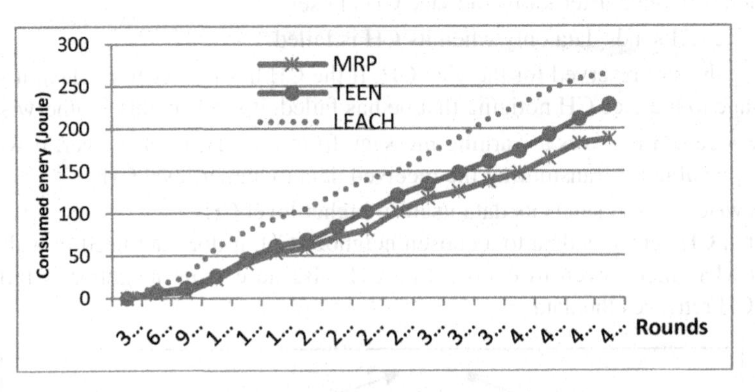

Fig. 4. Average energy consumed

The figure 4 shows the average energy consumed by the three protocols and highlights that our MRP outperforms the two referenced well known protocols.

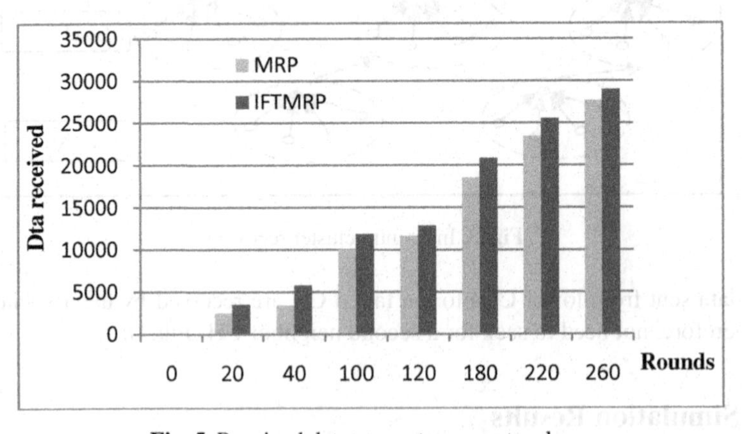

Fig. 5. Received data amounts per protocols

In the proposed protocol IFTMRP, only data of CHs are recovered in case of CH failures, on the other hand, data of cluster members are lost. For this reason, there is no great difference in the number of received data (Fig. 5) between the two proposals. These lost data will be recovered in the second solution.

We notice in figure 6 that the number of received data increases in IIFTMRP compared to IFTMRP and that is due to the data recovered by the vice CHs in the protocol tolerating faults. In this solution a vice CH gets back data of CHs and data of cluster members.

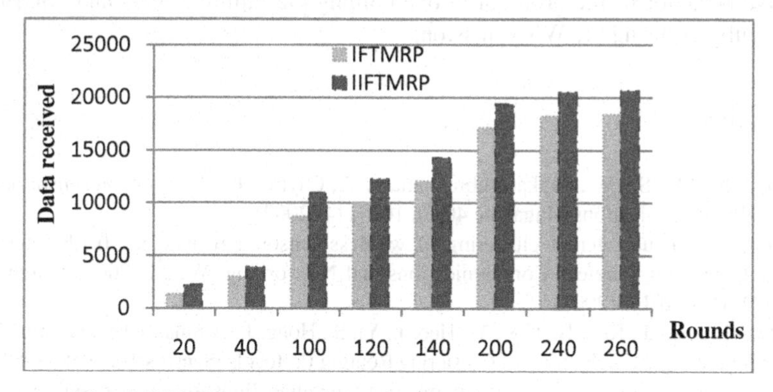

Fig. 6. IFTMRP vs. IIFTMRP regarding received data amounts

The figure 7 shows the proportion of clusters disconnected in WSN. When the cluster of the level 1, which is the nearest to the base station dies, all the clusters of lower levels linked to it, will be disconnected in MRP. Hence, sensed data cannot be forwarded to the base station. In IFTMRP, only the failed cluster will be disconnected. On the contrary, none of the clusters is disconnected when using IIFTMRP.

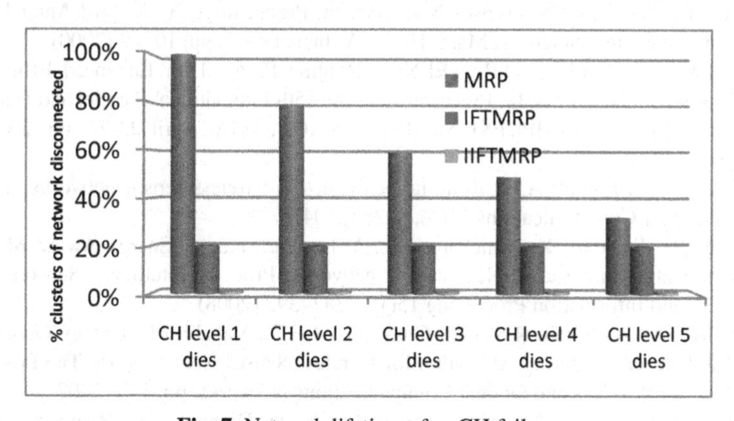

Fig. 7. Network lifetime after CH failure

8 Conclusion

Due to the deployment environment riskiness, failures of sensor nodes are very probable in WSNs particularly when the density of nodes is important. For this raison, the major services of a WSN must be designed in such a way they have to survive to breakdowns. Among these critical services, communication through routing protocols is one of them.

In this context, we have proposed in this paper a Fault Tolerant Multi-level Routing Protocol in which each clusterhead has to elect one of its cluster members as an assistant (vice). In case of a CH failure occurrence, the associate assistant takes over, so relaying CH activities. The results provided by the simulations showed an effective behavior of the protocol in overcoming the failure effects and contribute to successfully ensuring the WSN mission.

References

1. Akyildiz, I.F., Su, W., Sankarasubramaniam, Y., Cayirci, E.: A survey on sensor networks. IEEE Communications Magazine 40(8), 102–114 (2002)
2. Gupta, G.: Fault-tolerant clustering of wireless sensor networks. In: IEEE international Conference on Wireless Communications and Networking, WCNC 2003, March 20, pp. 1579–1584. IEEE (2003)
3. Min, H., Jung, J., Kim, B., Cho, Y., Heo, J., Yi, S., Hong, J.: A Smart Checkpointing Scheme for Improving the Reliability of Clustering Routing Protocols. Sensors 10, 8938–8952 (2010)
4. Abbasi, A.A., Younis, M.: A survey on clustering algorithms for wireless sensor networks. Comput. Commun. 30, 2826–2841 (2007)
5. Aliouat, Z., Aliouat, M.: Effective Energy Management in Routing Protocol for Wireless Sensor Networks. In: Proc. IEEE/IFIP 5th International Conference on New Technologies, Mobility and Security (NTMS), Istanbul Turkey, pp. 1–5 (2012)
6. Liu, X.: A Survey on Clustering Routing Protocols in Wireless Sensor Networks. Sensors 12, 11113–11153 (2012)
7. Heinzelman, W.R., Chandrakasan, A., Balakrishnan, H.: Energy-Efficient Communication Protocol for Wireless Microsensor Networks. In: Proceedings of the 33rd Annual Hawaii Int. Conf. on System Sciences, Maui, HI, USA, January 4-7, pp. 10–19 (2000)
8. Manjeshwar, E., Agrawal, D.P.: TEEN: A Routing Protocol for Enhanced Efficiency in Wireless Sensor Networks. In: Proceedings of the 15th International Parallel and Distributed Processing Symposium (IPDPS), San Francisco, CA, USA, April 23-27, pp. 2009–2015 (2001)
9. Al-Karaki, J.N., Kamal, A.E.: Routing techniques in wireless sensor networks: a survey. IEEE Wireless Communications 11(6), 6–28 (2004)
10. Xinfang, Y., Jiangtao, X., Chicharo, J.F., Yanguang, Y.: An Energy-Aware Multilevel Clustering algorithm for wireless sensor networks. Proc. of Intelligent Sensors, Sensor Networks and Information Processing 15(18), 387–392 (2008)
11. Rasid, M.F.A., Abdullah, R.S.A.R., Ghazvini, M.H.F., Vahabi, M.: Energy Optimization with Multi-level Clustering Algorithm for Wireless Sensor Networks. In: The Proc. of the Int. Conf. on Wireless and Optical Communications Networks, pp. 1–5 (2007)
12. Aliouat, M., Aliouat, Z., Naidja, M.: Adaptative Nodes Diagnosis and Recovery for Wireless Sensor Networks. In: IEEE, Symposium on Computer Applications and Industrial Electronics, Kota Kinabalu, Malaysia, December 3-4 (2012)

A New Approach for QCL-Based Alert Correlation Process

Lydia Bouzar-Benlabiod[1], Salem Benferhat[2], and Thouraya Bouabana-Tebibel[1]

[1] LCSI laboratory, Ecole nationale Supérieure d'Informatique (ESI), Algiers, Algeria
l_bouzar@esi.dz, t_tebibel@esi.dz
[2] CRIL-CNRS, Université d'Artois, Lens, France
benferhat@cril.univ-artois.fr

Abstract. Intrusion Detection Systems (IDS) are very important tools for network monitoring. However, they often produce a large quantity of alerts. The security operator who analyses IDS alerts is quickly overwhelmed. Alert correlation is a process applied to the IDS alerts in order to reduce their number. In this paper, we propose a new approach for logical based alert correlation which integrates the security operator's knowledge and preferences in order to present to him only the most suitable alerts. The representation and the reasoning on these knowledge and preferences are done using a new logic called Instantiated First Order Qualitative Choice Logic (IFO-QCL). Our modeling shows an alert as an interpretation which allows us to have an efficient algorithm that performs the correlation process in a polynomial time. Experimental results are achieved on data collected from a real system monitoring. The result is a set of stratified alerts satisfying the operators criteria.

Keywords: IDS, alert correlation, QCL, preferences, knowledge.

1 Introduction

Intrusion Detection Systems (IDS) [1] [2] are important network security tools. They filter traffic data searching for malicious activities. A security operator has to supervise log alerts emanating from different IDS. When the data traffic is important and the number of suspicious activities is high the security operator is quickly submerged by a huge number of IDS alerts [3]. Alert correlation is a process that aims to reduce the produced alerts number. It analyses IDS alerts, gathers them into groups or clusters of similar alerts [4] and provides intrusion reports to security operators.

Different alert correlation methods have been proposed in the literature. A new approach to alert correlation has been proposed in [15]. This method uses an extension of the preferences logic, called Qualitative Choice Logic (QCL) [16], to integrate the security operator knowledge and preferences resulting from his experience, to the correlation process. QCL increases the propositional logic language with the ordered disjunction operator:"$\times$". This operator expresses preference information. $A \times B$ means that "if possible A, otherwise at least B".

A. Amine et al. (Eds.): *Modeling Approaches and Algorithms*, SCI 488, pp. 125–134.
DOI: 10.1007/978-3-319-00560-7_17 © Springer International Publishing Switzerland 2013

This paper is in the same spirit as the one proposed in [15]. Our aim is to define a new method for alert correlation which consists in inserting a filter created by the operator to the alert correlation process. The filter is based on the operator knowledge and preferences related to the monitored system. The output of our model is a set of alerts that satisfy the security operator's criteria. Our proposed algorithm is based on an extension of QCL language. This paper contains three important contributions: (1) We first propose a framework for dealing with preferences using IFO-QCL language. (2) We then propose an algorithm for alerts correlation that takes into account the security operator knowledge and preferences. We simply view an alert as an interpretation. Hence the satisfaction of an interpretation, with respect to the security operator preferences, directly induces a ranking of alerts. (3) Lastly, we present experimental studies on real alerts issued from PLACID[1] project. The paper is organized as follows: Section 2 briefly describes existing alert correlation approaches. Section 3 and 4 detail our alert correlation model. Section 5 gives the main steps of our approach. Section 6 presents our experimentation results.

2 Related Works

Alert correlation is a multi-step transformation process. The inputs are IDS alerts. Alerts are analyzed, gathered and then returned to the operator as an intrusion report. One of the main alert correlation objectives is the reduction of the number of alerts. Among alert correlation methods, we can distinguish: (1) Correlation approaches based on similarity between alerts' attributes [5][6][7] and alert aggregation mechanisms [17] [18]. These approaches are based on the fact that alerts belonging to the same attack have similar attribute values (source IP address, etc.). An alert-filter is set up in order to determine the attribute values similarities. (2) Correlation by pre-defined attack scenarios [8][9][10]. It is an explicit correlation approach whose objective is to detect complex attack scenarios. These scenarios are stored in signature bases. The bases can be either obtained from historical data or specified by users. (3) Preconditions and postconditions alert correlation methods [11] [12] [13]. Each action is associated with a set of preconditions (needed for launching an action), and a set of postconditions representing consequences of the action. These methods correlate alerts if postconditions of present alerts may be correlated with preconditions of some future alerts.(4) Alerts prioritization based on weighted logics [15][23]. A fuzzy-logic based technique for scoring and prioritizing alerts was proposed in [23], where weights associated with alerts represent the seriousness of alerts.
Some methods combines different approches tomake a strong alert correlation mecanisms [24].

The prioritization method proposed in [15] integrates the security operator knowledge and preferences. This approach is based on a logic of preferences called Qualitative Choice Logic (QCL)[16]. This logic, and its extensions, have been used to represent security operator knowledge and preferences. In [15], first

[1] http://placid.insa-rouen.fr/

order logic predicate, denoted by "show-to-operator (x)", has been introduced to indicate whether an alert x should be presented to a security operator or not. The main problem with this approach is its high computational complexity, since the decision problem: "should an alert be presented to the security operator" is at least a Δ_p^2 complete problem.

We also propose a prioritization approach based on expert knowledge and preferences. We use a different modeling logical language from the one used in [15]. This language is more flexible than propositional QCL [16], and less expressive than the FO-QCL used in [15]. The defined logic is called Instantiated First Order Qualitative Choice Logic (IFO-QCL) and is more appropriate for alerts correlation. Contrarily to the modeling proposed in [15], our approach for alerts correlation can be achieved in a polynomial time.

3 Instantiated First Order Qualitative Choice Logic IFO-QCL

QCL [16] is a compact logic for representing and dealing with simple and complex knowledge and preferences. QCL increases the expressive power of the propositional logic with a new connective operator called ordered disjunction operator denoted by: $\times$. This operator expresses users' preferences.

The extension to full first order logic language is not needed for alert correlation. As we will show, a small extension (at least from the representation point of view) of propositional logic is enough. More precisely, we propose to use Instantiated First Order QCL (IFO-QCL).

In the following, we will present our language, which is in fact built over three encapsulated languages:(1) IFO formulas which will only be used to express knowledge or pieces of information that each alert should satisfy in order to be considered by a security operator. (2) IFO-BCF formulas (BCF for Basic Choice Formulas) which will be used to express simple forms of preferences. (3) IFO-GCF formulas (GCF for General Choice Formulas) which will be used to express general preferences. Our framework is parameterized with a normalized function that transforms any IFO-GCF formula to IFO-BCF one, this transformation can be achieved in a polynomial time. The vocabulary used to define IFO formulas and IFO-BCF formulas is defined by:

- A set of constant symbols: $\{c_1, c_2, \ldots, c_n\}$.
- A set of predicate symbols:P $\{P_1, P_2, \ldots, P_n\}$. Each predicate has an arity m representing the number of arguments it can support.
- A set of propositional operators $\neg, \vee, \wedge$ and new connector called ordered disjunction $\times$.
- The usual separators (,).

In our application, the predicate symbols will represent alerts' attributes such as kinds of used IDS, operating systems, etc. Constants represent instances of alerts' attributes.

3.1 Instantiated First Order (IFO) Formulas

Let P be a set of predicate symbols. The following defines the language composed of instantiated first order formulas

- (1) if P is a predicate symbol of arity m and $\{c_1, c_2, \ldots, c_m\}$ is a set of constants then $P(c_1, c_2, \ldots, c_m)$ is an IFO formulas.
- (2) if P and Q are IFO formulas then (P∧Q), (P∨Q), (¬P) are IFO formulas.
- IFO formulas are only obtained by applying items (1) and (2) a finite number of times.

In fact, IFO language is composed of instantiated first ordered formulas where all terms are constants. The following extended languages allow the representation of simple and complex preferences (respectively IFO-BCF, IFO-GCF).

3.2 IFO Basic Choice Formulas (IFO-BCF)

IFO Basic Choice Formulas (IFO-BCF) are ordered disjunctions of IFO formulas. They offer a simple way to order available alternatives. The language composed of IFO-BCF formulas is defined as follows:

- If ϕ is a IFO formula then ϕ is a IFO-BCF.
- If P and Q are two IFO-BCF formulas then P×Q is a IFO-BCF.
- Every IFO basic choice formula is only obtained by applying the two rules above a finite number of times.

Intuitively, P×Q means "satisfy either P or Q with a preference to P". Namely, solutions where P is true are preferred to the ones where ¬ P and Q are true. The language of IFO-BCF formulas will be simply denoted by IFO_{BCF}.

3.3 IFO General Choice Formulas (IFO-GCF)

IFO General Choice Formulas (IFO-GCF) allow to represent general forms of formulas. They are defined as follow :

- If ϕ is a IFO-BCF then ϕ is a IFO-GCF.
- If P and Q are two IFO-GCF then (P∧Q), (P∨Q), (¬P), (P×Q) are a IFO-GCF.
- Every IFO general choice formula is only obtained by applying the two rules above a finite number of times. As we will see later, every IFO-GCF can be reduced to an IFO-BCF by some normalization function "f".

The language of IFO-GCF formulas, will be referred to by IFO_{GCF}.

Example 1. Suppose that a security operator first prefers analyzing IDS alerts which have a 'high' 'severity' value. He, afterwards, chooses to analyze alerts having the 'medium' value and the last option is to analyze 'low' severity alerts. This preference is expressed using IFO_{BCF} as follows: Severity(high) × Severity(medium) × Severity(low). The predicate symbol is 'Severity' and the constant symbols are: "high", "medium" and "low".

3.4 Normalization Function: From IFO_{GCF} to IFO_{BCF}

Normalization functions transform IFO-GCF formulas to IFO-BCF ones. The advantage of such a transformation is to reuse efficient definitions of satisfaction degree of IFO-BCF for IFO-GCF.
$f_{IFO-QCL}$ is the normalization function which transforms every IFO-GCF formula to an IFO-BCF one. This normalization function considers that all preferences have the same level of importance. It is defined using the following three items:
1. If ϕ is a IFO-BCF then $f_{IFO-QCL}(\phi) = \phi$.
2. $f_{IFO-QCL}$ is decomposable with respect to negation, conjunction, disjunction and ordered disjunction. If ϕ and φ are two IFO-GCF (but either ϕ or φ is not IFO-BCF) then:

- (a) $f_{IFO-QCL}(\phi \wedge \varphi) \equiv f_{IFO-QCL}(f_{IFO-QCL}(\phi) \wedge f_{IFO-QCL}(\varphi))$
- (b) $f_{IFO-QCL}(\phi \vee \varphi) \equiv f_{IFO-QCL}(f_{IFO-QCL}(\phi \vee f_{IFO-QCL}(\varphi))$
- (c) $f_{IFO-QCL}(\phi \times \varphi) \equiv f_{IFO-QCL}(f_{IFO-QCL}(\phi) \times f_{IFO-QCL}(\varphi))$
- (d) $f_{IFO-QCL}(\neg\phi) \equiv f_{IFO-QCL}(\neg f_{IFO-QCL}(\phi))$

3. Let $\phi = a_1 \times a_2 \times \ldots \times a_n$ and $\varphi = b_1 \times b_2 \times \ldots \times b_m$ be two IFO-BCF formulas.

- (a) $f_{IFO-QCL}(\phi \wedge \varphi) \equiv c_1 \times c_2 \times \ldots \times c_k$
  $c_i = \bigvee (a_j \wedge b_l)$ with $1 \leq j \leq n, 1 \leq l \leq m$ and $j + l = i + 1$.
- (b) $f_{IFO-QCL}(\phi \vee \varphi) \equiv c_1 \times c_2 \times \ldots \times c_k$ where k=max (m,n) and

$$c_i = \begin{cases} a_i \vee b_i \text{ if } i \leq min(m,n). \\ a_i \text{ if } m \prec i \leq n. \\ b_i \text{ if } n \prec i \leq m. \end{cases}$$

- (c) $f_{IFO-QCL}(\neg\phi) \equiv \neg a_1 \times \neg a_2 \times \ldots \times \neg a_n$

Property "1" describes the fact that the normal form of an IFO-BCF ϕ is the IFO-BCF ϕ. Property "2" (from(a) to (d)) expresses that the normalization function is decomposable with respect to negation, conjunction, disjunction and ordered disjunction. Property "3" (from (a) to (c)) defines the conjunction, disjunction and negation of IFO-BCF formulas.

Example 2. Let ϕ: OS(Unix) $\times$ OS(Mac-OS) $\times$ OS(Windows) and φ: Alert-Source(External) $\times$ Alert-Source(Internal), be two IFO-BCFs.
$f_{IFO-QCL}(\phi \wedge \varphi) \equiv$ (OS (Unix) $\wedge$ Alert-Source (External)) $\times$ ((OS(Unix) $\wedge$ Alert-Source (Internal)) $\vee$ (OS (Mac-OS) $\wedge$ Alert-Source (External))) $\times$ (OS (Mac-OS) $\wedge$ Alert-Source (Internal)) $\vee$ (OS (Windows) $\wedge$ Alert-Source (External)) $\times$ (OS (Windows) $\wedge$ Alert-Source (Internal)).

4 Semantics and Satisfaction Degrees

This section defines the concept of interpretation in our framework and attributes a satisfaction degree to each preference with respect to an interpretation.

4.1 IFO-QCL Interpretation

An interpretation is a pair I=(D,I_v) where D is the domain of the interpretation and I_v is a function that assigns to each constant C an element of D, and to each predicate symbol P (of arity n) a subset of D^n. Intuitively, I_v(P) represents the set of all n-tuples that make P true in I.

Example 3. Let I_1 =(D,I_v) be an interpretation with D={Unix, Mac-OS, Windows } and OS be a predicate symbol with arity 1. I_v(OS)=Unix intuitively means that within the interpretation I_1 the Operating System OS has the value UNIX.

4.2 Satisfaction Relation in IFO-QCL

Let I=(D, I_v) be an interpretation, P and Q be two predicates and $\{c_1, c_2, \ldots, c_m\}$ be a set of constants. Let ϕ and φ be two IFO formulas.

We now define the concept of satisfaction of each interpretation with respect to a given IFO-BCF and IFO formula.

This satisfaction relation will be denoted by :$I \models_k \phi$ where I=(D,I_v) is an interpretation,ϕ an IFO or an IFO-BCF formula and k is an integer that represents the satisfaction degree of the formula ϕ in the interpretation I.

Let us start with the case of IFO formulas that do not involve the preference operator $\times$. In this case k can only be either equal to 1 or equal to 0.

4.3 Satisfaction of IFO Formulas

If P is a predicate symbol of arity m and $\{c_1, c_2, \ldots, c_m\}$ are a set of constant then:

- $I \models_1 P\{c_1, c_2, \ldots, c_m\}$ iff $\{I_v(c_1), I_v(c_2), \ldots, I_v(c_m)\} \in I_v(P)$ and $I \models_0 P\{c_1, c_2, \ldots, c_m\}$ iff $\{I_v(c_1), I_v(c_2), \ldots, I_v(c_m)\} \notin I_v(P)$
- $I \models_k \phi \wedge \varphi$ iff $I \models_i \phi$ and $I \models_j \varphi$ and $k = \max(i, j)$.
- $I \models_k \phi \vee \varphi$ iff $I \models_i \phi$ and $I \models_j \varphi$ and $k = \min(i, j)$.
- $I \models_k \neg\phi$ iff $I \models_{1-k} \phi$

In the following, for IFO formulas, we simply write $I \not\models \phi$ iff $I \models_0 \phi$, and $I \models \phi$ iff $I \models_1 \phi$.

4.4 Satisfaction of IFO-BCF Formulas

In this case, the satisfaction degree can be greater than 1 since an IFO-BCF involves different choices or options. More precisely, if $\phi = \varphi_1 \times \varphi_2 \times \ldots \times \varphi_n$ is an IFO-BCF formula (φ_i's are IFO formulas) then $I \models_k \varphi_1 \times \varphi_2 \times \ldots \times \varphi_n$ iff $\exists j$, such that $I \models \varphi_j$ and $I \not\models \varphi_i$ for all $i \prec j$. When $\forall \varphi_i, I \not\models \varphi_i$ then we simply write $I \not\models \phi$

Example 4. Let Severity and IDS-Type be two predicates, I_1 =(D, I_{v1}) and I_2 =(D, I_{v2}) be two interpretations such that D={high, medium, low, snort, Brother}, I_{v1}(IDS-Type) = {Snort}, I_{v1}(Severity) = {high}, I_{v2}(IDS-Type) = {Bro} and I_{v2}(Severity) = {Low}.
Let ϕ and φ be two preferences formulas: ϕ_1 Severity (high)$\wedge$ IDS-Type(snort)$\times$ Severity (low)$\wedge$ IDS-Tye(snort).
ϕ_2 IDS-Type(snort)$\times$ IDS-Type(bro).
$I_1 \models_1 \phi_1$, $I_1 \models_1 \phi_2$, $I_2 \not\models \phi_1$, $I_2 \models_2 \phi_2$.

5 Main Steps of the Proposed Approach

The inputs, outputs and steps of our method are enumerated below:

5.1 Set of Alerts A

The input of our correlation alert process is a set of alerts A issued from different IDS. Alerts are provided in IDMEF format (International Data Model Exchange Format) [21]. An IDMEF alert contains information about alerts such as: source, target, analyzers, alert detection time, etc. In our model, an alert will be represented as an interpretation where the domain of each attribute is simply the list of the different attribute values present in the alert. Each alert attribute is in predicate form, for instance: Severity(medium), AnalyzerID(27896), AnalyzerName(prelude-lml).

5.2 A Knowledge Base K

This knowledge base encodes the security operator's knowledge. It is a set of IFO formulas. The knowledge formulas detail the attributes values that the operator does not want to see in the resulting alert set. An example of a knowledge formula is: OS (Windows)$\wedge$Classification (http-Inspect). It encodes the fact that if the Operating System of the machine is "Windows" then the security operator refuses to consider the "http-inspect" alerts.

5.3 A Set of Preferences P

It is a set of the security operator's preferences encoded using IFO-GCF formulas. An example of preference can be: (AnalyzerName(snort) $\wedge$ (Severity(high)$\vee$ Severity (medium))) $\times$ (AnalyzerName(snort) $\wedge$ Severity(low)). This formula means that: "snort" alerts which have a high or a medium severity are preferred to low severity "snort" alerts.

5.4 Output

The output of our application is a subset of the initial alerts filtered by the knowledge and preferences sets. The resulting alerts are stratified in accordance with the security operator's knowledge and preferences.

5.5 Algorithm

The inputs are: the set of alerts, the knowledge and preferences sets (K and P). The results are given by the following algorithm.

1. Eliminate alerts that satisfy the knowledge formulas. We denote A' the resulting alerts set.
2. transform the preference formulas according to the normalization function.
3. compute the alerts satisfaction degree using the preference formulas.
4. rank alerts according to their satisfaction degree.
5. return the max preferred alerts to the security operator (the alerts having the minimum degree of satisfaction).

6 Experimentation Using Real Data

This section details the experimental studies on data collected within the PLACID project. This set of data is composed of more than 1,099,302 alerts.

The knowledge and preferences model used in tests is shown below. This model is extracted from the analysis of IDS returned alerts. Tests are done using rather restrictive knowledge and preferences set.

Knowledge Base
Classification(successful) $\wedge$ AnalyserName (prelude-lml)).
The knowledge expresses the fact that the security operator assesses that alerts coming from the HIDS (Host IDS) "prelude-lml" with a classification value "successful" are probably false positive alerts. Indeed, "prelude-lml" reports all the users logins. Then such alerts must be eliminated.

Preferences Base
p_1 : Severity(high) $\times$ Severity(medium) $\times$ Severity (low);
p_2 : AnalyserName(local-snort) $\times$ AnalyserName (public-snort).
The security operator expresses two preferences formulas. The first concerns the alert severity: alerts with a high severity are more important to treat than medium severity ones which are more important than low severity ones.

The second preference formula is about the IDS which generates the alert. the security operator assesses that "local-snort" alerts are more dangerous than the "public-snort" ones. Hence, "local-snort" monitores the local network traffic.

The interface of the developed application is intuitive for the security operator. Thus, he just enters the attributes values corresponding to his knowledge and/or his preferences in their dedicated fields, and runs the IFO-QCL process.

The most preferred alerts are those having a high severity value and coming from the local-snort IDS.

The results shows that from 1099302, only 27810 are important for the security operator. Thus only 2.53 % of the initial alerts are important for the security operator. This rate can be explained by the fact that the security operator knows well the monitored network, its weaknesses and the most dangerous attacks. Thus he expresses knowledge and preferences that allows to return only the most dangerous alerts.

The remaining alerts are not lost, they are ranked in decreasing order of importance (satisfaction). These alerts are shown in a second result application window and can be viewed by the security operator if necessary. Thus the security operator is more efficient if he has to analyse 2.53 % of the issued alerts.

The preferred alerts do not contain any false positive in the security operator point of view.

Our algorithm result (the preferred alert rate) depends on the security operator knowledge and preferences, more they are restrictive less the resulted alert set is voluminous.

7 Conclusion and Perspectives

We propose, in this paper, a new logic to deal with preferences called Instantiated First Ordered Qualitative Choice Logic (IFO-QCL). This logic treats the IDS alerts using the security operator's knowledge and preferences, in a polynomial time. An algorithm was built upon the defined logic. The application takes the security operator's knowledge and preferences and returns a set of stratified alerts. We test our application on real data using the whole PLACID dataset with a set of knowledge and preferences. The result is a set of filtered alerts which represent 2.53% from the initial ones. Other tests will be done with different of knowledge and preference sets (more restrictive and more passive) in order to establish a link between the weankness of these sets and the resulting alert number rate.

References

1. Anderson, J.: Computer security threat monitoring and surveillance. Technical report. James P. Anderson Company, Fort Washington, Pennsylvania (April 1980)
2. Axelsson, S.: Intrusion Detection Systems: A Survey and Taxonomy. Technical Report No 99-15, Dept. of Computer Engineering, Chalmers University of Technology, Sweden (March 2000)
3. Manganaris, S., Christensen, M., Zerkle, D., Hermiz, K.: A Data Mining Analysis of RTID Alarms. Recent Advances in Intrusion Detection Systems 34(4), 571–577 (2000)
4. Chifflier, P., Tricaud, S.: Intrusion Detection Systems Correlation: a Weapon of Mass Investigation, CanSecWest, Vancouver (March 2008)
5. Cuppens, F.: Managing alerts in a multi-intrusion detection environment. In: Proc. 17th Computer Security Applications Conference, pp. 22–31 (December 2001)
6. Julisch, K.: Clustering intrusion detection alarms to support root cause analysis. ACM Transactions on Information and System Security (TISSEC) 6(4), 443–471 (2003)
7. Valdes, A., Skinner, K.: Probabilistic alert correlation. In: Lee, W., Mé, L., Wespi, A. (eds.) RAID 2001. LNCS, vol. 2212, pp. 54–68. Springer, Heidelberg (2001)
8. Eckmann, S.T., Vigna, G., Kemmerer, R.A.: STATL: an attack language for state-based intrusion Detection. Journal of Computer Security 10(1-2), 71–103 (2002)

9. Morin, B., Debar, H.: Correlation of intrusion symptoms: An application of chronicles. In: Vigna, G., Kruegel, C., Jonsson, E. (eds.) RAID 2003. LNCS, vol. 2820, pp. 94–112. Springer, Heidelberg (2003)
10. Cuppens, F., Ortalo, R.: LAMBDA: A language to model a database for detection of attacks. In: Debar, H., Mé, L., Wu, S.F. (eds.) RAID 2000. LNCS, vol. 1907, pp. 197–216. Springer, Heidelberg (2000)
11. Cuppens, F., Miege, A.: Alert Correlation in a Cooperative Intrusion Detection Framework. In: Proceedings of the 2002 IEEE Symposium on Security and Privacy, p. 202 (May 2002)
12. Ning, P., Cui, Y., Reeves, S.: Constructing attack scenarios through correlation of intrusion alerts. In: CCS 2002: Proceedings of the 9th ACM Conference on Computer and Communications Security, pp. 245–254. ACM, New York (2002)
13. Templeton, S.J., Levitt, K.: A requires/provides model for computer attacks. In: NSPW 2000: Proceedings of the 2000 Workshop on New Security Paradigms, pp. 31–38. ACM, New York (2000)
14. Benferhat, S., Kenaza, T., Mokhtari, A.: A Naive Bayes Approach for Detecting Coordinated Attacks. In: COMPSAC 2008, pp. 704–709 (July-August 2008)
15. Benferhat, S., Sedki, K.: Two alternatives for handling preferences in qualitative choice logic. Fuzzy Sets and Systems Journal (FSS 2008) 159(15), 1889–1912 (2008)
16. Brewka, G., Benferhat, S., Le Berre, D.: Qualitative Choice Logic. Artificial Intelligence Journal (AIJ) 157(1-2), 203–237 (2004)
17. Cuppens, F.: Managing Alerts in a Multi-Intrusion Detection Environment. In: Proceedings of Recent Advances in Intrusion Detection, Davis, CA, USA, pp. 22–31 (October 2001)
18. Debar, H., Wespi, A.: Aggregation and Correlation of Intrusion-Detection Alerts. In: Proceedings of Recent Advances in Intrusion Detection, Davis, CA, USA, pp. 85–103 (October 2001)
19. Qin, X., Lee, W.: Attack Plan Recognition and Prediction Using Causal Networks. In: ACSAC 2004, pp. 370–379 (2004)
20. Geib, C., Goldman, R.: Plan Recognition in Intrusion Detection Systems. In: Proceeding of DARPA Information Survivability Conference and Exposition (DISCEX), vol. 1, pp. 46–55 (June 2001)
21. Debar, H., Curry, D., Feinstein, B.: The Intrusion Detection Message Exchange Format (IDMEF), Network Working Group, Request for Comments (RFC): 4765, Category: Experimental, SecureWorks, Inc. (March 2007)
22. Ranum, M.J.: False Positives: A Users Guide to Making Sense of IDS Alarms, ICSA Labs IDSC, white paper (2003)
23. Alsubhi, K., Al-Shaer, E., Boutaba, R.: Alert Prioritization in Intrusion Detection Systems. In: The 11th IEEE/IFIP Network Operations and Management Symposium (NOMS 2008), pp. 33–40 (April 2008)
24. Tabia, K., Benferhat, S., Leray, P., Me, L.: Alert correlation in intrusion detection: Combining AI-based approaches for exploiting security operators knowledge and preferences. In: Association for the Advancement of Artificial Intelligence (2011)

A Pragmatic and Scalable Solution for Free Riding Problem in Peer to Peer Networks

Mourad Amad[1], Djamil Aïssani[1],
Ahmed Meddahi[2], and Abdelmalek Boudries[3]

[1] Laboratory LAMOS, University of Bejaia, Algeria
amad.mourad@gmail.com, lamos_bejaia@hotmail.com
[2] Telecom Lille 1, France
ahmed.meddahi@telecom-lille1.eu
[3] University of Bejaia, Algeria
abdelmalek.boudries@yahoo.fr

Abstract. Peer-to-Peer (P2P) systems are increasingly popular as an alternative to the traditional client-server model for many applications, particularly file sharing. Free-riding is one of the most serious problems encountered in Peer-to-peer (P2P) systems like BitTorrent.

Free riding is a serious problem in P2P networks for both Internet service providers (ISPs) and users. To users, the response time of query increases significantly because a few altruistic peers have to handle too many requests. In P2P file sharing systems, free-riders who use others' resources without sharing their own cause system-wide performance degradation. Therefore, the rational P2P application developers should develop some anti-free riding measures to deal with it.

In this paper, we propose a pragmatic and scalable peer to peer scheme based on a utility function that uses two parameters: collaboration of peers and detection of tentative of free riders. Performance evaluation of the proposed solution shows that results are globally satisfactory.

Keywords: Free riding, Free rider, Peer-to-Peer (P2P) network, Utility function.

1 Introduction

The decentralized peer-to-peer model (P2P) of communication has emerged as a viable alternative to the traditional client server based centralized model in the large scale distributed systems. There are mainly three P2P architectures: centralized, decentralized and hybrid. The first architecture distributes the storage services but the resource search process is already centralized in like server. Napster is a representative example of such system. The second architecture is divided on two types; structured and unstructured architecture. Decentralized structured architecture such as HPM [2] is cost effective in terms of routing (*optimized and greedy*). However, the semantic aspect and the maintenance cost are not well considered. Decentralized unstructured architecture is opposite to the

A. Amine et al. (Eds.): *Modeling Approaches and Algorithms*, SCI 488, pp. 135–144.
DOI: 10.1007/978-3-319-00560-7_18 © Springer International Publishing Switzerland 2013

former one, the lookup service is not greedy (*flooding or probabilistic*). However, the topology maintenance and the semantic search are well considered. As an example of theses architectures, we can cote: Gnutella . The later architecture (hybrid) consists of a set of super peers organized in decentralized architecture (*structured or unstructured*). Each super peer manages a small network in centralized manner such as in Napster. KaZaad and Skype are representative examples of this architecture.

P2P networks are confronted to many problems: routing performance, semantic search, security aspect, firewall detection and traversal, anonymity of users, free riding, network management performance, etc. In this paper, we study the free riding problem. Free riding means exploiting P2P network resources (*through searching, downloading objects, or using services*) without contributing to the P2P network at desirable levels.

Free-riding problem is not just observed in P2P systems. This type of rational behavior also appears in the context of strategic network formation, selfish routing, mobile ad hoc networks and congestion control. However, the prevalence of free-riding behavior in P2P systems can be attribute d to some specific characteristics, including decentralization, high churn rate (*users dynamically joining and leaving the system*), availability of cheap identities (*pseudonyms*), hidden actions (*e.g. nodes not forwarding messages to reduce computing and communication cost*), and collusion (*users forming groups to maximize benefit*) [9]. In other words, free rider peers in P2P networks are already high consumers, and low or never producers.

Within a short time of its release, the Gnutella network found about 75% of its users free-riding, i.e., downloading files from the network without uploading any content [1]. In the same study, they found that only 10% of peers provide almost 90 % of services in the P2P network. Most peers share no more than ten files or some unpopular files, which are not queried or downloaded by others at all [1].

Many proposed free riding solutions have been proposed in the literature. However, most of them are much cost in terms of implementation or suffer from the scalability problem. In this paper, we try to give a new solution to free riding problem in P2P networks, considering both cited obstacles.

The rest of this paper is organized as follow: next section gives a brief overview of peer to peer networks and related works on free riding in such networks. In section 3, we introduce our proposed model to reduce and isolate the free rider peers from our network. The proposed solution is programmatic and scalable. Performance evaluation of the proposed solution is illustrated in section 4. Finally, we conclude and give some perspectives for future works.

2 Background and Related Works

In this section, we give a background on P2P networks, and then we explain the most important solutions that have been given to address the free riding problem in such networks.

2.1 Background on P2P Networks

It is widely believed that the success of P2P file sharing systems depends upon the quality of service (QoS) offered by such systems. Many types of applications are investigated by P2P technology such as: file sharing, parallel and distributed computing, collaboration and communication and recently Internet telephony. However, file sharing applications are the most used. Many studies show that the most important part of Internet flows are from P2P file sharing applications. Hybrid P2P architecture is well adapted to file sharing applications. The maintenance cost is reduced and the scalability is considered. However, mobility and time connectivity of super peers are a challenge. In general, we choose peer with low mobility and long time connectivity as a super peer.

For our proposed model against free riding problem, we use the hybrid P2P architecture represented on Figure 1 as administration and interconnexion architecture. In this architecture, we distinguish two types of peers: ordinary peer and super peer. Each set of ordinary peers is managed by a super peer. This later is principally characterized by high speed, long time connectivity and high storage capability.

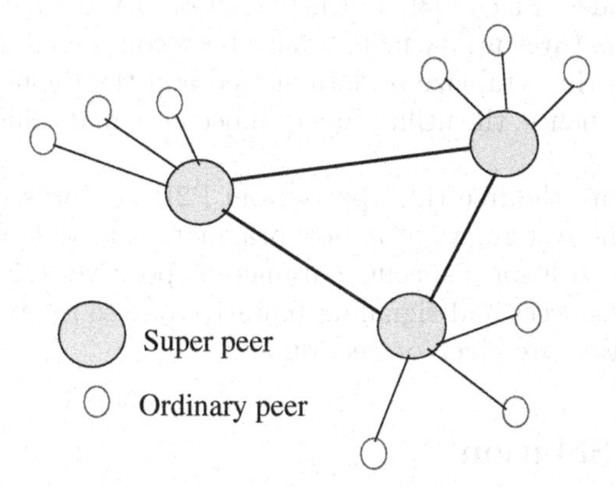

Fig. 1. Hybrid P2P architecture

Each peer (*ordinary peer or super peer*) shares their own resources and can search or request other resources. Ordinary node searches resources by asking its super peer. Super peer searches resources by asking its neighboring peers. Peer that searches resources and does not offering is called free rider.

2.2 Related Works on Free Riding

A number of solutions have been proposed to address the free-riding problem in peer-to-peer systems, particularly for file sharing applications or streaming applications [8]. The authors of [3] propose to introduce a utility function to capture contributions made by a user and an auditing scheme to ensure the integrity of a utility function's values. The principal problem of this solution is the choice of the utility function. In [7], Give-to-get algorithm has been proposed; it encourages peers to upload data to neighbors which prove to be the best forwarders. This policy hurts free-riders any time their neighbors have the opportunity to forward to well-behaving peers instead. The problem of this solution is the scalability. Like shown on [6], when the peers are behind Nats/Firewalls, it is very difficult to prevent free riding problem. In their paper, the authors prove that it is impossible to prevent free-riding when more than half of the peers are firewalled. In [12], the authors show by experimentation using PlanetLab, a new aspect of the free-riding problem in BitTorrent. When a client obtains a larger than normal view of the BitTorrent swarm, it increases its chances to become unchoked by lechers and to discover seeders. Consequently, it is able to attain good download rates without uploading. A preventive solution of free riders in BitTorrent called reat-Before-Trick is then proposed in [11].

Authors of [13] propose an anti-free riding approach based on migration and workload balancing. The heart of this mechanism is to migrate some shared files from the overloaded peers to the neighboring free riders automatically and transparently, which enforces free riders to offer services when altruistic peers are heavily overloaded. Like in [3], a utility function based scheme is proposed in [10]. In this scheme, three utility factors have been considered: the total number of the shared files, the total size of data shared and the popularity of the data shared. As a consequence, the utility function becomes more efficient as the basic one.

In order to isolate the free rider peers from P2P networks without affecting the scalability of the system, we propose a pragmatic and scalable solution based on utility function with some specific parameters: positive collaboration of each peer by offering resources and signaling tentative of free riders. Most details of the proposed solution are given on section 3.

3 Proposed Solution

The aim of our approach is to encourage resources sharing by forcing each peer to contribute in order to be served. For this, hybrid P2P architecture is used. Peer which offers more services (*resources*) is well served. However, peer egoist is progressively suspected.

3.1 Functional Principle

The most important aspect of the proposed solution is the collaboration between peers. All peers should contribute efficiently in resources sharing. Otherwise, they will be progressively suspected. Notations and parameters used in this paper are illustrated on table 1.

At the join operation, the new node (*peer*) initializes its parameters as follow: *Serv =0*, *Dect=0*, *FR=0*. When this peer gives a positive response to any requestor, the super peer on which is attached increments the corresponding value

Serv. When a peer receives many request from the same requestor, it signal this anomaly to it super peer. This later increments the value of *Dec* corresponding to this peer. After each period *T*, the super peer examines both values *Serv* and *Dect* for each attached peer. For each unchanged couple (*Serv, Dect*), the super peer increments the value of *FR*. When *FR* is sufficiently incremented, the corresponding peer will be suspected (*considered as a free rider*).

When a super peer receives a request from neighboring peers, it makes it in the queuing, it serves it when it becomes in the head of queuing, after applying the priority function as shown in table 1.

Table 1. Notations

Symbol	Definition
N	Total number of peers in the network
Serv	Number of positive requests given by the current peer
Dect	Tentatives of free riding signaling by current peer
FR	Status of current peer, it is incremented if during period *T*, both values *Serv* and *Dect* are already the same.
Prior	*Prior=f(Serv, Dect)= Serv+Dect* is a priority function

Figure 2 shows the functional principle of the proposed anti-free riding scheme.

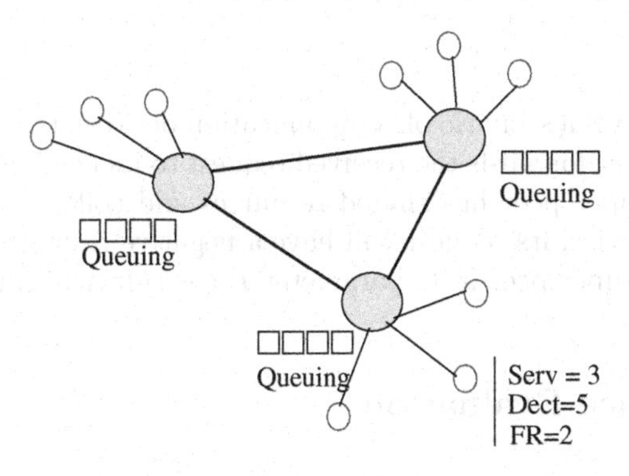

Fig. 2. Functional principle of the proposed scheme

3.2 Scenarios of Communication

Figure 3 shows a scenario of communication between ordinary node and its super peer. If ordinary node does not collaborate by offering resource, their requests will not be satisfied. It will be suspected by its super peer when its parameter *FR* is sufficiently incremented.

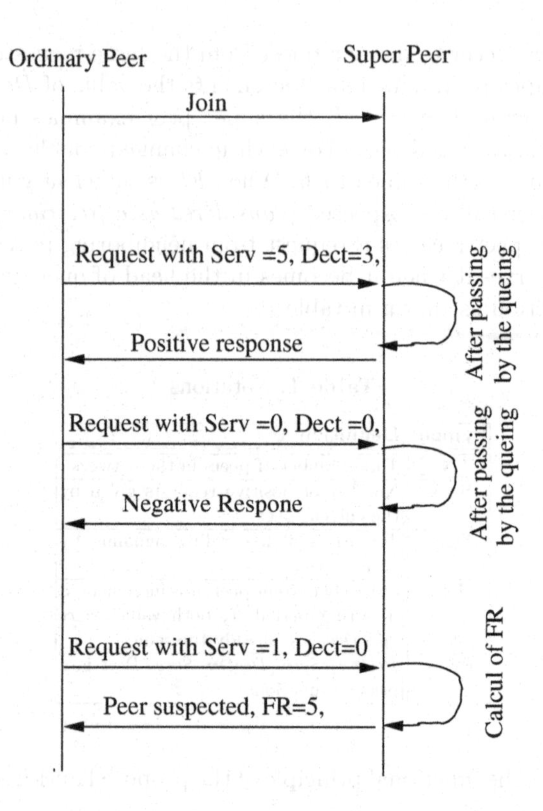

Fig. 3. Scenario of communication between ordinary peer and its super peer

Figure 4 illustrates a scenario of communication between two super peers. The requested super peer forwards the received request to its neighboring super peers if the requestor super peer has shared resources and collaborates positively in the system. Otherwise, its request will have a negative response. It is suspected by a neighboring super peer if its parameter *FR* is sufficiently incremented.

4 Performance Evaluation

In this section, we describe the simulations we performed and the corresponding results. We don't consider the mobility, join and leave when staring simulation.

4.1 Simulation Model

We generate randomly a network with 50 nodes, 7 of them are super nodes. We execute the application in two minutes; we generate randomly requests at arbitrary nodes.

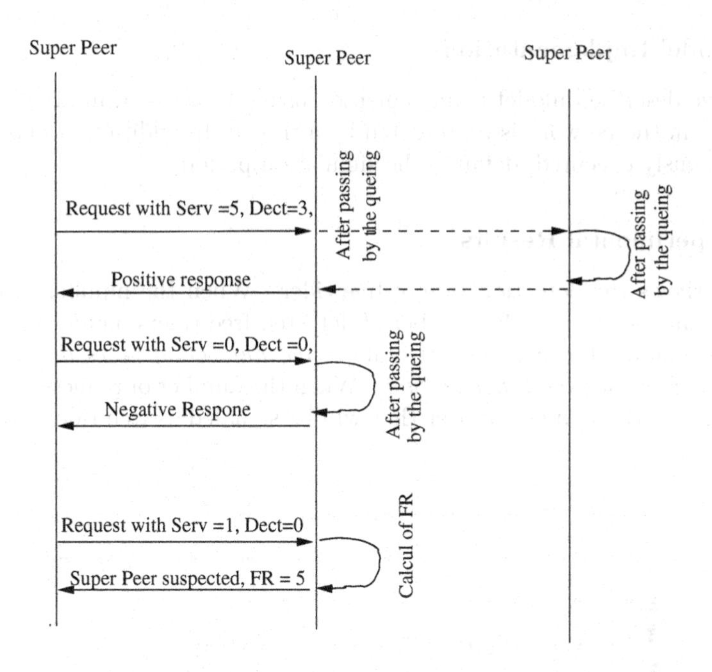

Fig. 4. Scenario of communication between two super peers

4.2 Simulation Parameters and Performance Metrics

As in [4], the following qualitative metrics can be considered for evaluating an anti-free riding scheme:

1. **Quality of service:** QoS parameters may include search time spent in resolving queries, high quality and quantity, and download time.
2. **Availability:** availability of content and services in a P2P network can be increased by forcing free riders to contribute.
3. **Scalability and load sharing:** when the cooperation in a P2P system increased, the load on those peers can be also shared by other peers that would otherwise be free riders. This will help the system to be more scalable where important number of download operations and search queries can be successfully executed on the system.
4. **Robustness:** mechanisms against free riding can make a P2P network more robust against disconnections operation and legal attacks of malicious nodes, which will increase network population in terms of available content and also in terms of number of nodes that are reachable.

As in [5], a number of quantitative parameters can be also considered: the number of downloaded files, the number of unsuccessful downloads, number of uploads, download cost, number of generated messages and fairness.

4.3 Model Implementation

The above described model is implemented using Java programming language. Each peer in the network is represented by a thread. In addition, a main thread is continuously executed, defining the simulation period.

4.4 Experimental Results

Figure 5 shows the detection rate of free riders. When the number of requests increment in the system, the number of detected free riders increments also.

Figure 6 shows the false negative alerts(*the number of peers that have been signaled as free riders but they are not*). When the number of request increments in the system, the number of peer that will be signaled as free riders with error is reduced.

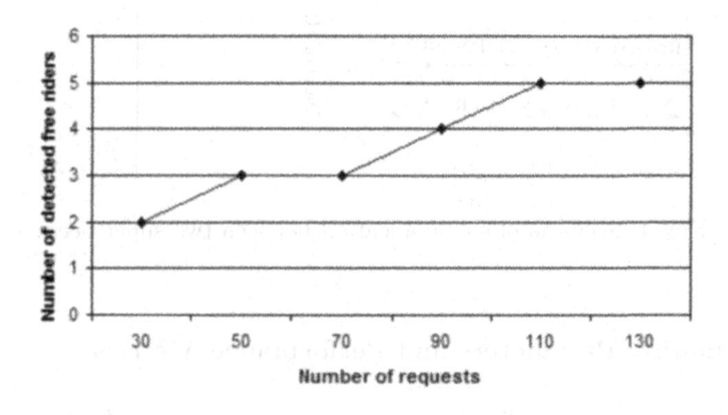

Fig. 5. Free riders detection rate

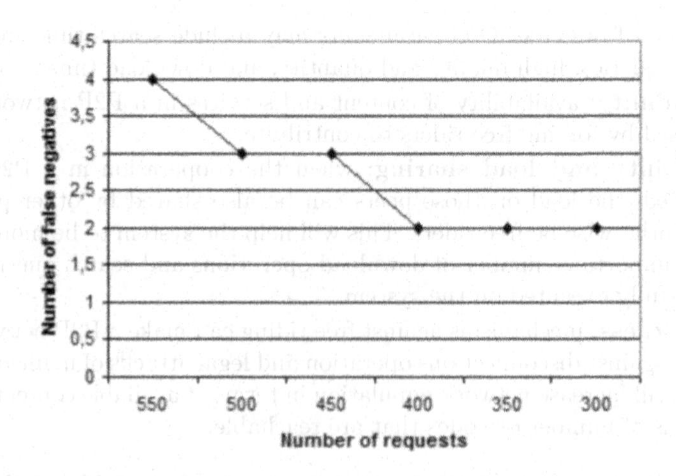

Fig. 6. False negative alerts

5 Conclusion and Perspectives

Free riding is a great challenge to the development and maintenance of Peer-to-Peer (P2P) networks, particularly for file sharing applications.

The free riding problem affects the system in two significant ways. First, the number of files in the system grows very slowly or becomes limited. The number of popular files may become even smaller as the time goes by. This adversely affects user's interest in the system and they eventually pull out of the system. When users who share popular files pull out of the system, the system becomes poorer in terms of the amount of files shared. This is a vicious cycle and it may eventually lead to the collapse of the system. Second, if only a few peers share popular files, all the downloading requests are directed towards those peers. This causes those peers to become hot spots, or exposed to DoS attacks, and overloading their machines and causing congestion on their network. Peers frequently experiencing CPU overloads or network congestion due to the P2P system may exit the system if it affects their other routine activities.

In this paper, we have proposed a new scheme for suspecting free riders from P2P network. The proposed solution is principally based on a specific utility function with priority operation. Peer with high collaboration with upload operation will be good served. Performance evaluation of the new scheme shows that is scalable and easy to implement.

In terms of future work, we envision to generalize and ameliorate our solution in order to take into consideration the free riding problem with two axes: lookup and file sharing.

Acknowledgement. The authors would to thank Dr. Soufiene DJAHEL from Dublin University (*Ireland*) for their constructive remarks.

References

1. Adar, E., Huberman, B.A.: Free Riding on Gnutella. First Monday 5(10) (2000)
2. Amad, M., Meddahi, A., Aîssani, D., Zhang, Z.: HPM:A Novel Hierarchical Peer-To-Peer Model for Lookup Acceleration With Provision of Physical Proximity. Journal of Network and Computer Applications (JNCA) 35, 1818–1830 (2012)
3. Ham, M., Agha, G.: ARA: A Robust Audit to Prevent Free-Riding in P2P Networks. In: 5th IEEE International Conference on Peer-to-Peer Computing (2005)
4. Karakaya, M., Korpeoglu, I., Ulusoy, Ö.: A Distributed and Measurement-based Framework Against Free Riding in Peer-to-Peer Networks. In: Peer-to-Peer Computing, pp. 276–277 (2004)
5. Karakaya, M., Körpeoglu, I., Ulusoy, Ö.: Counteracting free riding in Peer-to-Peer networks. Computer Networks 52, 675–694 (2008)
6. Mol, J.J.D., Pouwelse, J.A., Epema, D.H.J., Sips, H.J.: Free-riding, Fairness, and Firewalls in P2P File-Sharing. In: Eighth International Conference on Peer-to-Peer Computing (2008)
7. Mol, J.J.D., Pouwelse, J.A., Meulpolder, M., Epema, D.H.J., Sips, H.J.: Give-to-Get: An Algorithm for P2P Video-on-Demand. In: Proc. SPIE 6818 Multimedia Computing and Networking (2008)

8. Qin, F., Ge, L., Liu, Q., Liu, J.: Free Riding Analysis of Peer-to-Peer Streaming Systems. Journal of Computational Information Systems 7(3), 721–728 (2011)
9. Rahman, M.R.: A Survey of Incentive Mechanisms in Peer-to-Peer Systems. Technical Report CS-2009-22, University of Waterloo
10. Ramaswamy, L., Liu, L.: Free Riding: A New Challenge to Peer-to-Peer File Sharing Systems. In: HICSS (2003)
11. Shin, K., Reeves, D.S., Rhee, I.: Treat-Before-Trick: Free-riding Prevention for BitTorrent-like Peer-to-Peer Networks. In: The 2009 IEEE International Symposium on Parallel and Distributed Processing (IPDPS 2009), pp. 1–12 (2009)
12. Sirivianos, M., Park, J.H., Chen, R., Yang, X.: Free-riding in BitTorrent Networks with the Large View Exploit. In: IPTPS (2007)
13. Yu, Y., Jin, H.: Discourage free riding in Peer-to-Peer file sharing systems with file migration and workload balancing approach. Front. Comput. Sci. China 1(4), 436–449 (2007), doi:10.1007/s11704-007-0042-z

Cooperative Strategy to Secure Mobile P2P Network

Houda Hafi[1] and Azeddine Bilami[2]

[1] Computer science department, University of Kasdi Merbah-Ouargla-Algeria
hafi.houda@gmail.com
[2] LaSTIC Laboratory, computer science department, U.H.L-Batna-Algeria
abilami@yahoo.fr

Abstract. Mobile peer-to-peer networking (MP2P) is a relatively new paradigm compared to other wireless networks. In the last years, it has gained popularity because of its practice in applications such as file sharing over Internet in a decentralized manner. Security of mobile P2P networks represents an open research topic and a main challenge regarding to their vulnerability and convenience to different security attacks, such as black hole, Sybil...etc. In this paper, we analyze the black hole attack in mobile wireless P2P networks using AODV as routing protocol. In a black hole attack, a malicious node assumes the identity of a legitimate node, by creating forged answers with a higher sequence number, and thus forces the victim node to choose it as relay. We propose a solution based on a modification of the well-known AODV routing protocol and taking into account the behavior of each node participating in the network. Performances of our proposal are evaluated by simulation.

Keywords: Mobile wireless network, P2P, security, black hole.

1 Introduction

Traditionally, the exchange of services between computers is based on client / server model [1], this architecture has shown its limits in terms of performance, amount of processing data and network costs. The peer to peer architecture occurs as an alternative solution to the architecture client / server by providing a breakdown of traffic and load, fault resistance and anonymity. The term P2P refers to a distributed model in which entities called peers play dual role as client and server to provide to a community a service in a decentralized manner [2]. The concept behind this term is simple, its purpose is to bypass the central node in the network and to obtain a completely distributed model in which peers are considered volatile i.e. they join and leave the network unpredictably. The main consequence is the lack of confidence given to a connection between two peers. The model can be divided into three categories: centralized, hybrid or pure [3], according to the mechanism used to search and locate resources in the network. Among its many applications, we quote file sharing, distributed computing and collaborative work.

A. Amine et al. (Eds.): *Modeling Approaches and Algorithms*, SCI 488, pp. 145–154.
DOI: 10.1007/978-3-319-00560-7_19 © Springer International Publishing Switzerland 2013

P2P networks can be supported by wired networks as wireless networks. However, wireless P2P networks claim specific requirements in terms of security, because of their characteristics namely: unreliable links and more vulnerable to various attacks, mobile nodes powered by batteries, bandwidth much less than wired network, limited computing power. Currently very few researches have been conducted for solving security problems of wireless P2P networks; so far there are few scientific proofs on the subject. To protect the network against various malicious actions, most of these researches use traditional security mechanisms such as encryption, sealing, digital signature…

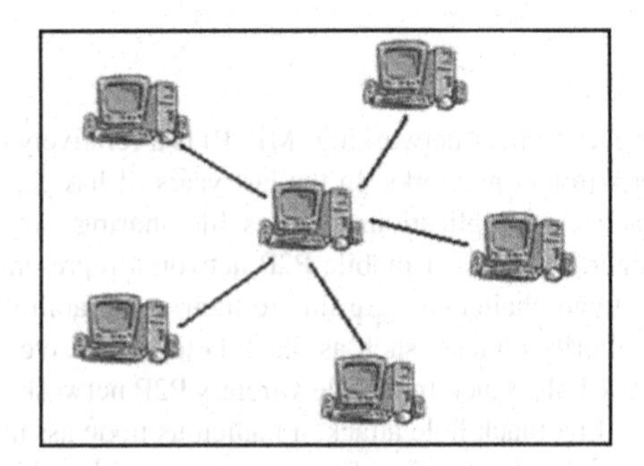

Fig. 1. Client/server model

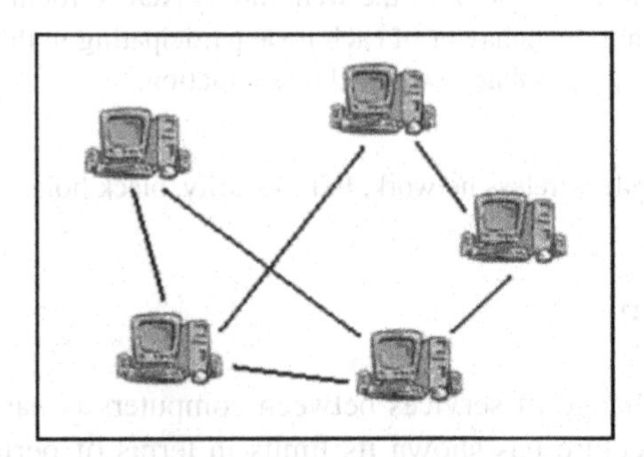

Fig. 2. Peer to Peer model

In this paper, we propose a new protocol based on the use of a trust model able to ensure the secure exchange in mobile wireless P2P networks, while taking into account the characteristics of these networks. We focus on wireless mobile ad hoc networks, where a collection of mobile entities are interconnected by a wireless technology, forming a temporary network without using any administration or any fixed support.

This paper is organized as follows, after the introduction, we present in section 2, the current state of the art of the proposed solutions described by different research teams in the literature. Our proposal is discussed in section 3. In the fourth section, we analyze the protocol performances under different simulation conditions. Finally, we conclude with some future directions for this research.

2 Security in Wireless P2P Network

Most of security mechanisms for P2P systems are developed under the assumption of a large amount of resources, almost unlimited (CPU, memory, energy), high reliability connections through wired links, absence of node mobility, unlimited scope. For all these reasons, most of the solutions adopted in wired networks are not directly transferable to wireless P2P.

The P2P architecture is more appropriate in ad hoc environment, because it is not necessary to maintain persistent connectivity with a server; however the lack of a centralized administration makes these networks more vulnerable to different attacks. Ad hoc networks and P2P architecture share a number of similarities namely: decentralization, self-organization, the volatility (arrival and departure of nodes), the scalability and anonymity [4]. They have a relatively close structure. The security problems in both systems are similar because they have common sources and thus securing ad hoc networks especially by using distributed solutions which are not based on any central point may provide a solid approach in order to secure mobile P2P network against various malicious actions such as the Sybil attack, the Eclipse attack, Man in the middle... etc. Below we cite the main solutions currently proposed for ad hoc networks.

In [5], the authors discuss a method which mitigates the effects of black hole attack; the proposed protocol requires intermediate nodes to include information on the next hop in the RREP packet. After receiving the packet by the source node, the intermediate node sends an additional route request (FREQ) to the next node to verify that the target node (i.e. the node that just returns the RREP packet) has a route to the intermediate node and the destination. When the next hop receives a Further Request, it sends a Further Reply which includes the check result to the source node. Based on the information contained in Further Reply the source rules on the validity of the route. In this protocol, the RREP control packet is modified to contain information about the next hop. After receiving RREP, the source node sends a RREQ again to the specified node as next hop in the RREP received. Clearly, this protocol increases the routing overhead and End to end delay. Also an intermediate node must send a RREP packet twice for the same route request.

In [6] the authors have devised a new method to detect the attack black hole: it is DPRAODV "A Dynamic Learning System Against. Black hole Attack in AODV ", which tends to isolate the malicious node network. The source stores the sequence number of destination (DSN) of RREP received in the routing table and checks to see if the RREP_Seq_No is greater than the threshold value. In this protocol, the threshold value is dynamically updated at each time interval. If the value of RREP_Seq_No is found above the threshold value, the node is suspected of being malicious and is added to a blacklist. It also sends an ALARM packet to its neighbors.

The solution allows participating nodes to realize that one of their neighbors is malicious, then the node is isolated from the network and it is no longer allowed to participate in the operation of packet forwarding.

In [7], the authors describe a protocol in which the source node verifies the authenticity of a node that initiates RREP, by finding more than one route to the destination. When the source receives RREP, and if the existing routes to the destination have shared hops, the source node can then recognize a safe route to the destination.

An algorithm presented in [8] called Pre_Process_RREP, detects the black hole attack in a MANET. The Process continues to accept RREP packets and calls a process called Compare_Pkts (packet p1, packet p2), which compares the destination sequence number of the two packets and selects the packet with higher destination sequence number in case of the difference between the two numbers is not significantly high. Packet containing exceptionally high destination sequence number is suspected to be a malicious node, an ALERT message containing the node identification is broadcasted to neighbor nodes.

3 The Proposed Model

The notion of trust has been applied in telecommunications with the notion of prior knowledge of identities. But today, the development of new communication models such as ad hoc networks; mobile and wireless P2P networks make this vision of confidence obsolete [9]. In addition of that the trust is not a technical problem; it is a social problem to oppose the notion of security, we need confidence when security is not sufficient.

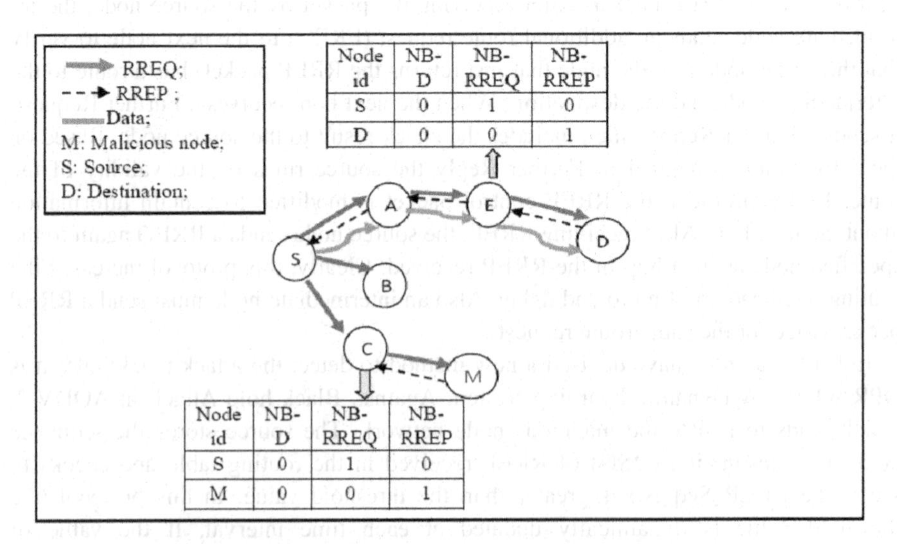

Fig. 3. Example of Diagram depicting the Detection of malicious node

In our model, in order to evaluate the node's confidence level in the network, each one maintains an activity table which stores the identifier of a node, the number of data packets, the number of route request packets (RREQ), and the number of reply packets (RREP) received from that node. When a legitimate node receives a packet (fig. 3), depending on the type of the received packet, it increases the number of received packets in the activity table. If the received packet is RREP, then it consults its activity table to check one of the following equations, based on the stored values in this table, it decides whether the node is an intruder one or not.

Whenever a black hole node receives a data packet, it deletes it directly and when it receives a RREQ packet, it responds by sending a false RREP without consulting its routing table. It avoids broadcasting the RREQ to other nodes. Based on this behavior, a legitimate node will not receive data and RREQ packet from a malicious node. It receives only the response packets.

Therefore, if we assume that:

- **NB-D:** the number of data packets received from a node X
- **NB-RREQ:** the number of RREQ packets received from a node X
- **NB-RREP:** the number of RREP packets received from a node X

1) If **(NB-D+NB-RREQ > NB-RREP)** then: X is trusted node
2) If **((NB-D+NB-RREQ! =0) and (NB-RREP >NB-D+NB-RREQ))** then: X is known node
3) If **(NB-D+NB-RREQ=0)** then: X is unknown node and it can be a black hole node

We present below the pseudo-code of our proposed algorithm:

Step 1:
The source node S starts the route discovery process

Step 2:
Each intermediate node receives a RREQ stores the sequence number of the source (SSN)

Step 3: When an intermediate node receives a RREP, first, it checks the existence of the node in the blacklist, if the condition is true, it deletes it directly. Otherwise it goes to step 4.

Step 4: In this step it checks a bit added to the RREP packet format, to prevent that multiple nodes check several times the same packet.

```
At step 3:
If (packet==RREP) then
    If (id-node ∈ {blacklist}) then
        Delete packet
    Else
        go to step 4
    End if
End if

At Step 4:
If (bit = 1) then
        The RREP was already checked by a node and the
        next node will not need to recheck the packet,
        in this case the node is judged to be trusted
        or known (node A in figure 3).
        Rebroadcast RREP to the source.
Else (bit =0) (node E and C in figure 3)
      Switch (state of node) {
              Case 1: the node is trusted
                      Put bit = 1
                      Rebroadcast RREP to the source
              Case 2: the node is known
                      Put bit = 1
                      Rebroadcast RREP to the source
              Case 3: node is unknown (The route is
              not secure, and the node can be a black
              hole (node M in figure 3)
                      If (DSN>>SSN) (to be sure) then
                       It doesn't refer to the source
                        Add the node to the blacklist
                      End if
                                  }
End if
```

Fig. 4. Pseudo code of intrusion detection algorithm

4 Simulation and Evaluation of the Proposed Approach

The implementation of malicious node behavior can be made at different levels of the protocol layer (Physical, MAC, Routing, and Application), due to different attack scenarios and vulnerabilities. In this paper a single attack on the routing layer is discussed. The implementation of malicious behavior is made on the routing layer. It can also be made by modifying existing routing protocols such as DSR, AODV etc ... therefore; the intruder detection algorithm was also implemented in routing layer.

In order to estimate the effect of our proposal on the network level, we study the behavior of a mobile P2P network using different versions of AODV (the original AODV (without attack) and AODV with Black hole Attack “BHAODV”).

To implement our protocol, we have used the NS2 simulator [10] and have made several changes at several levels; first we have implemented the attack and then have integrated the proposed protocol which is a modified version of AODV. To analyze the results, we have used awk scripts. Table 1 shows the simulation parameters used in our experiment

Table 1. Simulation parameters

Parameter	Value
Time	from 0 to 100s
MAC	802-11
Number of nodes	20
Traffic	CBR
Pause time	2(s)
Packet Size	512 octets

We measure and compare performances in terms of significant metrics such as:

- *Packet delivery ratio*: the ratio of the number of data packets received by destinations nodes to those generated by the source nodes.
- *Normalized Routing Load:* is the ratio of the number of control packets (RREQs, RREPs, and RERRs) generated by the routing protocol to the number of well received data packets.
- *Throughput:* is the amount of information transmitted per unit time.
- *The data packets received by destinations*

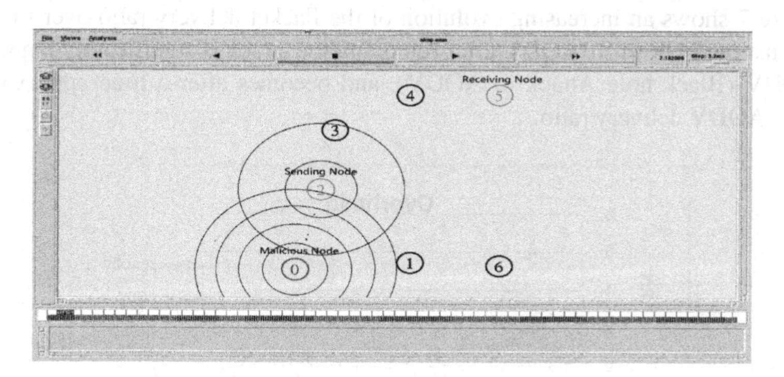

Fig. 5. Simulation of Black hole attack on mobile P2P Network under AODV

Figure 5 and figure 6 show respectively the scenario of the network simulation with the attack, and the scenario after integrating the solution. We notice in fig.6, that the malicious node has no more effect.

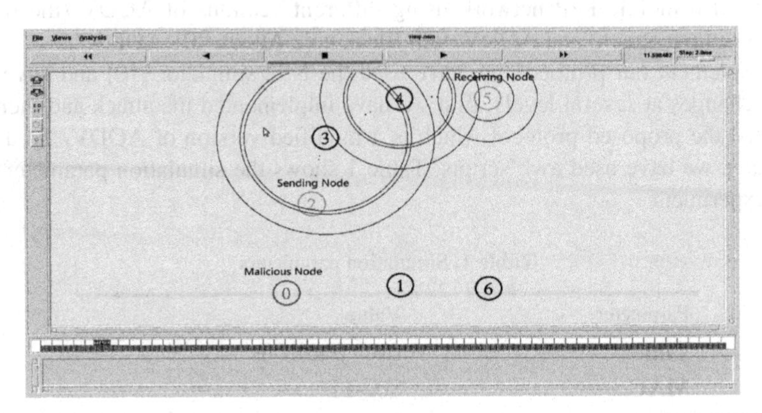

Fig. 6. Simulation of Black hole attack on mobile P2P Network under SBAODV

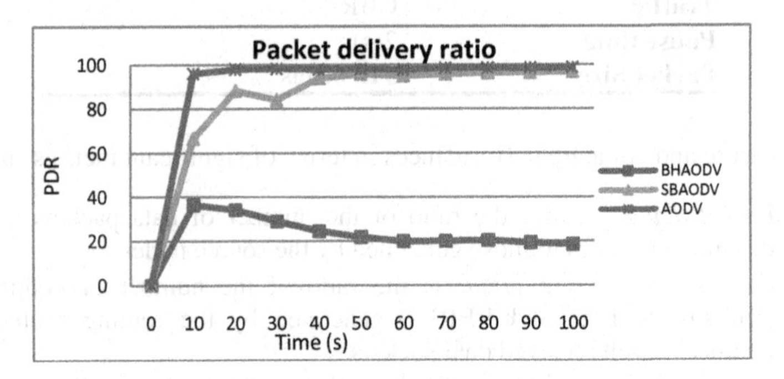

Fig. 7. Graph of Comparison Results for Packet delivery ratio

Figure 7 shows an increasing evolution of the Packet delivery ratio over time. We notice that the PDR of SBAODV after removing the attack is significantly higher than BHAODV (Black hole Attack in AODV), and becomes after a time approximately equal to AODV delivery ratio.

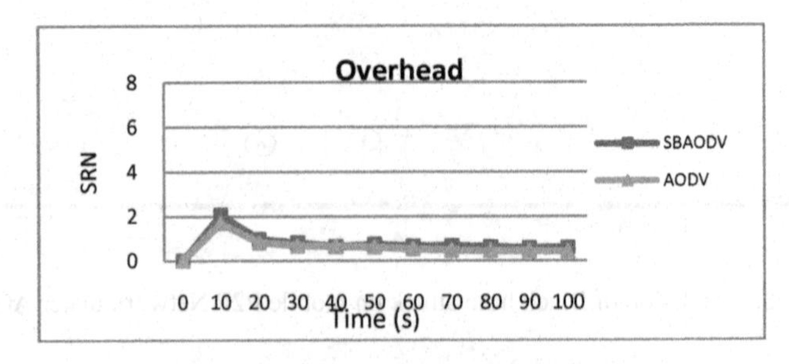

Fig. 8. Overhead using AODV and SBAODV

Figure 8 shows the routing load versus time, it is slightly more than the AODV.

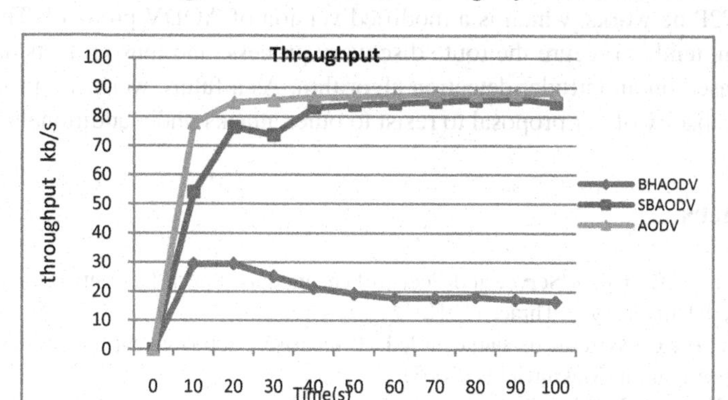

Fig. 9. Comparison Results for Throughput

Figure 9 shows the throughput versus time, for the protocol BHAODV it is very low, after the integration of our module, the throughput starts to increase gradually.

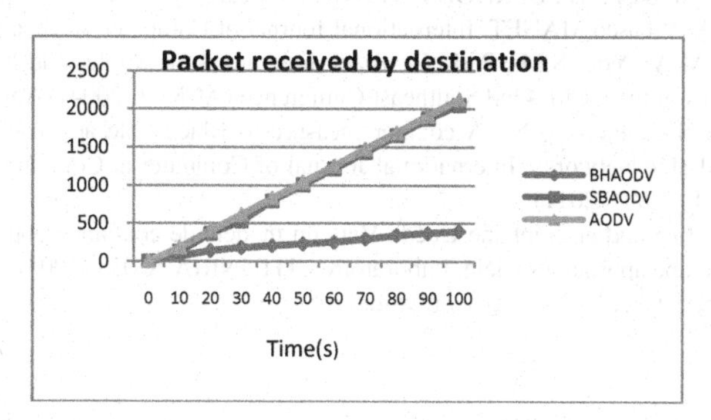

Fig. 10. The number of data packets received by destinations

In Figure 10, we calculate the number of data packets sent by the legitimate nodes and received by their actual destinations. The curve of the AODV under attack is very low compared to the others. Indeed, we clearly see that the attacker was able to isolate legitimate nodes and absorb traffic, packets received in this case are those nodes that are far from the malicious node, since we used 20 nodes, and if we reduce the number of nodes (to 7 for example), we notice that the curve of BHAODV tends to 0.

5 Conclusion

The security of mobile wireless P2P network presents a great challenge because of its autonomous and decentralized nature. Traditional security mechanisms based on the server provide an inappropriate means to protect a purely decentralized network, in such situations the distributed and cooperatives solutions are the most appropriate, because they respond suitably to the requirements of such networks.

In this work, we have implemented a new security protocol dedicated to mobile wireless P2P networks, which is a modified version of AODV protocol. The proposed mechanism tends to secure the route discovery process, and thus protects data transfer process based on an intruder detection algorithm. As a future work, we project to evaluate the capacity of our proposal to resist to other attacks under additional conditions.

References

1. Dunaytsev, R.: Client/Server and Peer-to-Peer models. Space Internetworking Center Democritus University of Thrace (2012)
2. Jarraya, H.: Un système de sauvegarde P2P sécurisé s'appuyant sur une architecture AAA, Maryline Laurent-Maknavicius (2008)
3. Shen, X., et al.: Handbook of peer to peer networking. Springer (2010)
4. Klemm, A., et al.: A Special-Purpose Peer-to-Peer File Sharing System for Mobile Ad Hoc Networks. Department of Computer Science, University of Dortmund
5. Deng, et al.: Routing "Security in Wireless Ad-hoc Networks". IEEE Communications Magazine (2002)
6. Raj, P.N., Swadas, P.B.: DPRAODV: A Dynamic Learning System Against Black hole Attack in AODV based MANET. International Journal of Computer Science (2009)
7. Shurman, M.A., Yoo, S.M., Park, S.: Black hole attack in wireless ad hoc networks. In: Proceedings of the ACM 42nd Southeast Conference (ACMSE 2004) (April 2004)
8. Mandhata, S.C., Patro, S.N.: A counter measure to Black hole attack on AODV- based Mobile Ad-Hoc Networks. International Journal of Computer & Communication Technology (IJCCT) 2(VI) (2011)
9. Véronique Legrand et Stéphane Ubéda Vers un modèle de confiance pour les objets communicants: une approche sociale, Laboratoire CITI, INRIA ARES (2004)
10. Ns-2, http://www.isi.edu/nsnam/ns/

An Efficient Palmprint Identification System Using Multispectral and Hyperspectral Imaging

Abdallah Meraoumia[1], Salim Chitroub[2], and Ahmed Bouridane[3]

[1] Université Kasdi Merbah Ouargla, Laboratoire de Génie Électrique
Faculté des Sciences et de la Technologie et des Sciences de la Matière, Ouargla, 30000, Algérie
[2] Signal and Image Processing Laboratory, Electronics and Computer Faculty, USTHB
P.O. Box 32, El-Alia, Bab-Ezzouar, 16111, Algiers, Algeria
[3] School of Computing, Engineering and Information Sciences, Northumbria University, Pandon Building, Newcastle upon Tyne, UK
Ameraoumia@gmail.com, S_chitroub@hotmail.com, Bouridane@qub.ac.uk

Abstract. Ensuring the security of individuals is becoming an increasingly important problem in a variety of applications. Biometrics technology that relies on the physical and/or behavior human characteristics is capable of providing the necessary security over the standard forms of identification. Palmprint recognition is a relatively new one. Almost all the current palmprint-recognition systems are mainly based on image captured under visible light. However, multispectral and hyperspectral imaging have been recently used to improve the performance of palmprint identification. In this paper, the MultiSpectral Palmprint (*MSP*) and HyperSpectral Palmprint (*HSP*) are integrated in order to construct an efficient multimodal biometric system. The observation vector is based on Principal Components Analysis (*PCA*). Subsequently, Hidden Markov Model (*HMM*) is used for modeling this vector. The proposed scheme is tested and evaluated using 350 users. Our experimental results show the effectiveness and reliability of the proposed system, which brings high identification accuracy rate.

Keywords: Biometrics, Palmprint, *MSP*, *HSP*, *PCA*, *HMM*.

1 Introduction

An accurate automatic personal identification is critical in a wide range of application domains such as access control, surveillance systems, and physical buildings. Traditional knowledge-based or token-based personal identification have both become inefficient and expensive as they are now incapable of meeting the needs of our fast-paced society [1]. Biometric recognition is emerging as a powerful means for automatically recognizing a person's identity with a higher reliability. As one of the developing biometric techniques, palmprint identification is becoming a popular and convincing solution for identifying persons'

A. Amine et al. (Eds.): *Modeling Approaches and Algorithms*, SCI 488, pp. 155–164.
DOI: 10.1007/978-3-319-00560-7_20 © Springer International Publishing Switzerland 2013

identity since palmprint is proved to be a unique and stable personal physiological characteristic [2]. Thus, majority of studies on palmprint identification are mainly based on image captured under visible light. However, multispectral and hyperspectral imagings [3,4], which give different information from a variety of spectral bands, have been recently used to improve the performance of palmprint identification because each spectral band highlights specific features of the palm, making it possible to collect more information to improve the accuracy and anti-spoofing capability of palmprint systems [5]. These bands provide different and complementary information on the same palmprint. In these techniques, an acquisition device, to capture the palmprints under different light resulting into several bands, is used. The idea is to employ the resulting information in these bands to improve the performance of palmprint identification system.

In this paper, the *MSP* and *HSP* are integrated in order to construct an efficient multimodal biometric system. In this method, the spectral band is compressed using *PCA* and some of principal components are selected for representing the feature vector. These vectors (for all spectral bands) are fused, based on feature extraction level, in order to obtain an observation vector whose represent the entire palmprint. Subsequently, *HMM* model is used for modeling this vector. Finally, log-likelihood scores are used for palmprint matching.

The remainder of this paper is organized as follows. The proposed scheme for *MSP*/*HSP* identification is exposed in section 2. Section 3 gives a brief description of the proposed feature extraction method. This section includes also an overview of *PCA* technique and *HMM* method. The Matching and fusion techniques are presented in section 4. The experimental results, prior to fusion and after fusion, are given and commented in section 5. Finally, section 6 is devoted to the conclusion and future work.

2 System Design

The proposed multimodal palmprint identification system is composed of two different sub-systems that exchange information at matching score level. Each sub-system, (see Fig. 1), consists of the following steps: preprocessing, feature extraction (observation vector), modeling, matching process and decision making. To enroll into the system database, the user has to provide a set of training palmprint modalities. Typically, a feature vector is extracted from each modality using *PCA*. This feature vector describes certain characteristics of the palmprint images that will be modeled using *HMM*. Finally, the model parameters are stored as model references. For identification, the same observation vector are extracted from the test palmprint and the log-likelihood is computed using all of reference models in the database. For the multimodal system, each subsystem computes its own matching score and the set of the individual scores are combined into a final score (using fusion at the matching score level). This final score is used for decision making to accept or reject the user.

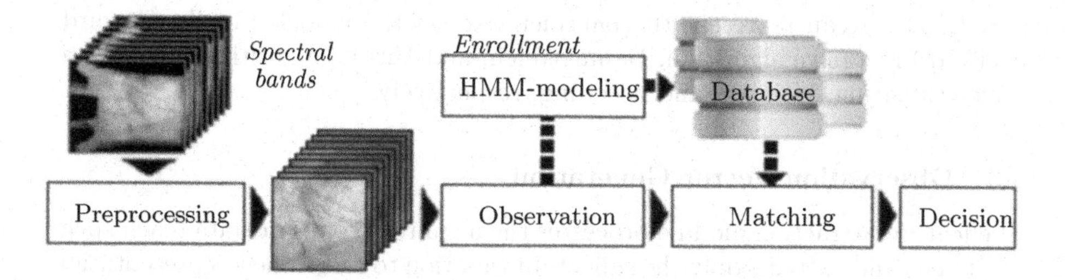

Fig. 1. Block-diagram of the *MSP/HSP* identification system based on *HMM*.

3 Feature Extraction

Before a further feature extraction, the palm area {Region Of Interest (*ROI*)} has to be located [6]. With referencing to the locations of the gaps between fingers, the palm is duly rotated and the maximum palm area is then located. At the features-extraction stage the features are generated from the *ROI* sub-images by *PCA* technique. This technique has been widely used for pattern recognition, as well as in the field of biometrics.

3.1 Principal Component Analysis

PCA is a powerful transformation encoding method, which transforms the original data into the feature vectors of smaller dimensionality where the variance of the feature data is maximal. *PCA* is a standard technique used to approximate the original data with lower dimensional feature vectors. The methodology for calculating principal component is given by the following method [7]:
Given W samples of a process, $\mathcal{O}$=$[o_1$,o_2,$o_3 \cdots o_W]$, the first step in *PCA* is to obtain an estimate of the mean of the process, $\widehat{\mathcal{O}} = \frac{1}{W}\sum_{i=1}^{W} o_i$. After that, each sample differs from the average samples $\widehat{\mathcal{O}}$, by $\varphi_j = \mathcal{O}_j - \widehat{\mathcal{O}}$ then a covariance matrix is constructed, $\widehat{\mathcal{C}} = \sum_{i=1}^{W} \varphi_i \varphi_i^T$. Then, eigenvectors, v_k and eigenvalues, λ_k with symmetric matrix $\mathcal{C}$ are calculated. v_k determine the linear combination of W difference samples with φ to form the eigensamples, $\vartheta_l = \sum_{k=1}^{W} v_{lk}\varphi_k$. Finally, from these eigensamples, K eigensamples, with $K < M$, are selected to correspond to the K highest eigenvalues.

3.2 Hidden Markov Model

An *HMM* is a collection of finite states connected by transitions. Each state is characterized by two sets of probabilities: a transition probability, and either a discrete output probability distribution which, given the state, defines the condition probability of emitting each output symbol from a finite alphabet. Thus, an *HMM* is characterized by [8]: a state transition probability matrix (A), an initial state probability distribution (π) and a set of probability density functions associated with the observations for each state (B). A compact notation

$\lambda = (A, B, \pi)$ can represent the parameter set of the model. Finally, forward-backward recursive algorithm, Baum-Welch, and Viterbi algorithm are used to solve evaluating, training, and decoding, respectively.

3.3 Observation Vector Generation

The feature extraction module processes the acquired biometric data (each spectral band) and extracts only the salient information to form a new representation of the data. In our method, the spectral band is typically analyzed using *PCA* to construct observation vectors. Since an *HMM* model of each observation vector is constructed.

Let I_X represent a $H \times W$ input *ROI* sub-image, *X*≡*HSP* or *X*≡*MSP*,

$$I_X = [x_1, x_2, x_3, \cdots x_W], \tag{1}$$

Applying the *PCA* transform on I_X, we decorrelate the W columns (W one-dimensional vectors) and concentrate the information content on the first components of the transformed vectors.

$$V_X = PCA(I_X) = [o_1, o_2, o_3, \cdots o_W], \tag{2}$$

Then, some of components, *K*, are selected for representing the feature vector.

$$O_X = [o_1, o_2, o_3, \cdots o_K], \tag{3}$$

Finally, in order to obtain the observation vector for representing the entire palmprint. The feature vectors extracted from different spectral bands are fused based on feature extraction level. Feature vectors combination is the main idea of fusion at the feature level. So, all vectors (for all spectral bands) are fused, based on feature extraction level. As results, the entire palmprint as a single template (observation vector) as follows:

$$\nu_X = [O_{X1}, O_{X2}, \cdots O_{Xn}] = [v_1, v_2, v_3, \cdots v_\eta], \tag{4}$$

The size of ν_X is $[H, \eta]$, were $\eta = K \times n$, with n represent the band number.

4 Matching, Normalization and Fusion

4.1 Matching Strategy

In the matching process, the observation vector corresponding to the test palmprint is compared with a particular user's biometric model. This typically results in an observation vector score which is somehow related to how likely it is that this particular user is the source of that vector. *HMM* classifiers will use a likelihood-based score. Thus, the probability of the observation sequence given a *HMM* model is computed via a Viterbi recognizer [9]. The model with the highest log-likelihood is selected and this model reveals the identity of the unknown palmprint. Thus, during the identification process, the characteristics of the test image are extracted. Then the log-likelihood score of the observation vectors given each model, $P(\nu_{Xj}|\lambda_i) = \ell(\nu_{Xj}, \lambda_i)$, is computed. Therefore, the score vector is given by:

$$\mathfrak{L}(\nu_{Xj}) = [\ell(\nu_{Xj}, \lambda_1) \quad \ell(\nu_{Xj}, \lambda_2) \quad \ell(\nu_{Xj}, \lambda_3) \cdots \ell(\mathcal{O}_i^X, \lambda_d)] \tag{5}$$

where d represents the size of model database. Finally, a threshold T_o regulates the system decision. The system infers that pairs of biometric samples generating score, D_o, higher than or equal to T_o are belong to the same person. Consequently, pairs of biometric samples generating score lower than T_o are they belong to different persons.

4.2 Fusion Strategy

Several contributions have shown that fusion of scores obtained from various unimodal biometrics identification systems often enhances the overall system performance. Fusion at the matching-score level consists in combining the scores, which describe the similarities between the biometrics acquired and their models, obtained by each biometric system [10]. During the system design we experiment five schemes: SUM-score (*SUM*), MIN-score (*MIN*), MAX-score (*MAX*), MUL-score (*MUL*) and sum-WeigHTing-Score (*WHT*).

5 Experimental Results and Discussion

To evaluate the performance of the proposed multimodal identification scheme, a database containing *MSP* and *HSP* was required. In this work, we construct a multimodal database for our experiments based on *MSP* and *HSP* from the Hong Kong Polytechnic University (PolyU)[11]. The database consists of 12 palmprint images per person with total of 350 persons. Each *MSP* contain 4 spectral bands. Thus, for the *HSP*, the number of spectral bands is equal to 69 .

Identification occurs when the biometric system attempts to determine the identity of an individual. A biometric is collected and compared to all the models in a database. Identification is *closed-set* if the person is assumed to exist in the database. In *open-set* identification, the person is not guaranteed to exist in the database [12]. In our work, the proposed method was tested through the both modes. To evaluate the efficiency of this proposed method, the experiments were designed as follow: three palmprints of each person were randomly selected for enrollment, and the rest, nine, palmprints are used as test images for identification, respectively. The test phase is divided into two parts. First part presents the performance of the proposed palmprint identification algorithm for different palmprint image types (*MSP* and *HSP*). However, the second part presents the performance of palmprint identification with the fusion of *MSP* and *HSP*.

5.1 Unimodal System Identification

HSP test results. The principal components (*PCs*) vectors reflect the compact information of different column vectors (V_X) for each spectral band. Most of

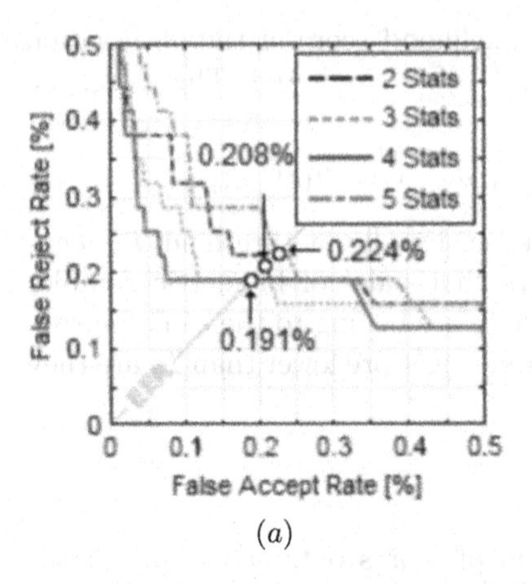

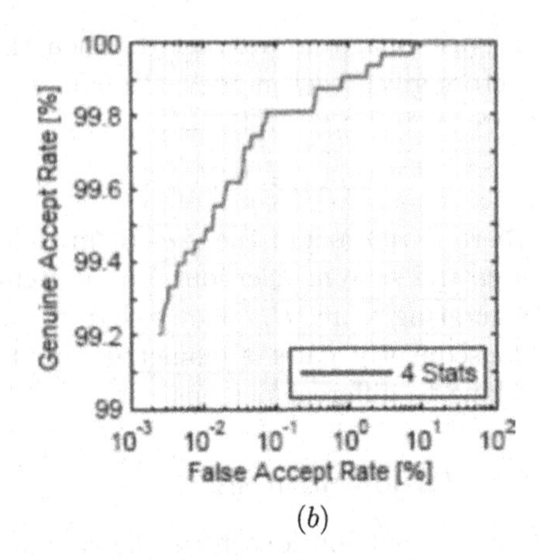

Fig. 2. *HSP*-based unimodal *open set* identification test results. *(a) ROC* curves with respect the number of states and *(b) ROC* curve for the best case.

these vectors become negligible, as result; the vector derived from the initial vectors computation (V_X) is limited to an array of summed vectors within all components. The test was repeated for various numbers of *PCs* vectors in each V_X, only 15 *PCs* vector (> 90 % of the total information) for each spectral band are enough to achieve good identification rate.

A series of experiments were carried out using the *HSP* database to select the best number *HMM*-states. This is carried out by comparing all N states, with $N = 2$ to 5 that gives the best *open set* identification rate. The problem we address is as follows: we want chosen the number of *HMM*-states such that the Equal Error Rate (*EER*) is minimized. Thus, in Fig. 2.*(a)*, we plot the system performance as a function of various numbers of states in the *HMM*. The reason Fig. 2.*(a)* was generated to show how the number of numbers of states in the *HMM* used might have an effect on the performance of our system. Therefore, the system can achieve higher accuracy at three and four states compared with the other number of states in terms of the *EER*, it can achieve an *EER* equal to 0.191 % at a threshold T_o equal to 0.7601 and 0.7420, respectively. Fig. 2.*(b)* shows the Receiver Operating Characteristics (*ROC*) curve, which is a plot of Genuine Acceptance Rate (*GAR*) against False Acceptance Rate (*FAR*), for all possible thresholds, when the number of *HMM*-states equal to 4. Finally, the performance of the *open set* identification system under different number of *HMM*-states is shown in Table 1.

Table 1. *HSP/MSP* based unimodal *open set* identification system performance

Modalities	2 States		3 States		4 States		5 States	
	T_o	EER	T_o	EER	T_o	EER	T_o	EER
HSP	0.7757	0.208	0.7601	0.191	0.7420	0.191	0.8041	0.224
MSP	0.7713	3.781	0.8022	2.311	0.7985	2.275	0.7915	2.530

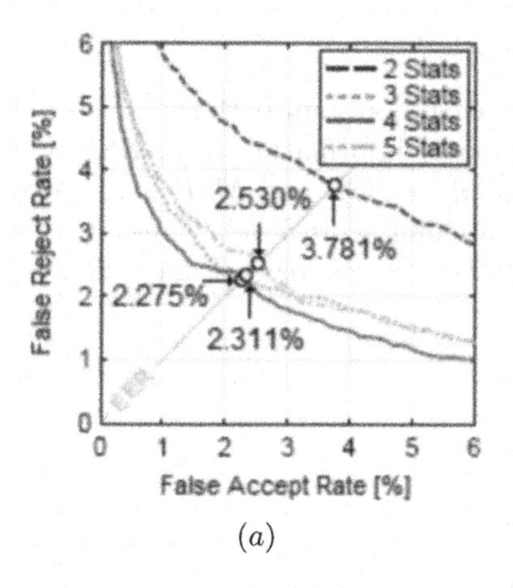

(a)

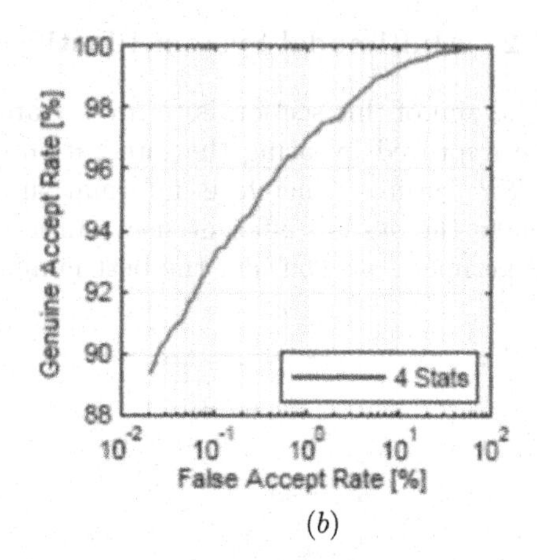

(b)

Fig. 3. *MSP*-based unimodal *open set* identification test results. *(a) ROC* curves with respect the number of states and *(b) ROC* curve for the best case.

MSP Test Results. The aim of this section is to investigate whether the system performance could be improved by using the *MSP* imaging. Therefore, information presented by different spectral bands is fused to make the system efficient. Thus, in this experiment, always the number of *PCs* in each band is equal to 15. In order to see the performance of the *open set* identification system, we usually present, in Fig. 3.*(a)*, the results for all number of *HMM*-states. This figure shows that four states offers better results in terms of *EER*. Our *open set* identification system can achieve an *EER* of 2.275 % at threshold $T_o = 0.7985$. Finally, the results expressed as a *ROC* curve, which is a plot of *GAR* against *FAR*, obtained by the four *HMM*-states is plotted in Fig. 3.*(b)*. Finally, Table 1 showing the results for all number of *HMM*-states.

In the case of a *closed set* identification, a series of experiments were carried out to select the best number of *HMM*-states. This has been done by comparing all number of *HMM*-states and finding the number that gives the best identification rate. Table 2 present the experiments results obtained for *HSP* and *MSP* types. For the *HSP* case, from Table 2, the best results of Rank-One Recognition (*ROR*) produce 99.250 % with lowest Rank of Perfect Recognition (*RPR*) of 189 with four *HMM*-states. Also, in the case of *MSP*, the best result of *ROR* is given as 89.821 % with lowest *RPR* of 349 for a number of *HMM*-states equal to 4.

Table 2. *HSP/MSP* based unimodal *closed set* identification system performance

Modalities	2 States		3 States		4 States		5 States	
	ROR	RPR	ROR	RPR	ROR	RPR	ROR	RPR
HSP	98.857	143	99.071	162	99.250	189	98.750	242
MSP	83.571	331	88.143	332	89.821	349	89.464	308

5.2 Multimodal System Identification

The aim of this section is to investigate whether the system performance could be improved by using the integration or fusion of information from *HSP* and *MSP* features. Therefore, information presented by *HSP* and *MSP* is fused to make the system efficient using fusion at matching score level. Thus, in this experiment, we consider the best number of *HMM*-states for each modality.

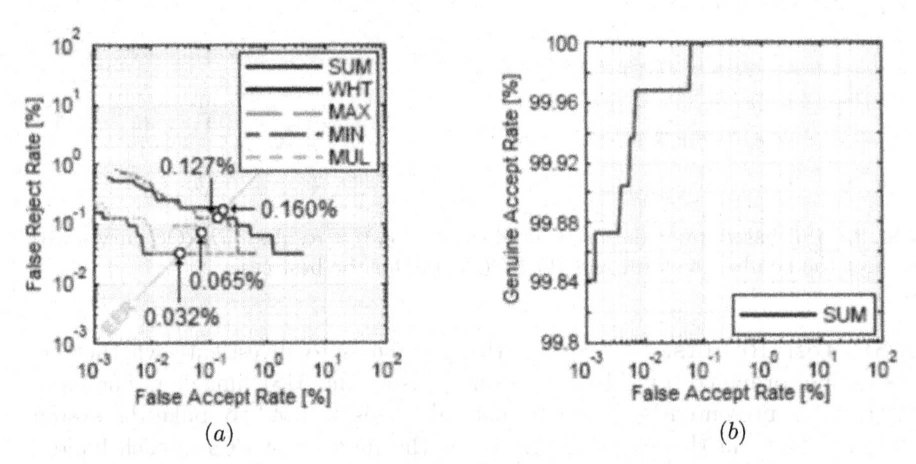

Fig. 4. Multimodal palmprint *open set* identification test results. *(a)* The *ROC* curves with respect fusion rules and *(b)* The *ROC* curve for the best case (*SUM* rule).

At the matching score level fusion, the matching scores output by multiple matchers are integrated using several fusion rules. For that, a series of experiments were carried out to selection the best fusion rule that minimize the *EER* using the best unimodal result for the two modalities. In order to see the performance of the *open set* identification system, graphs showing the *ROC* curves were generated (see Fig. 4.*(a)*). This figure shows that the *SUM* and *MIN* rules offers better results in terms of the *EER*. For example, if *WHT* rule is used, we have $EER = 0.160$ %. In the case of using *MUL* rule, *EER* was 0.065 %. Using *MAX* rule, *EER* was 0.127%. A *SUM* and *MIN* rules, both improves the result (0.032 % at $T_o = 0.8044$ and 0.7084, respectively) for a database size equal to 350. Therefore, the system can achieve higher accuracy at the fusion of the two matching score compared with a single matching score. Finally, the results expressed as a *ROC* curve obtained with *SUM* rule fusion are plotted in Fig. 4.*(b)*. Table 3 presents the obtained results for all fusion rules for 350 users.

Table 3. Multimodal *Open set* identification test result

DB size	SUM		WHT		MUL		MAX		MIN	
	T_o	EER	T_o	RPR	ROR	EER	T_o	EER	T_o	EER
350	0.8044	0.032	0.7321	0.160	0.9541	0.065	0.7583	0.127	0.7084	0.032

To validate our idea we have run other tests for the *closed set* identification. Thus, to determine the best fusion rule, a table contains the results for all fusion rules can be established (see Table 4). The experiments described in table 4 suggest that the *SUM* rule performs better than that the others fusion rules. The best result of *ROR* is given as 99.786 % with lowest *RPR* of 16.

Table 4. Multimodal *Closed set* identification test result

DB size	SUM		WHT		MUL		MAX		MIN	
	ROR	RPR	ROR	RPR	ROR	RPR	ROR	RPR	ROR	RPR
350	99.786	16	99.357	160	92.536	11	99.179	138	99.714	35

To find the better methods graphs showing the *ROC* curves for the *open set* and *closed set* identification using unimodal and multimodal system, were generated (see Fig. 5.*(a)* and Fig. 5.*(b)*). By the analysis of those plots, it can be observed that the performance of the system is significantly improved by using the fusion technique. In the case of *open set* identification, the *SUM* rule gives the best result (*EER* = 0.032 %). The best result (*ROR* = 99.786 %) in the case of *closed set* identification is given always when *SUM* rule fusion is used. However, the experimental results show that the identification rate of multimodal system based on fusion of *HSP* and *MSP* modalities is higher than that of the identification rate adopting the unimodal system.

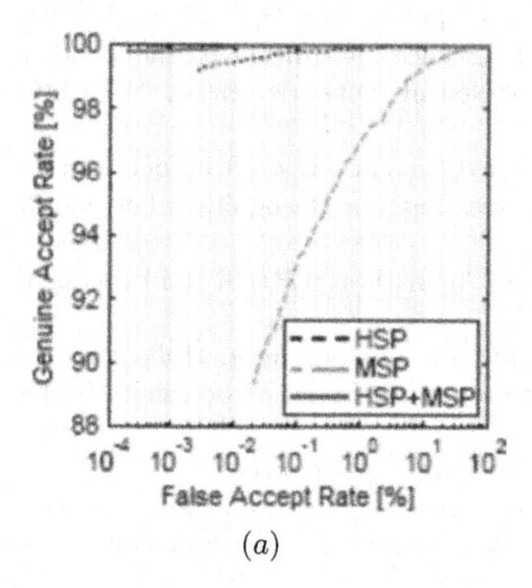

(a)

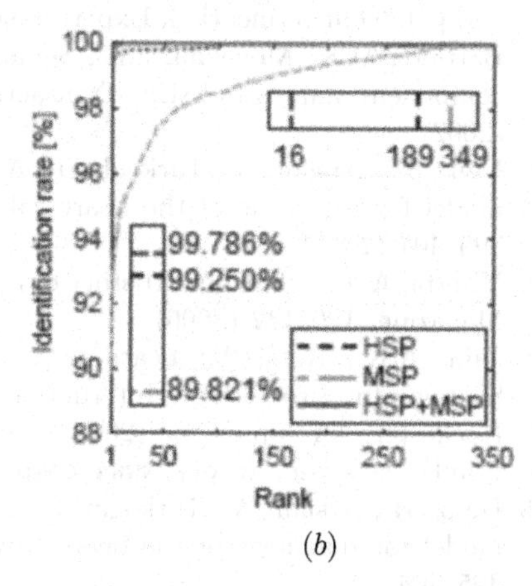

(b)

Fig. 5. Comparison of the unimodal and multimodal systems. *(a) Open set* identification systems and *(b) Closed set* identification systems.

6 Conclusion and Further Work

In this paper, an identification system based on *HSP* and *MSP* modalities and by using *HMM* method has been proposed. Fusion is performed at the matching score level to generate a unique score that is used for recognizing a palmprint data. The proposed system is evaluated using the PolyU database. The obtained experimental results show that the combination of *HSP* and *MSP* data performs best, in both *open set* and *closed set* identification, and is much better than person identification using only single palmprint data. For further improvement of the proposed system, our future work will focus on the performance evaluation using large size database and using other feature extraction techniques such as Gabor filter [13] and discrete cosine transform [14].

References

1. Wayman, J., Jain, A., Maltoni, D., Maio, D.: Biometric Systems, Technology, Design and Performance Evaluation. Springer, London (2005)
2. Yue, F., Zuo, W., Zhang, D., Li, B.: Fast palmprint identification with multiple templates per subject. Pattern Recognition Letters 32, 1108–1118 (2011)
3. Zhang, D., Guo, Z., Lu, G., Zhang, L., Zuo, W.: An Online System of Multispectral Palmprint Verification. IEEE Transactions on Instrumentation and Measurement 59(2) (February 2010)
4. Guo, Z., Zhang, D., Zhang, L., Liu, W.: Feature Band Selection for Online Multispectral Palmprint Recognition. IEEE Transactions on Information Forensics and Security 7(3) (June 2012)
5. Zhang, D., Guo, Z., Lu, G., Zhang, L., Zuo, W.: An online system of multi-spectral palmprint verification. IEEE Trans. Instrumentation Measurement 59(3), 480–490 (2010)
6. Zhang, D., Guo, Z., Lu, G., Zhang, L., Liu, Y., Zuo, W.: Online joint palmprint and palmvein verification. Expert Systems with Applications 38, 2621–2631 (2011)
7. Bartlett, M.S., Movellan, J.R., Sejnowski, T.J.: Face recognition by independent component analysis. IEEE Transactions on Neural Networks 13(6), 1450–1464 (2002)
8. Uguz, H., Arslan, A., Turkoglu, I.: A biomedical system based on hidden Markov model for diagnosis of the heart valve diseases. Pattern Recognition Letters 28, 395–404 (2007)
9. Viterbi, A.J.: A Personal History of the Viterbi Algorithm. IEEE Signal Processing Magazine, 120–142 (2006)
10. Jain, A.K., Ross, A.: Learning User-Specific Parameters in a Multibiometric System. In: Proc. IEEE International Conference on Image Processing (ICIP), Rochester, NY, pp. 57–60 (2002)
11. PolyU Database, `http://www4.comp.polyu.edu.hk/~biometrics/`
12. Uguz, H., Arslan, A., Turkoglu, I.: A biomedical system based on hidden Markov model for diagnosis of the heart valve diseases. Pattern Recognition Letters 28, 395–404
13. Dey, S., Samanta, D.: Iris Data Indexing Method Using Gabor Energy Features. IEEE Transactions on Information Forensics and Security 7(4), 1192–1203 (2012)
14. Meraoumia, A., Chitroub, S., Ahmed, B.: Gaussian Modeling And Discrete Cosine Transform For Efficient And Automatic Palmprint Identification. In: IEEE International Conference on Machine and Web Intelligence-ICMWI, USTHB, Algiers, pp. 121–125 (2010)

RDAP: Requested Data Accessibility Protocol for Vehicular Sensor Networks

Mansour Louiza and Moussaoui Samira

Computing Departement
University of Sciences and Technology Houari Boumediene
Algiers, Algeria
{Mansour.louiza,Moussaoui_samira}@yahoo.fr

Abstract. Vehicular Sensor Networks (VSNs) are an emerging paradigm in vehicular networks. This new technology uses different kind of sensing devices available in vehicles, to gather information in order to provide safer, efficient and comfort for roads users. One of the VSNs challenges is how to deal with dynamic data collection. To achieve this, an efficient collaboration between sensors and vehicles is required. This paper proposes a new multi-hop data collection and dissemination scheme based on data replication on VSNs in an urban scenario. The aim of our proposal scheme is to achieve a high accessibility to a requested data while maintaining a low level of channel utilization. The simulation results show that this protocol can achieve significant performance benefits.

Keywords: Vehicular Sensor Network, Data Collection, Data Dissemination, Data Replication.

1 Introduction

A Vehicular Sensor Network VSN can be built on top of a Vehicular Ad hoc NETworks VANET [1, 2] by equipping vehicles with sensing devices. Compared to traditional static sensors, a sensor node is not subject to processing, storage, and energy limitations. Moreover, a mobile sensor node improves sensing coverage with low costs. The two primary distinct features of vehicle networks are that i) vehicles can be highly mobile so an intermittent connectivity , and ii) their mobility patterns are more predictable than those of nodes in Mobile Ad hoc NETworks MANET [3] due to the constraints imposed by roads, speed limits, and commuting habits. These specific characteristics generate a number of new research challenges (e.g. data dissemination, data aggregation, data collection, security and authentication ...) that need to be addressed for VSNs to be widely deployed. In this work, we focus on one of the most important topics in VSNs: data collection. The typical scale of a VSN over a wide geographic area, the volume of generated data, important mobility of vehicles and the limited bandwidth make it infeasible to adopt traditional sensor network solutions [4]. The challenge is processing queries in this highly mobile environment and avoids broadcast storm problem [5] to overcome delay, overhead and accuracy. Some

A. Amine et al. (Eds.): *Modeling Approaches and Algorithms*, SCI 488, pp. 165–173.
DOI: 10.1007/978-3-319-00560-7_21 © Springer International Publishing Switzerland 2013

approaches have been proposed to deal data collection in VSNs. In [1], an opportunistic dissemination is used to harvest meta-data. A cluster based approach is used in [6], and in [7] a data replication technique is used. These approaches have some constraints, as they limit their data harvesting schemes on limited area [1], use centralized access point [6] and a high-cost band for communication [2].

We propose a new scheme for disseminating and harvesting data in a VSN using multi-hop communication and data replication. To reduce overhead, our protocol takes advantage of the vehicle's motion by using a communication between vehicles.

The aims of our proposition is to satisfy the requesting vehicle (collect a requested data), and make this data available for other future requesting vehicles.

The rest of this paper is organized as follows. Section II presents the related works focused on data dissemination and data collection on VSNs. Section III describes our proposed protocol. The performance evaluation of this protocol is presented in Section IV. Finally, we conclude the paper and give some perspectives to our work in Section V.

2 Related Work

Several dissemination protocols were proposed for VANETs. They could be sorted into two classes: (i) protocols for infotainment services [1] that have constraints related to the bandwidth, and (ii) protocols for emergency services [8] that have end-to-end delay and delivery ratio constraints.

To collect sensed data, a number of models have been widely used in wireless sensor networks [9, 10]. However, these approaches are less efficient in VSNs because of i) the typical scale of a VSN, ii) the volume of generated data, and iii) vehicles mobility.

CarTel [6] is a mobile sensor computing system designed to collect, process, deliver, and visualize data from sensors located on vehicles. Each CarTel node uses a geographical dissemination to gather data; process sensor readings and delivering them to a central station (portal). CGP (Clustered Gatheatring Protocol) [2] is a cross-layered gathering, dissemination and aggregation protocol, based on a geographical clustering in a hybrid vehicular architecture (V2V and V2I). It collects information from nodes, aggregates them and sends them to a provider via high cost links. The MDHP (MobEyes Data Harvesting Protocol) [1] proposes a solution to disseminate and harvest data. Private vehicles (regular nodes) opportunistically and autonomously spread summaries of sensed data. A harvesting protocol is used by police agents (authority nodes) to build a distributed index of the mobile storage of sensed data. This index allows law enforcement agents to querying a huge database without centralization. MDHR (Multi-hop Data Harvesting with Replicas in Vehicular Sensor Networks) [7] presents a multi-hop geographic data dissemination method using data replication technique to collect sensor data in VSNs adopting a V2I communication. In MDHR, to achieve the desired data to the requested vehicle, the only way to communicate is between static nodes, so we can consider the environment used like a traditional WSN. In Road Probing [5], the Road Side Unit (RSU) initiates the probing process and selects the passing by vehicles as probes to collect traffic and environment information. The selected vehicles sense the desired data and forward it back to the RSU in a multi-hop fashion.

The aim of our proposal is to develop an efficient strategy to collect a required data. By efficient we mean that the process has to be able to collect data, with a low bandwidth communication cost. RDAP optimizes the bandwidth usage by minimizing the amount of vehicles that have to rebroadcast packets by using the same principle as DDT (Distance Defer Transfer Protocol) [11]. In our approach, we try to satisfy a requesting vehicle and offer a high accessibility to the requested data by using data replication. We take advantage of the vehicles' mobility, thus enable a collaborative harvesting by mobile and static sensor nodes.

3 RDAP: Requested Data Harvesting Protocol in VSNs

Currently, some critical issues in VSNs, such as data transmission delay, packets overhead, and network connectivity are not completely solved. The presented protocols have some restrictive assumptions, as they suppose a data collecting schemes to only single-hop or short distance transmissions [1], or use relay stations and cellular communications to support multi-hop collection [2, 6], which can be very costly and complex. However many of reported works are solutions to specific applications. The presented protocols use either a V2V communication [1] which can induce the broadcast storm problem, or a V2I communication [6, 7] which it's scalable but expensive.

RDAP is a new scheme for disseminating and collecting data in a VSN. It works on demand and stops when the data collection is achieved. Using multi-hop communication and data replication, the aim of **RDAP** is to satisfy vehicular node's request and also allow accessibility to the data for other potential requesting vehicular nodes.

3.1 Protocol Environment

We consider a hybrid vehicular sensor network using both V2V and V2I communication in an urban area. The ITS (Intelligent Transports Systems) [12] applications aimed by this proposed protocol are general information services that's provides conviviality like recognize the nearby road situation or traffic monitoring. On this environment, we assume that the IEEE 802.11p standard is used and the nodes are equipped with a GPS device.

3.2 Protocol Overview

In this section, we detailed the proposed protocol from the request transmission to the data reception. The scheme will be divided into four parts: (i) Data Requesting Phase, (ii) Data Delivery Phase (iii) Data Replication Phase, and (iv) Data Sharing Phase.

Each node (sensor nodes and vehicular nodes) creates and maintains a neighborhood list via beacon messages, and a received replicas table.

Data Requesting Phase

When a vehicle node needs to collect a data, it sends a request packet. This packet contains vehicle's information (identifier, velocity v (m/s) and coordinates).

Three different cases can be observed:

- The required data is available on a neighborhood vehicular node: Upon receiving a query, the vehicular node tries to perform it locally in its database. In case of success, the data packet will be automatically sent to the source node. In this phase, the query spreads only because of the vehicle mobility. This communication decreases the packet overhead since there is a probability that another vehicle has before made the same request. After a timeout, if there is no response, the request is sent to a sensor node.

- The Requested Data is available on a neighborhood static sensor node: In this case this sensor node sends straightforwardly the Data Packet to the vehicular node.

- The Requested Data is available on a non-neighborhood static sensor node: In this case, the Requested Data is sent to the farthest static node. This static node (RS: Requesting Sensor) adds its identifier and its coordinates to the packet, and forwards it. The RS initiates a timer "Replication Timer" RepT (seconds) which will be used later for supervising the Data Replication process. The value of RepT will be the estimated time to retrieve the Data and to deliver it to the RS node. The Requested Data will be forwarded, until it reaches the node holding the requested data.

In an urban environment, and during the request propagation, the request packet may probably cross an intersection. To deal with this case, in RDAP a set of nodes is defined as intersection nodes. When the packet gets to an intersection node, it will be broadcasted (flooding).

To improve RDAP, another step is added where each intersection node collects periodically data in its road segment. When a vehicular node request for the data, it can found it more easily, because it's probably already collected by at least one of the intersection node.

Data Delivery Phase

When the node holding the Requested Data receives the Requested Packet, it sends the Requested Data to the previous node on the Requested Packet path.

Data Replication Phase

The replication process depends on the requesting vehicle velocity $\boldsymbol{v}$, the size of the Data $\boldsymbol{p}$ and the sensors transmission rate $\boldsymbol{r}$. When the RS receives the Data, it broadcasts the packet to its neighbors, creating replicas of the Data. For restricting the flooding overhead, we use a Time-To-Live value TTL. Formula (1) gives the number of hops that the RV has moved while receiving the Data.

$$\boldsymbol{v \times RepT \div d} \tag{1}$$

Where $\boldsymbol{d}$ is the distance separating two sensor nodes.

While formula (2) gives the distance (by hops) the RV has traveled while the Data is transmitted to the vehicle.

$$v\times(p\div r)\div d \tag{2}$$

The two equations (1) and (2) will be bring together and rounded up to determine the appropriate TTL value.

$$TTL = [v \times RepT \div d + v \times (p \div r) \div d]$$

Depending on the TTL value, RS broadcasts the Data Packet to its neighbors and sends a unicast message to the farthest static node on the requester vehicular node's direction. This node will do as the RS until expiration of the TTL. When the Data reaches a sensor node close to the initiator vehicular node RV, it sends it to the RV.

Data Sharing Phase

In an urban environment there is a high probability that a number of vehicular nodes request for the same data. In this phase, we use the replicas created in the previous phase. After the replicas are made by a request from the first vehicular node; other following vehicular nodes can benefit from the data availability and access it directly.

4 Performances Evaluation

We evaluate RDAP performance through extensive simulations using ns-2 [13]. For realistic mobility generation, we use SUMO [14] and MOVE [15]. The simulations consider a vehicular network with a number of vehicles between 80 and 120 and 120 sensor nodes distributed uniformly on each side of the road in a 2100*2110 m2 area. The distance between each node is set to 200 m. The generation frequency of beacon messages is set to 10 messages /s. The average speed of vehicles v is between 0 m/s and 30 m/s and the size of the data p is set to 1000 bytes and these of the request and beacons is set to 200 bytes.

The data is supposed existed in one sensor node. The distance separating the vehicular node originator of the request and the sensor node holding the data is set between 45m and 1800m.

To evaluate RDAP performances, we mainly focused on (i) the number of hops it's take to collect the data, (ii) the Response Time and (iii) the Packets Overhead.

In figures 1, 2 and 3, the x-axis was configured to the number of vehicles requesting for the data, which varies from 1 to 6.

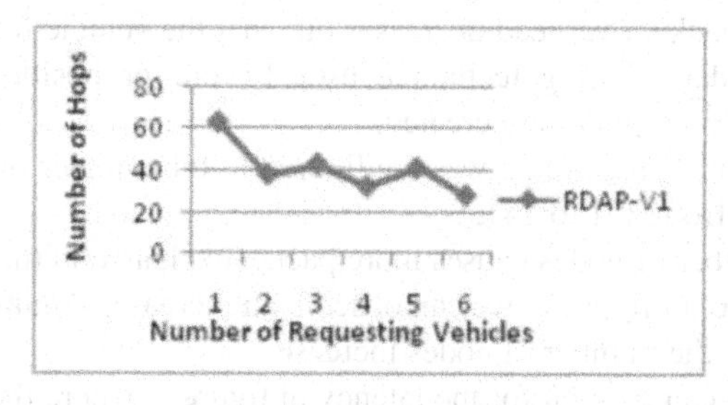

Fig. 1. Numbers of hops / Number of vehicles requesting for data

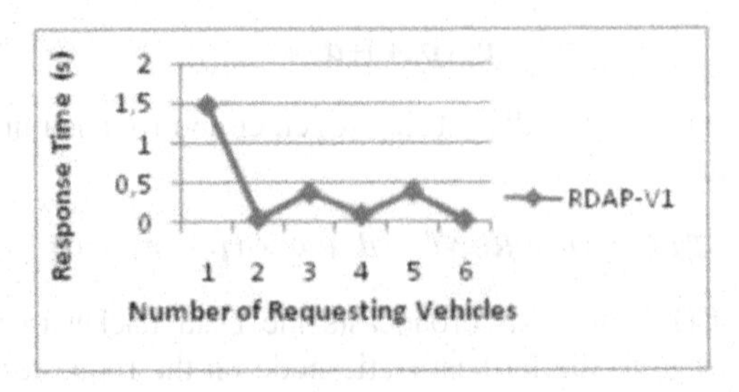

Fig. 2. Time of Response / Number of vehicles requesting for data

Figure 1 shows the number of hops it takes to deliver the requested data depending on the number of requesting vehicles.

It can be seen that in the proposed scheme more the number of vehicle requesting the data is important, more the number of hops of data delivery decreases. This is the benefit of data replication, where vehicles can access to the data more easily.

Figure 2 shows the response time to deliver the data from sending the first request packet to receiving the last data packet depending on the number of vehicles requesting the data. It can be seen that more the number of vehicle requesting the data is important, more the delivery time decrease. This is due that vehicles can reach the data more quickly, because of data replication process and communication between vehicles.

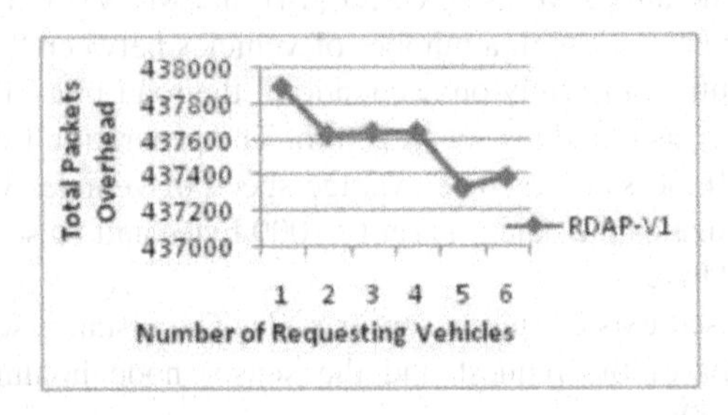

Fig. 3. Total Packet Overhead / Number of vehicles requesting for data

Figure 3 shows the total Packet overhead generated by our proposition. We can observe that more the number of vehicles requesting for the data increase, more the packet overhead is important due to the beaconing messages, till it reach the sixth vehicle, where the packet overhead decreases due that the vehicle found the data more quickly cause the data sharing technique used by our proposition reduces further packet overhead, as no replicas are created.

In figures 4, 5 and 6, the x-axis was configured to the number of nodes in the network, which varies from 201 to 241.

Increase the number of nodes causes more packet exchange in the network, causing collisions. Therefore, in figure 4, we can observe an increase of number of hops in the proposed scheme as the number of nodes increase.

The same results can be seen for the latency in figure 5, where the time of response increases as the number of nodes increase.

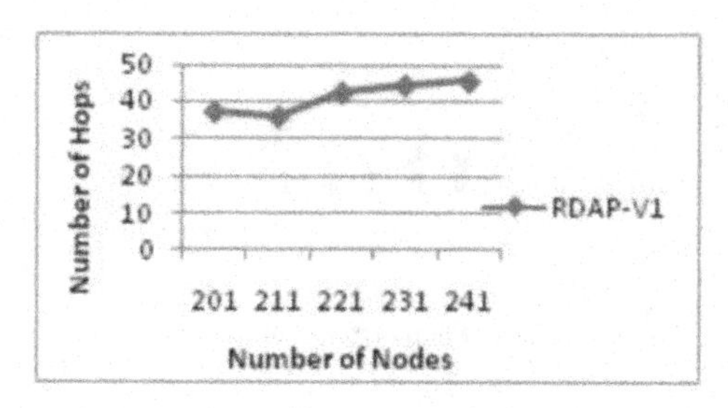

Fig. 4. Number of hops / Number of Nodes

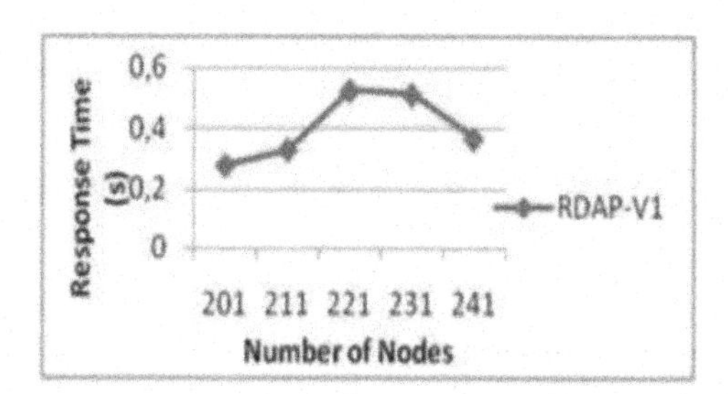

Fig. 5. Time of Response / Number of Nodes

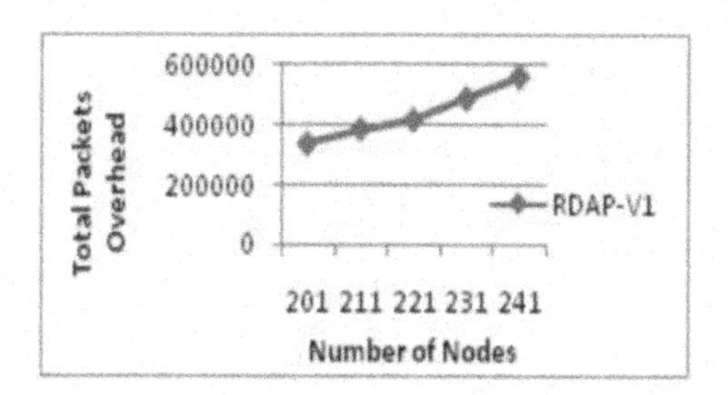

Fig. 6. Total Packet Overhead / Number of Nodes

Caused by the beaconing messages, the total Packet overhead generated by our proposition increase gradually as the number of nodes increase 6.

In figures 7, 8 and 9, we compare RDAP performance's with MDHR. The x-axis was configured to the number of requesting vehicles in the network, which varies from 1 to 6.

Figure 7 shows the comparison of the number of hops it takes from sending the request packet to receiving the data packet. It can be seen that the proposed scheme has advantages over the MDHR protocol. This is can be explained that the data collection strategy used in MDHR can collect the data where ever the data is, whereas, in MDHR, even if the data is on a neighborhood vehicular node, it's cannot be delivered in a single hop, and a new costly process of research and delivery of the data is initiated.

Figure 8 demonstrates the advantage of our solution over MDHR in term of time of response. These is because of the data replication mode used where the data is copied in a larger area (far static nodes are reach) compared to MDHR.

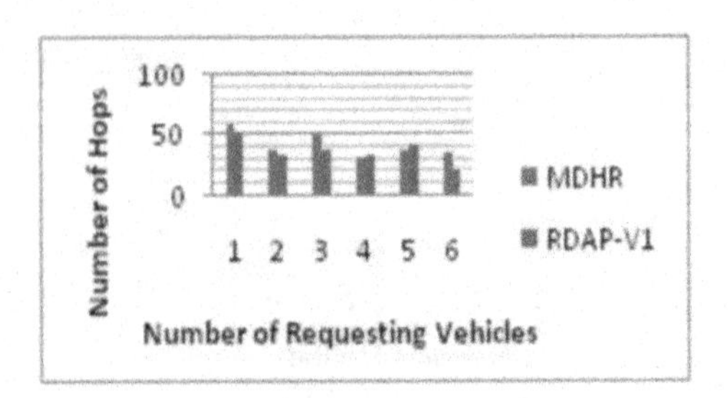

Fig. 7. Number of hops / Number of vehicles requesting data

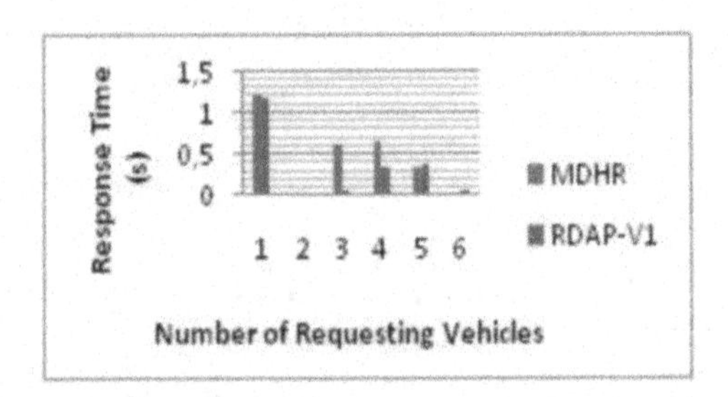

Fig. 8. Response Time / Number of vehicles requesting data

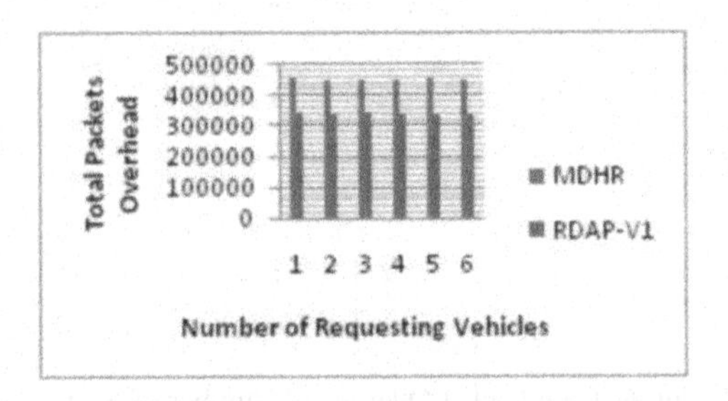

Fig. 9. Total Packets Overhead / Number of Nodes

In figure 9, the packets overhead is compared. We can see that RDAP outperform MDHR, this is due that in RDAP, the dissemination/ replication mode used minimize the collisions and congestions problems (Broadcast Strom Problem), whereas, the data replication mode used in MDHR is flooding, which can generate an important congestions and collisions.

The data collection technique used in MDHR can affect the collect time, packets overhead and also the efficiency (rate of successful requests) if the distance separating between the node holding the data and the RS node is important.

5 Conclusion

In this paper, a new data dissemination and gathering system RDAP based on geographical dissemination and data replication mechanism on vehicular sensor networks is proposed. Designed for hybrid VSN architecture, it allows communication between vehicles and between vehicles and road infrastructure. RDAP use a hybrid data collect, where a continuous data collection is used by the intersection nodes and a request driven collection is used by vehicular nodes. The aim of RDAP is to deliver a requested data and make this data accessible for other requesting vehicles. Simulation results of RDAP demonstrate the advantage of using the data replication process, the communication between vehicles and the dissemination strategy used to overcome collisions and overhead in the network.

References

1. Lee, U., Magistretti, E., Zhou, B., Gerla, M., Bellavista, P., Corradi, A.: Efficient Data Harvesting in Mobile Sensor Platforms. In: IEEE PerSeNS 2006 Workshop, Pisa, Italy (March 2006)
2. Salhi, I., Cherif, M.O., Senouci, S.M.: A New Architecture for Data Collection in Vehicular Networks. 978-1-4244-3435-0/09/$25.00 ©2009 IEEE (2009)
3. Ni, S.Y., Tseng, Y.C., Chen, Y.S., Sheu, J.: The Broadcast storm problem in a mobile ad hoc networks. In: Proc. 5thAnnual ACM/IEEE International Conference on Mobile Computing and Networking, Seattle, Washington (August 1999)
4. Kheroua, L., Moussaoui, S., Mansour, L.: An Agent based Rumor Dissemination for Routing in Wireless Sensor Networks. In: ISPS, Algeria (2011)
5. Yang, L., Xu, J., Wu, G., Guo, J.: Road Probing: RSU Assisted Data Collection in Vehicular Networks. IEEE (2009)
6. Hull, B., Bychkovsky, V., Zhang, Y., Chen, K., Goraczko, M., Miu, A., Shih, E., Balakrishnan, H., Madden, S.: CarTel: A Distributed Mobile Sensor Computing System (2006)
7. Lim, K.W., Jung, W.S., Ko, Y.-B.: Multi-Hop Data Dissemination with Replicas in Vehicular Sensor Networks. In: Proceeding Vehicular Technology Conference, VTC Spring 2008, Singapore, May 11-14, pp. 3062–3066. IEEE (2008) ISSN: 1550-2252, ISBN: 978-1-4244-1644-8
8. Seth, A., Darragh, P., Liang, S., Lin, Y., Keshav, S.: An Architecture for Tetherless Communication. In: DTN Workshop (2005)
9. Bononi, L., Di Felice, M.: A Cross Layered MAC and Clustering Scheme for Efficient Broadcast in VANETs. In: IEEE MASS 2007, Pisa, Italy (October 2007)
10. Wang, F., Zeng, D., Yang, L.: Smart cars on smart roads: an IEEE intelligent transportation systems society update. IEEE Pervasive Computing 5(4) (2006)
11. Piran, M., Murthy, G.: A Novel Routing Algorithm for Vehicular Sensor Networks. Internation Journal of Wireless Sensor Networks (WSN), 919–923 (2010)
12. Wang, F., Zeng, D., Yang, L.: Smart cars on smart roads: an IEEE intelligent transportation systems society update. IEEE Pervasive Computing 5(4) (2006)
13. ns-2 (The Network Simulator), `http://www.isi.edu/nsnam/ns`
14. Krajzewicz, D., Rossel, C.: Simulation of Urban Mobility (SUMO). German Aerospace Centre (2007), `http://sumo.sourceforge.net/index.shtml`.
15. MOVE (MObility model generator for VEhicular networks): Rapid Generation of Realistic Simulation for VANET (2007), `http://lens1.csie.ncku.edu.tw/MOVE/index.htm`.

Improvement of LEACH for Fault-Tolerance in Sensor Networks

Mohamed Lehsaini[1] and Herve Guyennet[2]

[1] STIC Laboratory, Faculty of Technology, University of Tlemcen, 13000Tlemcen, Algeria
m_lehsaini@mail.univ-tlemcen.dz
[2] LIFC Laboratory, University of Franche-Comte, 25030 Besancon, France
herve.guyennet@femto-st.fr

Abstract. In wireless sensor networks, failures occur due to energy depletion, environmental hazards, hardware failure, communication link errors, etc. These failures could prevent them to accomplish their tasks. Moreover, most routing protocols are designed for ideal environment such as LEACH. Hence, if nodes fail the performance of these protocols degrade. In this context, we propose two improved versions of LEACH so that it becomes a fault-tolerant protocol. In the first version, we propose a clustered architecture for LEACH in which there are two cluster-heads in each cluster: one is primary (CHp) and the other is secondary (CHs). In the second version, we propose to use the checkpoint technique. Finally, we conducted several simulations to illustrate the performance of our contribution and compared obtained results to LEACH protocol in a realistic environment.

Keywords: LEACH, FT1-LEACH, FT2-LEACH, Fault-tolerance, Checkpoint, WSN.

1 Introduction

Wireless sensor networks (WSN) consist of a large number of low-cost and low-powered sensor devices, communicating with each other through wireless links and collaborating to accomplish a common task. Sensors can be deployed over a geographical area for monitoring physical phenomena like temperature, humidity, vibrations, seismic events, and so on [1]. Now, WSN are permeating a variety of application domains such as avionics, environmental monitoring, structural sensing, telemedicine, space exploration, and command and control.

WSN should have a long lifetime to accomplish the application requirements. However, In addition to resource constraints in WSN, the failure of sensor nodes is almost unavoidable due to energy depletion since they have been usually deployed in hostile environments and their batteries cannot be recharged or replaced, hardware failure, communication link errors, and so on [2,3,4]. Therefore, in WSN, fault tolerance has become as important as other performance metrics such as energy efficiency, latency and accuracy.

A. Amine et al. (Eds.): *Modeling Approaches and Algorithms*, SCI 488, pp. 175–183.
DOI: 10.1007/978-3-319-00560-7_22 © Springer International Publishing Switzerland 2013

In general, the consequence of these failures is that a node becomes unreachable, violates certain conditions that are essential for providing a service or returns false readings which could cause a disaster especially in critical applications. Furthermore, the above fault scenarios are worsened by the multihop communication nature of WSN. It often takes several hops to deliver data from a source node to the remote base station; therefore, failure of a single node or link may lead to missing reports from the entire region of WSN.

Therefore, since sensors are prone to failure, fault tolerance should be seriously considered in many sensing applications which are generally required to be fault-tolerant, where any pair of sensors is usually connected by multiple communication paths. Recently, several studies have dealt with fault tolerance in WSN, particularly in the routing process. Moreover, these works focus on the detection and recovery of failures in WSN.

We evaluate LEACH in a realistic environment in which sensor nodes can fail and links may be lost. Then, we propose two improved versions FT1-LEACH and FT2-LEACH of LEACH so that it becomes a fault tolerant protocol. FT1-LEACH involves two cluster-heads in each cluster: one is primary and the other secondary and FT2-LEACH use the checkpoint technique. Moreover, cluster-heads are elected based on their capabilities and in clusters, main cluster-heads and their vice cooperate with each other to reduce extra costs by sending only one copy of sensed data to the sink.

FT1-LEACH and FT2-LEACH could tolerate the failures of links and therefore guarantee routing reliability in WSN while dissipating less extra energy and time. Finally, we conducted several simulations to demonstrate the performance of our contribution and we compared obtained results with those of LEACH [5] in a realist environment.

The rest of this paper is organized as follows: Section 2 presents briefly the protocol LEACH; in Section 3, we propose two improved versions of LEACH; Section 4 illustrates performance analysis of LEACH and the proposed schemes in a realistic environment. Finally, we conclude our paper and discuss future research work in Section 5.

2 Presentation of LEACH

LEACH (Low Energy Adaptive Clustering Hierarchy) is a hierarchical cluster-based routing protocol for wireless sensor networks which partitions the nodes into clusters. In each cluster a dedicated node called cluster-head (CH) and other nodes are cluster members. CH is responsible for creating and manipulating a TDMA schedule and sending aggregated data from nodes to a remote base station using CDMA technique. Moreover, this protocol is divided into rounds and each round consists of two phases:

2.1 Setup Phase

During this phase, cluster formation is carried out. In which, each sensor node decides independently of other nodes if it will become a CH or not. This decision takes into account when the node served as a CH for the last time i.e. the node that has not been a CH for long time is more likely to elect itself as a CH.

Once CHs are elected, they inform their neighborhood with an advertisement packet that they become CHs. Each non-CH node picks the advertisement packet with the strongest received signal strength and it sends the message "Join Packet" to request to join its corresponding CH.

After this process, the CH knows the number of member nodes and their IDs. Based on all messages received within the cluster, the CH creates a TDMA schedule and broadcasts it to its cluster members. Then, it picks a CDMA code randomly to avoid interference when transmitting data to the base station.

2.2 Steady-State Phase

During this phase, data transmission begins. Sensor nodes send their data collected during their allocated TDMA slot to their respective CHs. The radio of each non-CH node can be turned off until the nodes allocated TDMA slot, thus minimizing energy dissipation in these nodes. When all the data has been received, the corresponding CH aggregates these data and sends them to the remote base station as presented by Fig. 1.

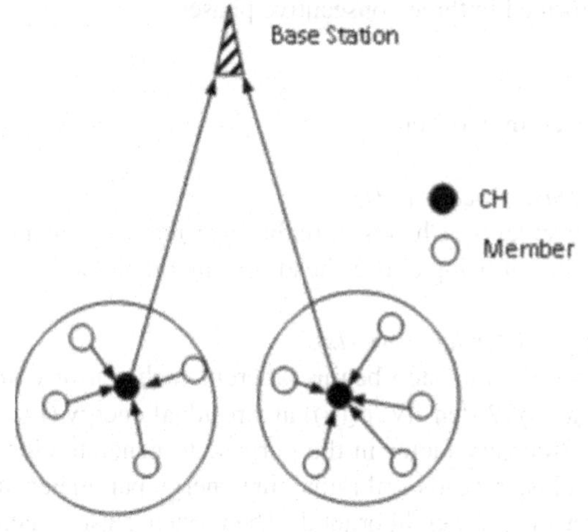

Fig. 1. Cluster formation in LEACH

LEACH is able to perform local aggregation of data in each cluster to reduce the amount of data transmitted to the base station. Although LEACH protocol acts in a good manner, it suffers from many drawbacks such like;

- Cluster-heads are randomly selected without take into account their ability to play this role and CHs can quickly deplete their batteries and hence they stop working and cause holes in the target area.
- It cannot cover a large area when some sensor nodes fail.

Since LEACH has many drawbacks, many works have been done to make this protocol performs better but to the best our knowledge there is no work dealing with fault-tolerance in LEACH.

3 Contribution

In most routing protocols, fault-tolerance was not considered particularly in LEACH. In this context, we propose two improved versions of LEACH protocol so that it becomes a fault-tolerant protocol. In the first version called FT1-LEACH, we propose a clustered architecture for LEACH in which there are two cluster-heads in each cluster: one is primary (CH_p) and the other is secondary (CH_s). In FT1-LEACH, we used the same process for the formation of clusters as in LEACH and we tried to introduce minor changes so that LEACH becomes a fault-tolerant protocol. In the second version, we propose to use the checkpoint technique to make LEACH as a fault-tolerant protocol.

3.1 Contribution 1: FT1-LEACH

FT1-LEACH is performed in three consecutive phases:

Cluster formation
This phase is performed in two steps:

Election of primary cluster-heads (CHp).
The election of primary cluster-heads is realized in the same manner as in LEACH based on the probability of being cluster-heads during this period.

Election of secondary cluster-heads (CHs).
After the election of CH_p, this step begins wherein each sensor computes its weight which is a combination of 2-density (ρ_2(u)) and residual energy (E(u)) as presented in Eq. (1) We involve 2-density factor in the purpose to generate clusters whose members are linked with cluster-heads and remaining energy parameter to select the nodes with more energy in their 2-neighborhood. Then, each cluster member generates a 'Hello' message including two extra fields addition to other regular contents: Weight and CH_p and broadcasts it as well as it eavesdrops its neighbor's 'Hello' message. The node with the largest weight in each cluster is elected as secondary cluster-head (CH_s).

$$Weight(u) = \alpha * \rho_2(u) + \beta * E(u) \quad \wedge \quad \alpha + \hat{a} = 1 \tag{1}$$

The values of α and β are chosen depending on the application. For example if we want to favor the node that has more energy as cluster-head we would attribute great value to β.

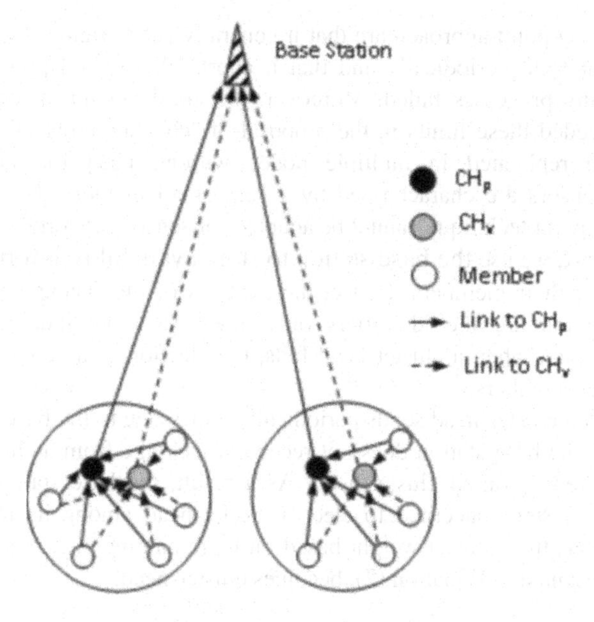

Fig. 2. Cluster formation in FT1-LEACH

Routing Paths

When a node detects a relevant event, it sends an alert message to its corresponding cluster-heads (CH_p and CH_s). If the primary cluster-head does not send a message to the remote base station within a threshold time interval, the secondary cluster-head considers it that is down and it sends data to the base station.

There are two kinds of routing:

Intra-cluster

Cluster members do not transmit data collected directly to the base station but they send them to their respective cluster-heads (CH_p and CH_s).

CH-to-BS

In each cluster, primary cluster-head is responsible to transmit the aggregated data to the remote base station and if they do not perform this task within a threshold time interval the secondary cluster-head considers that the primary cluster-head is down and it sends collected data to the base station.

3.2 Contribution 2: FT2-LEACH

Our second contribution is based on the checkpoint technique which is considered the most typical approach to tolerate failures in parallel and distributed systems. By writing checkpoints into stable storage periodically, the checkpoint approach is able to tolerate the failure of the whole network. The advantage of this approach is that it is very general and is able to tolerate the failures of the whole network. However, the limitations of checkpoint approach are that it generally needs stable storage to save a global consistent state periodically and that it aborts all survival processes even if only one of many processes failed. Moreover, the latest version of checkpoint approach has exceeded these limits in the proposal of "checkpointing Diskless" where checkpoints are replicated in multiple nodes without need for stable storage. However, the sensors are characterized by a very small memory. Hence this latest version of checkpoint technique cannot be adapted for sensor networks.

In this context, we use the base station to store availability information about cluster-heads and their members. Hence, the contents of the checkpoint represented by the base station are the identifiers of cluster-heads and their corresponding members. Therefore, when a cluster-head fails, the election of a new cluster-head is done among their members.

Moreover, each cluster-head sends periodically a message to the base station and if during a period, the base station does not receive a message from a cluster-head, the latter is considered as failed cluster-head. As a result, the base station transmits a message to the cluster concerned to elect a cluster-head among its members. The member which has the greatest weight based on its remaining energy and its density parameter as presented in Equation (2), becomes cluster-head.

$$Weight(s) = \frac{1}{2}\,Energy(s) + \frac{1}{2}\,density(s) \quad (2)$$

4 Evaluation and Simulation Results

In our experiments, we conducted extensive simulations to evaluate FT1-LEACH and FT2-LEACH performance and compare them with LEACH in terms respectively of the ratio of successful reception at the base station during the network's lifetime and energy consumption. To achieve these goals, the simulations have been performed in NS-2[6] using the MIT_uAMPS [7]. We have carried out these simulations with the same scenario used to evaluate the performance of LEACH in order to illustrate the performance of our contribution. Hence, we considered a network topology with 100 non-mobile sensor nodes with a sensing range of 25 meters. Sensor nodes are placed randomly in a 100 m×100 m square area by using an uniform distribution function, and the remote base station is located at position x = 50, y = 125, i.e. the base station was placed 75 meters outside the area where the sensor nodes were deployed. At the beginning of the simulation, all the sensor nodes had an equal amount of energy i.e. the sensor nodes started with 2 Joules of energy. We note that system lifetime is defined as the time when last sensor dies in the sensor network.

The simulations were performed until all the sensors in the network consumed their energy and the average values were calculated after each round whose duration is 20 seconds. This duration represents the cluster timeout. It is used to prolong network lifetime and balance energy deviation among all its sensors. On expiry of this period, FT1-LEACH and FT2-LEACH triggered the cluster-head's election process again. Moreover, we used the same energy parameters and radio model as discussed in [8], wherein energy consumption is mainly divided into two parts: receiving and transmitting messages. The transmission energy consumption requires additional energy to amplify the signal according to the distance from the destination. Thus, to transmit a k-bit message to a distance d, the radio expends energy (E_{Tx}) as described by the formula (3), where ε_{elec} is the energy consumed for radio electronics, $\varepsilon_{friss\text{-}amp}$ and $\varepsilon_{two\text{-}ray\text{-}amp}$ for an amplifier. We distinguish two kinds of energy consumption during transmission according to the distance between the cluster-heads and the base station.The reception energy consumption is $E_{Rx} = \varepsilon_{elec} \times k$.

$$\boldsymbol{E_{Tx}} = \begin{cases} \mathring{a}_{elec} * k + \mathring{a}_{friss-amp} * k * d^2 & if\ d < d_{Crossover} \\ \\ \mathring{a}_{elec} * k + \mathring{a}_{two-ray-amp} * k * d^4 & if\ d \geq d_{Crossover} \end{cases} \tag{3}$$

Simulated model parameters are set as shown in Table 2. The data sizes were 500 bytes/message plus a header of 25 bytes. The message size to be transmitted was:

$$k = (500\ bytes + 25\ bytes) \times 8 = 4\ 200\ bits$$

Table 1. Model Parameters

Parameter	Value
Network grid	(0,0) x (100,100)
Base Station	(50,125)
ε_{elec}	50 nJ/bit
$\varepsilon_{friss\text{-}amp}$	10 pJ/bit
$\varepsilon_{two\text{-}ray\text{-}amp}$	0.0013 pJ/bit
$d_{Crossover}$	87 m
Data packet size	500 bytes
Packet Header size	25 bytes
Intial energy per node	2 J
Number of nodes (N)	100

Fig.3 shows that the ratio of successfully received packets to the base station is higher than in LEACH because in LEACH, if a cluster-head stops working information will not be forwarded to the base station while in FT1-LEACH if the main cluster-head is down its vice transmits the information to the base station. However, in FT2-LEACH, a cluster-head may fail before the end of the alert period of the base station.

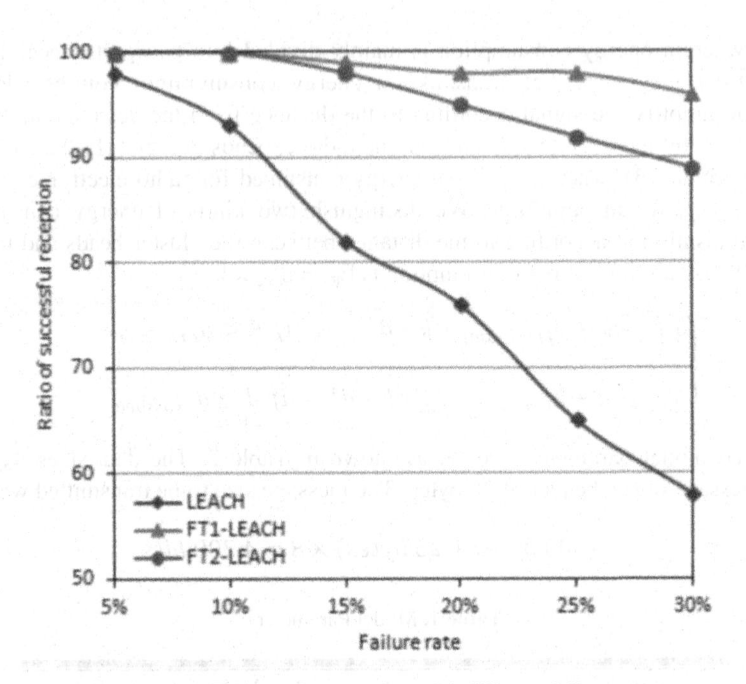

Fig. 3. Ratio of successful reception with different failure rates

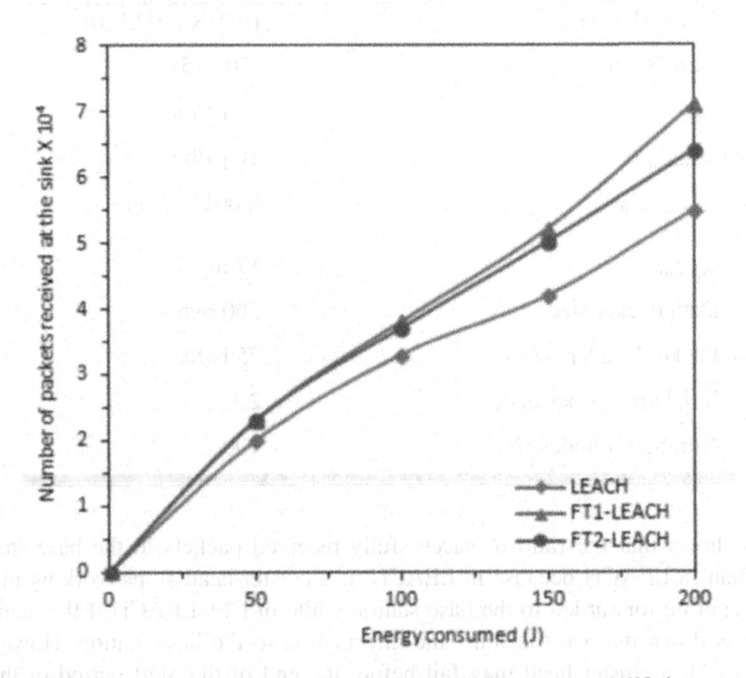

Fig. 4. Amount of packets received at the sink during network lifetime

Fig.4 illustrates that in FT1-LEACH and FT2-LEACH, the amount of packets received at the base station during network lifetime is higher than in LEACH because in LEACH, if a cluster-head stops working information will not be forwarded to the base station and hence the energy will be lost without sending information caused by the redundant transmissions. Furthermore, in FT1-LEACH and FT2-LEACH if the main cluster-head is down its vice transmits the information to the base station and hence the energy consumed reflects a transmission.

5 Conclusion

The evaluation of LEACH in a realistic environment showed that LEACH loses performance when some nodes fail. In this context, we have proposed two improved versions of LEACH so it becomes a fault-tolerant protocol.

Simulation results showed that our contributions have improved the performance of LEACH in terms of number of packets successfully received at the base station and the energy to send these packets to the base station.

References

1. Akyildiz, I.F., Su, W., Sankarasubramaniam, Y., Cayirci, E.: Wireless sensor networks: a survey. Computer Networks 38(4), 393–422 (2002)
2. Woo, A., Tong, T., Culler, D.: Taming the underlying challenges of reliable multihop routing in sensor networks. In: Proceedings of the 1st International Conference on Embedded Networked Sensor Systems, pp. 14–27 (November 2003)
3. Paradis, L., Han, Q.: A Survey of Fault Management inWireless Sensor Networks. Journal of Network and Systems Managemen 15(2), 171–190 (2007)
4. Ammari, H.M., Das, S.K.: Fault tolerance measures for large-scale wireless sensor networks. Journal ACM Transactions on Autonomous and Adaptive Systems (TAAS) 4(1), 1–28 (2009)
5. Heinzelman, et al.: Energy-efficient communication protocol for wireless microsensor networks. In: IEEE Proceedings of 33rd Annual Hawaii International Conference on System Sciences (HICSS 2000), Maui, Hawaii, USA, vol. 2 (January 2000)
6. NS-2, Network Simulator, World Wide Web, http://www.isi.edu/nsnam/ns/ns-build.htmlMIT_uAMPSLEACHNS2Extensions
7. http://www.ece.rochester.edu/research/wcng/code/index.html
8. Heinzelman, et al.: An application-specific protocol architecture for wireless microsensor networks. IEEE Transactions Wireless Communications 1(4), 660–670 (2002)

Locally Distributed Handover Decision Making for Seamless Connectivity in Multihomed Moving Networks

Zohra Slimane, Abdelhafid Abdelmalek, and Mohammed Feham

STIC Laboratory, University of Tlemcen, Algeria
{z_slimani,a_abdelmalek,m_feham}@mail.univ-tlemcen.dz

Abstract. Providing seamless connectivity emerges nowadays as a key topic in internet mobility management. NEMO BS Protocol was recently proposed by IETF for managing mobility of moving networks. The protocol is based on hard handover (Brake-Before-Make). The results of NEMO handover analyzes in literature prove that, the support of mobility such as it is defined, does not make it possible to ensure seamless connectivity. To overcome limitations of NEMO BS protocol, several optimizations were proposed. The solutions suggested are based on the optimization of each component of the handover using cross-layer design, network assistance and multihoming. Unfortunately, these optimizations remain insufficient to meet the needs for the applications to critical performance such as handover delay and packet loss. This paper proposes an intelligent support to manage mobility in a locally distributed manner for a multihomed moving network. The task is distributed between mobile routers who cooperatively perform the decision making and execution of the handover based on information gathered by MIH services. NS2 Simulations experiments were investigated to validate the proposed model. The results show that the solution provides excellent performance meeting the needs of time-sensitive applications.

1 Introduction

The management of the mobility of an entire mobile network embarking several equipments able to communicate with the Internet aroused a great interest for the researchers and the industrialists who wish to deploy such networks in the public transport and moving systems. The problem can be solved by using MIPv6 [5] for example or one of its improvements, this would require on the one hand that each device be MIPv6 capable, and on the other hand that each embarked device carries out the functionalities of the MIPv6 protocol leading thus to an overload of traffic. Another solution NEMO BASIC Support [6] based on the extension of MIPv6 was proposed recently by IETF working group NEMO (Network Mobility). This basic solution does not take unfortunately charges all the functionalities and several issues such as Routing Optimization (RO), Multihoming, Handover performance and security were raised by the IETF.

In particular, the optimization of the handover, as well in terms of protocols as of decisional algorithms remains up to now an open issue. In the basic specification of

A. Amine et al. (Eds.): *Modeling Approaches and Algorithms*, SCI 488, pp. 185–194.
DOI: 10.1007/978-3-319-00560-7_23 © Springer International Publishing Switzerland 2013

protocol NEMO, the mobile router can use only one IPv6 address to attach the access networks. The problem is that during a change of network, the mobile router must carry out a hard handover so that a loss of connectivity is inevitable. This kind of problem is not at all acceptable in the case of time-constraint applications. All research tasks aiming at improving the performances of handover NEMO BS [2,3], [7-11] remain ineffective to meet the needs of these services, so that further optimizations are needed.

The objective of this paper is to suggest a new mobility support (NEMO BS protocol extension) for the context of multiple mobile routers multihomed NEMO network [1]. Such context means that NEMO network has several mobile routers (MR) of which each one presents a single external interface and admits a Home Agent (HA), but only one MNP prefix is used. Our aim is to develop an infrastructure independent solution. For that, we move towards an intelligence located in the centre of NEMO network itself. To take the totality of mobility decisions and to avoid single point of failure, we distribute the task of handover decision making and execution among all mobiles routers in the moving network. The mobile routers (MRs) are connected to each other and to MNNs via switchs. In the proposed solution, we will adopt a proactive approach which consists at executing soft handovers (Make Before Brake) conducting to the availability of several tunnels simultaneously. Thus, the breakdown of an active link for a given MR causes only a switching in the traffic via another MR having a previously established tunnel. We will use the standard IEEE 802.21 [4] especially remote MIH services to manage the mobility.

The remainder of the paper is organized as follows. In section 2, we give a brief description of our proposed mobility support. Section 3 is devoted to the specification of our protocol. In section 4, simulation results are presented. Finally, section 5 concludes the paper.

2 Locally Distributed Mobility Management Support (LDMMS)

We consider a multihomed moving network (NEMO network) which changes its point of attachment to internet via multiple mobile routers (MRs). The MRs are domiciled to different Home Agents (HAs), but only one prefix is delegated. NEMO network accepts a primary MR mobile router and one or several secondary MRs. The HA of the Primary MR is called the Primary HA.

The MNP prefix delegated to the primary MR is announced to MNNs nodes inside the NEMO network. The main idea behind our proposal is, if possible, to establish tunnels in advance between each secondary MR and the Primary HA, and save them for use when the primary MR (or a secondary MR) with an operational open tunnel becomes out of service or loses connectivity to the access network (Fig. 1). So, once a failure within the actual used path to internet is detected, one of the registered MRs possessing a tunnel to the primary Home Agent should replace the MR which link with the access network is broken down (the MR for the replacement is called substituent).

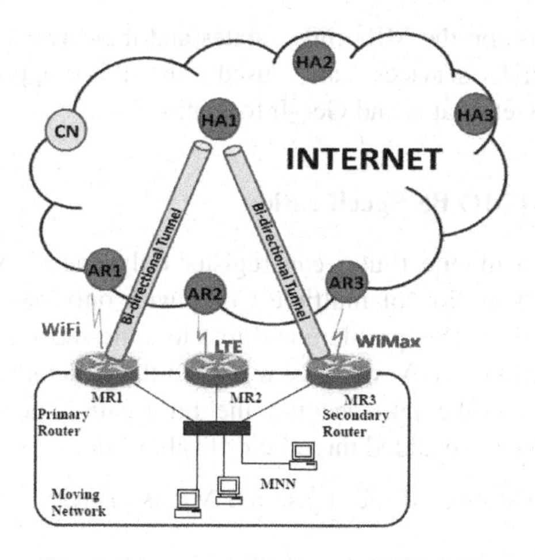

Fig. 1. Diagram architecture of LDMMS

2.1 Architecture of the MR Mobility Stack

Each MR could choose and set out the MR router substituent on the base of the data it keeps concerning all the MRs. But, at a given moment of the system operation only one MR can take this decision. We assign this task to MR having an actual open tunnel with the primary HA (we call this MR "Active MR"). For this purpose, we plan a module to be implemented at each MR level which we call DMMS (Distributed Mobility Management Support). DMMS collects information about local links and others MRs' links (Fig.2).

The MIH services of IEEE 802.21standard allow us to assist the NEMO in its tasks such as environment detection. We will use the local and remote MIES and MICS services at the level of each MR. The local services allow detecting the changes in MRs links' states, to activate the triggers associated. The remote MIES services allow MRs the notification of MR about the events associated to links occurring at their level; in the other hand the "Active MR" may use the remote MICS services to obtain

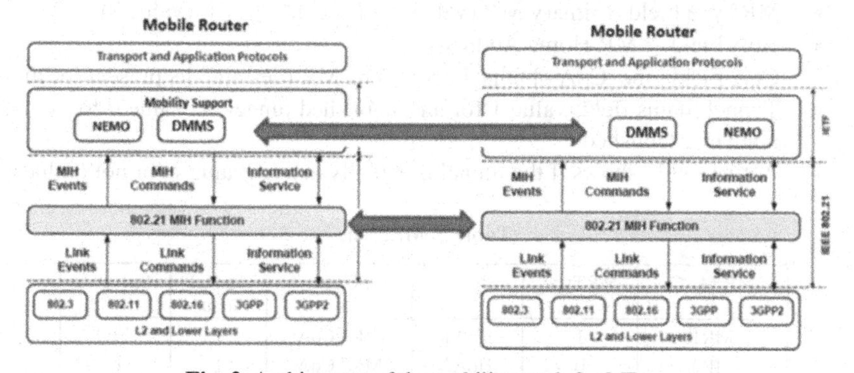

Fig. 2. Architecture of the mobility stack for MRs

the needed information on the MRs links' states and their quality and characteristics. The following MIH services are used in this approach: Link_Down, Link_Going_Down, Get_Status and Get_Information.

2.2 Changes in NEMO BS Specification

In NEMO BS [14], a mobile router can register only one CoA address at its HA. Consequently, the registration of multiple CoAs with only one HoA address is not possible. In order to settle this problem and to allow the secondary mobile routers to be registered at the primary HA, we have modified the structure of the binding cache of the primary HA to take into account the information on the possible mobile substituent routers. We have added three fields (Table 1):

- Type: This indicates if the registered MR is primary (value 1) or secondary (value 0).
- Tunnel: This indicates if this tunnel is open (value 1) or closed (value 0) to the data traffic. Once established, the tunnel is by default open to the signalization traffic and closed to the data traffic.

Table 1. New Binding Cache for the CoAs multiple registration support

Prefix	CoA	Type	Tunnel
MNP	MR1 CoA	1	1
MNP	MR2 CoA	0	0
---	---	---	---

2.3 Definition of a New Cache at MR Level

We have defined in the module DMMS a cache named MRcache (Table 2) to maintain the necessary information concerning all MRs in NEMO network. Each cache entry contains the following fields:

- MR MAC field: the ID of the MR (for example, the physical address of the ingress interface)
- MRType Field: Primary MR (value 1) or secondary MR (value 0)
- HoA Field : MR Home Address
- CoA Field : the CoA obtained address by MR (defining its present location)
- Tunnel Status field: value 1 for an established tunnel or value 0 for a tunnel non yet established
- Active field : shows if the tunnel is actually active (value 1) or not (value 0)

Table 2. MRcache

MR MAC	MRType	HoA	CoA	Tunnel Status	Active
MR1	1	HoA1	MR1 CoA	1	1
MR2	0	HoA2	MR2 CoA	1	0
MR3	0	HoA3	MR3 CoA	0	0

2.4 Link Break Down Detection

We use the approach based on MIH_Link_Going_Down service for the breakdown detection of MR links presenting open tunnels with the primary HA. In this approach, in order to activate the tunnel of a MR substituent before the activation of LD trigger, we use the Link_Going_Down (trigger LGD) event allowing the prediction of the links break down. The LGD trigger will be used with a preset threshold:

$$P_{rx} < \alpha_{LGD} P_{th}$$

where:

P_{rx}: is the received signal strength (RSS)
P_{th}: is the threshold strength below which the link is considered broken down
α_{LGD}: is the LGD coefficient slightly superior to1 (typical values: 1.05, 1.10, ...)

So, if a LGD event is generated by the external interface of a MR, the MIHF of this MR sends a remote MIH_Link_Going_Down to the MIHF of the Active MR. This latter notifies immediately the DMMS module.

3 Handover Operations

The handover consists of two phases: (i) Registration of a substituent MR and (ii) Tunnel switching.

3.1 Phase 1: Registration of a Substituent MR

Let's consider (Fig.1) where Home network is the WiFi network and MR1 is the primary router. Actually, MR1 is the "Active MR". Assuming the secondary mobile router MR2 got a temporary CoA address on the LTE access network and established a tunnel with its HA2 Home Agent. The mobile router MR2 then notifies the "Active MR" (MR1) of this tunnel establishment by using a New_tunnel_Notification message, which is sent by DMMS module at MR2-level to the DMMS module at MR1-level. This latter updates MR2's entry of its MRcache. Then, it sends to its Home Agent HA1 a Sub_Reg_Request message to carry out the registration of MR2 as a substituent of MR1. Upon receiving this message, the HA1 sends to MR2 a message of invitation to registration (Sub_Reg_Invite). This message arrives at MR2 via its Home Agent HA2. MR2 answers HA1 with a SBU message (Substituent Binding Update). HA1 registers MR2 as a substituent for MR1 in its binding cache and sends to MR2 a message SBA (Substituent Binding Acknowledgement). In parallel, HA1 sends a Substitute_Reg_Notification message to MR1 to inform it that MR2 was registered as a substituent. The tunnels established with HA1 are either on open state or a closed state according to whether the tunnel is currently used to convey the traffic or not. The new tunnel established between MR2 and HA1 is initially put on the closed state, expecting an order of switching to make it swing to the open state.

3.2 Phase 2: Tunnel Switching

Let's consider again (Fig.1). Assuming that NEMO network movement makes MR1 at the point of losing the connectivity with its actual access network, when the MR1 detects this event by means of DMMS module (Link_Going_Down Event), it looks immediately in its MRcache for the available substituent MRs (Tunnel Status = 1, Active = 0), it selects one following a predefined policy (for example according to bandwith) and sends it a Tunnel_Activation_request message asking it to request the HA1 (the primary HA) the opening of the associated tunnel. Assuming that MR1 choses MR2 to replace it. MR1 sends, therefore, the message Tunnel_Activation-Request to MR2, this latter sends in its turn a Tunnel-Opening_Request message to HA1 indicating the parameters and the reason of this request (tunnel MR1_HA1 inaccessible). At the reception and the validation of this message : (i) HA1 opens the tunnel HA-MR2 in both direction by which the traffic coming from or going to NEMO network is immediately transmitted, (ii) HA1 sends to MR2 a Tunnel-Opening_Replay message. Once MR2 received a Tunnel_Opening-Replay message (Successful Operation), it replies to MR1 with a Tunnel_Activation-Replay message.

The HA1 will use simultaneously the tunnels HA1-MR1 and HA1-MR2 until the confirmation of the HA1-MR1 tunnel breakdown. HA1 will employ a special process: the tunnel survivability test; it sends to MR1 a message Tunnel_Alive_Request. If the MR1 has not lost its link with the access network, it replies by the message Tunnel_Alive_Replay.

4 Simulation Experiments

The implementation is done under NS-2.29 version [15] including the MIH package of NIST [16]. In one hand, we created a C++ DMMS_Agent inheriting from the Agent class of NS-2 [14], to assume the mobility management at each MR level.

In addition, we have brought modification on the NEMO agent implemented at the MR and HA levels to assume the registration mechanism of multiple CoAs of secondary routers and tunnel switching.

The topology of the simulated network is presented in Fig.3, where a hierarchical addressing [14] is adopted:

- Node 0: a router (0.0.0) presenting four wired interfaces
- Node 1: a corresponding node CN (3.0.0).
- Node 2: the HA1 Home Agent (4.0.0) of the primary router MR1.
- Node 3: a base station IEEE 802.11 (access router AR1 (1.0.0)) with a coverage of 100m.
- Node 4: a base station IEEE 802.16 (access router AR2 (2.0.0)) with a coverage of 1000m.
- Node 5: the primary mobile router MR1 (4.1.0) presenting two interfaces: an external interface (802.11) and an internal interface (802.3) linked to MNN.

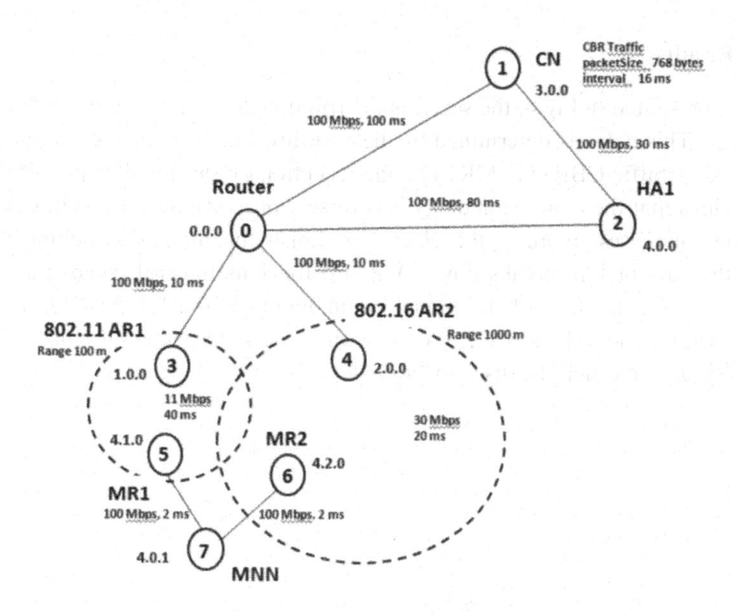

Fig. 3. Topology of simulated NEMO network

- Node 6: a secondary mobile router MR2 (4.2.0) presenting two interfaces: an external interface (802.16) and an internal interface (802.3) linked to MNN.
- Node 7: MNN (4.0.1)

The characteristics of the links and those of the bandwidth as well and the delay are referred in Fig.3. A streaming traffic CBR is sent by the CN towards a node MNN of the NEMO network (PacketSize=768 bytes, Interval=16 ms). The simulation time is fixed at 20s. The NS-2 simulator doesn't withstand the mobility of a full network. For this reason we have emulated the movement of NEMO network by reducing the emission power of the base station IEEE 802.11 (access router AR1 (1.0.0)). Following the importance of the caused power, we activate either LGD trigger or LD trigger.

We consider the following simulation plan: at the beginning of the simulation (from t=0 to t=5s), NEMO network has a single link with the internet network, the link with the base station 802.11(AR1), the base station 802.16 being deactivated. Consequently, a single tunnel is available, the one established between MR1 (the primary router) and its HA. At t= 5s, the base station 802.16 is active, a second tunnel is established between the secondary router MR2 and the HA, MR2 is registered in HA as a substituent of MR1. At t=10s, we activate the LGD trigger at MR1 level, and this by reducing suitably the emission power of the base station IEEE 802.11. Then, at t=10 s to t = 10.5 s with a step equal to 0.1 s, we activate the LD trigger to scan the range [0,500 ms] of the possible values for the TimeInterval parameter associated to LGD trigger. This manipulation will allow us to determine the threshold for which it is possible to ensure a seamless connectivity.

4.1 Results

We have presented in Fig.4, the services interruption delay according to Time-Interval parameter. This delay is determined by the past time between the reception of the last data packet (traffic CBR) via MR1 and the reception of the first data packet via MR2. It is obvious that the connexion delay is inversely proportional to TimeInterval, more this latter increases, more is the chance to finish the tunnels switching operation before the current link breaks down. Value 0 for TimeInterval means that the used trigger for the activation of the tunnel commutation is the Link_Down, i.e. we are dealing with a hard handover. We note the threshold value of 250 ms for the TimeInterval, for which the disconnection delay is null.

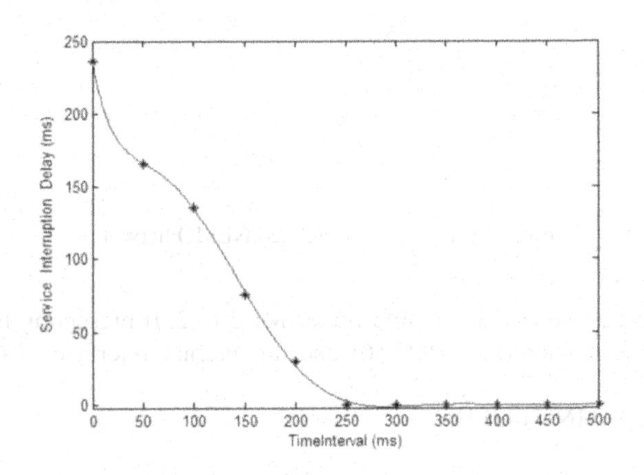

Fig. 4. Services interruption delay vs. TimeInterval

For applications such as the vision conference and the voice on IP (VoIP), a delay of 50 ms is tolerable [12], thus a threshold of 200 ms for TimeInterval can be kept for anticipation. Notice that we have used a RTT maximal value between AR2 and HA given by [13] that is 200ms. Assuming a log-distance path loss model with conservative path loss exponent (β=4, see[17]), we give in Fig.5 for different network speeds the optimal values for LGD factor.

Fig.6 represents the number of the lost packets according to TimeInterval. We can determine this by making the difference between the sequence number (TCP packet) of the first packet of the received data after re-establishing the connection and the sequence number of the last data packet received before disconnecting.

We notice a maximal loss of 15 packets for the case TimeInterval = 0. If we make a comparison with the NEMO BS protocol, with the same application CBR (PacketSize=768 bytes, Interval=16 ms), handover of this latter for an average time of 1250 ms will generate a loss of about 78 packets.

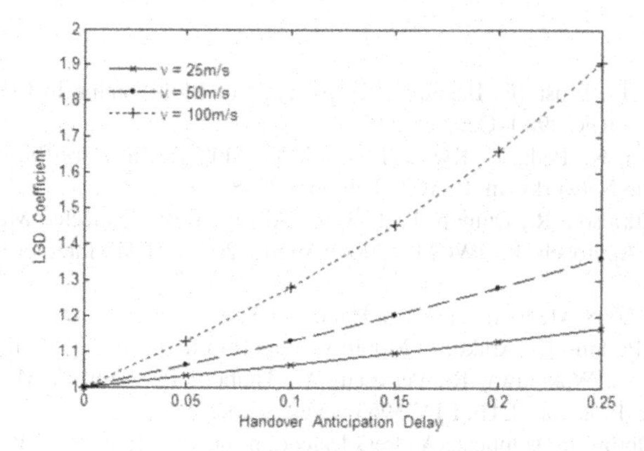

Fig. 5. LGD factor vs Anticipation Time

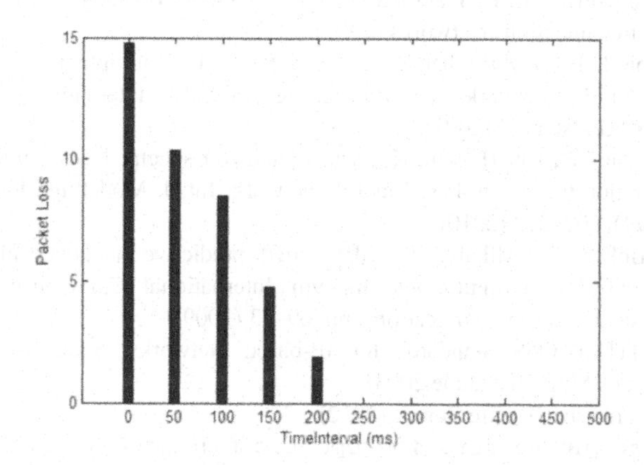

Fig. 6. Packet Loss vs TimeInterval parameter (LGD trigger)

5 Conclusion

In this paper, we have described a new distributed approach for mobility management in multiple mobile routers multihomed NEMO context. The proposed solution doesn't require any change in Internet architecture, while ensuring seamless connectivity to meet ITU-T QoS Standards for IP-based Networks. We considered a multihomed NEMO network with several MRs domiciled in different HAs. This architecture is based on the introduction of the notion of primary and secondary router, and the distribution of the task of mobility management among all MRs using local and remote MIH IEEE 802.21 services. We extended the NEMO BS protocol with the registration of substituent MRs and the tunnel switching supports. A prospect as continuity with this work would be the definition of a routing policy, load sharing or even bandwidth aggregation schemes if several tunnels are simultaneously open.

References

1. Ng, C., Paik, E., Ernst, T., Bagnulo, M.: Analysis of Multihoming in Network Mobility Support. IETF, RFC 4980 (October 2007)
2. Park, C., Choi, N., Paik, E., Kwon, T., Choi, Y.: Multiple Interface/Prefix Selection for Virtual Mobile Networks. In: ICACT (February 2008)
3. Wang, Q., Atkinson, R., Dunlop, J.: Towards Always Best Connected Multi-Access: the MULTINET Approach. In: IWCMC 2008 (August 2008), IETF, Internet Draft (February 2004)
4. IEEE 802.21-2008, Media Independent Handover Services (2008)
5. Johnson, D., Perkins, C., Arkko, J.: Mobility Support in IPv6. RFC 3775, IETF (2004)
6. Devarapalli, V., Wakikawa, R., Petrescu, A., Thubert, P.: Network Mobility (NEMO) Basic Support Protocol. IETF, RFC 3963 (January 2005)
7. Moore, N.: Optimistic Duplicate Address Detection for IPv6. IETF RFC 4429 (April 2006)
8. Mussabbir, Q.B., Yao, W.: Optimized FMIPv6 using IEEE 802.21 MIH Services in Vehicular Networks. IEEE Transactions on Vehicular Technology. Special Issue on Vehicular Communications Networks (2007)
9. Lin, H., Labiod, H.: Hybrid handover optimization for multiple mobile routers-based multihomed NEMO networks. In: Proceedings of IEEE International Conference on Pervasive Service, Istanbul (2007)
10. Huang, Z., Yang, Y., Hu, H., Lin, K.: A fast handover scheme based on multiple mobile router cooperation for a train-based mobile network. Int. J. Modelling, Identification and Control 10(3/4), 202–212 (2010)
11. Jeney, G., Bokor, L., Mihaly, Z.: GPS aided predictive handover management for multihomed NEMO configurations. In: 9th International Conference on Intelligent Transport Systems Telecommunications, pp. 69–73 (2009)
12. Neal seitz, ITU-T QoS Standards for IP-based Networks. Standards Report. IEEE Communications Magazine (June 2003)
13. North Americ Internet Traffic Report (2012), http://www.internettrafficreport.com/namerica.htm
14. The VINT Project. The ns Manual, http://www.isi.edu/nsnam/ns/ns-documentation.html
15. NS-2 Network Simulator, http://www.isi.edu/nsnam/ns
16. The Network Simulator NS-2 NIST add-on—IEEE 802.21 model (based on IEEE P802.21/D03.00), National Institute of Standards and Technology (NIST) (January 2007)
17. Rappaport, T.S.: Wireless Communication: Principles and Practice. Personal Education International (2002)

A New Hybrid Authentication Protocol to Secure Data Communications in Mobile Networks

Mouchira Bensari and Azeddine Bilami

LaSTIC Laboratory, Computer Science Department, UHL Batna, Algeria
bensari2010@gmail.com, abilami@yahoo.fr

Abstract. The growing area of lightweight devices, such as mobile cell phones, PDA conduct to the rapid growth of mobile networks, they are playing important role in everyone's day. Mobile Networks offer unrestricted mobility and tender important services like M-business, M-Learning, where, such services need to keep security of data as a top concern. The root cause behind the eavesdroppers in these networks is the un-authentication. Designing authentication protocol for mobile networks is a challenging task, because, mobile device's memory, processing power, bandwidths are limited and constrained. Cryptography is the important technique to identify the authenticity in mobile networks. The authentication schemes for this networks use symmetric or asymmetric mechanisms. In this paper, we propose a hybrid authentication protocol that is based on Elliptic Curve Cryptography which is, actually, the suitable technique for mobile devices because of its small key size and high security.

Keywords: Authentication, Asymmetric algorithms, ECC, Energy saving, Mobility, Mobile networks, RSA, Security.

1 Introduction

The proliferation of mobile devices leads to the growing popularity of mobile networks. A mobile network can be defined as an autonomous collection of mobile nodes using diverse devices: mobile phones, PDA, tablet PC… which communicates over a wireless link with infrastructure [1]. These mobile devices are constrained in terms of bandwidth, energy, storage capacity, computing power… Since mobile networks are based on cellular network technology, the coverage area is generally carved on geographically limited small surfaces called cells. Each cell is covered by a base station (BTS) which allocates resources for launching services to mobile nodes inside the cell. The cell is sliced into zones of localizations where each zone is managed by a switch (MSC) recording information about the users in a database of visitor (VLR) and the database of permanent users of the network (HLR). [2]

There are different generations of mobile networks. From a generation to the next one, there is an improvement of the voice /data services that are offered by the network. [3] The last generations (since 3G) are not limited to vocal services, but they also offer access to the Internet, and new forms of important services in our

A. Amine et al. (Eds.): *Modeling Approaches and Algorithms*, SCI 488, pp. 195–204.
DOI: 10.1007/978-3-319-00560-7_24 © Springer International Publishing Switzerland 2013

day life such as: M-Learning, M-Tickets, M-Media, M-Banking, M-Payment, M-Commerce and M-business [4], where the security represents a critical focus.

Security in the mobile networks must ensure the basic conditions of security such as confidentiality, authentication, integrity, availability and no-repudiation. [5] These conditions have different degrees of importance according to the used context in the mobile network. Evidently, the authentication symbolizes the base of security in any crucial application using these networks.

Designing authentication protocol is one of many important security issues in mobile networks. Because of the low-power computing capability of mobile devices, most security technologies, which require expensive computations, are not fully applicable to mobile networks. According to these constraints and those related to the use of a wireless medium, the design of authentication protocols that are suitable for mobile networks, is a challenge task. Several researches are carried out in this field and several protocols are conceived. The existing protocols are based either on symmetrical, or on asymmetrical cryptography or both. The symmetrical protocols have several security weaknesses. Although they are compatible with the constraints of mobile devices, the asymmetrical protocols allow more protection than the symmetrical ones, but they require costly computation operations. The hybrid protocols consist in providing a compromise between cost and security. Currently, the authentication protocols are directed towards the elliptic curve cryptography (ECC).

Elliptic curve cryptography has emerged as an alternative to the traditional asymmetrical cryptographic systems. Several researches showed that this technique is well adapted to mobile devices with limited resources [6].

In this paper, we propose a new authentication protocol for mobile networks. The proposed protocol pursues a hybrid scheme starting with a full authentication. Our proposal inherit from ECC technique, the benefit of higher computational speed, lower power consumption, smaller bandwidth requirement and smaller size message exchanges between the communicating parties.

The remainder of this paper is organized as follows. In Section 2, we review the existing authentication protocols for mobile networks. In Section 3, we give the basics of ECC. In Section 4, we describe the proposed protocol. Section 5 analyzes the security and efficiency of our proposal and finally section 6 concludes the paper.

2 Existing Authentication Protocols for Mobile Networks

At the beginning of each confidential communication, a correspondent must identify his interlocutor and check that it is indeed the supposed one. This is the principle of the authentication which permits to confirm something (or someone) like authentic. It interposes between two important phases, the identification and the attribution of the rights. It completes the process of identification by proving a declared identity. So, it puts in failure all the attacks proceeding by usurpation of the identity or the role [7].

Recent years have seen an explosive growth of interest in securing data communications in mobile networks, especially by authentication protocols. We can classify these authentication protocols in three kinds: symmetric, asymmetric and hybrid protocols.

The first authentication protocols designed for mobile networks are symmetrical; the authentication procedure is carried out using symmetric-key cryptography such as the Advanced Encryption Standard (AES).

Several symmetrical authentication protocols were presented like the first protocols designed to GSM systems [8]. The authentication in these protocols is only unilateral. Furthermore, the user's identity and location are not anonymous.

The enhanced protocols such as in [9] provide more security functions like identity, confidentiality and mutual authentication, but they all need a third trust part such as HLR or the old VLR [10].

These protocols seem suitable to mobile environments. However, the authentication by these protocols is often vulnerable and needs a third part for the construction and the distribution of keys. The solution is the sensationalism to the asymmetric protocols.

The asymmetric protocols are based on asymmetric key cryptography and use different keys i.e. public key and private key, such as RSA and ECC.
Several certificate-based protocols have been proposed for mobile networks like the one proposed by [8] for the GSM and The ASPeCT protocol [11] for UMTS. They are the most significant examples of asymmetric protocols. The basic idea behind these protocols consists in reducing the frequency of long-distance signaling across domains when the roaming across an area becomes more frequent. The drawback of these schemes consists in the incorrect use of certificates which may generates security flaws [10].

These protocols are more effective and less vulnerable than the symmetrical protocols. However, these protocols use keys of big sizes and require a longer computing time and more resources comparing to the symmetrical protocols, because of the complexity of the executed operations. Therefore, it is not encouraging to use these protocols in a mobile environment in spite of their effectiveness in security.

The adequate solution tends to use Elliptic Curve Cryptography because of its high security and its compatibility with mobile networks constraints. The protocols defined in [12] are good examples in this context.

Other proposed protocols have proposed a hybrid scheme while using a combination of both symmetric and asymmetric key cryptographies.
A series of authentication protocols described in [13] are suitable for the GSM architecture using the hybrid approach. The protocols proposed by Beller et al. presented in [8], were built using the Rabin crypto-system based on the difficulty of computing a square root modulo a composite integer. The Rabin crypto-system was specifically chosen because of its very efficiency and its appropriateness for implementations in mobile stations (MS).

3 Elliptic Curve Cryptography

Elliptic curve cryptography (ECC) was proposed by Victor Miller and Neal Koblitz in the middle of the 1980 years. Currently ECC is a wall crypto-system; it is an attractive alternative to RSA, particularly, to devices with limited resources. It offers equivalent security with smaller key sizes resulting in faster computations, lower power consumption, as well as memory and bandwidth savings [6].

The conventional crypto-systems operate directly on long entireties where ECC operates on points belonging to an elliptic curve defined over a finite field **Fp**.

Let p be a prime number, and let Fp denote the field of integers modulo p. An elliptic curve E over Fp is defined by an equation of the form:

$$y^2 = x^3 + ax + b \tag{1}$$

Where a and b, in Fp, satisfy $4a^3 + 27b^2 \neq 0 \pmod{p}$, if Fp is a Prime Field. Else (Binary Field):

$$y^2 + xy = x^3 + ax^2 + b \tag{2}$$

Where $b \neq 0$.

A pair (x, y), where x, y in Fp, is a point on the curve if (x, y) satisfying the equation (1) or (2) [7], [12].

ECC's main cryptographic operation is scalar point multiplication, which computes $\boldsymbol{Q = kP}$ where **P** is a point; $\boldsymbol{k}$ is an integer and **Q** is also a point of the curve. Scalar multiplication is performed through a combination of point additions and point doublings.

The force of ECC relies on the difficulty of solving the Elliptic Curve Discrete Logarithm Problem (ECDLP), which states that given **P** and **Q**, it is hard to find $\boldsymbol{k}$ in ECC, this problem is indissoluble for great values of $\boldsymbol{k}$. The integer $\boldsymbol{k}$ acts as a private key, while the result of the multiplying **Q** serves as the corresponding public key [12].

In some curves, ECDLP may be solved efficiently, for that; a set of curves possessing the necessary security properties has been published. The use of these curves is also recommended as a means of facilitating interoperability between different implementations of a security protocol.

The ECC crypto operations are adopted and standardized by NIST, ANSI, IEEE and IETF. Currently, there are three applications of ECC in cryptography, the Elliptic Curve Integrated Encryption Scheme (ECIES), which is a variant of the ElGamal encoding scheme, the Elliptic Curve Digital Signature Algorithm (ECDSA), which is the elliptic curved variant of the DSA and the Elliptic Curve Diffie-Hellman (ECDH), which is the typical diagram of key exchange based on Diffie-Hellman mechanism applied to the elliptic curves [7].

4 HAPMON Protocol

We propose a new hybrid authentication protocol for mobile networks, which is based on ECC. Our protocol called HAPMON (Hybrid Authentication Protocol for MObile Networks) provides a compromise between high security and the constraints of mobile environment. It is a protocol suitable composed of two parts: certificate-based authentication protocol (CBA); and the ticket-based authentication protocol (TBA) (Figure 1), where, each one of these protocols provides a mutual authentication.

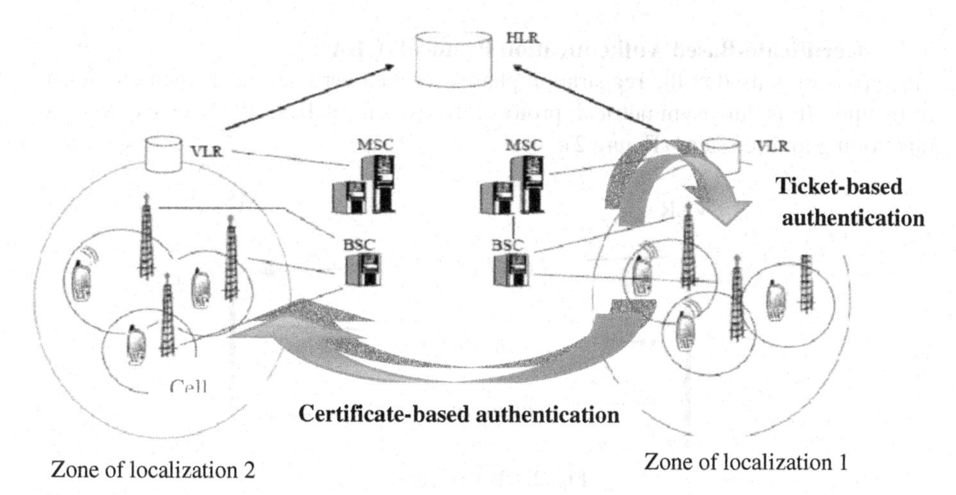

Fig. 1. HAPMON protocol

Like other authentication protocols for mobile networks, HAPMON has two phases, the bootstrapping and the authentication phase.

4.1 Bootstrapping Phase

In this phase (registration phase), the nodes that ask for services are provided with some data (bootstrapping material) and must prove the knowledge of the appropriate information in the authentication phase like a proof of eligibility. The responsible part in this phase is the HLR where a public key infrastructure (PKI) is employed.

Before the generation of X.509v3 certificate, the HLR chooses an elliptic curve (**E**) defined over a finite field **Fq** such as the ECDLP, which is more difficult to be solved. It chooses also a reference point **(P)** on E. Then it chooses a secret number **d** and calculates **Q = dP.**

The HLR generates after the pair key, public key (**KU**), KU= Q, and the private key (**KR**), KR= d. It generates, also, for MS a key **ki** used for coding the data before the checking of the certificate, so that nothing is sent clearly on the network. This key is equivalent to the **ki** of GSM network. **E**, **Fq**, **P** and **KU** are public, where the **KR** and **ki** are private.

4.2 Authentication Phase

After the first phase was completed successfully, the authentication phase starts. In this phase, the mobile node and the VLR exchange mutually the evidence that has been provided by HLR in the first phase. The authentication phase is a suite of CBA protocol and TBA protocol. The choice of one protocol depends mainly on MS localization at the time of requesting services and other considerations (certificate and ticket validity).

A. Certificate-Based Authentication Protocol (CBA):

This protocol is used at the registration phase, at handover phase and when the ticket is invalid. It is an asymmetrical protocol based on ECIES. We can explain its functioning in three steps (Figure 2):

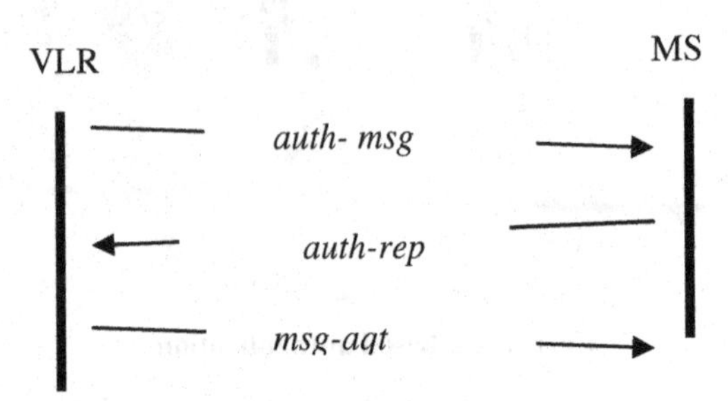

Fig. 2. CBA protocol

Step 1:

When the MS receives or asks for service, the VLR in which MS resides uses the temporal identifier of MS (**TID**) to access a secret database and extract the MS's data that are **E**, **P** and **ki**. Then, the VLR generates an authentication message auth-msg(CertVLR;R1)$_{ki}$ which contains a random point R1 = *a*P (*a* selected randomly from Fq) and their certificate CertVLR. This message is encrypted using ***ki*** then sent to MS. At the reception of auth-msg, MS decrypts it using ***ki*** then, it checks the validity of CertVLR by the public key of HLR (**KU_{HLR}**), if it is valid, MS calculates the shared key with the VLR (**kc**),

$$kc = KR_{MS}KU_{VLR} \quad (3)$$

Where KU_{VLR} is consulted from CertVLR. Then, MS calculates R1' = R1 + kc, subsequently, it chooses an *b* from Fq and calculates *b*P = R2. To answer the auth-msg, MS constructs an authentication response message auth-resp(CertMS;R1';R2)$_{KUVLR}$, which contains their certificate CertMS;R1', and R2. This message is encrypted using the KU_{VLR} and then sent to VLR.

Step 2:

VLR decrypts the message using its private key (**KR_{VLR}**), it checks the validity of CertMS using **KU_{HLR}**, if it is valid, it calculates the shared key with MS,

$$kc = KR_{VLR}KU_{MS} \quad (4)$$

KU_{MS} is consulted from CertMS. Then, it calculates R1, R1= R1`- kc, if the calculated R1 is equivalent to the sent one in auth-msg, then this implies that MS's secret key is valid, (it isn't a case of identity usurpation). Thus MS is well authenticated to VLR. VLR will generate a ticket (Figure 3) to MS and calculate the key of the next session **K1**.

The ticket is a data structure containing: the MS's ID, the ticket's issues, the end date, and the signature of the VLR. Then, It also calculates R2`= R2 + kc.

Ms-id: F904.6001.B270.9845
***The issue date*:** 19/03/2013
***The end-date*:** 15/14/2013
***VLR-sign*:** D0F4

Fig. 3. Ticket

Finally VLR will send an authentication acknowledgment message msg-aqt (ticket, $R2`)_{KUMS}$ to MS. the key of the next session **K1** is calculated with ECDH ;

$$K1 = aR2 \tag{5}$$

Step 3:
After the reception of msg-aqt, MS decrypts it by its KR_{MS}, it calculated R2, R2= R2`- kc. If R2 corresponds to the sent one in auth-resp, then it implies that VLR's secret key $\mathbf{KR_{VLR}}$ is valid. Thus, the VLR is well authenticated to MS.

The MS will register the ticket and calculates the key of the next session based on ECDH using R1. **k1=** $\boldsymbol{b}$**R1.**

B. Ticket-Based Authentication Protocol (TBA)
This protocol is used when the MS stays at the same zone (address to the same VLR) and requests services several times, and the ticket is valid. It is a symmetrical protocol based on AES. The TBA protocol (Figure 4) is described as follows:

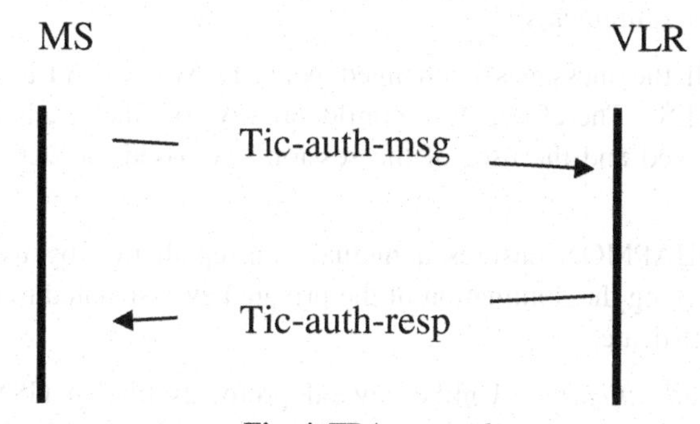

Fig. 4. TBA protocol

Step 1:
With each request of service, MS constructs a ticket authentication message, Tic-auth-msg(ticket, R1 $)_{k0}$, containing R1= aP and the ticket. The message is encrypted using the generated key during the last session ($\mathbf{k_0}$) and sent to VLR.

After receiving the Tic-auth-msg, the VLR uses the **TID** of MS to find the **k0** and the ticket of MS. Then VLR checks the ticket, if it is legal, it decrypts the message.

If the ticket is the same in Tic-auth-msg so MS has a valid ticket, thus MS is authenticated successfully. The VLR responds MS with ticket authentication response, Tic-auth-resp(R1, R2)$_{k0}$, which contains R2= *b*P (*b* selected randomly from Fq) and R1 (consulted from Tic-auth-msg). Encoding is also carried using **k0**. VLR finally generates the key of the next session **K1** on base ECDH, K1 = *b*R1. If the ticket is invalid, the VLR asks MS to reapply the CBA protocol.

Step 2:
When receiving Tic-auth-resp from VLR, MS uses the **k0** to decrypt it, then it checks if R1 is the one in Tic-auth-msg , if it is, that means that VLR has the correct **k0**, so, the MS well authenticates the network. Therefore, it generate the key of the next session **K1** based on ECDH, K1 = *a*R2.

5 Analysis

HAPMON is implemented in J2ME platform, which is appropriate to test applications for mobile devices, where the ECC operations are offered by Bouncy castle library. We have assigned the VLR as server and the MS as client and the communication is established via Sockets. In this section, we discuss the security and efficiency of our proposal and compare it with other schemes. We compare it to the well-known elliptic and non-elliptic authentication schemes for mobile networks.

5.1 Security Analysis

The performances of HAPMON are evaluated and compared while considering the following security requirements:

Confidentiality: all the messages exchanged between MS and VLR are encrypted by ECIES or AES. Therefore, the confidentiality of the exchanged data is completely guaranteed and the uses of the session key based on ECC augments the confidentiality.

Non-repudiation: HAPMON ensures a mutual non-repudiation by using electronic certificate and checking the domination of the private key associated to the public key indicated in the certificate.

Explicit mutual authentication: Unlike several protocols like in GSM, HAPMON ensures a mutual authentication. It is explicitly expressed. MS and VLR are mutually authenticated, if each one of them proves to the other that it has the correct **Kc**.

Resistance to the dictionary attack: HAPMON is able to ensure protection against the dictionary attack upon the shared key. Even with the knowledge of **E**, **P** and **KU** of MS and VLR, an adversary cannot carry-out the dictionary attack in order to obtain the session key, because the time necessary to find the session key (the resolution time of ECDLP), is considered higher than the desired one.

5.2 Efficiency Analysis

We compare our proposal (192-bits) with non elliptic [8] (1024-bits) protocol, Beller-Chang-Yacobi protocol (1024-bits) and the elliptic Aydos [13] protocol (160-bits). The parameter lengths, the computational load, the bandwidth and the storage requirements on the user side represent the comparison criterions.

From Figure 5, we can observe the great difference between HAPMON and the non-elliptic protocols. We observe also a small difference between CBA and Aydos protocol, because this last one is based on the digital signatures which are not considered as complex certificates. Our protocol requires less bandwidth, low storage requirements on the user side, especially the TBA, which is the most frequently used .This efficiency, is a result of ECC utilization.

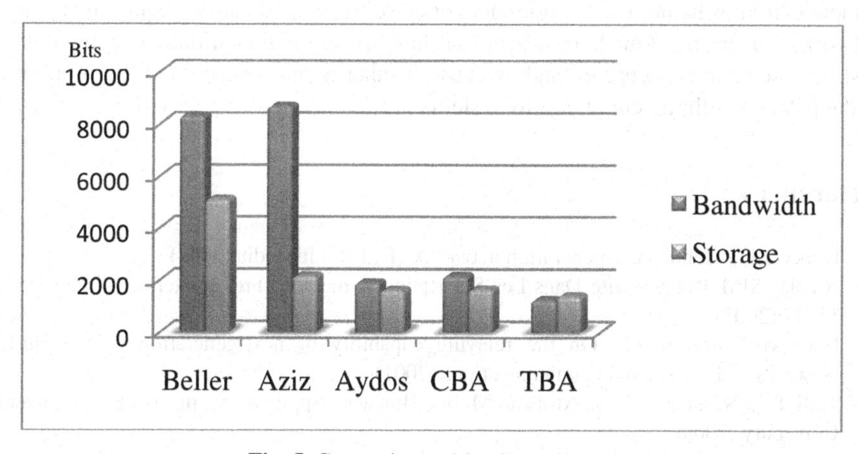

Fig. 5. Comparison with other schemes

The analysis of the computational load in these protocols gives the table below.

Table 1. Computational load analysis

protocol		computational load
Beller		2 PKE + 1PKD + Pre-computation
Aziz		3 PKE + 2 PKD
Aydos		1 eP + 1 ECD-SAV + 2 SKE +1 SHA
HAPMON	CBA	1 ECAES dec + 2 ECIES
	TBA	1 ECAES enc + 1 ECAES dec

PKE: Public Key Encryption. PKD: Public Key Decryption. eP : Point Multiplication. ECAES: elliptic curve AES. SKE: Secret Key Encryption or Decryption. ECDSAV: Elliptic Curve Digital Signature Algorithm Verification. ECIES: Elliptic Curve Integrated Encryption Scheme

From the table 1, we remark that our scheme has a low computational load (ECAES and ECIES operations in CBA, 2 ECAES operations in TBA), since it is based on ECC. The other protocols need complex operations, because they are based on the traditional cryptography (RSA, Rabin, SHA), they demand even Pre-computation (Beller). So, they have a complex computational load.

6 Conclusion

The difficulty to secure mobile networks is due to the limited capacity of the nodes. Authentication protocol for mobile networks was tackled by many researchers, but the results didn't reach the awaited ones. In this paper, we have presented a new authentication scheme for mobile networks respecting security requirements and networks constraints. Our hybrid proposal has low communication and computation cost. Our scheme is secure enough because it inherits the security and the properties of the powerful elliptic curve crypto-systems.

References

1. Huber, J.F.: Mobile next-generation networks. IEEE MultiMedia (2004)
2. Nicolas SIMON : Sécurité Dans Les Smartphones, memoir. Free university of Bruxelles, 11–37 (2007)
3. Neonakis Aggelou, G.: On the relaying capability of next-generation GSM cellular networks. IEEE Personal Communications (2001)
4. Tatli, E.I.: Security in Context-awareMobile Business Applications, pp. 11–15. Mannheim University (2008)
5. Beghriche, A., Bilami, A.: Modélisation et Gestion de la Confiance dans les Réseaux Mobiles Ad hoc, CIIA (2009)
6. Hankerson, D., Menezes, A., Vanstone, S.: Guide to Elliptic CurveCryptography. Springer (2004)
7. Rajeswari, G., Thilagavathi, K.: A Novel Protocol For Indirect Authentication. In: Mobile Networks Based On Elliptic Curve Cryptography, JATIT (2009)
8. Aydos, M., Yanık, T., Koc, Ç.K.: An High-Speed ECC-based Wireless Authentication Protocol on an ARM Microprocessor *. IEE Pro.: Comms (2001)
9. Molva, R., Samfat, D.: Tsudik: Authentication of Mobile Users. IEEE Network 8(2), 26–34 (1994)
10. Stojmenovic, I.: Handbook of Wireless Networks and Mobile Computing, 87–109 (2002)
11. Günther, H., Kheith, M., Chiris, J.: Authentication protocols for mobile network environment value-added services. IEEE Transactions on Vehicular Technology 2 (2002)
12. Rajeswari, G., Thilagavathi, K.: An Efficient Authentication Protocol Based on Elliptic Curve Cryptography for Mobile Networks. IJCSNS 9(2) (2009)
13. Tzeng, Z.-J., Tzeng, W.-G.: Authentication of Mobile Users in Third Generation Mobile Systems. National Chiao Tung University (2001)

Mapping System for Merging Ontologies

Messaouda Fareh[1], Omar Boussaid[2], and Rachid Chalal[3]

[1] LRDSI Laboratory, Blida University, Algeria
farehm@gmail.com
[2] ERIC Laboratory, Lyon 2 University, France
omar.Boussaid@univ-lyon2.fr
[3] LMCS Laboratory, ESI, Oued-Smar, Algeria
r_chalal@esi.dz

Abstract. In this paper we present a new mapping system for merging OWL ontologies. This work is situated in the general context of stored information heterogeneity in a decisional system such as data, metadata and knowledge, for cohabitation and reconciliation of these information by mediation. Our Mapping approach focuses on computing semantic similarity between concepts of ontologies to merge, it is based on a weighted combination of computing similarity methods, we use syntactic, lexical, structural, and semantic technics. The proposed mapping process makes use of several types of information in a manner that increases the mapping accuracy.

Keywords: Mapping, ontologies, similarity measure, semantic similarity, structural similarity.

1 Introduction

This work is situated in context of information heterogeneity in decisional systems, such as data, metadata and knowledge. Before cohabiting these different forms of information, an algorithm for merging ontologies must be developed to integrate them. The aim of this paper is to develop a mapping method for merging ontologies.

The ontologies merger is an interesting perspective in the design of ontology, by these significant advantages in saving time and adding clarity and optimality in the representation of domain. However, it has not yet been adequately explored because of the merger process complexity and the lack of help mechanisms to the merger.

The matching process consists in finding relationships or correspondences between entities belonging to different ontologies. That is, given two ontologies, one should be able to map concepts found in the first ontology onto the ones found in the second one.

Figure 1 schematized the matching process. The matching can have other parameters as input (e.g., similarity threshold) and other semantic resources in order to enrich and explicit the semantics of ontologies to be matched (e.g.: thesaurus like WordNet [16], DBpedia [1] and Yago [24] or other kind of semantic that can be helpful for matching).

A. Amine et al. (Eds.): *Modeling Approaches and Algorithms*, SCI 488, pp. 205–216.
DOI: 10.1007/978-3-319-00560-7_25 © Springer International Publishing Switzerland 2013

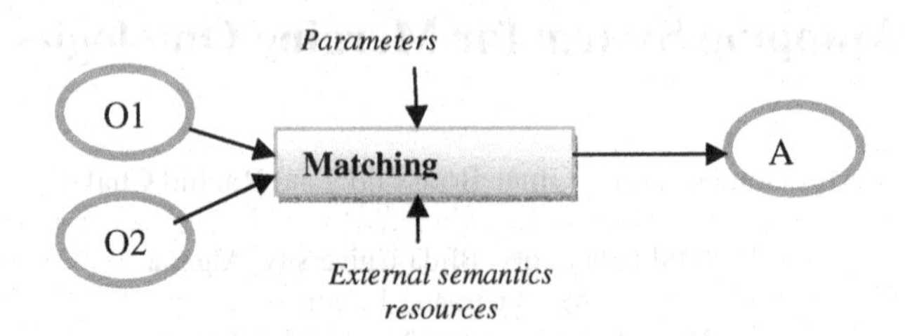

Fig. 1. Matching process

In the literature, there exist several methods of ontologies mapping, but for merging ontologies, there are few methods have been proposed, which we quote [23] [6] [17] [7] [15] and [12]. The purpose of this paper is to develop a mapping system for merging ontologies.

Most approaches to merge/map ontologies are limited to the use of semi-automatic algorithms, and user intervention for semantic conflict resolution, as [14] [23] [25] [6] [17] [7] [15] and [12]. Unlike these approaches, our method allows a full automation of the map process.

These approaches [17] [23] [4] and [14] are based on syntactic similarity between the concepts. However, two concepts can have the same syntax, although their semantics are different. Our method is distinguished by the introduction of a semantic dimension in the mapping process through the use of WordNet. On the other hand, the semantic dimension is based on the structural position of the entities to be compared, but also on the combination of similarity measures in order to obtain a value very fine degree of similarities between these entities.

We have organized our paper as follows: Section 2 discusses the limitations of existing merging/mapping techniques and describes related work. Section 3 presents the proposed mapping system. An evaluation study is presented in Section 4. Finally, Section 5 gives concluding remarks and future work.

2 Related Work

Merging ontologies is a process for creating a single ontology that encompass the knowledge of two or more existing ontologies, which describe the same subject, or are in the same application domain. So, we create a new ontology, called the merged ontology capturing the knowledge of the original ontologies. The challenge is then to ensure that all the correspondences and differences between ontologies, are properly reflected in the resulting ontology [3].

The merged ontology unifies and replaces the original ontologies. This definition does not specify how the resulting ontology is linked to the originals ontologies, to let open the problem of the choice of merging method. The most common approaches use the union or intersection operators.

Therefore, several mechanisms for merging/ mapping ontologies are developed: ONIONS [7], Chemaira [15], FCA-Merge [23], MOMIS [6], PROMPT [17], OLA

[4], S-Mach [8], H-Mach[2], RiMOM [14] and MLMA [12]. For merging ontologies, there are a limited number of methods and tools that are proposed, by cons, for mapping ontologies, the amount of existing methods is very important, we decided to make a selection of the most common methods to illustrate the variety of approaches.

ONIONS (Ontological Integration Of Naive Sources) provides guidelines to analyse and merge existing ontologies, and stresses the reuse of terminologies domain [7]. It has been developed since the early 90s, to reflect the problem of conceptual heterogeneity. ONIONS uses similarity measures terminology then it exploits the structural features.

Chimaera is a semi-automatic environment for merging and testing (diagnosing) large ontologies [15]. Matching in the system is performed as one of the major subtasks of a merge operator. Chimaera searches for merging candidates as pairs of matching terms, involving term names, term definitions, possible acronym and expanded forms, names that appear as suffixes of other names. This tool calculates similarities using simple methods, then it can redefine the obtained hierarchy after the merger, as well as edit it manually.

FCA-Merge (Formal Conceptual Analysis Merge) made a comparison based on instances: it uses technical language; key terms connected with instances of ontology concepts (use a dictionary). The main problem of FCA-Merge method is its side semi-automatic because it needs the intervention of a human, for the validation of the construction process. Another side which doesn't focus too much on semantics.

MOMIS (Onvironment Mediator for Multiple Information Sources) has been established to address, the problems of heterogeneous information sources presented in the web [6]. MOMIS enrolled as a mediator for the extraction and integration of information structured and semi-structured. MOMIS is based on the merger process, a set of tools that facilitates the creation of the merged ontology, such as: WordNet for the discovery of lexical relations between extracted terms, and ARTEMIS allows the grouping class size.

PROMPT performs some tasks automatically and guides the user to accomplish other tasks, for which his intervention is necessary [17]. PROMPT begins with searching for possible matches linguistic similarity, but is based primarily on the structure, it is semi-automatic in that it only proposed alternatives that cannot be validated only by human intervention. In PROMPT, the user is assisted and guided throughout the process.

S-Match [8] is a system for searching semantically matches. S-Match takes two graph-like structures, e.g., classifications, XML schemas, or ontologies, as input and returns logical relationships, e.g., equivalence and subsumption, found between the nodes of the graphs. Ontology entities are converted to logical formulas. Then, the match manager uses various basic element-level matchers and logic provers to find relationships between these formulas, which in turn correspond to the relationships between entities.

OLA (OWL Lite Aligner) [4] is a system that takes an equal contribution of each component of the ontologies, e.g., classes, instances … etc in order to find the matching entities of the input ontologies. It considers ontologies as graphs, and determines the similarity of the graph nodes based on string, language, and structure based similarities. These similarities are aggregated. For computing these similarities, OLA starts with base distance measures computed from labels and concrete data types. Then, it iterates a fixpoint algorithm until it no longer yields an improvement.

H-Match [2] takes OWL ontologies as its input. Internally, these input ontologies are represented by graphs using the H-model representation. Moreover, H-Match computes two types of similarities: linguistic and contextual. These are then combined using weighting schemas to yield a final measure, called semantic similarity. In determining the contextual similarity, H-match considers neighboring concepts, e.g., linked through the taxonomy of the actual concept.

RiMOM (Risk Minimization based Ontology Mapping) [14] integrates multiple strategies, such as edit distance, statistical learning, and three similarity propagation-based strategies. Then, it applies a strategy selection method in order to decide on which strategy it will rely more. As a result, RiMOM combines the conducted alignment. RiMOM offers three possible structural propagation strategies: concept-to-concept (CCP), property-to-property (PPP), and concept-to-property (CPP) propagation strategy. To choose between them, RiMOM uses heuristic rules. For instance, in CCP, similarities of concept pairs are propagated across the concept hierarchy structure.

(MLMA+) Multi-Level Matching Algorithm [12] is a strategy benefits from existing ontology match techniques and combines their match results to provide enhanced ontology matching results. The string and linguistic based techniques evaluate the given entities by analyzing their names, labels and comments. They consider both the lexical and linguistic features as terms of comparison. Moreover, the structure-based techniques take into account the structural layout of the ontologies considered.

Umer and Munday (2012) [25] propose a framework of ontology integration which supports and uses a hybrid approach of concept matching in order to combine the advantages of different concept matching approaches. The proposed concept matching technique mostly relies on the semantics of the concepts. User confirmation is also involved in the concept matching process for complicated cases in terms of their semantics and hierarchy. The proposed approach of comparing concepts based on their semantics and structure along with their syntax is a better way of concept comparison.

After analyzing these approaches, we note that the majority of the mentioned approaches use heuristics, propositional logic or probability. Some approaches are based on using of computing syntactic similarity to identify correspondences between concepts, but rarely take into account their semantics.

The cited techniques above, agree on some properties but they have limitations:

- Most of these approaches are limited to the use of semi-automatic algorithms for merging/mapping ontologies.
- To link the concepts of different ontologies, some approaches are based on syntactic similarities. However, two concepts can have the same syntax, while their semantics are different because it's placed in different contexts. Conversely, two concepts can have the same semantics as they are described by different syntaxes.

- Among these methods, some can discover similarities that can be ignored by others, and vice versa. More complex methods most frequently based on the use of information provided by simpler methods.
- The greatest need in relation to methods of mapping is the effective combination of several complementary methods rather than creating a single method fits all situations.
- The automation level is frequently limited to a partial autonomy, requiring validation of expert proposals that are made. Some ontology mapping tools are fully automatic and produce all their matches without external intervention, but they are not intended for merging ontologies. By cons, for merging ontologies methods require user intervention.

3 Proposed Approach to Ontologies Mapping

To overcome these significant limitations, as to the relevance and flexibility in the ontologies mapping, we have proposed and developed the system that focuses on:

- Adding a semantic dimension by prior enrichment of ontologies, which is done by defining a set of metadata annotations, to the original concepts of two ontologies.
- The method is based on computing semantic similarity between concepts of ontologies. It is based on a weighted combination of methods for calculating similarities existing with different specificities.
- The automation of the mapping process between ontologies to unload as much as possible the user.

Before ontology is able to be mapped via our system, it must be the subject of enrichment by expert. During the enrichment, the concepts of the ontologies will be located in their context. That is to say that the expert provide a definition and synonyms that stick as much as possible to the area to which refers their ontologies, via the help of a lexical database WordNet, or by referring to the expert knowledge.

If the system is faced with a compound or complex concept, which Wordnet offers no definition, the expert will give his own definition (always in relation to the field). This enrichment will have a major impact on computing similarity between concepts from different ontologies.

Once the ontologies are enriched, the system can proceed by the different technics of similarity calcula.

Euzenat and Shvaiko (2007) [5] provide a comprehensive state of the art of different alignment techniques. These methods are based on similarity measures. In order to improve the alignment results, several measures can be used simultaneously, which brings up an inevitable conflict between the different results produced by each of these measures.

In this section, we calculate various measures of similarity, which will allow us to make correspondences between the two original ontologies. To do this, we chose a combined approach of computing measures. This will be done in three steps:

3.1 Computing Terminological Similarity

Measures based on syntactic and lexical comparison, are applied to pairs of concepts to measure their terminological similarity. Syntactic methods [16]: they are based on the comparison of words, strings or texts. Linguistic or lexical methods using external resources (dictionnaire, taxonomy…): the similarity between two entities, represented by terms, is calculated using semantic links, existing in external resources.

In our system we used the Jaro distance [10] for the syntactic similarity and WordNet for lexical similarity.

- **Syntactic similarity:** There are many similarity measures of this type, to estimate the proximity of two words based on the letters they have in common.

The Levenshtein distance (1966) [13] was one of the first to take account of this notion. This distance between two strings is the minimum number of insertion, deletion, or substitution of a single character needed to transform one string into the other. It is called edit distance, it is mainly used in the context of spelling corrections, to estimate the most probable term that a user wanted to use.

Several measures are based on the Levenshtein distance, as metric Smith & Waterman (1981) [22], which was applied in the biomedical field to find similar regions in DNA strands. Also citing, Gotoh (1982) [9] proposes an extension of the work of Smith and Waterman (1981), or Monge & Elkan (1996) [16] that divide the words into sub-strings. Elkan distance has better performance in terms of results in several experiments; it can be used in several areas. Finally, citing the Jaro distance (1989) [10] and that of Winkler (1999) [26] which is an extension of the previous one. These two distances take into account in the comparison of two character strings, on the one hand, the number of characters in common, and also the order of the characters. The Jaro/Jaro-Winkler distance has almost the same performance, that Monge-Elkan results, but much faster.

The Jaro distance measures the similarity between two strings. More Jaro distance between two chains, is high, more they are similar. This measure is particularly adapted to the treatment of short chains, such as names or passwords. The result is normalized, so as to have a measure between 0 and 1, zero representing the absence of similarity.

Jaro distance between strings ***s1*** and ***s2*** is defined by:

$$\boldsymbol{SimSyn} = \frac{1}{3}\left(\frac{m}{|s1|} + \frac{m}{|s2|} + \frac{m-t}{m}\right)$$

where : m is the number of corresponding characters, and t is the number of transpositions. Two identical characters of $s1$ and $s2$ are considered as corresponding, if their distance (ie the difference between their positions in their respective chains) does not exceed:

$$\left(\frac{\max(|s1|, |s2|)}{2}\right) - 1$$

The number of transpositions is obtained by comparing the ith character corresponding to ***s1***, with the corresponding character in th ***s2***. The number of times these characters are different, divided by two, gives the number of transpositions.

- **Lexical similarity:** The lexical methods require the use of external resources. Several types of resources can be used, our choice is WordNet.

WordNet is a lexical resource, in English, available on the Internet, which groups of words (nouns, verbs, adjectives and adverbs) into sets of synonyms called synsets. A synset contains all the words denoting a given concept. To calculate the linguistic similarity, the function Syn(*c*) calculating the set of synsets of WordNet of concept ***c***, let S = Syn (*c1*) ∩ Syn (*c2*) the set of common sense between ***c1*** and ***c2*** to compare, the cardinality of S is:

$$\lambda (S) = | Syn (c1) \cap Syn (c2) |$$

Let min (|Syn (*c1*) |, | Syn (*c2*) |): the minimum of the cardinalities of two sets Syn (*c1*) and Syn (*c2*), then the similarity between two concepts ***c1*** and ***c2*** is defined as follows:

$$SimLex(c1, c2) = \lambda (S) / min(Syn (c 1), Syn (c 2))$$

This measure return 1.0 if at least ***c1***, is the only synonym of ***c2*** or ***c2*** is the only synonym of ***c1***. If ***c1*** and ***c2*** are not synonyms, and have no relation (hyponymy, antonyms, etc ...) we will measure 0.

These two measures will be combined by the formula:

$$SimTer(c1, c2) = \frac{(SimLex(c1, c2) * CoeffLex) + (SimSyn(c1, c2) * CoeffSyn))}{CoeffLex + CoeffSyn}$$

Where: Coeff: is a numerical coefficient which is given for each measurement. It is calculated as follows: $Coeff = Exp^{Sim}$

3.2 Computing Structural Similarity

Structural methods: These methods, which deduced the similarity of two entities, using structural information, when the entities in question are linked to others by semantic links, forming a hierarchy of entities. The internal structural methods: they calculate the similarity between two concepts, using the information on their internal structure, and external structural methods or conceptual: they use the hierarchical structure of ontology, and are based on counting technics arc to determine the semantic similarity between two entities [21].

(Rada et al 1989) [20] suggest that to measure the distance between two ontological concepts, based on the arcs minimum number to traverse, for go from one concept to another. The authors note that this proposition is valid for all hierarchical links (is-a, kind-of, part-of, ...) but must be adapted for other types of links (cause, etc..).

Resnik measure [21] is based on the informational content notion. It uses jointly ontology and corpus. The informational content of concept reflects the concept relevance in the corpus, taking into account the frequency of its occurrence in the corpus, and the occurrence frequency of the concepts it subsumes. Resnik defines the semantic similarity between two concepts by the information quantity they share. This shared information is equal to the informational content of the lowest Common

Ancestors (LCA), This measure depends only on the lowest Common Ancestors, is therefore rather summary because we have LCA(a, b) = LCA (d, e) while d and e are nearer to the LCA as a and b. The Jiang & Conrath measure [11] overcomes the limitations of Resnik measure, combining the informational content of the LCA, to those of concepts. It also takes into account the arcs number.

Wu and Palmer [27] define the similarity, in function of the distance which separates two ontological concepts in the hierarchy, and also by their position relative to the root. The similarity is defined relative to the distance between two concepts in relation to their lowest common ancestors and the root of the hierarchy.

Structural similarity is calculated by measuring the Wu and Palmer:

$$Sim_{wp}(c1,c2) = \frac{depth(LCA(c1,c2))}{depth(c1) + depth(c2)}$$

Where: LCA (***c1***, ***c2***) is the common ancestor (Lowest Common Ancestors of ***c1*** and ***c2*** and depth (LCA (***c1***, ***c2***)) is number of edges between LCA (***c1***, ***c2***) of the root, and depth (***c1***) is the path length between ***c1*** root through LCA (***c1***, ***c2***). We base for the calculation, on the Wordnet hierarchy [28].

3.3 Computing Semantic Similarity

Semantic similarity is calculated by the combination of terminological and structural similarity. These will be combined by the following formula:

$$SimSem(c1,c2) = \frac{\sum_{(rc1,rc2)\in VR(c1,c2)} SimTer(rc1,rc2) * SimWp(c1,c2) * (1 - |d1 - d2|^2)}{|VR(c1,c2)|}$$

Where: **rc**$_i$: the near relative (father or son) of this concept c_i..

$$VR(c1,c2) = \{(rc1,rc2) | rc1 \in V(c1), rc2 \in V(c2) \text{ and } Sim(rc1,rc2) \geq thresold\}$$

Such that we must take the couple: rc_1 and rc_2 (father-father, son-son). $d_i = Sim_{wp}(c_i, rc_i)$. If a concept is not a neighbor relative to check the above condition, the semantic similarity is calculated by the following formula:

$$SimSem(c1,c2) = \begin{cases} \frac{(SimTer * coeffTer) + (SimStr * coeffStr)}{coeffTer + coefStr} & \text{if } SimTer \geq threshold, SimStr \neq 0 \\ SimTer & \text{if } SimTer \geq threshold, SimStr = 0 \\ 0 & \text{in other cases} \end{cases}$$

After these steps we obtain a matrix of similarity measures " correspondence matrix ".

4 Experimentation

To evaluate the performance of our System, a prototype is realized in Java. Our tool supports input two ontologies described in OWL, given that OWL is a standard for representing ontologies, and then comparing the correspondence obtained by our tool and those by a manual mapping. By computing performance measures used to evaluate the quality of the correspondence produced between the concepts of ontologies, they are primarily measures of calculating the pertinence in information search, such as precision and recall. We conducted a series of tests is provided in some tests using the Benchmark database available to Test library of Ontology Alignment Evaluation Initiative OAEI [19].

The following table summarizes the results obtained by our Mapping system, based on the values of the metrics alignment quality (measures of precision, recall and Fmeasure):

Table 1. Performance measures of our mapping system

Tests	*Precision*	*F-measure*	*Recall*
101-207	1	0.786	0.648
101-204	1	1	1
101-209	0.882	0.844	0.81
101-304	0.871	0.931	1
101-301	0.882	0.9	0.937
101-247	1	1	1
101-303	0.952	0.952	0.952

In this table we note that in some tests such as 101-204, 101-247 and 101-207, the value of precision is 1, that means that the results of our mapping tool are the same given by an expert, and for the other tests, the precision value is between 0.882 and 0.969, therefore, our system gives good results which are encouraging.

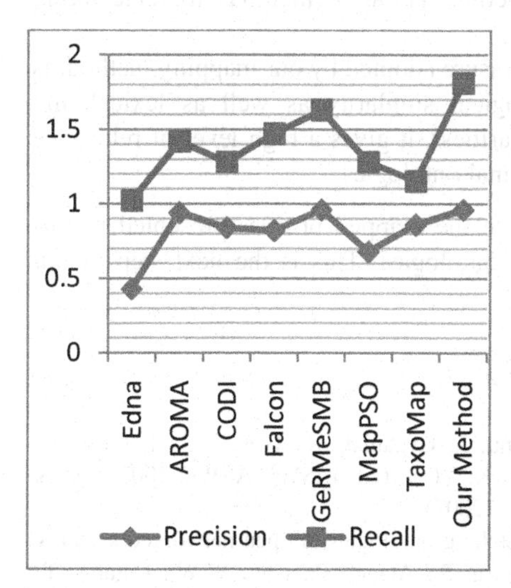

Fig. 2. Comparison curve of matching quality results in 2010 with our Method

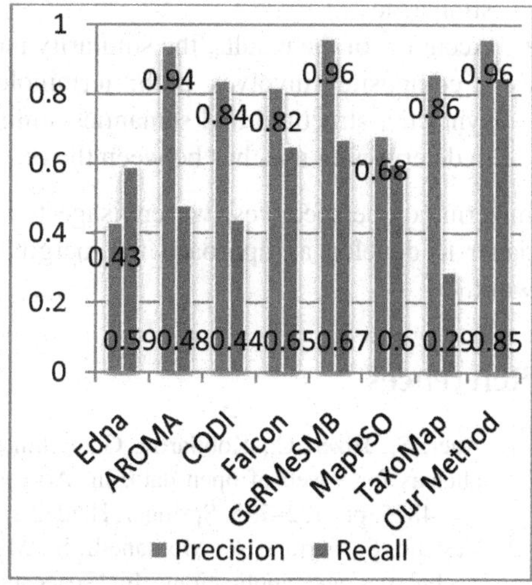

Fig. 3. Comparison histogram of matching quality results in 2010 with our Method

After computing precision and recall values, for our Mapping system, we compared these values with those of the following systems: Edna, AROMA, CODI, Falcon, GeRMeSMB, and MapPSO TaxoMap. The precision and recall values of comparison systems are presented in summarized results provided by OAEI 2010 [18].

A comparative summary of the results of OAEI on the benchmarks is shown in Figures 2 and 3. These Figures provide the consolidated results by groups of tests. We display the results of participants as well as those given by some simple edit distance algorithm on labels (edna). Two systems are ahead: our system and GeRMeSMB, with AROMA as close follower, while CODI, Falcon, MapPSO, had presented intermediary values of precision and recall. No system had strictly lower performance than edna, specially in terms of precision, because edna is a simple edit distance algorithm on labels, which is used as a baseline.

5 Conclusion

Our objective is to develop a mapping method for merging ontologies, as part of the reconciliation of stored information in decisional system.

Our research and findings, we resulted in the development of a novel mapping method for merging ontology. It is based a prior enrichment ontologies to add a semantic dimension to mapped process. The originality of our method resides in the combination of the following:

- Automation of the mapped process, the user need only select ontologies that he wants to map.
- The method is based on computing semantic similarity between concepts of ontologies. It is based on a weighted combination of methods for calculating similarities.
- Accuracy of the results, the similarity measure adopted by our mapping method, is a composite (involves both: terminological similarity as well as lexical and syntactic, structural and semantics similarities), it gives a high level of relevance in determining matches between the original ontologies.

In terms of perspectives, we envisage to use the mapped process presented in this paper to develop an approach for merging ontologies. This is the next step of our work.

References

1. Auer, S., Bizer, C., Kobilarov, G., Lehmann, J., Cyganiak, R., Ives, Z.G.: DBpedia: A nucleus for a web of open data. In: Aberer, K., et al. (eds.) ISWC/ASWC 2007. LNCS, vol. 4825, pp. 722–735. Springer, Heidelberg (2007)
2. Castano, S., Ferrara, A., Montanelli, S.: Matching ontologies in open networked systems: Techniques and applications. In: Spaccapietra, S., Atzeni, P., Chu, W.W., Catarci, T., Sycara, K. (eds.) Journal on Data Semantics V. LNCS, vol. 3870, pp. 25–63. Springer, Heidelberg (2006)

3. Elbyed, A.: ROMIE, une approche d'alignement d'ontologies à base d'instances. Doctoral thesis. National Institute of telecommunications (2009), `http://tel.archives-ouvertes.fr/docs/00/54/18/74/PDF/TheseELBYED.pdf`
4. Euzenat, J., Valtchev, P.: Similarity-based ontology alignment in OWL-lite. In: Proc. 15th European Conference on Artificial Intelligence (ECAI), Valencia, ES, pp. 333–337 (2004)
5. Jerome, E., Shvaiko, P.: Ontology Matching. Springer, Heidelberg (2007)
6. Fergnani, A.: Ontology dynamics for Semantic Web: the MOMIS approach (2002), `http://www.dbgroup.unimo.it/tesi/fergnani.pdf`
7. Gangemi, A., Steve, G., Giacomelli, F.: ONIONS: An Ontological Methodology for Taxonomic Knowledge Integration. In: ECAI 1996 Workshop on Ontological Engineering, Budapest (1996)
8. Giunchiglia, F., Shvaiko, P.: Semantic matching. The Knowledge Engineering Review 18(3), 265–280 (2003)
9. Gotoh, O.: An improved algorithm for matching biological sequences. Journal of Molecular Biology 162(3), 705–708 (1982)
10. Jaro, M.A.: Advances in record linking methodology as applied to the 1985 census of Tampa Florida. Journal of the American Statistical Society 84(406), 414–420 (1989)
11. Jiang, J., Conrath, D.: Semantic similarity based on corpus statistics and lexical taxonomy. In: Proceedings on International Conference on Research in Computational Linguistics, Taiwan (1997)
12. Ahmed, K.A.: A Multi-Matching Technique for Combining Similarity Measures in Ontology Integration, A Thesis, Concordia, University,Montréal, Québec, Canada (2010)
13. Levenshtein, V.: Binary Codes Capable of Correcting Deletions, Insertions and Reversals. Soviet Physics Doklady 10, 707 (1966)
14. Li, Y., Zhong, Q., Li, J., Tang, J.: Results of ontology alignment with RiMOM. In: Proc. International workshop on Ontology Matching (OM), Busan, Korea, November 11, pp. 227–235 (2007)
15. McGuinness, D.L., Fikes, R., Rice, J., Wilder, S.: An environment for merging and testing large ontologies. In: Proceeding of KR, pp. 483–493 (2000)
16. Monge, A., Elkan, C.: The field-matching problem: algorithm and applications. In: Proceedings of the 2nd International Conference on Knowledge Discovery and Data Mining, pp. 267–270 (1996)
17. Noy, N.F., Musen, M.A.: The prompt suite: interactive tools for ontology merging and mapping. Int. J. Hum.-Comput. Stud. 59(6), 983–1024 (2003)
18. Ontology Alignment Evaluation Initative, Benchmarks results (2010), `http://oaei.ontologymatching.org/2010/results/benchmarks/index.html`
19. Ontology Alignment Evaluation Initiative Test5 library (2011), `http://oaei.ontologymatching.org/2011/benchmarks`
20. Rada, R., Mili, H., Bicknel, E., Blettner, M.: Development and application of a metric on semantic nets. IEEE Transaction on Systems, Man, and Cybernetics 19(1), 17–30 (1989)
21. Resnik, P.: Semantic similarity in a taxonomy: An information based measure and its application to problems of ambiguity in natural language. Journal of 2. Artificial Intelligence Research 11, 95–130 (1999)
22. Smith, T., Waterman, M.: Identification of common molecular subsequences. Journal of Molecular Biology 147, 195–197 (1981)
23. Stumme, G., Maedche, A.: FCA-merge: bottom-up merging of ontologies. In: 17th IJCAI, Seattle (WA US), pp. 225–230 (2001)

24. Suchanek, F., Kasneci, G., Weikum, G.Y.: A Large Ontology from Wikipedia and WordNet. Elsevier Journal of Web Semantics 6(3), 203–217 (2008)
25. Umer, Q., Mundy, D.: Semantically Intelligent Semi-Automated Ontology Integration. In: Proceedings of the World Congress on Engineering, WCE 2012, London, U.K., July 4 - 6, vol. II (2012)
26. Winkler, W.E.: The State of Record Linkage and Current Research Problems. Rapport interne, Statistical Research Division, U.S. Census Bureau (1999)
27. Wu, Z., Palmer, M.: Verb Semantics and Lexical Selection. In: Proceedings of the 32nd Annual Meetings of the Associations for Computational Linguistics, pp. 133–138 (1994)
28. Zahaf, A., Malki, M., Fellah, A.: Alignement des ontologies: utilisation de WordNet et une nouvelle mesure structurelle. In: Proceeding CORIA, pp. 401–408 (2008)

80 Gb/s WDM Communication System Based on Spectral Slicing of Continuum Generating by Chirped Pulse Propagation in Law Normal Dispersion Photonic Crystal Fiber

Leila Graini and Kaddour Saouchi

Laboratory of Study and Research in Instrumentation and Communication
of Annaba, Badji Mokhtar University, P.O. Box 12, Annaba 23000, Algeria
graini_leila@yahoo.fr

Abstract. In this paper, the combination of the initial positive chirp parameter of picosecond pulse, and the high non-linearity and low normal dispersion of photonic crystal fiber is studied to optimize the spectral flatness of continuum source for WDM system application. This continuum source is defined on the C-band of optical communication window and is capable of providing all the necessary channels after spectral slicing by an optical demultiplexer. If the initial source is delivered at repetition rate of 10 GHz, the obtained continuum allows to generate more than 32 channels spaced 100 GHz all centered at 1550 nm, where each channel is a pulses train has the same repetition rate than the initial source and suitable to modulate with modulation rate of 10 GHz to achieve data transfer rate of 10 Gb/s, the type of modulation format RZ-OOK was used for channels coding. In this context, the transmission chain of WDM communication system at 8.10 Gb/s is demonstrated.

Keywords: initial positive chirp parameter, parabolic pulse, WDM system, photonic crystal fiber, continuum source.

1 Introduction

In a classical WDM link (Wavelength Division Multiplexing), the transmitter part requires a set of laser diodes emitting at different wavelengths but close enough (in the vicinity of wavelength of 1550 nm), this technique is simple but its disadvantage is the laser noise; these fluctuations impose an ultimate limit to the performance of any optical system communications. This noise is caused especially by spontaneous emission in the laser diode. In addition, laser chirp is the major disadvantage in the WDM directly modulated system performance, and it increases when the bit rate increased, which limits the transmission at bit rate less than 5 Gb/s [1, 2].

In order to avoid these disadvantages, the continuum source technique used to replaces laser diodes, which is subsequently sliced in the spectral domain to create multi-wavelength light sources used as WDM transmitter [3, 4, 5], For this technique, optimizing the spectral density for a given pump power whilst maintaining a flat spectral

A. Amine et al. (Eds.): *Modeling Approaches and Algorithms*, SCI 488, pp. 217–225.
DOI: 10.1007/978-3-319-00560-7_26 © Springer International Publishing Switzerland 2013

profile is critical. In normally dispersive fibers, the combination of dispersion and self-phase modulation (SPM) can allow the generation of flatter spectrum [6]. The main limit to spectral pulse quality in this regime is the spectral ripple that arises from SPM of short pulse lasers. Such effects can be avoided by using chirped pulses [7]. Recently however we numerically studied the impact of initial chirp parameter on the nonlinear propagation of chirped picosecond pulse through a highly nonlinear and normal dispersion fiber for the continuum source generation, we found that the continuum spectrums generated by positive chirped pulses are ultra flat in the wavelength bands that are of utmost interest in optical communication systems. Moreover, positive initial chirp makes the required fiber length shorter for effective continuum [7].

In this paper, we present results in terms of the extent of the reduction of spectral ripple relative to the case for unchirped Gaussian shape pulses and then we go on to demonstrate the benefits of this technique can provide for spectral slicing source applications at WDM system. In this aim, we propose to generate a single continuum source capable of providing all the necessary channels for the efficiency of the WDM system. The continuum source we used is based on the propagation of chirped picosecond pulse in special highly nonlinear and law normal dispersion fiber PCF (Photonic Crystal Fiber) [8], and is defined on the C-band (1530 *nm* - 1565 *nm*) compatible with the spectral band of optical amplifiers currently used, with ultra flat power spectral density. The obtained continuum allows to generate more than 32 channels spaced 100 GHz all centered at 1550 nm. If the initial source is delivered at repetition rate of 10 GHz we can achieved a rate of 320 Gb / s.

2 Continuum Source Generation

The numerical model of the pulse propagation in PCF is the well-known generalized nonlinear Schrödinger equation (GNLSE) that is suitable for studying the evolution of ultrashort pulse in nonlinear media [9]:

$$\frac{\partial A}{\partial z}+\frac{\alpha}{2}A+i\frac{\beta_2}{2}\frac{\partial^2 A}{\partial t^2}-\frac{\beta_3}{6}\frac{\partial^3 A}{\partial t^3}=i\gamma(|A|^2 A+\frac{i}{\omega_0}\frac{\partial}{\partial t}(|A|^2 A)-T_R A\frac{\partial |A|^2}{\partial t}) \quad (1)$$

Equation (1) given for pulse shorter than $5ps$. Where A is the slowly varying amplitude of the pulse, α is the attenuation of the PCF, β_2 and β_3 are the coefficients of second and third-order dispersion respectively, γ is the nonlinear coefficient, and T_R=3fs at wavelength λ=1550 *nm*.

It is possible to solve equation (1) numerically by using the split step Fourier method (SSF) [9], the split step Fourier method used extensively to solve the pulse propagation problems in nonlinear dispersive media.

Using the real cross-section of the PCF used in [8], the chromatic dispersion has been computed for a wavelength range extending from 1060 to 1680 *nm*. The calculations were performed by means of a full vectorial finite element method [10]. The fiber shows ultra flattened chromatic dispersion of $D \pm 0.4$ *ps/(nm.Km)* from 1060 to 1680 *nm* wavelength range (D=0.8 *ps/(nm.Km)* at 1550 *nm*), has a high nonlinear coefficient $\gamma=51[W.Km]^{-1}$ at 1550 *nm*, and a third order dispersion β_3=-0.01 ps^3/Km.

For the PCF mentioned above, with low normal group velocity dispersion, small β_3, and high nonlinearity, it would be perfectly suit for required continuum spectrum for WDM application, when the flat spectrum are interesting because it can generate channels with the same power level.

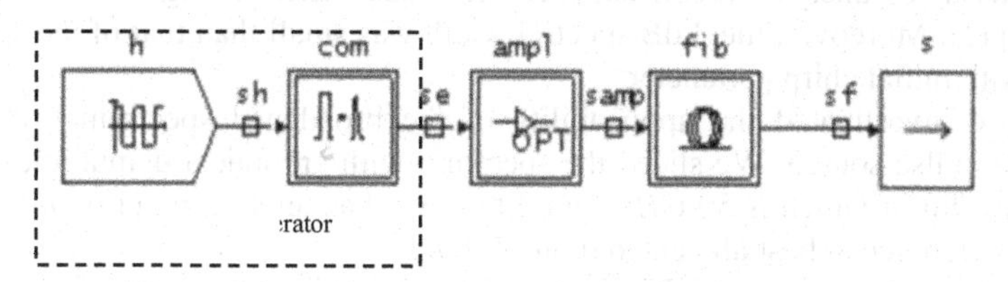

Fig. 1. Diagram of continuum generation

We assume the incident pulse, to be of chirped Gaussian shape, the electric field *A (0, t)* with chirp parameter c corresponding to such a pulse can be expressed in form:

$$A(0,t) = \sqrt{P_0}\exp\left(-\frac{(1+ic)t^2}{2T_0^2}\right) \tag{2}$$

Where P_0 is the power of the pulse and T_0 is the input pulse width, and it is related to the full wide at half maximum (FWHM) of the input pulse by $T_{FWHM} \approx 1.665T_0$. The specific values of parameters used in simulation are given as follows: $P_0 = 0.2$ W, $T_{FWHM} = 2.4$ *ps*, the central wavelength λ_0=1550 *nm* and the initial chirp parameter C=10 [7].

Fig.1 is the diagram of continuum generation that we are going to simulate. The proposed system components are simulated using COMSIS software.

The input signal is formed of clock representing binary states (0 or 1) of the message to be transmitted. For each binary state 1, one pulse is generated. The maximum of the pulse is located at the centre of bit time (optical short pulse generator of chirped Gaussian type at repetition rate of 10GHz). The optical amplifier of 30 dBm allows achieving levels of high peak power required for the non-linear effects to continuum generation. We obtained after optical amplification a pulses train at repetition rate of 10 *GHz*, which allows us to obtain after propagation in the non-linear fiber, a pulses train of 10 *GHz* having for each spectrum our continuum.

During the propagation of the chirped Gaussian pulse through the PCF, the pulse waveform becomes parabolic due to the interaction of the linear frequency chirp induced by the SPM effects, and the normal dispersion. The frequency chirp induced by a parabolic waveform is linear, and the accumulation of such chirp with the initial chirp parameter results in flat spectrum [7].

From Fig .2 we can see that the output spectrum of chirped Gaussian pulse shows a widening of 100 nm (bleu trace) with good spectral flatness over a range of 40 nm. the spectrum is considerably ultra flat in the central region of the pulse spectrum, this can be explain by that the sign of frequency chirp induced by the SPM effects varies to same as that of the initial chirp parameter. The continuum centered on the wavelength λ = 1550 nm and has spectral width of 40 nm can cover the C-band is suitable for WDM systems.

To compare the performance with that of the unchirped pulse, we plotted the curve of spectrum of pulse without initial chirp parameter and generated a broad spectrum from the same pulse source (Fig. 2: green trace). As shown in Fig. 2, the spectrum has a small oscillatory structure which could compromise the quality of the processed signal in terms of unequal WDM channel amplitudes after slicing of the continuum spectrum [11]. Moreover, the 3 dB spectral width was small than that of the spectrum of pulse with initial chirp parameter.

Next, we investigated the applicability of the broadened spectrum to a multi-wavelength pulse source. We sliced the spectrum with an optical demultiplexer. The total bandwidth of which is 50 *GHz* with 200 GHz channel spacing (1.6 *nm*) in order to limit interference at best all centered on 1550 *nm*.

We show a superposition of the spectra of 16 channels in the output of the optical demultiplexer (Fig.3). The spectral width of the channels is taken less than or equal to that of the filter of the optical demultiplexer. We obtained sub-band spectral of 1 *nm*. The channels are generated in the 1540- 1558.2 *nm* wavelength range and have the same power level.

In the following, we consider the propagation of eight channels at 10 GHz repetition rate with an inter-channel separation of 200 GHz. For ease of interpretation of results, the number of channels has been kept low. It is possible to analyze more number of channels since the design can support these possibilities.

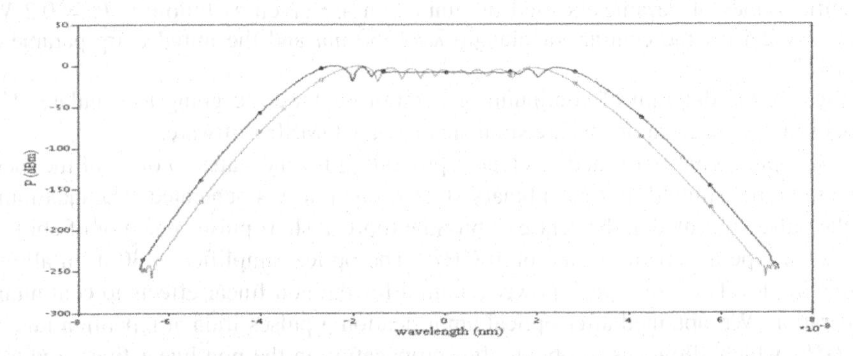

Fig. 2. Spectrum of chirped pulse (bleu trace), spectrum of unchirped pulse (green trace)

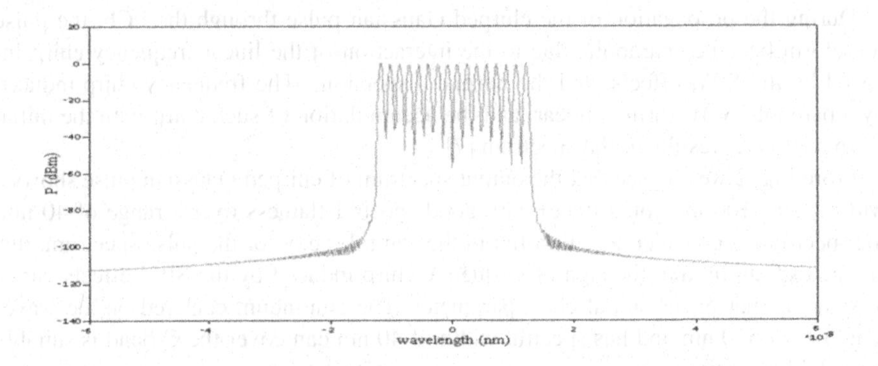

Fig. 3. Spectrum obtained after spectral slicing by optical demultiplexer

3 8-Channel WDM System Simulation

The simulation setup of the WDM system is shown in Fig.4. This is essentially the same as proposed in our earlier work [11]. The transmitter was based on the bock diagram represented in Fig. 1 and produced a 10 GHz train of 2.4 ps chirped pulses at a wavelength of 1550 nm. These pulses were then used for generating continuum sources, and are subsequently sliced in the spectral domain to create an 8-channel WDM multi-wavelength source (wavelength 1-8). The wavelengths range from 1548 nm to 1551.2 nm, with 1.6 nm wavelength spacing (200 GHz) i.e. carrier signal and this carrier signal is used to modulate the pseudo random bit sequence data of the user to compose a 10 Gbit/s RZ transmitter. An intensity modulator which is Mach-Zehnder is used on-off keying modulation (OOK) to modulate each channel according to the RZ electrical data. The encoded data from all users are multiplexed by optical multiplexer and then passed through a 20 km span of standard single mode optical fiber (SMF) followed by a 2 Km span of dispersion compensating fiber (DCF) and a loss compensating optical amplifier which is EDFA. The multiplexing signal finally arrives at optical receiver. The receiving block is designed to convert the optical signal that carries the information into electrical pulses. It is composed of demultiplexer of 50 *GHz* bandwidth and 200 GHz channel spacing used to separate different channels, PIN photodiodes, and decision circuits [12].

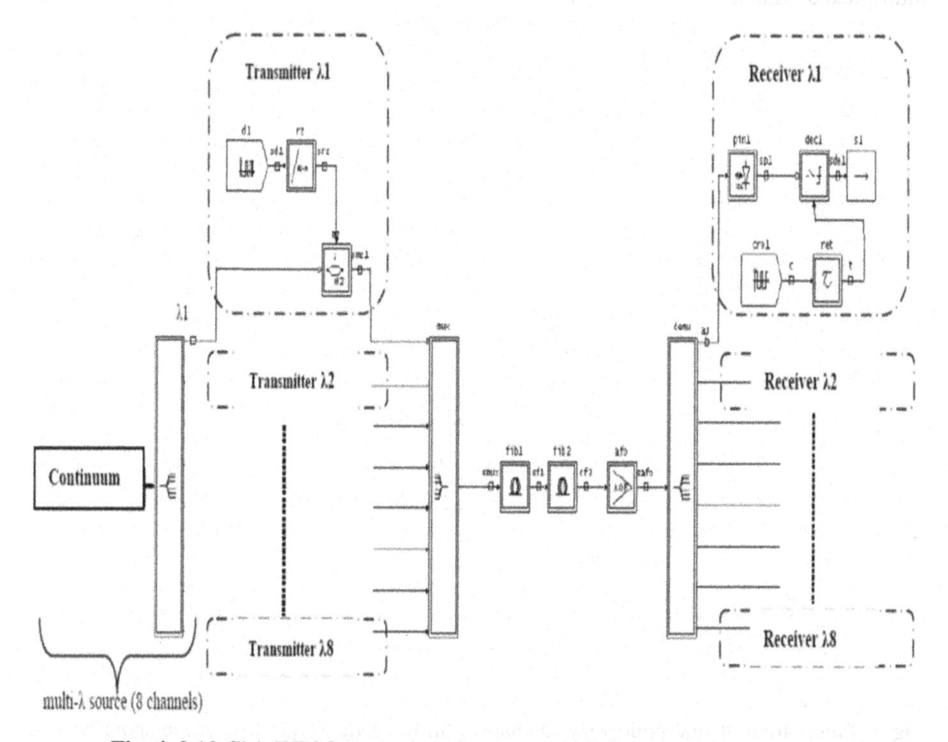

Fig. 4. 8.10 Gb/s WDM system based on spectral slicing of continuum source

Once the link is built, and according to the result of simulations, we visualized the signals at various points of the link. This allows to visualizing the successive transformations of the signal during his career as well as the behavior and influence of each block.

On the transmission side of the system, the pulses train of the first optical channel centered in the 1550 *nm region* is plotted in Fig.5.a. This region was chosen to minimize influence of fiber loss and to have a potential opportunity to use EDFAs for spans longer than 10 *km*. Each channel modulated with modulation rate to achieve data transfer rate of 10 Gb/s. the Return to Zero (RZ) code format was used for signal coding (Fig.5.b). The temporal waveform of modulated sliced signal is shown in Fig.5.c. The resulting signal is RZ-OOK (Return to Zero – On –Off Keying) type format and whose pulse duration is 0.33 % of the bit time. The pulse widths products are almost constant at ~7*ps* across all channels, as determined mainly by the demultiplexer characteristics.

For multiplexing the 8- modulated channels into a single fiber, the optical multiplexer has the same parameters as the demultiplexer used for slicing the continuum in terms of spectral bandwidth and channel spacing is used to reduce cross-talk between adjacent channels. The system was designed so that to minimize noise from the adjacent channels. The spectrum of the 8-optical channels multiplexed signal at output of multiplexer is shown in Fig. 6.a. The optical spectrum of the signal from the multiplexer allows to illustrating the superposition of channels, which is operated by wavelength multiplexing (power spectral density of the optical signal at the output of multiplexer). The spectral width of the optical signals is visible in the spectrum of the multiplexed signal.

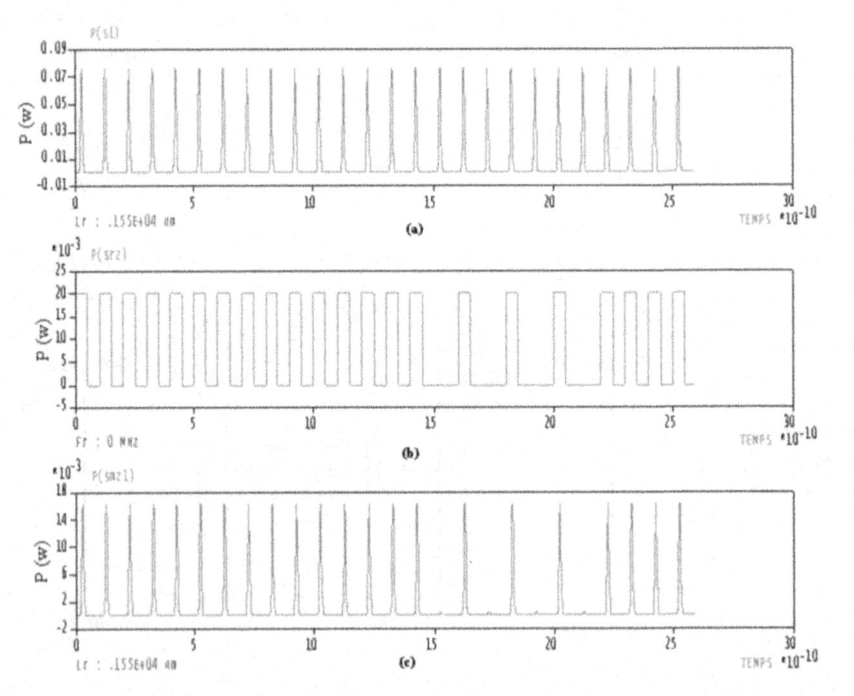

Fig. 5. Pulses train of first optical sliced channel (a) RZ code format (b), and temporal waveform of first modulated channel (c)

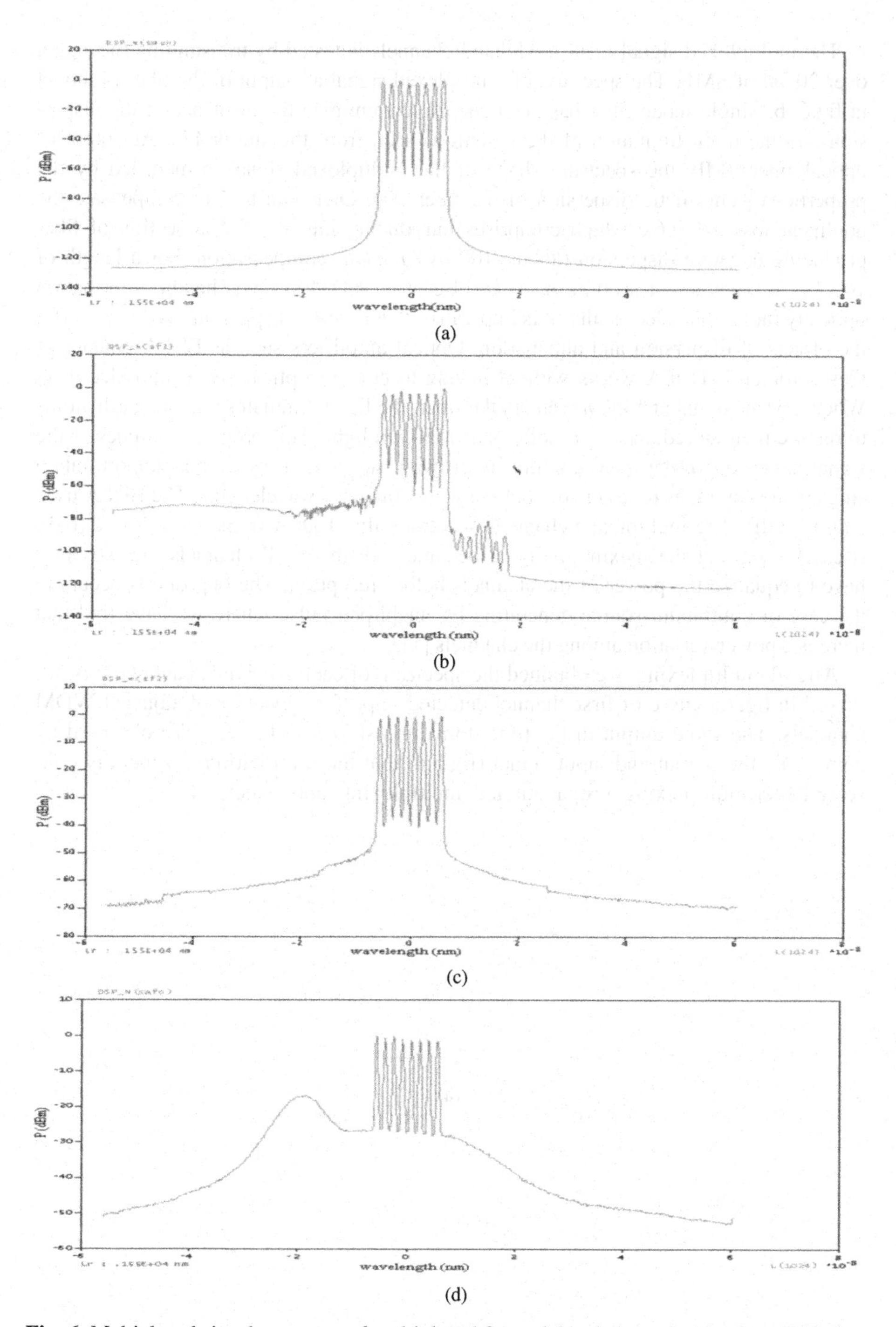

Fig. 6. Multiplexed signal spectrum of multiplexed 8- modulated channels (a), after: SMF (b), DCF (c), and EDFA (d)

The multiplexed signal of 8- modulated channels is tested by transmitting the signal over 20 *km* of SMF. The spectrum of multiplexed signal at output of the fiber is shown in fig.6. b. Single mode fiber has been chosen to minimize the influence of the dispersion and the main limitation of the system as seen from the simulation. At output of optical fiber SMF, the spectrum shape of the multiplexed signal is modified by the properties of chromatic dispersion of the fiber. The DCF is a tool to compensate the nonlinear losses Fig.6.c which comprises introducing into the link a section of fiber producing negative dispersion (about -100 *ps / nm.km*) compensation. For a length of 100 *km*, it takes about section of 10 *km*. Because WDM systems handle information optically rather than electrically, it is imperative that long-haul applications do not suffer the effects of dispersion and attenuation. Optical amplifiers such as EDFA counteract these problems. EDFA works without having to convert optical signal into electrical. When a weak signal at 1550 *nm* enters the fiber, the light stimulates the rare earth atoms to release their stored energy as additional 1550 *nm* light. This process continues as the signal passes down the fiber, continually growing stronger. They do not support only a single wavelength, as repeaters do, but the whole range of wavelengths. The EDFA used gain is 30dB. The multiplexed channels spectrum after EDFA is shown in figure (6.d). It can be see that the maximum signal becomes 0dBm for all channels, so, we don't have to equalize the power of the channels before reception. The opposite is occurs in the case of continuum source generating by unchirped pulse where we have find that there is a power variation among the channels [11].

After demultiplexing, we obtained the spectrum of each received channels. It is observed in fig.7.a curve of first channel detected imperfect rejection of adjacent WDM channels. The same output in the time domain is shown in Fig.7.b. We observed by comparing the output and input signal (fig.5.c) that the information is generally conserved after multiplexing, propagation, demultiplexing, and detection.

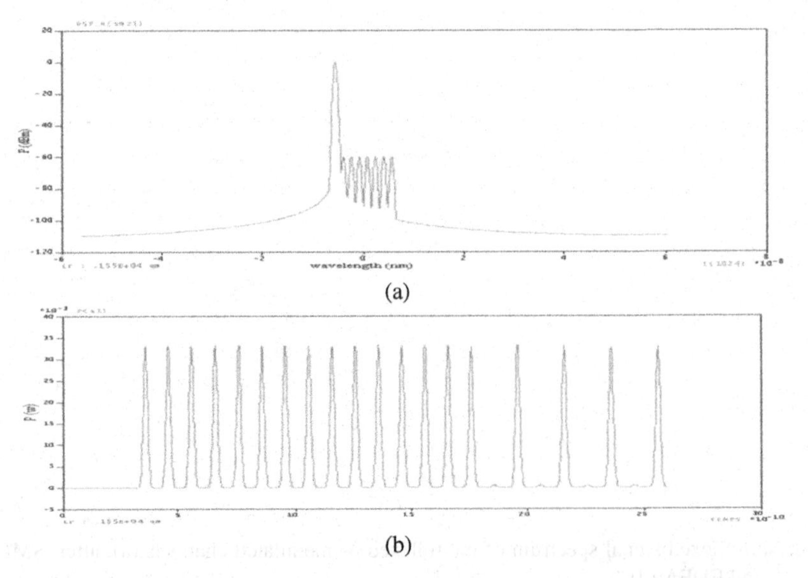

Fig. 7. Spectrum (a) and temporal waveform of first detected channel (b)

4 Conclusion

We have developed a single optical source to achieve WDM system by exploiting the combined effects of initial positive chirp parameter of picosecond pulse and the law normal dispersion PCF on the continuum source generation. The obtained continuum source is an ultra flat continuum covering the whole C-band of optical communication window and is capable of providing all the necessary channels with the same power level after spectral slicing by an optical demultiplexer of 50 GHz bandwidth and 200 GHz (1.6 nm) channel spacing suitable to modulate by user data. In order to well understand the proposed method, the 8-channel WDM telecommunication system at 8.10 Gb/s is demonstrated. The performance of a multi-channel derived from a continuum source will be numerically studied.

References

1. del Río Campos, C., Horche, P.R., Martín Minguez, A.: Effects of Modulation Current Shape on Laser Chirp of 2.5 Gb/s Directly Modulated DFB-Laser. In: Proceedings of the Third International Conference on Advances in Circuits, Electronics and Micro-electronics, CENICS 2010, Venice, Mestre, Italy, July 18-25, pp. 51–55 (2010)
2. del Río Campos, C., Horche, P.R., Martín Minguez, A.: Interaction of semiconductor laser chirp with fiber dispersion: Impact on WDM directly modulated system performance. In: Proceedings of the Fourth International Conference on Advances in Circuits, Electronics and Micro-electronics, CENICS 2011, Nice, France, August 21-27, pp. 17–22 (2011)
3. Takushima, Y., Kikuchi, K.: 10-GHz, over 20-channel multiwavelength pulse source by slicing super-continuum spectrum generated in normal-dispersion fiber. IEEE Photon. Technol. Lett. 11, 322–424 (1999)
4. Ozeki, Y., Taira, K., Aiso, K., Takushima, Y., Kikuchi, K.: Highly flat super-continuum generation from 2 ps pulses using 1 km-long erbium-doped fibre amplifier. Electron. Lett. 38(25), 1642–1643 (2002)
5. Ozeki, Y., Takushima, Y., Aiso, K., Taira, K., Kikuchi, K.: Generation of 10 GHz similariton pulse trains from 1.2 Km-long erbium doped fiber amplifier for application to multi-wavelength pulse sources. Electronics Letters 40(18) (2004)
6. Xu, Y., Ren, X., Wang, Z., Zhang, X., Huang, Y.: Flatly broadened supercontinuum generation at 10 Gbits using dispersion flattened photonic crystal fibre with small normal dispersion. Electronics Letters 43(2) (2007)
7. Graini, L., Saouchi, K.: Effect of initial chirp parameter on the continuum generation for WDM application. In: International Conference on Embedded Systemsin Telecommunications and Instrumentation (icesti 2012), annaba, algeria, November 5-7 (2012)
8. Begum, F., Namihira, Y., Kinjo, T., Kaijage, S.: Supercontinuum generation in photonic crystal fibers at 1.06, 1.31, and 1.55 um wavelengths. Electronics Letters 46(22) (2010)
9. Agrawal, G.P.: Nonlinear Fiber Optics, 3rd edn. Academic, New York (2001)
10. Zghal, M., Chatta, R., Bahloul, F., Attia, R., Pagnoux, D., Roy, P., Melin, G., Gasca, L.: Full vector modal analysis of microstructured optical fiber propagation characteristics. In: SPIE Proc., vol. 5524, pp. 313–322 (2004)
11. Graini, L., Saouchi, K.: Simulation of DWDM communication system at 8.10 gbit/s using continuum source in the transmitter. Accepted for published in Acta Technica Napocensis Electronica-Telecomunicatii 54(1) (2013)
12. Verneuil, J.-L.: Simulation of de telecommunications systems of optical fiber at 40 Gb/s. PhD thesis, Limoges university (2003)

Fuzzy Logic Based Utility Function for Context-Aware Adaptation Planning

Mounir Beggas[1,4], Lionel Médini[2], Frederique Laforest[3], and Mohamed Tayeb Laskri[4]

[1] Université de Lyon, CNRS, INSA-Lyon - LIRIS UMR5205
[2] Université de Lyon, LIRIS UMR5205 CNRS
[3] Université de Lyon, LT2C,Telecom Saint Etienne
[4] Department of Computer Science, University of Annaba, Algeria
{mounir.beggas,lionel.medini}@liris.cnrs.fr,
frederique.laforest@telecom-st-etienne.fr,
laskri@univ-annaba.org

Abstract. Context-aware applications require an adaptation phase to adapt to the user context. Utility functions or rules are most often used to make the adaptation planning or decision. In context-aware service based applications, context and Quality of Service (QoS) parameters should be compared to make adaptation decision. This comparison makes it difficult to create an analytical utility function. In this paper, we propose a fuzzy rules based utility function for adaptation planning. The large number of QoS and context parameters causes rule explosion problem. To reduce the number of rules and the processing time, a rules-utility function can be defined by a hierarchical fuzzy system. The proposed approach is validated by augmenting the MUSIC middleware with a fuzzy rules based utility function. Simulation results show the effectiveness of the proposed approach.

Keywords: context-awareness, middleware, adaptation planning, QoS, fuzzy logic.

1 Introduction

Mobile context-aware service based applications involve two kinds of parameters: context parameters and service QoS parameters. The adaptation planning selects the service variant that best fits the current context state. Rules based systems are used for adaptation planning [1] [2]. In such systems, the application designers need to explicitly describe the rules defining adaptive reaction of the system in response to context change. Utility function based adaptation planning [3] is proposed to overcome theses limitations. Utility function is defined independently of existing service variants. It is defined by an analytical model that uses service QoS parameters to score service variants according to the current context sate. Formulating such analytical model is a complex task [3–5] and it has always been a source of confusion for the designers.

A fuzzy control system is a computing with words manner [6] that maps the input space to the output space using fuzzy sets and human-like if-then

A. Amine et al. (Eds.): *Modeling Approaches and Algorithms*, SCI 488, pp. 227–236.
DOI: 10.1007/978-3-319-00560-7_27 © Springer International Publishing Switzerland 2013

fuzzy rules. Fuzzy control systems can define nonlinear systems by manipulating imprecise data [7]. It has been successfully applied to various application-specific network QoS management systems [8], service QoS management systems [4] [9]. Fuzzy logic controllers can be used for adaptation planning design, allowing context and application QoS parameters to be expressed with fuzzy sets and linguistic terms.

The large number of QoS and context parameters in a mobile operating environment causes the rule explosion problem. Hierarchical fuzzy systems [10] [11] were introduced to reduce the number of rules using hierarchical fuzzy control, in which correlated linguistic variables are hierarchically inferred and grouped into abstract linguistic variables.

In this paper, we propose a fuzzy rule based utility function for adaptation planning in context-aware middleware. The proposed approach combines the advantages of rule based system and utility based systems. In addition it can manipulate imprecise data. To address the problem of the explosion in the number of rules, we propose to use hierarchical fuzzy control techniques. The proposed approach is successfully applied to augmenting the MUSIC[1] middleware by a fuzzy logic based utility function.

The remainder of this paper is structured as follows. Section 2 describes the related work. An introduction to context-aware middleware and the MUSIC middleware is presented in Section 3. In section 4, we present a definition and a modeling of context and QoS parameters. In Section 5, we present the proposed approach for adaptation planning based on fuzzy adaptation rules. Section 6 provides experimental results and evaluates the efficiency of the proposed approach. Section 7 concludes the paper and opens perspectives for the future work.

2 Related Work

Researchers have proposed several techniques to simplify service based application reconfiguration. In [1] znd [2], the adopted adaptation strategy is based on situation-action rules. Each rule is formulated in terms of conditions and actions. An obvious limitation of rule-based approaches is the imposed binary decision logic. Another limitation of rule-based approaches for SBA adaptation is that they do not allow to dynamically add new variants without redefining the whole set of rules. Furthermore, application designers need to explicitly describe the adaptive reaction of the system in response to any relevant context change.

To overcome these limitations, [3] and [12] use a utility function for adaptation decision for context aware applications. Utility function is also applied to adaptation planning for home automation applications [13]. Utility functions are mathematical artifacts that map each possible system state, or alternative implementation, to a real scalar value. For SBA reconfiguration, the highest utility score indicates the most suitable variant for the current context. In MADAM and MUSIC [3] [14], the middleware uses utility functions to calculate utility scores

[1] MUSIC, acronym of Self-Adapting Applications for Mobile USers In ubiquitous Computing environments. `http://ist-music.berlios.de`

for each application variant. Defining utility functions, especially for complex applications with multiple QoS and context parameters is a difficult task [3–5]. Utility functions are based on variant properties prediction, current context situation and user preferences. Therefore, application developers need to formulate every possible adaptation aspect according to different context states in mathematical equations.

3 Middleware for Context Management

In context-aware SBAs, the adaptation control loop is controlled by a middleware [4] [14] where the application management is based on IBM's MAPE-K (Monitor, Analyze, Plan, Execute, Knowledge) reference model for autonomic control loops [15]. The middleware manages a collection of service based applications and seeks to maximize the overall utility. Figure 1 illustrates the main components of the middleware. The context manager monitors and analyzes the environment context; on context changes, an adaptation planning process is started. The adaptation planning deduces the suited adaptation plan for the current context situation. The reconfiguration engine reconfigures application services by executing the adaptation plan. The reconfigured service based application is composed of selected service variants and the necessary bindings between them.

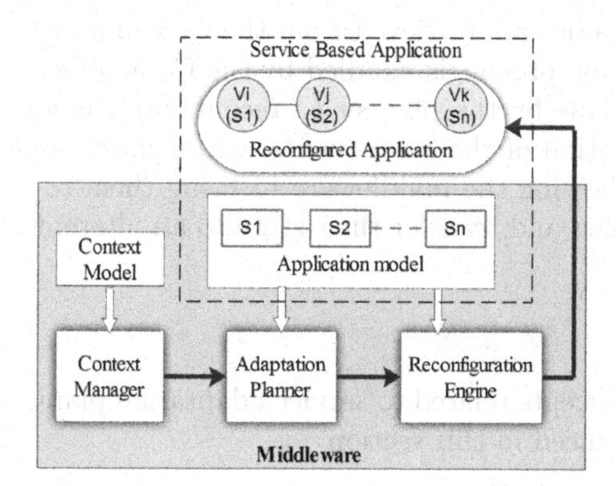

Fig. 1. Main components of the middleware

3.1 MUSIC Middleware

The MUSIC Middleware provides an execution environment managing the self-adaptation of service oriented mobile applications in ubiquitous environments. The middleware manages a collection of active applications and seeks to maximize the overall utility in different context conditions [16].

Figure 2 depicts the component-based architecture of the MUSIC platform. The planning is triggered by context changes detected by the Context Manager.

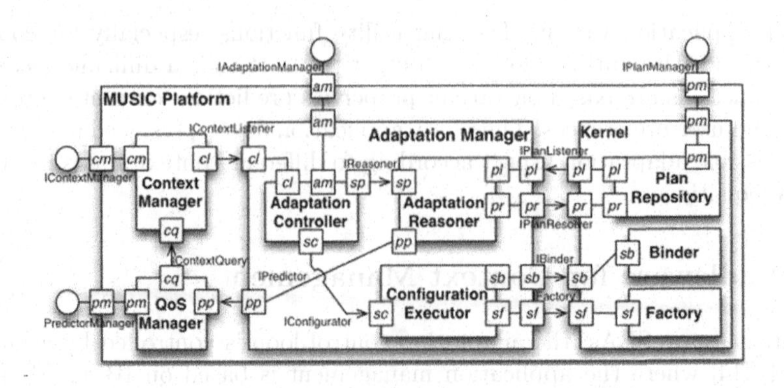

Fig. 2. Architecture of MUSIC middleware [16]

The Adaptation Controller coordinates the adaptation process. The Adaptation Reasoner supports the execution of the planning heuristics, which is driven by metadata included in the plans (service variant parameters). The Plan Repository provides an interface IPlan Resolver to the adaptation reasoner allowing for the recursive retrieval of variant plans associated to a given service. The adaptation reasoner builds a valid application configuration and discards those whose dependencies are unresolved. Then, the heuristics ranks the application configurations by evaluating their utility based on the computation of the predicted properties, whose values are retrieved from the QoS Manager.

The reconfiguration process is handled by the Configuration Executor, which uses the selected plans by the planner to reconfigure the application. This requires the collaboration of the components, which must implement a reconfiguration interface allowing the middleware to bring them to a state where they can be safely replaced and transfer their state to an alternative component.

4 Definitions

The definition of concepts related to service adaptation planning and their fuzzy extension are introduced in this section.

Definition 1 (Service). *A service is an abstract component that defines a class of service variants offering similar functionalities. It comprises functional and extra-functional parameters. Service parameters are defined as follows:* $S = (F, QoS)$. *The functional interface* $F(S) = (I, O)$ *specifies the service inputs and outputs. Extra-functional parameters denoted* $QoS(S)$ *are detailed below.*

Definition 2 (Service variant). *A service variant* $V_i(S)$ *or simply a variant is an instance of the service* S. *All variants of* S *share the same functional interface and are defined with specific values of QoS parameters.*

$$V(S) = \{V_i(S),\ V_i(S) = (F(S),\ QoS(V_i(S))) \mid i = 1 \ldots k\}, \tag{1}$$

with k being the number of variants of the service S. $QoS(V_i(S))$ is defined below.

Definition 3 (QoS parameters). *QoS parameters of the service S are defined as follows:*

$$QoS(S) = \{Q^j \mid j = 1 \ldots n_Q\}, \tag{2}$$

where n_Q is the number of quality parameters for the service S and Q^i is a class of QoS parameters defined on a domain $dom(Q^i)$.

The set of quality parameters of the ith variant of the service S is represented as: $QoS(V_i(S)) = \{q_i^j \in dom(Q^j) \mid j = 1 \ldots n_Q\}$.

Definition 4 (Context parameters). *The context is modeled through a finite set of context parameters. In particular, for a given application, we define its contextual environment as a set of n_c context parameters $\{C_K \mid K = 1 \ldots n_c\}$.*

Definition 5 (Context state). *A context state corresponds to an assignment of values to context parameters at some point in time t. In particular, a context state $W_t = \{c_k \in dom(C_k) \mid K = 1 \ldots n_c\}$.*

Definition 6 (Service Based Application). *A service based application is modeled by a composition of services and the necessary binding between them. Let application services be defined as: $A = \{S_i \mid i = 1 \ldots n_s\}$, with n_s being the number of application services. The running application is performed by a dynamic binding between service variants. In a given context state W_t, the running application is: $A(W_t) = \{V_j(S_i) \in V(S_i) \mid S_i \in A \text{ and } i = 1 \ldots n_s,\ j \in [1 \ldots k_i]\}$, with k_i being the number of variants of the service S_i.*

Definition 7 (Adaptation planning). *An adaptation planning is the decision making process that maps the current context state to the suitable service variants by selecting the variants maximizing the application utility. The application overall utility is a function of the form:*

$$U_c = f(U_{V_{j1}(S_1)}, U_{V_{j2}(S_2)}, \ldots, U_{V_{jn}(S_n)}, U_p),$$

where, U_p is the contribution of parameters, that are non-related to the individual components, rather related to the composition, communication among components etc., to the utility.

To calculate the application overall utility, we use the following utility function proposed for MUSIC in [17]:

$$U_c = \sum_{i=1}^{n} W_i \, U_{V_{ji}(S_i)} \; + \; W_{n+1} \, U_p \;,$$

where, $\sum W_i = 1.0$ and each W_i indicates the relative importance (weight term) of a service within a composition.

Each variant utility function has as input variant $QoS(V_i)$ and context parameters W_t. It is defined by the following function:

$$U_{V_i(S_j)} = f\left(W_t, QoS(V_i)\right). \tag{3}$$

5 Fuzzy Rules Based Utility for Adaptation Planning

In this work, we propose to use fuzzy logic controller to define a utility-rule based adaptation planning; where the adaptation control loop follows the MAPE-K reference model presented in Section 3. A service variant has a certain utility for a particular context state based on its QoS-properties. Utility can be evaluated at run-time by a designer-defined fuzzy rule-based function. For that, fuzzy sets and fuzzy terms are defined for each context and QoS parameters. The proposed adaptation planner is depicted in Figure 3.

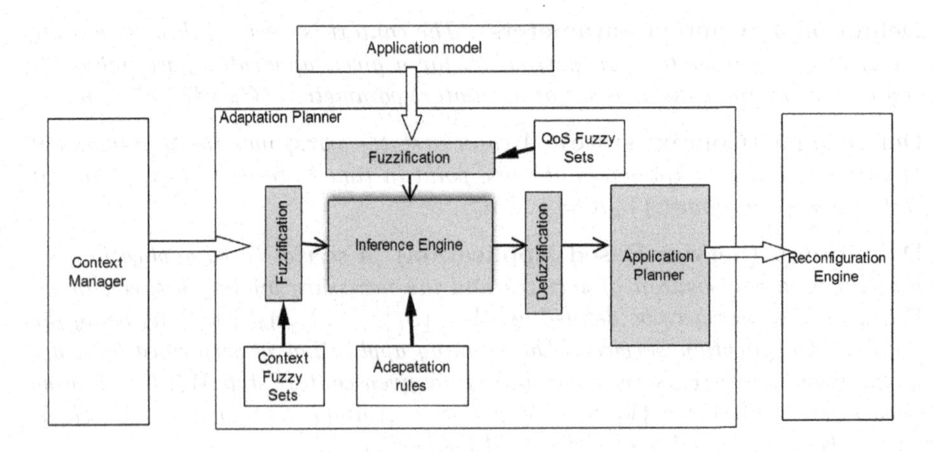

Fig. 3. Fuzzy control based adaptation planner

The adaptation planner receives current context state values from context manager. Each context parameter is fuzzified to the relevant fuzzy terms based on the predefined context fuzzy sets. Also, variants' QoS parameters are fuzzified based on the predefined QoS fuzzy sets. For each variant, a fuzzy reasoning process is applied by the inference engine to deduce its utility in the current context state. The application planner, for each application service, selects the variant that best satisfies the current context state. The adaptation plan is then sent to the reconfiguration manager for plan realization. Next sections detail the proposed planning function.

5.1 QoS and Context Fuzzy Sets

Fuzzy sets represent the fuzzy meaning of a given term. The defined terms are used by the designer to define the adaptation rules. For each QoS and context parameter, a set of suitable fuzzy sets are defined. Fuzzy sets can be defined for binary or multi-value parameters. The binary parameter can be defined by a singleton fuzzy sets. For example location (indoor/outdoor) can be described by singleton fuzzy sets of Figure 4(a). The multi-value parameter can be defined by continuous fuzzy set membership function. For example, the battery-level is a value between $0 - 4\,ah$ that can be Low, Medium or High level (Figure 4(b)).

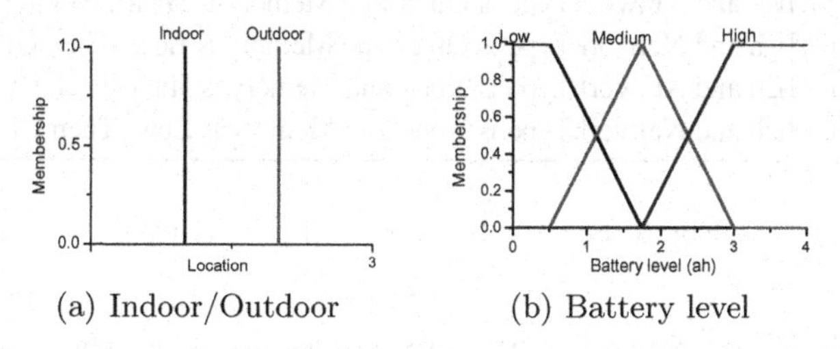

(a) Indoor/Outdoor (b) Battery level

Fig. 4. Linguistic terms and fuzzy sets membership function examples

Having defined the fuzzy set membership functions, the mapping between the context and QoS parameters, and utility of each variant is done by defining a list of if-then adaptation rules.

5.2 Utility Adaptation Rules

Fuzzy logic adaptation decision is mainly based on if-then adaptation rules. Adaptation rules use the defined QoS and context linguistic terms to make antecedent propositions, and utility terms for consequent propositions. Adaptation rule antecedent propositions concern context and QoS parameters, for instance *(NetworkType is WiFi and Memory is High)*. During rule evaluation, all the rules are applied in parallel and their order is not significant. For fuzzy rule examples, see Figure 5. The rule based utility function is evaluated by the fuzzy inference process [7].

To compare our proposition with analytical utility function, wa take an example of utility function applied for MUSIC application from [17]:

$$
\begin{aligned}
Ut =&0;\ if\ EnvNoise = LOW\\
&((1.0;\ if\ context.Memory \geq 100\\
&\quad 1.0 - (100 - context.Memory)/100;\ otherwise;\\
&) + (1.0;\ if\ context.NetworkType = WiFi\\
&\quad 0.0;\ if\ context.NetworkType = None\\
&\quad 0.5;\ otherwise\\
&))/2.0;\ otherwise
\end{aligned}
$$

This function is difficult to define, and difficult to update [3–5]. The proposed fuzzy rule based utility function that replaces this function is depicted in Figure 5. Compared to analytical utility function, fuzzy rules based utility function is simple to define and to update, and intuitive to understand as fuzzy rules follow human logic [18].

IF EnvNoise is Low Then Ut is VeryLow
IF EnvNoise is High and NetworkType is WiFi and Memory is High Then Ut is High
IF EnvNoise is High and NetworkType is WiFi and Memory is Low Then Ut is MedHigh
IF EnvNoise is High and NetworkType is Other and Memory is High Then Ut is MedHigh
IF EnvNoise is High and NetworkType is Other and Memory is Low Then Ut is MedLow
IF EnvNoise is High and NetworkType is None and Memory is High Then Ut is MedLow
IF EnvNoise is High and NetworkType is None and Memory is Low Then Ut is Low

Fig. 5. Fuzzy rules based utility function

A hierarchical fuzzy controller [10] is introduced to reduce the number of rules and as a consequence the processing time. Adaptation rules can be transformed into a hierarchical adaptation rules by hierarchically inferring and grouping correlated terms into abstract terms for the input of higher level rules.

6 Evaluation

In order to evaluate our approach, we have carried out two evaluation studies. First of all, we evaluated the feasibility of the proposition by simulation of the adaptation process with random generated data. We simulate the inference process in two manners: (1) fuzzy control system and (2) hierarchical fuzzy control system. The effectiveness of the proposed approach is evaluated by a comparison between the results of both systems with respect to execution time and number of rules. Second, we augment the MUSIC middleware by a the proposed fuzzy rules based adaptation planning. The proposed approach is validated based on different basic examples provided with the MUSIC platform in the web page: `http://ist-music.berlios.de/site/documents.html`.

We have implemented and tested our proposed approach using the fuzzy control language and tool JFuzzyLogic[2].In our experiments, we used the Mamdani type [19] for rules and inference models. The experiments have been conducted on a Pentium Core i5 with 6 GByte RAM.

6.1 Performance Evaluation

To evaluate the effectiveness of the proposed approach we have varied the number of variants from 10 to 500 and compare the execution time in both approaches. Figure 6 illustrates the obtained results. To compare the number of fuzzy rules of both approaches, we vary the number of QoS and context parameters from 3 to 8. This comparison is shown in Figure 7.

[2] `http://jfuzzylogic.sourceforge.net`

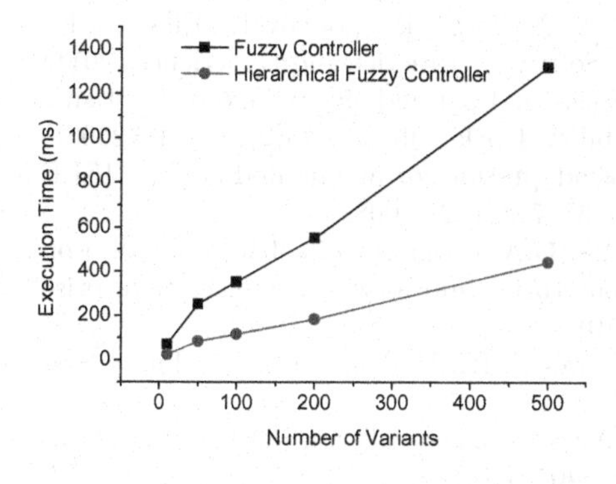

Fig. 6. Execution time according to the number of variants

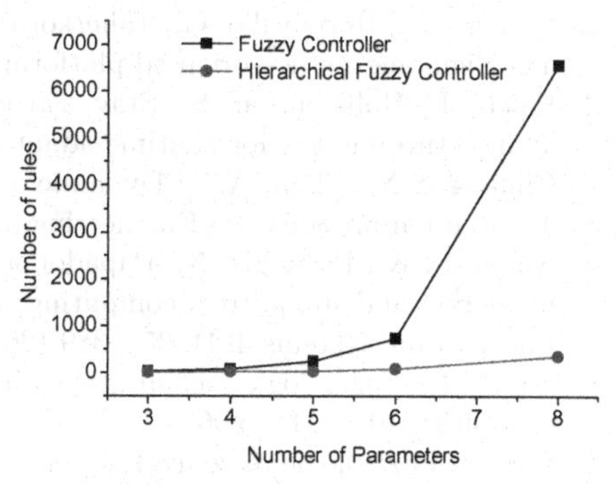

Fig. 7. Number of rules vs number of QoS and context parameters

In Figure 6, we notice the difference of the execution time of both approaches. As shown in Figure 7, the number of rules increases linearly as a function of the number of parameters when applying the hierarchical fuzzy control. Although, it increases exponentially when applying the non hierarchical fuzzy control system.

7 Conclusion and Future Work

We have described an adaptation planning approach for middleware dealing with context-aware applications. This approach is based on the fuzzy logic theory. It uses a fuzzy logic control system to deal with the adaptation decision problem in a human like if-then rules. To reduce the number of rules and the processing time, we show that the adaptation rules can be described hierarchically.

As future work, many challenges still need to be addressed. For example, we plan to incorporate more optimization in the decision making process. It can be done by defining a well adapted service composition overall utility. This requires the definition of more complex rules to represent the dependencies between parameters of different variants. Another problem that could be studied concerns the integration of services with the same functional parameters but with some differences in the types of extra-functional parameters. This requires developing a new adaptation strategy to make decision.

Acknowledgements. Many thanks to Mohammad Ullah Khan of the MUSIC project for his friendly help on the MUSIC system.

References

1. Garlan, D., Cheng, S., Huang, A., Schmerl, B., Steenkiste, P.: Rainbow: Architecture-based self-adaptation with reusable infrastructure. Computer 37(10), 46–54 (2004)

2. Romero, D., Hermosillo, G., Taherkordi, A., Nzekwa1, R., Rouvoy, R., Eliassen, F.: The digihome service-oriented platform. Software - Practice and Experience (2011)
3. Floch, J., Hallsteinsen, S., Stav, E., Eliassen, F., Lund, K., Gjorven, E.: Using architecture models for runtime adaptability. IEEE Software 23(2), 62–70 (2006)
4. Chuang, S.N., Chan, A.T.: Dynamic qos adaptation for mobile middleware. IEEE Transaction on Software Engineering 34(6), 738–752 (2008)
5. Kakousis, K., Paspallis, N., Papadopoulos, G.A.: A survey of software adaptation in mobile and ubiquitous computing. enterprise information systems. Enterprise Information Systems 4(4), 355–389 (2010)
6. Zadeh, L.: Fuzzy logic= computing with words. IEEE Transactions on Fuzzy Systems 4(2), 103–111 (1996)
7. Lee, C.: Fuzzy logic in control systems: fuzzy logic controller. i. IEEE Transactions on Systems, Man and Cybernetics 20(2), 404–418 (1990)
8. Tsang, D., Bensaou, B., Lam, S.: Fuzzy-based rate control for real-time mpeg video. IEEE Trans. Fuzzy Systems 6(4), 504–516 (1998)
9. Pernici, B., Siadat, S.H.: A fuzzy service adaptation based on qoS satisfaction. In: Mouratidis, H., Rolland, C. (eds.) CAiSE 2011. LNCS, vol. 6741, pp. 48–61. Springer, Heidelberg (2011)
10. Yager, R.: On the construction of hierarchical fuzzy systems models. IEEE Transactions on Systems, Man, and Cybernetics 28(1), 55–66 (1998)
11. Lee, M.L., Chung, H.Y., Yu, F.M.: Modeling of hierarchical fuzzy systems. Fuzzy Sets and Systems 138(2), 343–361 (2003)
12. Paspallis, N., Kakousis, K., Papadopoulos, G.: A multi-dimensional model enabling autonomic reasoning for context-aware pervasive applications. In: Proceedings of Mobiquitous 2008 (2008)
13. Bratskas, P., Paspallis, N., Kakousis, K., Papadopoulos, G.: Applying utility functions to adaptation planning for home automation applications. Information Systems Development, 529–537 (2010)
14. Geihs, K., Barone, P., Eliassen, F., Floch, J.: A comprehensive solution for application adaptation. Software Practice and Experience 39(4), 385–422 (2009)
15. Huebscher, M., McCann, J.: A survey of autonomic computing degrees, models, and applications. ACM Comput. Surv 40(3), 1–28 (2008)
16. Rouvoy, R., Barone, P., Ding, Y., Eliassen, F.: Music: Middleware support for self-adaptation in ubiquitous and service-oriented environments. Software Engineering for Self-adaptive Systems, 164–182 (2009)
17. Khan, M., Reichle, R., Wagner, M., Geihs, K., Scholz, U., Kakousis, C., Papadopoulos, G.: An adaptation reasoning approach for large scale component-based applications. Electronic Communications of the EASST 19 (2009)
18. Yang, Q., Yao, D., Garnett, J., Muller, K.: Using a trust inference model for flexible and controlled information sharing during crises. Journal of Contingencies and Crisis Management 18(4), 231–241 (2010)
19. Mamdani, E.H.: Applications of fuzzy algorithm for control of simple dynamic plant. Proceeding of the IEEE 121, 1585–1588 (1974)

Social Validation of Learning Objects in Online Communities of Practice Using Semantic and Machine Learning Techniques

Lamia Berkani[1], Lydia Nahla Driff[1], and Ahmed Guessoum[2]

[1] Department of Computer Science, USTHB University, Bab-Ezzouar, Algiers, Algeria
[2] Artificial Intelligence Laboratory (LRIA), Department of Computer Science, USTHB
l_berkani@hotmail.com, driff.nahla@gmail.com,
{lberkani,aguessoum}@usthb.dz

Abstract. The present paper introduces an original approach for the validation of learning objects (LOs) within an online Community of Practice (CoP). A social validation has been proposed based on two features: (1) the members' assessments, which we have formalized semantically, and (2) an expertise-based learning approach, applying a machine learning technique. As a first step, we have chosen Neural Networks because of their efficiency in complex problem solving. An experimental study of the developed prototype has been conducted and preliminary tests and experimentations show that the results are significant.

Keywords: online communities of practice, learning object, user profile, social validation, ontologies, machine learning technique.

1 Introduction

The integration of the "social" dimension in the web has created new ways of collaboration between people as well as exchange, creation and sharing of knowledge. Lytras and Ordóñez de Pablos (2009) discussed the evolution of the social web and explored the potential of web 2.0 and its synergies with the semantic web. This integration has encouraged the emergence of social networks and online communities where a large number of people work in a community to collaborate, share resources and experiences. However, one of the crucial problems of the current web is the problem of big data (Halevi and Moed, 2012). Indeed, a huge number of resources are made available every day through repositories, collections, and websites, etc. For this reason, this problem has attracted the attention of many researchers who are faced with different challenges, including (Manovich, 2011): capture, design, storage, search, sharing, analysis, and visualization.

In our research, we are interested in this problem with a particular focus on the validation dimension. We believe that with the amount of knowledge resources available on the web, it is crucial to get a fairly precise idea about the quality of the resource. The main objective is hence the assessment whether the resource is good or not and this can be represented by a degree of validity, provided preferably by an "expert" from the

A. Amine et al. (Eds.): *Modeling Approaches and Algorithms*, SCI 488, pp. 237–247.
DOI: 10.1007/978-3-319-00560-7_28 © Springer International Publishing Switzerland 2013

domain. Since experts are considered trusted sources, but which are not always available, it would be interesting to take advantage of their previous validations.

In order to support this process, we highlight in this paper the need to automate the social validation of knowledge resources using past experiences. Social validation may be seen as an important topic of the social web, it has nonetheless not yet been addressed widely. Very little research has addressed this issue. One can cite the work presented by Cabanac et al. (2010) about the social validation on collective annotations. The authors were based on the social theory of information and their main objective was how to leverage the readers' cognitive overload by characterizing relevant annotations. In this paper, we will try to automate the process of validation of Learning Objects (LOs) within an online Community of Practice (CoPs) of Teachers. We use both the members' assessments and an expertise-based learning approach. We use Machine Learning (ML) techniques in our problem of social validation of LOs; the intuition is to have automated learning from past experience of human's evaluations and the experts' assessments of the quality of these evaluations.

2 Problem Statement

A large number of LO repositories are made available to any user searching for educational content on various topics (Tzikopoulos et al., 2007). For instance, the ARIADNE[1] infrastructure provides access to some hundreds of thousands of LOs and MERLOT[2] includes more than 20,000 LOs. However, in order to support the selection of LOs and maintain quality control, feedback on LOs is needed.

The quality of LOs is considered as an important topic of metadata research (Vuorikari et al., 2007). According to Shreeves et al. (2006), the quality concerns both the metadata that describes the LO (i.e. the descriptive metadata) and the metadata that is generated by users in different formats of annotations, reviews and ratings (i.e. the evaluative metadata).

We are interested in our research in the validation of LOs within the framework of an online CoP of teachers. We address in this paper the need to support members of this community in the validation process of LOs. The main idea is to automatically assess the social validity of a LO based on the members' evaluations and using past experiences. In our context, past experiences refer to the validations manually done by experts in the respective LO domains of the members' evaluations.

3 Our Approach

In order to automate the validation process of LOs we take into account the existing experience. Our approach is based on two facets: (1) a semantic aspect; and (2) an expertise-based learning approach. The first facet is concerned with the semantic representation of domain knowledge, members' profiles and LOs. Two kinds of

[1] http://www.ariadne-eu.org
[2] http:// www.merlot.org

metadata are considered for a LO: the descriptive metadata that describe the LO and the evaluative metadata that contains a set of metadata to evaluate the LO. The second facet is concerned with the modeling of a ML technique.

3.1 Semantic Aspect

We present the domain knowledge ontology as well as a semantic representation of members' profiles and LOs.

3.1.1 Domain Knowledge Ontology (DKOnto)

We consider in our research a hierarchical ontology of the computer science domain which is derived from the well-known ACM (http://www.acm.org/class/1998/) taxonomy. However, taking into account that each domain has a degree of importance in the CoP, we have enriched this ontology with weights as presented in Fig.1. These weights reflect the importance of the corresponding domains and relevance to the exchanges within the CoP. The leaf nodes directly reflect this relevance while each parent's weight is obtained by summing up those of its children nodes (a parent has a higher weight as it covers more domains). The weights may be updated automatically by observing the members' interactions within the community. For instance, if members capitalize and validate a high number of LOs related to the domain D_i which may have a low weight at some specific time, then this domain will get its weight increased to reflect the relevance of D_i.

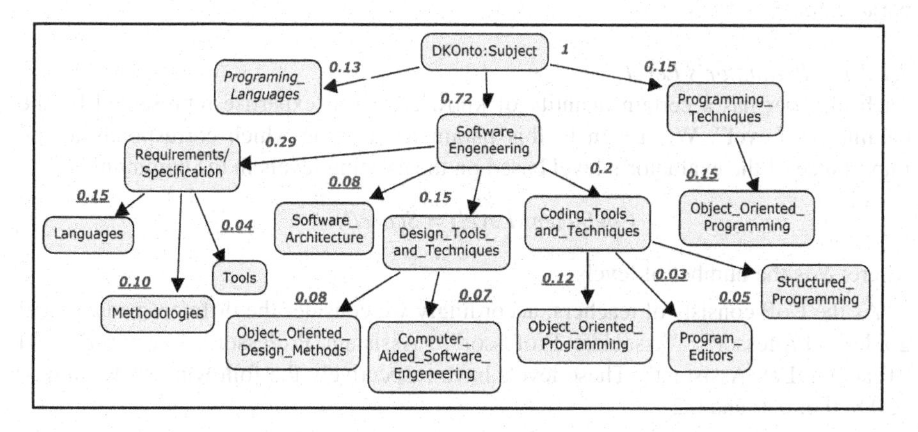

Fig. 1. Example of a domain knowledge ontology

3.1.2 Semantic Representation of a Member's Profile

A member's profile is based on two types of information: *(1) Static information:* includes personal characteristics (e.g. contact details, academic background, member's level, specialty, languages, etc.). *(2) Dynamic information*: different dimensions are identified such as: the "Expertise" and "Interest" about a specific domain, with a respective degree of expertise and interest. In this paper, we will consider all the knowledge of a member as a competence.

3.1.3 Semantic Representation of LOs

Fig. 2 illustrates the semantic representation of a LO. We focus on the evaluative metadata, i.e. the parameters which will be used in the automatic training phase.

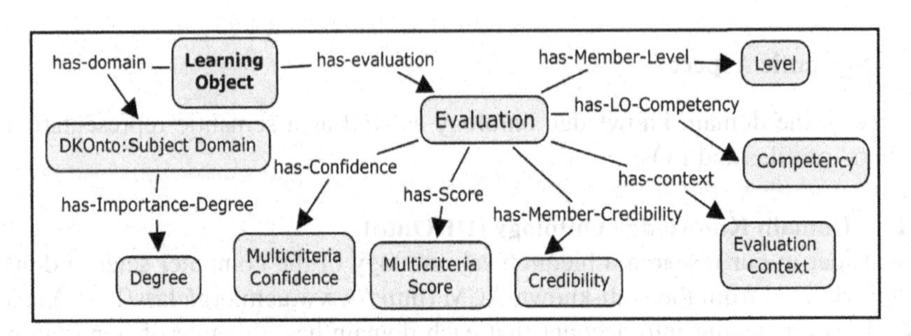

Fig. 2. Excerpt from the semantic representation of a LO

The prediction of the degree of validity for a given LO is based on the collective assessments of the LO as done by members. A set of parameters have been proposed to formalize each assessment. As illustrated in Fig. 2, the evaluation of a LO by a given member M_i is categorized by two classes of metadata: those related to the member (i.e. the level of the evaluator, his competence level with respect to the given LO and his credibility degree) and those concerning the resource (the assigned score, the confidence degree, the evaluation context). We explain below each of these parameters.

3.1.3.1 Evaluator's Level

Each member has a certain quantity of knowledge and expertise represented by the parameter "Level". We assign to this parameter a value which corresponds to the importance of the evaluator's level based on the existing levels in the community.

$$Coefficient(Level) = Score/N \quad (1)$$

where: N is the number of levels.

As the CoP consists of teachers, accordingly we consider the different pedagogical grades: "Professor", "Associate Professor", "Assistant Professor", "Lecturer", and "Teaching/Lab Assistant". These levels have respectively the following score values: 1; 0.8; 0.6; 0.4; and 0.2.

3.1.3.2 Multicriteria Score

Each evaluator will give a multicriteria score representing his opinion about the LO being evaluated using a five-point Likert scale (1-5), from very good to very bad. Different components are considered: "pedagogy", "didactics", "clarity", "relevance", "structure" and "originality" of the LO. The final score will be calculated using this formula:

$$Score = \frac{1}{N}\sum_{i=1}^{N} S_i P_i \quad (2)$$

where:

N is the number of criteria,
S_i is the given score for the ith criterion, and
P_i is the weight assigned to the ith criterion.

3.1.3.3 Multicriteria Confidence

A confidence degree is assigned by the member for each score component he gives using a five-point Likert scale (1-5), from a higher degree of confidence to a lower one. The final confidence degree will be calculated using this formula:

$$Confidence\ degree = \frac{1}{N}\sum_{i=1}^{N} Conf_i P_i \tag{3}$$

where:

N is the number of criteria,
$Conf_i$ is the given confidence degree for the ith criterion, and
P_i is the weight assigned to the ith criterion.

3.1.3.4 Evaluation Context

The assessment of the LO is based on a given context in which the evaluator is to give his evaluation. The importance of this reference is different based on each one's credibility. Accordingly, we assign a weight to each reference to indicate its reliability. Several formats can be distinguished: "tested results", "a similar LO", "an approved opinion", or "a new LO".

3.1.3.5 Competency Level

Given that each LO is related to certain domains of knowledge, then each member will have a certain level of knowledge called "skill" or "competency level" with respect to this LO. Three cases can be distinguished: (1) all the LO domains belong to the member's competence domains (high rate); (2) some domains belong to the member's competence domains (average rate); (3) the member has no knowledge about the domains but may have knowledge about other domains (low rate).

In order to calculate the competency level of a member about the domains of the LO which is under evaluation, it is necessary to calculate the similarity between each of his domains of competence and each of the domains of the LO. Table 1 describes rules that cover all the possible cases for the relative positions of two domains in the taxonomy.

Table 1. Similarity rules

	Description	Similarity
Rule1	D corresponds to D'	Sim = 1
Rule2	D (direct or indirect) parent of D'	Sim = 1
Rule3	D (direct/indirect) son of D'	$\mathrm{Sim} = \frac{\lvert \mathrm{weight(D)} - \mathrm{weight(D')} \rvert}{\Delta \mathrm{Depth}}$
Rule4	D and D' are independent	$\mathrm{Sim} = \begin{cases} \mathrm{weight}(D_c) & \text{if D and D' at the same depth} \\ \frac{\mathrm{weight}(D_c)}{\Delta \mathrm{Depth}} & \text{otherwise} \end{cases}$

where:

D is one of the member's domains of competence,
D' is one of the LO domains,
Dc is the closest (parent) domain common to D and D', and
ΔDepth is the difference between the depths of D and D' in the ontology.

The competency level of a member is calculated in relation to all of his domains of competence using the ontology DKOnto (see Fig. 1). We formulate the problem as follows:

Let M be a member and R a LO. Let M's domains of competence be defined as a vector $C_M = [C_1, C_2, \ldots, C_m]$. We associate with this vector a vector $D_M = [d_1, d_2, \ldots, d_m]$ (with same size as C_M) meaning that M has a degree of competence d_i in the domain C_i, where $0 \leq d_i \leq 1$. (It should be noted that 'm' can be strictly smaller than the total number of domains in the ontology.) Each LO resource has a set of relevant domains denoted by vector $D_R = [D_{R_1}, D_{R_2}, \ldots, D_{R_k}]$

The idea behind this formalization of the problem is that each domain of competence C_i of M is similar to a certain degree to each domain D_{R_j} of LO. Thus, we define the similarity matrix of M's competences with respect to LO as an (m x R_k) matrix as follows:

$$Similarity(M,R) = \begin{pmatrix} Sim(C_1, D_{R_1}) & \ldots & Sim(C_1, D_{R_k}) \\ \vdots & Sim(C_i, D_{R_j}) & \vdots \\ Sim(C_m, D_{R_1}) & \ldots & Sim(C_m, D_{R_k}) \end{pmatrix} \quad (4)$$

where:

$0 \leq Sim(C_i, D_{R_j}) \leq 1$ is the similarity between M's domain of competence C_i and the domain D_{R_j} of LO.

We know that each domain D_{R_j} of LO has a weight $0 \leq p_{R_j} \leq 1$ for this resource. Each member M has a degree of competence $0 \leq d_i \leq 1$ with respect to his domain of competence C_i. We define the degree of competence of *M* with respect to D_{R_j} (domain *j* of LO) as follows:

$$Competence(M, D_{R_j}) = \sum_{i=1}^{m} d_i \times Sim(C_i, D_{R_j}) \quad (5)$$

We finally define an overall degree of competence of *M* with respect to LO as being:

$$Competence\ (M,R) = \sum_{j=1}^{k} p_{R_j} \times Competence(M, D_{R_j}) \quad (6)$$

3.1.3.6 Credibility Degree

A member may have a high degree of credibility because of his correct assessments, i.e. he generally evaluates positively resources that are assigned a high degree of validity and negatively resources that obtain a low degree of validity. Thus, we assign to him a degree of credibility which represents the distance from one of his evaluations to the final validity of the LO according to the final score he gave it. This is captured in the following formula:

$$Credibility\ (S_i) = 1 - |Score - validity\ (S_i)| \tag{7}$$

The final member's credibility score is calculated taking into consideration the relative success of all the LOs he has evaluated during the whole period of his membership of the community:

$$Credibility = \frac{1}{N}\sum_{i=1}^{N}(Credibility(Si)) \tag{8}$$

3.2 Use of a Machine Learning Method

We apply a ML method on data using the representation of an evaluation. We have chosen Neural Networks (NN) (Muller and Reinhardt, 1991; Fausett, 1994) because of their efficiency for complex problem solving (e.g. classification, optimization problems). In our case, we have used this method to predict a degree of validity of a given LO. We have designed several models of Multi-Layer Perceptrons, tried various learning and activation functions, and varied the number of neurons in each situation. The final NN architecture is proposed as follows:

- Start with a multilayer Perceptron with six neurons on the input layer, twenty neurons on the hidden layer and one neuron on the output layer.
- Feed the input into the network with a set of evaluations on various LOs.
- Specify in the output the validity score specified by the expert member.
- Train the network, changing the architecture until the best learning is gradually obtained.
- Simulate the network, once a good learning has been achieved.

Note that we consider that the parameters do not have the same priority. We have given them different degrees of importance. For a given "score", the "confidence" and the "evaluation context" are parameters that give us more information about the score. As such, we give them the same (average) degree. On the other hand, the "competence", "credibility" and "level", are parameters that help us control the given score, but with different (high) degrees. We consider however that the competence about a LO will be given a higher degree than the credibility and that the latter will be given a higher degree than the level.

4 Results and Experimentations

4.1 A Prototype of a Social Validation System

Fig. 3 presents the architecture of the prototype of the social validation system. This system is proposed to be integrated within an online community platform. Different services are proposed, including mainly: adding a new LO or updating/deleting some existing ones; evaluating a LO; and manually validating a LO by experts (this validation is necessary since the generated data will be exploited during the learning phase). Furthermore, the system includes an automatic learning service that uses the evaluations made by members (represented according to the six parameters), as well as the validations made by experts. This service applies a NN method. Finally, the system offers an automatic validation service based on new members' evaluations and the data used during the learning phase.

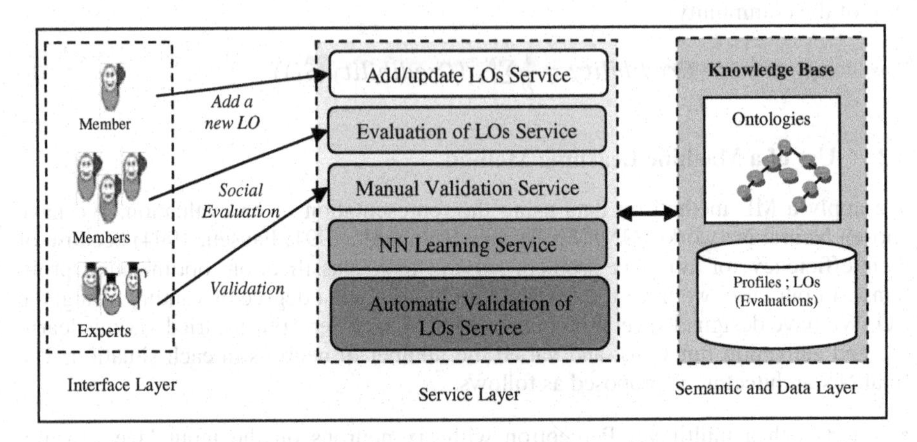

Fig. 3. A general architecture of the social validation system

We present in Fig. 4 an example of an evaluation of a LO made by a member *M1*.

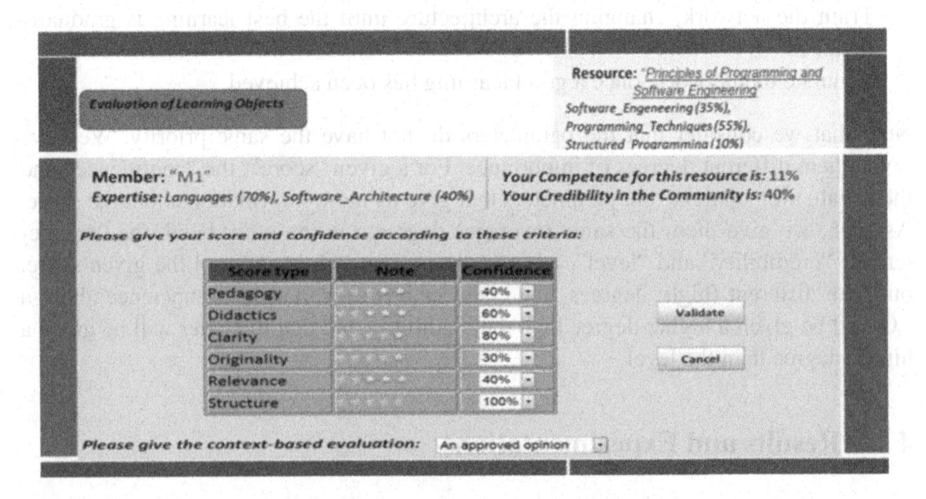

Fig. 4. Screenshot of the social validation system

The figure illustrates the different parameters, either extracted directly from the knowledge base (member's level), calculated (the member's competence with respect to the LO), or captured by the member (score, confidence, evaluation context).

4.2 Tests and Evaluation

Fig. 5 illustrates the results obtained after some tests using the NN method based on a large number of evaluations. To execute our learning process, we have used 1500 evaluations generated automatically using a supervised random algorithm and another 100 captured manually using the social validation system. For the test process, we have generated 800 instances automatically and 100 have been captured manually. The data generated automatically guarantees the inclusion of correct cases (e.g. a member with high degrees of "level", "competence" and "credibility" gives similar scores to those given by experts). However we have added some contradictory cases to test their influence.

As shown in Fig.5, the error rate decreases more and more with the increasing number of examples used as input during the NN learning phase.

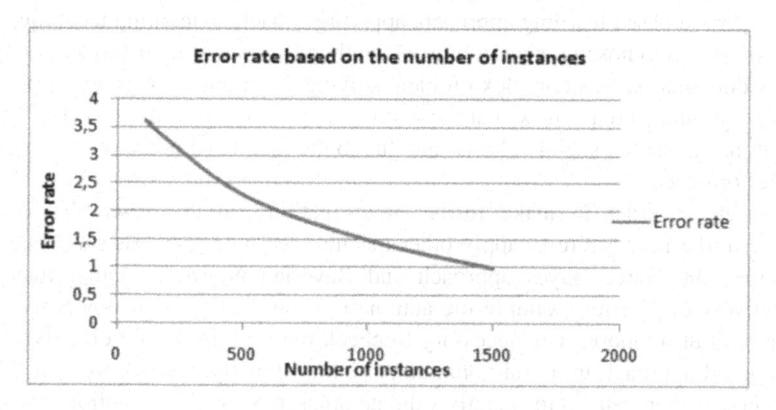

Fig. 5. Error rate of the NN method

Using a NN has various advantages, among which we can mention a fast simulation, a statistical treatment of the available instances, and giving an approximation if a case does not exist.

In order to analyze the results produced through the use of the NN method, we have studied the relationships between the different assessments. To this end, we have applied a Data Mining technique, using the Close algorithm (Pei et al., 2000), for mining association rules. A rule is a set of parameters with specific values generated (if some items frequently appear together) to predict the degree of validity. An example of a rule can be stated as follows:

$$\text{If (Level = Professor)and (Competence} \geq 0{,}7\text{)and (Credibility} \geq 0.7\text{) Then (Validity} \geq 0.8\text{)}$$

The Close algorithm is applied after coding the evaluations and the degrees of validity. A set of association rules has been generated. However, in order to check the reliability of a rule, it is necessary to predict a degree of validity using the NN with values that correspond to the values of the rule. After a series of tests, if the results obtained by the NN are close to the degrees proposed by the member for the respective evaluations, then we can confirm the reliability and applicability of the rule.

The main objective of this data mining step is an automatic exploitation of the association rules. Once generated, the rules can be used to generate various statistics. For instance, we can have the rate of members who adequately assess a LO with a high level; the degree of credible assessments, i.e. where the member's competence or credibility degree is high; etc.

5 Conclusion

This paper presents a novel approach for the validation of LOs within an online CoP of teachers. The originality of our contribution is in proposing an automation of the process of social validation of LOs. Our approach is based on two features: (1) the members' assessments, which we have formalized and represented semantically; and (2) an expertise-based learning approach, applying a machine learning technique. As a first step, we have chosen a neural network method because of their proven efficiency in many domains such as complex problem solving. Preliminary tests and experimentations on simulated data show that the results are promising because of the "correctness" of the calculated social validity and due to the fact that a neural network gives a good performance.

However, in order to make further investigations on this research work, we envisage in the near future to apply other machine learning methods such as genetic algorithms, the Naïve Bayes approach and Bayesian Networks. Other interesting perspectives concern for example the automatic control of the coherence of a given evaluation. Furthermore, it is necessary to check the usefulness and effectiveness of the proposed approach in a real community setting. Finally, we believe that this approach may be generalized to socially validate other items such as solutions (proposed either to the same problem or to different ones), artifacts, etc.

References

1. Cabanac, G., Chevalier, M., Chrisment, C., Julien, C.: Social validation of collective annotations: Definition and experiment. Journal of the American Society for Information Science and Technology 61(2), 271–287 (2010)
2. Fausett, L.: Fundamentals of Neural Networks: Architectures, Algorithms, and Applications. Prentice Hall (1994) ISBN: 0133341860
3. Halevi, G., Moed, H.: The Evolution of Big Data as a Research and Scientific Topic: Overview of the Literature. Research Trends, Special Issue on Big Data 30, 3–6 (2012)
4. Lytras, M.D., Ordóñez de Pablos, P.: Social Web Evolution. Integrating Semantic Applications and Web 2.0 Technologies, pp. 1–340. IGI-Global (2009)
5. Manovich, L.: Trending: The Promises and the Challenges of Big Social Data. Debates in the Digital Humanities (2011)
6. Muller, B., Reinhardt, J.: Neural Networks. Springer (1991) ISBN: 3540523804
7. Pei, J., Han, J., Mao, R.: Closet: An effcient algorithm for mining frequent closed itemsets. In: SIGMOD International Workshop on Data Mining and Knowledge Discovery (2000)

8. Shreeves, S.L., Riley, J., Milewicz, E.: Moving towards sharable metadata'. First Monday 11(8) (2006)
9. Tzikopoulos, A., Manouselis, N., Vuorikari, R.: An Overview of Learning Object Repositories. In: Northrup, P. (ed.) Learning Objects for Instruction: Design and Evaluation, pp. 29–55. Idea Group Publishing, Hershey (2007)
10. Vuorikari, R., Manouselis, N., Duval, E.: Using Metadata for Storing. In: Go, D.H., Foo, S. (eds.) Sharing and Reusing Evaluations for Social Recommendations: the Case of Learning Resources, pp. 87–108. Idea group Publishing, Hershey (2008)

Modelling Mobile Object Activities Based on Trajectory Ontology Rules Considering Spatial Relationship Rules

Rouaa Wannous[1], Jamal Malki[1], Alain Bouju[1], and Cécile Vincent[2]

[1] Univ of La Rochelle, L3i laboratory, France
[2] Univ of La Rochelle, LIENSs laboratory, France
{rwannous,jmalki,abouju,Cvincent}@univ-lr.fr

Abstract. Several applications use devices and capture systems to record trajectories of mobile objects. To exploit these raw trajectories, we need to enhance them with semantic information. Temporal, spatial and domain related information are fundamental sources used to upgrade trajectories. The objective of semantic trajectories is to help users validating and acquiring more knowledge about mobile objects. In particular, temporal and spatial analysis of semantic trajectories is very important to understand the mobile object behaviour. This article proposes an ontology based modelling approach for semantic trajectories. This approach considers different and independent sources of knowledge represented by domain and spatial ontologies. The domain ontology represents mobile object activities as a set of rules. The spatial ontology represents spatial relationships as a set of rules. To achieve this approach, we need an integration between trajectory and spatial ontologies.

Keywords: Trajectory data modelling, Modelling activities, Ontology rules, Spatial data modelling.

1 Introduction

Over the last few years, there has been a huge collection of real-time data on mobile objects. These data are obtained by many kind of systems like GNSS[1] (GPS[2] or ARGOS[3]), phone location or RFID[4]. This opens new perspectives for developing applications, such as bird migration monitoring [14], daily trips of employees [19], military application [13] and marine mammal tracking [8]. The raw data captured, commonly called trajectories, traces moving objects from a departure point to a destination point as sequences of pairs (sample points captured, time of the capture). In [14], the authors give a general definition of a trajectory: *A trajectory is the user defined record of the evolution of the position*

[1] GNSS : Global Navigation Satellite System.
[2] GPS: Global Positioning System.
[3] ARGOS: Advanced Research and Global Observation Satellite.
[4] RFID: Radio Frequency IDentification.

A. Amine et al. (Eds.): *Modeling Approaches and Algorithms*, SCI 488, pp. 249–258.
DOI: 10.1007/978-3-319-00560-7_29 © Springer International Publishing Switzerland 2013

(perceived as a point) of an object that is moving in space during a given time interval in order to achieve a given goal. Raw trajectories contain neither contextual information about the displacement of a moving object nor its accomplished activities [1]. Semantic trajectories can be seen as a high-level data layer on raw trajectories [19]. Furthermore, it becomes necessary to provide mechanisms for storage, modelling, efficient analysis and knowledge extraction from these data to enable interoperability between systems and services. Ontologies have been proposed as a solution for modelling data with their semantic information. To exploit raw trajectories, we need other data sources. Temporal, spatial and domain related information are fundamental sources. In the continuation of our previous work [8], a domain ontology was constructed to model semantic trajectory concepts and domain rules. We focused on semantic annotations for seal trajectories activities. We discussed the temporal data dimension of trajectories. This approach takes into account the temporal data features from low-level to high-level trajectory modelling. We give an evaluation of our approach on generated and real data. In this work, we are interested in modelling mobile object activities while considering the spatial relationships.

Raw trajectories can be captured as sample points given by their coordinates with the time of capture. So, a trajectory can be considered as spatio-temporal data. From this point of view, we can consider spatio-temporal data models to represent trajectory data. Nevertheless, these models do not propose specific support for a trajectory as a whole entity [5,12]. Trajectory can also be considered from the point of view of the moving object. Moving object data models have been defined to represent trajectories [6]. Our approach models a trajectory by an RDF graph combining features from domain application and both spatio-temporal and moving objects data models. In this paper, we illustrate our proposal to integrate these three data models.

This paper is organized as follows. Section 2 summarizes some recent related work on semantic trajectories. Section 3 represents the domain application considered in this work. Section 4 details independently the ontological modelling approach: domain seal trajectory and spatial ontologies. Section 5 introduces the domain and spatial rules. Section 6 illustrates the integration between seal trajectory and spatial ontologies. Finally, Sect. 7 concludes this paper and presents some future prospects.

2 Related Work

Data management techniques including modelling, indexing, inferencing and querying large spatio-temporal data have been actively investigated during the last decade [17,9,7]. Most of these techniques are only interested in raw trajectories [19,15,1]. The objective is to represent and query moving object trajectories. In [6], authors notice two data modelling points of view for trajectories: the conceptual modelling view and the moving objects view. Both need spatio-temporal data modelling and reasoning.

Projects like GeoPKDD [4] and MODAP [10] emphasized the need to address and to use semantic data about moving objects for efficient trajectory analysis.

For example, in [14], bird migration monitoring was analysed to get better understanding of bird behaviour. Scientists tried to answer queries such as: where, why and how long birds stop on their travels, the activities they engage during their stops, and which weather conditions the birds face during their flight. Considering these new requirements, new research has emerged offering data models that can be easily expanded taking into account semantic data. Thereby, a trajectory is seen as a user defined time-space function from a temporal interval to a space interval. To consider semantic trajectories, a conceptual view was defined by three main concepts: stops, moves, and begin-end of a trajectory [14]. Each part contains a set of semantic data. Based on this conceptual model of trajectories, several studies have been proposed such as [1,2]. Moreover, in [19], authors designed a conceptual model of trajectories from low-level real-life GNSS data to different semantically abstracted levels. Their application concerned daily trips of employees from home to work and back.

Using ontologies as a model for semantic spatio-temporal data is a recent research field. In [15], authors worked on marine mammal tracking with the objective of understanding the behaviour of the animal by studying its activities. To model semantic trajectories, an ontological approach was defined to represent trajectory concepts. The ontologies constructed are formalised in RDF and OWL languages. This approach takes into account thematic and temporal rules [8]. In consequence, the inference mechanism was based on domain rules in addition to temporal and spatial rules defined as entailments. Moreover, in [9], authors worked on a military application domain with complex queries that require sophisticated inference methods. For this application, they presented an upper-level ontology defining a general hierarchy of thematic and spatial entity classes and associating relationships to connect these entity classes. They intended for application-specific domain ontologies in the thematic dimension to be integrated into the upper-level ontology through subclassing of appropriate classes and relationships. Consequently, the inference mechanism was based on several domain specific table functions and used only RDFS rules indexes.

Correspondingly, an integration between application domain ontology and spatial ontology led to the discovery of more semantics on trajectories. Furthermore, an ontological framework was produced in [18], composed of a modular ontology and its three component modules. The three following ontologies were integrated into a unique ontology by setting up rules between them to get more semantics:

1. Geometric trajectory ontology is a generic ontology that describes in particular the spatio-temporal features of a trajectory;
2. Geographic ontology describes the geographic objects;
3. The domain application ontology describes the thematic objects of the application.

3 Domain Application

Our modelling approach considers trajectories of seals. The data comes from the LIENSs[5] (CNRS/University of La Rochelle) in collaboration with SMRU[6]. The captured spatio-temporal data of seal trajectories can be classified into three main states: haulout, cruise and dive [15]. In every state, there is a specific activity: resting, travelling and foraging, respectively. Based on these activities, we aim at answering queries, such as:

Example 1. seals foraging in a specific area

Analysing this query highlights the necessary of defining seal activities, such as `foraging`. Nevertheless, the spatial concepts representing `area` and the spatial relationship `contains` must be defined. Table 1 analyses the example query and illustrates the domain and spatial requirements.

Table 1. Domain, spatial concepts and rules needed for answering the example query

Concepts and rules			Description
Concepts	Domain	Dive	specific part of the seal trajectory
	Spatial	Polygon/Region	spatial concept for area
Rules	Domain	Foraging	seal activity
	Spatial	Contains	spatial relationship between the domain and spatial concepts

4 Ontology Based Modelling of Trajectory

The need of a spatial model and its relationships clearly appears from Table 1. In this section, we consider independently trajectory and spatial data models.

4.1 Trajectory Ontology Model

The seal trajectory ontology, called `owlSealTrajectory`, is a result of a model transformation like in model-driven engineering approaches. The input of this transformation is the semantic seal trajectory model represented by a UML class diagram. Figure 1 presents an extract of this ontology, where:

- `Seal` is a mobile object. It represents the animal equipped with a tag;
- `Sequence` is captured in the form of temporal intervals.
- `Trajectory` is a logical form to represent a set of sequences;
- `Activity` is the semantic part representing seal activities for a sequence;

[5] `http://lienss.univ-larochelle.fr`
[6] SMRU: Sea Mammal Research Unit- `http://www.smru.st-and.ac.uk`

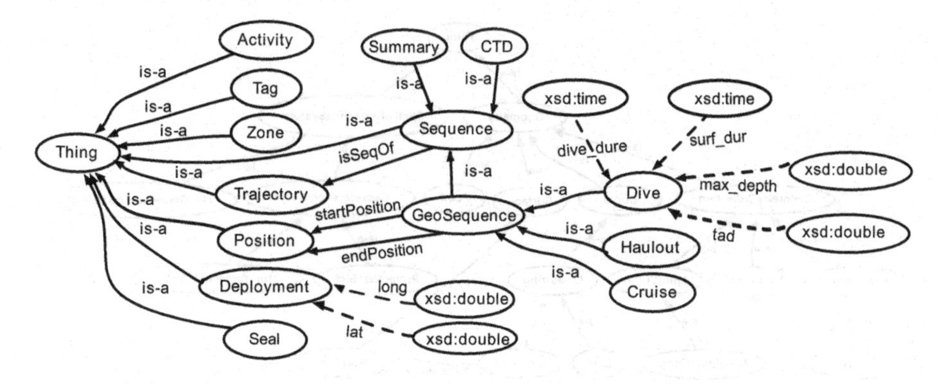

Fig. 1. An extract of `owlSealTrajectory` ontology

- `GeoSequence` is the spatial part in seal trajectory ontology and can be `Haulout`, `Cruise` or `Dive`;
- `Position` is a point location of a geosequence.

Besides these concepts, `owlSealTrajectory` defines relationships like:

- `seqHasActivity` is an object property between an activity and a sequence;
- `startPosition`, `endPosition` are object properties between a position and a geosequence. They represent start and end captured position of a geosequence;
- `long`, `lat` are data properties for the position of a captured point;
- `dive_dur`, `sur_dur` and `max_depth` are dive duration, surface duration and maximum depth of a dive, respectively;
- `TAD` is Time Allocation at Depth which defines the shape of a seal's dive [3].

4.2 Spatial Ontology Model

To model the spatial data dimension of a trajectory, we introduce a spatial ontology. In our approach, we choose `owlOGCSpatial` ontology developed by Malki [7]. This ontology is obtained by a model transformation. The input of this transformation is the spatial model represented by a UML class diagram proposed by Open Geospatial Consortium (OGC) [16]. An extract of the declarative part of this ontology is shown in Fig. 2.

5 Rule Definition in the Trajectory Ontology

Table 1 highlights the need of rules defined between ontology concepts: domain rules which are seal activities as well as spatial relationship rules.

5.1 Domain Seal Trajectory Rules

Throughout the rules associated with the domain seal trajectory, we focus on seal activities. With our domain expert, we define four seal activities during their

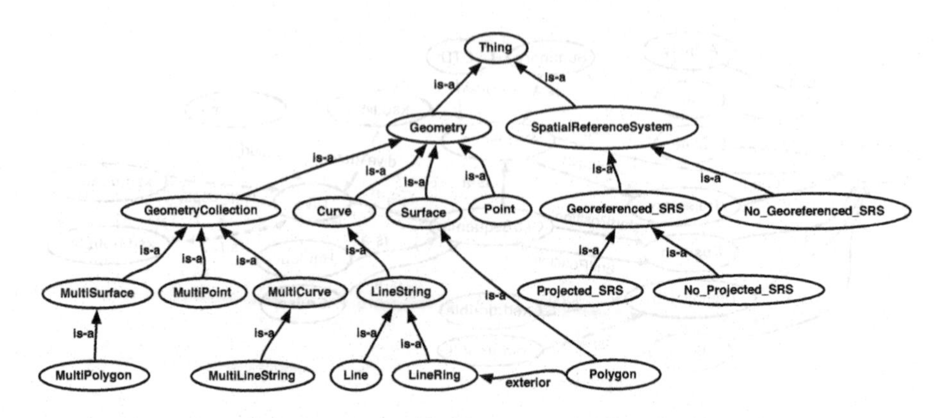

Fig. 2. A view of `owlOGCSpatial` ontology

trajectory: travelling, resting, foraging and travelling-foraging. Figure 3 shows the declarative part of these activities. We implement them as an object relationship `seqHasActivity` between the `sequence` and `activity` concepts. The implementation of these rules is based on Table 2. This decision table shows the classification of seal activities based on parameters and considerations established by the domain expert. We use Oracle Semantic Data Store to implement these rules. We create the rule base `sealActivities_rb` to hold this implementation. Code 1.1 shows the implementation of `foraging_rule`: where maximum dive depth is more than 3 meters, TAD is bigger than 0.9 and less than 1 and finally, surface duration divided by dive duration is smaller than 0.5.

Table 2. Decision table associated with seal activities

Rules	**Max dive depth** (meter)	**Dive shape** or TAD	**Surface ratio** = surface dur / dive dur
Resting	< 10	all	> 0.5
Travelling	> 3	$> 0 \ \& < 0.7$	< 0.5
Foraging	> 3	$> 0.9 \ \& < 1$	< 0.5
Travelling_Foraging	> 3	$> 0.7 \ \& < 0.9$	< 0.5

```
1  EXECUTE SEM_APIS.CREATE_RULEBASE('sealActivities_rb');
2  INSERT INTO mdsys.semr_sealActivities_rb
3  VALUES( 'foraging_rule',
4  '(?diveObject rdf:type  s:Dive)(?diveObject s:max_depth ?maxDepth)
5   (?diveObject s:tad      ?tad)  (?diveObject s:dive_dur ?diveDur)
6   (?diveObject s:surf_dur ?surfaceDur)',
7  '(maxDepth > 3) and (tad > 0.9) and (surfaceDur/diveDur < 0.5)',
8  '(?diveObject s:seqHasActivity ?activityProberty)
9  (?activityProberty rdf:type s:Foraging)',
10 SEM_ALIASES(SEM_ALIAS('s','http://l3i.univ-larochelle.fr/owlSealTrajectory#')));
```

Code 1.1. Implementation of `foraging rule`

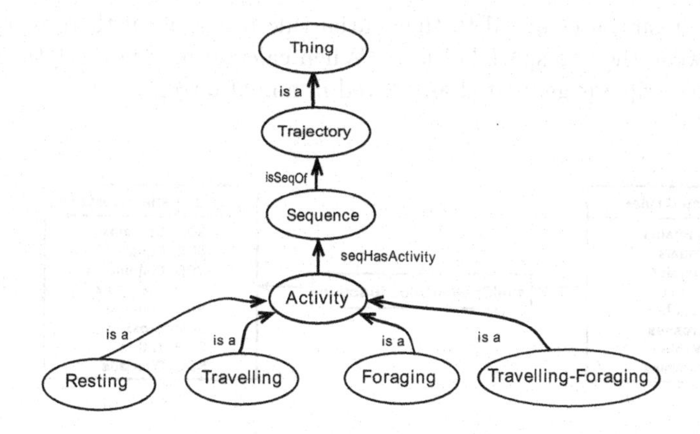

Fig. 3. Declarative part of seal activities

5.2 Spatial Relationship Rules

Spatial relationships are usually classified as topological, directional, and metric relationships. In this paper, we consider the topological relationships: `Equals`, `Disjoint`, `Intersects`, `Touches`, `Crosses`, `Overlaps`, `AnyInteract`, `Within`, and `Contains`. We use Oracle Semantic Data Store to implement these relationships. For each spatial relationship, we associate an ontology rule in the rule base `owlSpatialOnto_rb`. For example, Code 1.2 presents the implementation of the imperative part of `Contains_rule`. In line 6, the property `wkt` represents the coordinate of the spatial objects. While we consider these ontology rules based on Oracle spatial layer, we implement a PL-SQL function called `evalSpatialRules`. This function connects the spatial rules with the corresponding Oracle spatial operators, as shown in Fig. 4. Spatial Data Option `SDO_` is the prefix for the implementations of Oracle spatial operators [11]. Figure 6 illustrates the algorithm for calculating an inference for two spatial objects. For every two spatial objects, the inference procedure calls the spatial rules. The `evalSpatialRules` function calls the corresponding Oracle spatial

```
1  EXECUTE SEM_APIS.CREATE_RULEBASE('owlSpatialOnto_rb');
2  INSERT INTO mdsys.semr_owlSpatialOnto_rb VALUES(
3  'contains_rule',
4  '(?spObj1 rdf:type os:Geometry) (?spObj2 rdf:type os:Geometry)
5   (?spObj1 os:srid  ?sridSpObj1)    (?spObj2 os:srid   ?sridSpObj2)
6   (?spObj1 os:wkt  ?strSpObj1) (?spObj2 os:wkt   ?strSpObj2)',
7  '(evalSpatialRules(spObj1,strSpObj1,spObj2,strSpObj2,sridSpObj2,
8           ''CONTAINS'')=1)',
9  '(?spObj1 os:contains ?spObj2)',
10 SEM_ALIASES(SEM_ALIAS('os','http://l3i.univ-larochelle.fr/owlOGCSpatial#')));
```

Code 1.2. Implementation of `Contains` spatial rule

operator. The result is returned to the spatial rule to specify if there is a relationships between the two spatial objects. When calculating a new relationship, a new inference triple is generated and saved in an entailment.

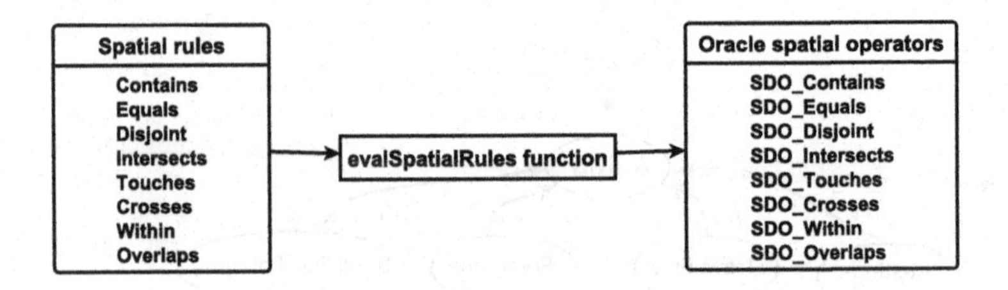

Fig. 4. Connect spatial rules with Oracle spatial operators

6 Integrating Trajectory and Spatial Ontologies

The need of a semantic integration is fundamental while considering different and independent sources of information. For this integration, we are based on `Position` and `GeoSequence` concepts in seal trajectory ontology as mentioned in Sect. 4.1. The integration process of `owlOGCSpatial` ontology with seal trajectory ontology follows these steps, as shown in Fig. 5:

1. `owlSealTrajectory:Position` is mapped by the OWL statement `owl:equivalentClass` to `owlOGCSpatial:Point`.
2. `owlSealTrajectory:GeoSequence` is mapped by OWL statement `owl:equivalentClass` to `owlOGCSpatial:Line`.

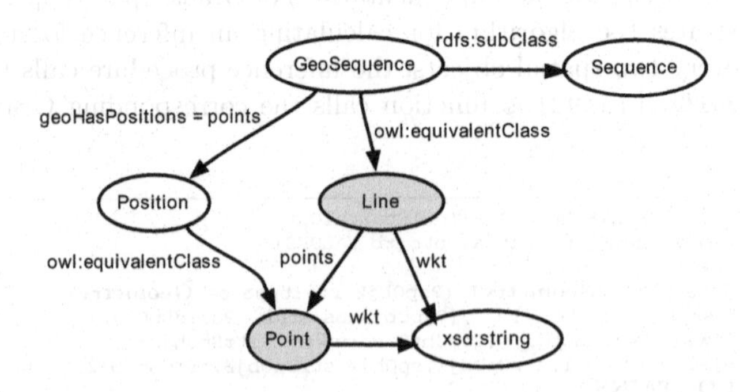

Fig. 5. Integrating `owlSealTrajectory` and `owlOGCSpatial` ontologies

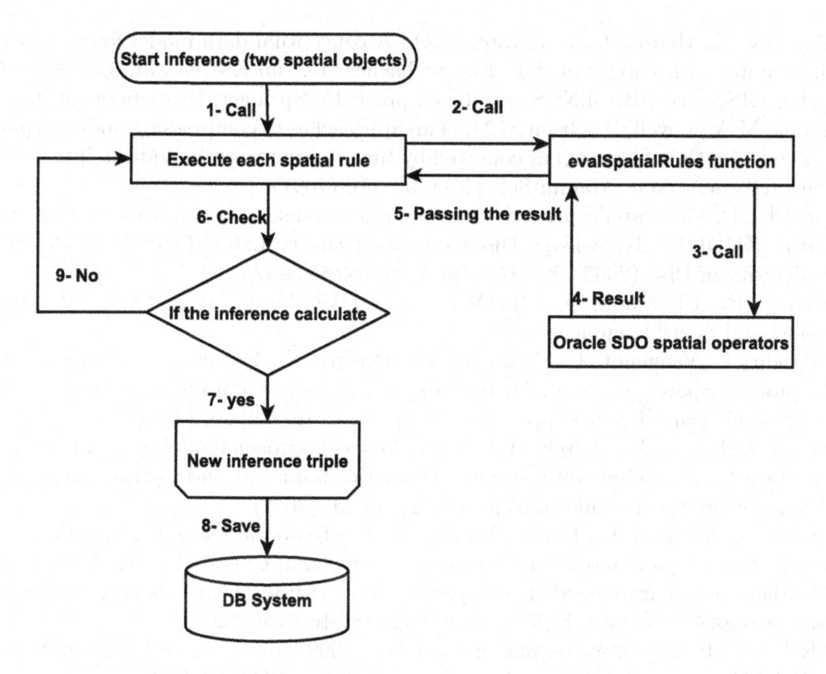

Fig. 6. Calculate the inference for two spatial objects

7 Conclusion and Future Work

Trajectories are usually available as raw data. Indeed, raw trajectories, collected by sensors, do not embed any kind of information about the travel of the moving object or about a possible interpretation of this travel. So, trajectory lacks semantics which is fundamental importance for its efficient use. In this work, we present an ontological approach for modelling semantic trajectories. This approach considers the spatial characteristics of semantic trajectories. Based on the principle of reusing existing ontologies and considering different and independent sources of knowledge, we define an ontological integration approach to connect the domain and spatial ontologies. Throughout the defined ontologies, we implement the domain rules and spatial relationship rules.

In our future work, we will evaluate this approach on real data and we will compare results with other approaches. Moreover, we are highly interested in defining new notions of semantic trajectories and the integration of data mining algorithms with ontological rules.

References

1. Baglioni, M., Macedo, J., Renso, C., Wachowicz, M.: An ontology-based approach for the semantic modelling and reasoning on trajectories. In: Song, I.-Y., et al. (eds.) ER 2008 Workshops. LNCS, vol. 5232, pp. 344–353. Springer, Heidelberg (2008)

2. Bogorny, V., Heuser, C.A., Alvares, L.O.: A conceptual data model for trajectory data mining. In: Fabrikant, S.I., Reichenbacher, T., van Kreveld, M., Schlieder, C. (eds.) GIScience 2010. LNCS, vol. 6292, pp. 1–15. Springer, Heidelberg (2010)
3. Fedak, M.A., Lovell, P., Grant, S.M.: Two approaches to compressing and interpreting time-depth information as collected by time-depth recorders and satellite-linked data recorders. Mar. Mamm Sci. 17(1), 94–110 (2001)
4. GeoPKDD. Geographic privacy-aware knowledge discovery and delivery. Coordinator: KDDLAB, Knowledge Discovery nad Delivery Laboratory, ISTI-CNR and University of Pisa (2005), http://www.geopkdd.eu/
5. Huang, B., Claramunt, C.: STOQL: An ODMG-based spatio-temporal object model and query language (2002)
6. Macedo, J., Vangenot, C., Othman, W., Pelekis, N., Frentzos, E., Kuijpers, B., Ntoutsi, I., Spaccapietra, S., Theodoridis, Y.: Trajectory data models. In: Mobility, Data Mining and Privacy, pp. 123–150. Springer, Heidelberg (2008)
7. Malki, J., Bouju, A., Mefteh, W.: An ontological approach modeling and reasoning on trajectories. taking into account thematic, temporal and spatial rules. TSI. Technique et Science Informatiques 31(1), 71–96 (2012)
8. Malki, J., Wannous, R., Bouju, A., Vincent, C.: Temporal reasoning in trajectories using an ontological modelling approach. Control and Cybernetics 41, 1–16 (2012)
9. Matthew, P.: A framework to support spatial, temporal and thematic analytics over semantic web data. PhD thesis, Wright State Univ. (2008)
10. MODAP. Mobility, data mining and privacy (2009), http://www.modap.org/
11. Oracle. Oracle spatial developer's guide 11g release 2 (11.2) (1996), http://docs.oracle.com/cd/E11882/
12. Parent, C., Spaccapietra, S., Zimanyi, E.: Spatio-temporal conceptual models: Data structures + space + time. In: Proceedings of the 7th ACM International Symposium on Advances in Geographic Information Systems, pp. 26–33. ACM (1999)
13. Perry, M., University, W.S.: A Framework to Support Spatial, Temporal and Thematic Analytics Over Semantic Web Data. Wright State University (2008)
14. Spaccapietra, S., Parent, C., Damiani, M., Demacedo, J., Porto, F., Vangenot, C.: A conceptual view on trajectories. Data and Knowledge Engineering 65(1), 126–146 (2008)
15. Wannous, R., Malki, J., Bouju, A., Vincent, C.: Time integration in semantic trajectories using an ontological modelling approach: A case study with experiments, optimization and evaluation of an integration approach. In: Pechenizkiy, M., Wojciechowski, M. (eds.) New Trends in Databases & Inform. AISC, vol. 185, pp. 187–198. Springer, Heidelberg (2012)
16. Werner, K., Martin, R.: Open GIS consortium, inc, openGIS simple features specification for SQL (1999)
17. Yan, Z., Chakraborty, D., Parent, C., Spaccapietra, S., Aberer, K.: SeMiTri: A framework for semantic annotation of heterogeneous trajectories. In: Proceedings of the 14th International Conference on Extending Database Technology, pp. 259–270. ACM (2011)
18. Yan, Z., Macedo, J., Parent, C., Spaccapietra, S.: Trajectory ontologies and queries. Transactions in GIS 12(s1), 75–91 (2008)
19. Yan, Z., Parent, C., Spaccapietra, S., Chakraborty, D.: A hybrid model and computing platform for spatio-semantic trajectories. In: Aroyo, L., Antoniou, G., Hyvönen, E., ten Teije, A., Stuckenschmidt, H., Cabral, L., Tudorache, T. (eds.) ESWC 2010, Part I. LNCS, vol. 6088, pp. 60–75. Springer, Heidelberg (2010)

A Model-Based on Role for Software Product-Line Evolving Variability

Yacine Djebar[1,2], Nouredine Guersi[1,3], and Mohamed Tahar Kimour[1,3]

[1] Badji Mokhtar University, Embedded System Laboratory –LaSE
[2] Department of Computer Science, University of 08 Mai 1945 Guelma, Algeria
[3] University of Badji Mokhtar-Annaba, Algeria
djebyac@yahoo.fr, {nguersi,mtkimour}@hotmail.fr

Abstract. Modeling evolving variability has always been a challenge for software Product line developers. Indeed, the most recent approaches discuss the problem with the architecture aspect through languages or models. Despite the contributions of these approaches, they have not discussed the possibility to represent the evolving Product line variability with the current UML role given that the latter was designed for a single software system. In this paper, we focused on the use of the concept of evolving role resulting from the adaptation of UML role to represent the evolving variability in the software product line.

Keywords: Evolving roles, product line, evolution diagram, UML.

1 Introduction

Most approaches that have studied the problem of evolving variability in the product line (PL) are based on codes and architecture models. The challenge is to seek mechanisms to manage automatically the evolving variability .The most recent is that of Seidl and al [1].The latter use a conceptual basis of a system for the evolution of model-based SPL which maintains consistency of models and feature mapping .However, despite the significant contribution of this work , authors provided only a partial solution to the problem and developers are forced to study the evolving variability manually or on using an appropriate modeling technique . Motivated by this constraint, we sought a mechanism that can provide developer a solution to model and control this aspect by using the concept of the role.

Considering a PL as a set of variable features playing roles with constraints can help to resolve this issue. However the role as defined in UML requires an adaptation in order to take into account the evolving aspect of PL variability.

Modeling variability by the evolving roles is a promising conceptual modeling which may be expressed in simple situations in which a variable feature can acquire and lose different roles through its life cycle without changing. However, when we use UML to model this aspect, we are compelled to use available UML associations, particularly the generalization. The existing associations limit the modeling process and may lead to ambiguous models that can affect the design step and even

A. Amine et al. (Eds.): *Modeling Approaches and Algorithms*, SCI 488, pp. 259–272.
DOI: 10.1007/978-3-319-00560-7_30 © Springer International Publishing Switzerland 2013

implementation of the whole domain. A detailed study on the UML roles and their problems has been achieved by Steinmann [2], [3]. Jodlowski and al [4] have tried to introduce roles to manage the dynamic objects in the UML class diagram by presenting appropriate object-oriented data models that can also solve problems related to multiple inheritance aspects. Dahchour and al [5], [6], [7] propose a gradual extension of UML with a new 'role-of' association. New meta-classes and OCL rules were added to the UML meta-model to capture the semantics of roles. Kazimierz and Subietay [8] discuss the concept of role that can support the object dynamic aspect as an alternative to a conceptual modeling and define a data structure - a sub-object of an object - to be implemented in an object-oriented or relational DBMS.

Other approaches have also modeled the dynamic aspect of objects. Chernuchin and al [9] presented a canonical extension of object oriented development through the introduction of roles in which they distinguish from a syntactic view point between classes and roles .The class contains roles and their dependencies. Zhu and al [10], [11], [12] used an oriented programming role (RBP) for representing the dynamic aspect of objects.

Unfortunately and despite the contributions of all these approaches for a single software system, no approach has discussed the possibility of representing the evolving PL variability by the role. In our model, variants are perceived in the class diagram as roles can be played by the variation points. Each variable point (optional, parameterized or with alternatives) is represented by a super class with a stereotype to define its type .the variants are represented by role subclasses. They were called by varying roles or unchanging roles. In addition, the model controls the variability evolution through the feature roles that we called f-roles.

This evolving aspect of roles is strengthened by an association of transition-state machine, defined for each variable entity. The rest of the article is organized as follows: Section 2 explains the problem and describes our model. We present in the section 3 the integration of the new concepts in the UML meta-models. Section 4 is devoted to an illustration of the model. Works related to the area of SPL evolution and conceptually closest to f-role are examined in Section 5. We conclude the paper in Section 6.

2 Modeling Variability

2.1 The Overall Model

The product line is a set of variable features playing roles that we called f-roles (Fig. 1). The variability is expressed through the variable features that correspond to optional variable points, variants or variation points with variable or mandatory alternatives. The relationship between each variable feature and its f-roles is achieved with a new association that we have called *'role'*. In the model, alternative variable features are represented as super-classes and their variants as f-role subclasses. The *"role"* association allows f-role subclasses to inherit their super class (parent class). Our model also supports multiple instantiation of the same class i.e. an f-role can

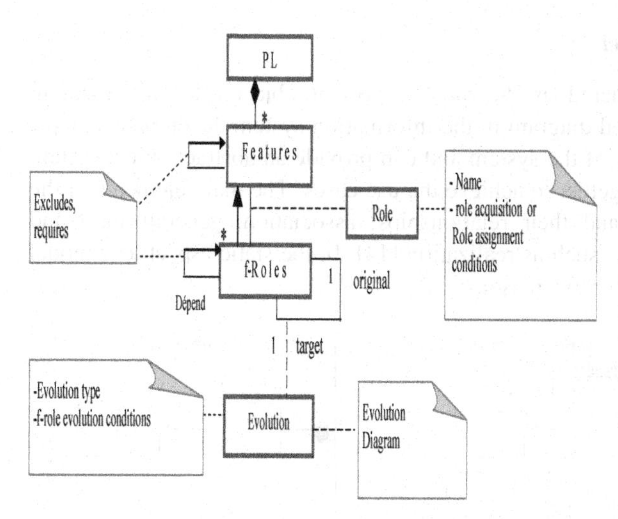

Fig. 1. The overall model

become a more than once instance in the same variable class while still belonging to this class. The mechanism of dependence between the f-roles is the same used by UML (requires, excludes ...).

The model provides a role *"original role"* under conditions to evolve (acquire or dispose of f-roles with keeping or losing the f-roles obtained previously) to another f-role *"target role"*. This evolution can be represented in two views, a static view by using class diagram and a dynamic view by using the state-transition diagram. Now we'll define the concepts we use in our model. The classes that correspond to variable points are called *"variable classes"* and variants of variation points are called *"f-role classes"*. To complete the modeling formalisms in the static model aspect, a listing of evolving f-role constraints can be attached to the model. These constraints may be expressed as predicates in the UML Object Constraint Language (OCL).

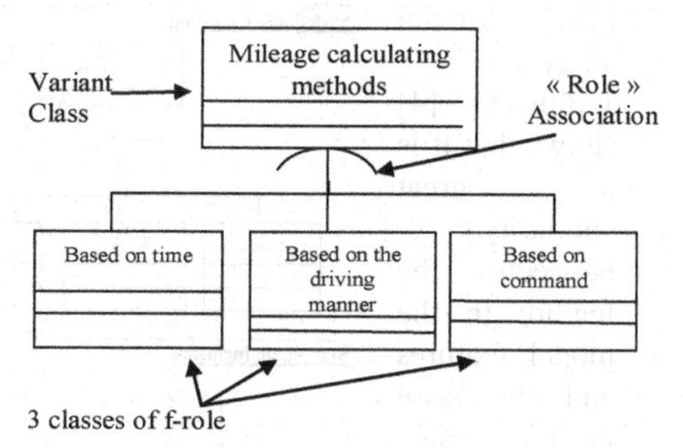

Fig. 2. Modeling of the role

2.2 Use of the "Dimension" Attribute

This new concept previously introduced into UML [12] allows to know the minimum and the maximum of the instances of any variable class that can be used to control the role alternative cardinality. The latter is very important to represent variability in PL.

In the rest of paper we will consider the possibility to express the variability through the use of the UML class diagram role associated with a state-transitions diagram adapted to the PL domain.

2.3 Concepts of the Model

The class diagram is considered as the most important object oriented modeling diagram .It is the only required diagram in the information system development steps. It shows the internal structure of the system and can provide an abstract object system representation that interact together to achieve the use cases. The main elements of the class view are the classes and their relationships: association, generalization and several types of dependencies, such as realization [14]. In the static aspect, our model focuses on the class diagram for two reasons:

- The main works inspired by feature models introduce variability in the first level of abstraction (The domain analysis). Jodlowski [4] shows that it is a great similarity between the feature in the model features and the classe in the object oriented-language .
- The second reason is more related to the fact that the class diagram is the main view of the UML notation.

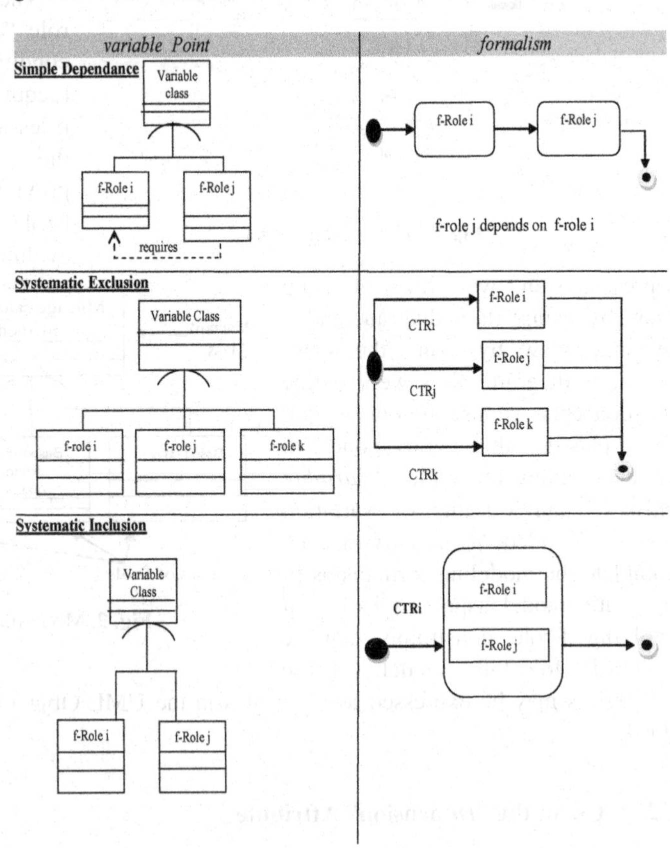

Fig. 3. Some modeling cases represented by the evolution diagram

Concepts and formalisms. Concepts and formalisms that allow the UML role to represent the evolving variability aspect are:

- The f-*role*. It is represented by a class (Fig. 2).
- The "*Role*" Association. it is represented by the symbol << ⌒ >> in the class diagram .It links a variable class that represent a variation point or an optional variable point to variant subclasses called "roles" which identify the variability.

The role class describes the evolving of roles for the variable class it is associated. This class expresses the variability through an optional class or a variant class (classes of alternatives).

- Instances of the role class are called *"f-roles"*.
- Role classes inherit the properties and operations of their variable parent class.
- All the roles (f-roles) played by the variable super-class, must be represented in the class diagram
- An optional role is a role that can be played if one condition at least is satisfied and it ceases to be played if the same condition and not another is not satisfied. The "optional role" is represented by a class identified in the class diagram by the dimension: "0,1" [13].
- A "mandatory role" is a role that must be always played. It does not depend on any condition imposed by its parent class. It also depends on other role. The class that corresponds to a "mandatory role" is identified in the class diagram by the dimension: "1,1" [13].

Instances of the variant class (with alternatives) express the variability by playing or stop playing roles if some conditions are satisfied. These conditions are called respectively *"acquisition"* and *"assignment"* constraints of roles (CTR i) which are then associated. These constraints can be defined as transition predicates and described in a constraint listing with UML OCL formal declarative language [14].

- Our model also supports multiple instantiation of the same class. Indeed, a role can become an instance more than once in the same class, while maintaining his membership in the class. For example, a "method based on the time" can be used with two different engines: petrol and diesel (Fig. 2). It will play two different roles for the same class.

The semantic of the "Role" association. In this part of our work, we are inspired by the 'role-of' association of Dahchour [6] that we adapted to product line engineering. Thus, the following formalisms must be observed:

- The "Role" association is similar to the generalization. It links a subclass to a parent class, but it differs in its evolving aspect. In the context of generalization, an object is a static instance of a class, contrary to "Role" association where the model takes into account the evolving changes of the variable super-class variants through role classes .We use the concept of association to represent the variability in the class diagram through the concept of role.
- The "Role" association (represented by V: v ← R) links a class called *"variable class"* or *"class of variants"* (V) with another class called *"the class of f-roles"* (R), which describes the evolving roles for variable class "v".
- The class "v" can be a class of roles for another class. For this reason it is so called *"variable f- role class"*.
- The variable class "v" instances play roles, which are instances of the role class (R).
- The role classes (f-roles) are used to model the evolving aspect of variable class.

- The no evolving entity (no evolving role and an optional class) is represented as respectively a permanent class or a super class.
- An evolving entity, in addition to be a permanent instance of a role class is represented by a set of role class instances. 8) When a variable class “v” takes a new role, a new appropriate instance of the role class (R) is created.
- If a variable class “v” loses or assigns a role, the instance of the corresponding role class (R) is not taken into account.

The Figure 3 shows three Role associations that link the class of “methods of mileage calculating” to three classes of roles (f-roles) : *"time-based method"*, *"based on the conduct type"* and *“based on the automatic command”*. The class *“methods of calculating mileage”* defines the permanent properties of three variants. The sub-class *“based on time”* ,*"based on the type of conduct"* and *“based on automatic control”* define some properties as a transitional method of calculation. The class *“method of mileage calculating*” has an optional role *“based on automatic control”* with transient properties as a method of calculating based on software code. The latter is attributes that contain different metrics.

Basic Scenarios of role. Consider a class of variants “V”. Variants of this class are subclasses of roles. f-role1, f-role2 represent two evolving role classes of the variable class V. The f-roles may evolve according to two different basic scenarios:

- *The acquisition*, denoted → R shows that a role can acquire an additional role of (R) with some conditions or automatically.
- *An assignment*, denoted R →, indicates that a “role” class can stop to play a role R automatically or under some conditions.

Scenarios of evolving role. Two scenarios can be developed:

- *The Mutation*. Allows to purchase a new f-role with losing of another. It is denoted R1 → - R2 for “role” classes R1 and R2 with R1 different from R2. This notation indicates that a “role” class playing f-role r1 (instance of R1) may stop play the f-role r1 and acquire the f-role r2 (instance of R2). In addition, the f-roles of R2 can not be created independently of the R1 f-roles. Such acquisition may be in one direction or both directions according to the required constraints (exclusive or Mutex dependencies).
- *The expansion*. Allows purchasing a new f-role without losing of another. It is denoted R1 → + R2, with R1 and R2 not necessarily different. Indicating that the role class plays the f-role r1 of R1 can acquire new f-role r2 of R2 by holding r1 (inclusive or).

Note that these two scenarios use the basic scenarios to achieve the role evolution. However, the four scenarios semantically cover all variability expression cases described in the previous section.

The Transition Constraints. Consider a class of variants "v" as a related class of f-roles R1 and R2. A transition predicate is associated with R1 to describe the conditions and / or sufficient on how instances playing the roles of R1 may explicitly or automatically acquire roles in R2. The transition predicates are described with UML declarative formal language OCL [14].

2.4 Use of the State-Transition Machine

Several situations relating to the feature variability are not clearly shown in the class diagram. These situations are:

- Formalisms of class diagram can not represent multiple Instantiating of the same class.

A *"method based on time"* (Fig. 2) can be used with two different engines : "diesel" and "fuel". It will play two different roles "diesel" and "fuel" for the same class.

- The class diagram can not represent explicitly the acquisition and assignment role conditions.

A class diagram becomes overloaded and unreadable .So, it require other visual mechanisms to be integrated to UML. To overcome this deficiency, we use two new associations in the class diagram the "Evolution" and the "Role" (Fig. 1) and we propose another diagram that supports this variability aspect. This is the state–transition diagram. Indeed, this diagram describes the internal behavior of an object that uses a finite state automaton. It shows the possible sequences of statements and actions that a class instance can be treated during its life cycle in response to discrete events (such as method invocations ...).

It is the only standard UML diagram, providing an all complete and unambiguous behavior elements to which it is attached. Indeed, an interaction diagram provides a partial view corresponding to a scenario without specifying how the different scenarios interact.

We use a state machine to represent the evolving roles of variable classes with the following specifications:

- The states of machine represent the roles (f-roles);
- The denominated "CTRi" transitions indicate the constraints faced by the variable class instances to evolve.
- An initial state and a final state are added to complete the diagram formalism.

The necessary "CTRi" constraints of original role to evolve into a target role can be described with the UML formal declarative language (OCL) [14]. The inter-dependencies constraints of the role are obtained by unidirectional evolution from an original role into a target role.

The unidirectional arrows show the dependencies between roles in the diagram. Unlike other roles, a role, concerned with such constraints will be removed if the original role is removed. Its existence also depends on the original role i.e. that this role cannot be created before the creation of an original role.

2.5 The Evolution Diagram

For a better variability evolving visibility, we propose a new diagram that combines the two views: a static (represented by the class diagram) and a dynamic (represented by the state machine). The diagram that we call "Evolution *Diagram*" has several levels, each of them correspond to a variable virtual partition of a class diagram.

This partition represents a part of the class diagram expressing the variability. The class diagram is structured into multiple virtual partitions according to the number of variation and optional points of PL. Each part may contain a variable class, the role subclasses and optional class related to the variable class. Each evolution diagram level is split into 2 parts, the first represents the roles of each variable class (f-roles) and the second (that corresponds to this variable partition) is a state machine that represents the interactions between evolving roles of the same level.

This state machine represents an evolving roles view. The figure 3 shows an example of this diagram. The role behavior is described by a set of states and transitions. The statements describe some conditions of the role and the transitions indicate possible role reactions with the conditions of the current state and the events that trigger these reactions. A transition is defined by:

- A transition trigger event.
- A transition guard condition.
- An action when the transition is verified.

In our model, we use a no hierarchical state machine models. The figure 3 shows some modeling situations of the variant-role evolution.

The formalisms used through the modeling situations are applicable in the evolution diagram state machine part.

3 Integration to UML

3.1 Integration of New Concepts Used in the Class Diagram

We present in this section the integration of the "Role" association, f-roles and evolving variability in the UML meta-models according to the specification formalisms presented in the previous section.

Integration of the "Role" association and roles. We'll define now the "Role" association Meta-Model (Fig. 4). The "Role" association is modeled as a direct subclass of relationship of the UML Core package [14].

We have created two Meta-Classes: “VariableElement”, to represent variable features and “RoleElement” to represent the roles (f-roles) of variable features. These two “Meta-Classes” are defined as direct “ModelElement” subclasses.

The Meta-Class “VariableElement” represent the classes of variable features involved by “Role” association.

The Meta-Class “RoleElement” represent the role class involved by a “Role” association.

The “Role” Meta-Class association connects “VariableElement” to “RoleElement”.

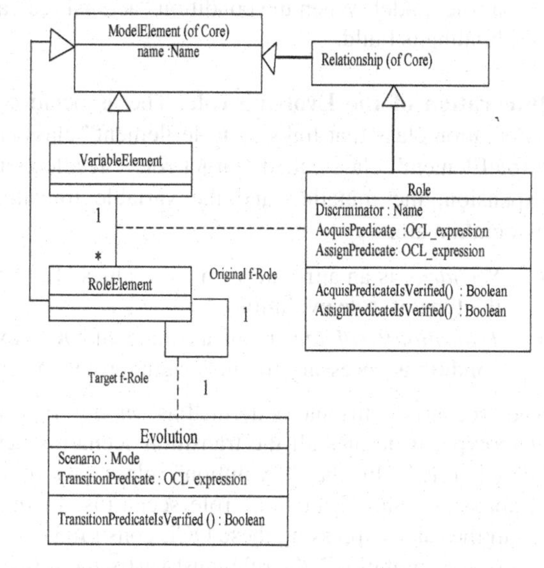

Fig. 4. Integration of the Model concepts to UML Meta-Models

The "Evolution" Meta-Class connects the “RoleElement” class called "original role” and another class of “RoleElement” called "target role”. It allows to specify the evolution concepts i.e. the “expansion” and “mutation” described in section 2.3.4 and the variable predicate transitions.

"Role" is a relationship between a variable class and role class .It contains the following attributes:

- *Discriminator:* is the name of a virtual variable partition in the class diagram (defined in section 2.5) in which the Role association belongs. It means all the links of roles (f-roles) that share the same variable class (it has the same name as the variable super class).

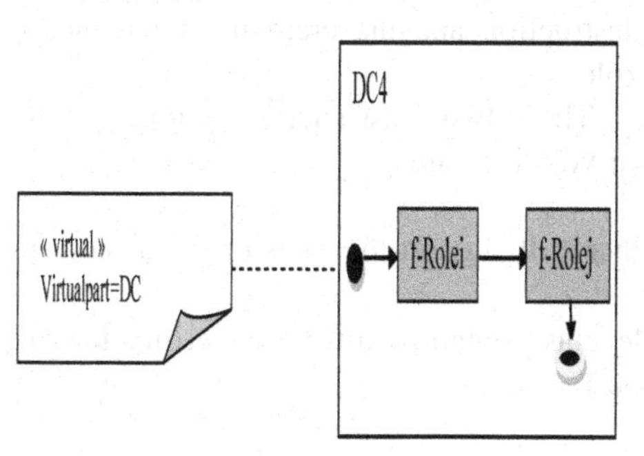

Fig. 5. The state Machine “Virtuality”

- *AssignPredicat:* its type is OCL_Expression and it describes the conditions and / or sufficient on how the variable class can explicitly or automatically assign roles.
- *AcquisPredicat*: its type is OCL_Expression and it describes the conditions and / or sufficient on how the variable class can explicitly or automatically acquire roles.
- The role assignment is controlled by the attribute “AssignPredicate” of a role model. When the condition “AssignPredicate” is verified, this binding role is deleted. The acquisition of a role is controlled by the attribute “acquisPredicate” of

a role model. When the condition “acquisPredicate” is verified, a new role for this binding is build.

Integration of the Evolving role. The association class "Evolution" is a recursive association class that links a “RoleElement” class called “original role” with another “RoleElement” class called “target role”. It allows to specify role evolutions i.e. the expansion, the mutation and the variable transition predicates. Attributes of this association are:

- *Scenario.* is an attribute with two values "Expansion" or "Mutation" that defines the type of role evolution .
- *TransitionPredicate.* is an attribute of OCL expression type that describes the conditions necessary for an “original role” instance to evolve into “target role” .

The “scenario” attribute is defined as “*mode”* type which is a type of “Enumeration” stereotype. It defines all the transition scenarios between the “original role” and the “target role”. In the “Evolution” class, we combine two rules managing the "Expansion" and "Mutation" role scenarios. In the interest of brevity, we will not detail the rules expressing these OCL constraints:

For the “mutation”, the rule must verify the following conditions:

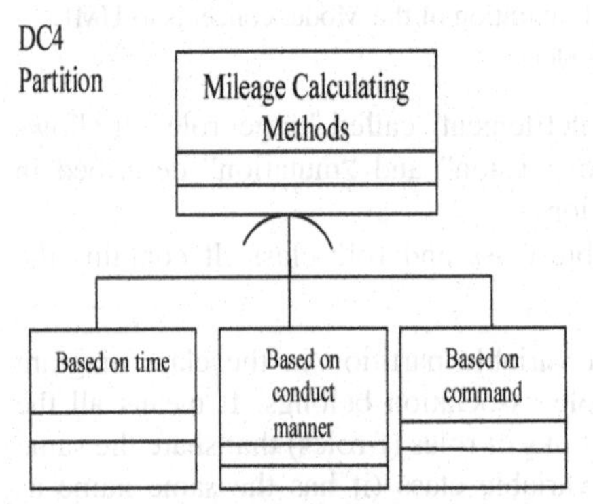

Fig. 6. Variability diagram of the << calculating methods >> partition

- The scenario of "Evolution" must be a “Mutation”.
- The original and target roles are classes of roles.
- Check “TransitionPredicate” evolution class predicate (T*ransitionPredicateIsVerified ()* function = true).

The result of the rule is the original role destruction and the creation of the target role.

The two associations related to “Evolution” are:

- *Original.* It refers to a class of roles that will support the rules of Expansion or Mutation.
- *Target.* It refers to a destination role class obtained after the verification of transition conditions (on the original role).

3.2 Integration of the New Concepts Used in the State-Transition Machine

The technique used to integrate the evolution diagram state machine in UML is the state machine “virtuality” [14].

This means that this state-machine behavior can be redefined by another refinement machine associated with a particular product representing specific state of the state-machine. This technique is used to parameterize a state machine in order to be applied for a specific state. In our model, the parameter is a role.

The concept of "Virtuality" is introduced by the <<virtual>> stereotype and virtualPart tagged value that indicates the occurrence of the virtual machine. In our model, the VirtualPart represents a virtual variable partition in the class diagram (defined in section 2.5). These partitions are named DC and numbered by i (DC1, DC2. ...) (Fig. 5).

4 Illustration

The Figure 6 shows an example of a evolution diagram level where a virtual variable partition (DC4) composed of three role associations links the *"methods of mileage calculating"* variable class to three f-role classes : "based on time", "based on the manner of conduct" and "based on the automatic command".

The *"method of mileage calculating"* class defines permanent properties of the method of variants with the following attributes: Fuel Type, Oil Index, Fuel Consumed Quantity, Fuel Available Quantity; Initialize Mileage Counter, Cumulative Mileage Counter; Last-date-measure-mileage, Date-early-use-brake-pads and Date-change-brake-pads....

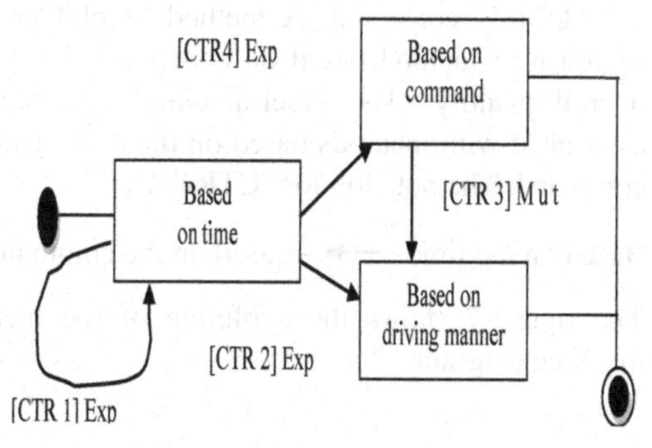

Fig. 7. The evolution of the << method of mileage calculating >>

The *"based on the time"* subclass defines some transient properties as a method based on the elapsed time by the vehicle with the follow attributes: "length between two successive final stops", "Number of vehicle stops" ...

The *"based on the conduct manner"* subclass defines some properties as a transitional method based on the manner of conduct .Its attributes are: "Period between gear changes", "Period between successive braking", "Maximal speed used".....

The *"Method of mileage calculating"* class has an optional role *"based on automatic command* "with transient properties (as a method based on a software command) are attributes that contain different measures (a measure of fuel consumption ...). Our role model supports multiple instantiation of the same class, the *"method based on time"* can be used with two different engines types like gasoline and diesel.

For "CTR1" constraint. A method that plays the role of a "calculation method based on the time" can be for "fuel" and "regular gasoline" for unleaded fuel at the same time". The predicate "number-type-fuel energy ≤ 2 *and* energy = gasoline" define the number "2" as the maximum of fuel type that the method can handle for the gasoline energy at the same time. The notation for CTR1 is:

Based on the time ⟶ + Based on the time.

For “CTR2” constraint. A method that plays the role of a “calculation method based on the time” can be both based on the “manner of conduct” if the engine's power is greater than or equal to 9 CV" as requiring an economy in fuel consumption. The notation of “CTR2” is:

Based on the time ⟶ + Based on driving manner.

For “CTR3” constraint". A method playing the role-based “method of automatic command” can switch to a “method based on the conduct manner” if the order status = 0" i.e. if the command fails. The predicate command status = 0 is associated with methods that switch to a calculation based on the manner of conduct. The notation of “CTR3” is:

Based on the command ⟶ - Based on driving manner.

For “CTR4” constraint. A method of playing the role method based on time may switch to a method based on command if fuel remaining quantity is less than 30% overall quantity" .The "fuel quantity” predicate remaining <30% overall quantity” is associated with methods based on the time switch to a calculation based on automatic command. The notation for “CTR4” is:

Based on the time ⟶ -Based on the command.

The figure 7 shows the evolution of the method of calculating through the state machine diagram.

5 Related Works

Several authors have discussed different areas of SPL evolution. We will discuss the conceptually closest approaches to the concepts used in our approach. One of the most recent of the latter is the work of Seidl [1]. Within this work, the feature model is the only target of evolution originating in the problem space. Other assets located in the problem space such as use cases and other requirement models are not considered. Five types of evolution are used for the problem space: ‘Duplicate Feature’ copies the selected feature (Original-Feature) and adds the clone (CopiedFeature) as sibling of the selected feature to the feature model. ‘Insert Feature’ creates an entirely new feature (NewFeature) in the feature model as child element of the currently selected feature. ‘Split Feature’ is used to model a feature in a more fine grained way than was originally intended by distributing its functionality to an arbitrary number of constituents. ‘Remove Feature’ can be applied to delete a selected feature (OriginalFeature) from the feature model. ‘Remove Feature and Owned Assets’ is a more complex evolution, which deletes a selected feature (Original-Feature) from the problem space and removes all solution space assets that were used exclusively by this feature but no other feature (OwnedAssets). The most advantage of this approach concerns the intra-spatial evolution that does not affect the feature mapping.

However, the transition constraints and the dependencies between features have not been considered. Furthermore, the approach does not support multiple instantiation of the same class. These limitations do not facilitate the evolution and can lead to understandable models.

Borba et al [15] have presented a general theory for software product line refinement describing when evolutions alter the semantic of a SPL. However, they do not take into account the semantic of individual solution space assets and do not offer real operations for evolution.

Anquetil et al [16] have presented an approach, which allows to trace links between SPL artifacts with four dimensions: the variability dimension, which relates problem to artifacts of solution space and relates these artifacts to the problem, the time dimension, which describes how an artifact changes through evolution. It can be used to recover the changes made by an evolution. The refinement dimension captures relations between artifacts of different abstraction levels and the similarity dimension captures relations between artifacts of the same abstraction levels. The procedure supporting evolution is fairly complex as requiring the management of several dimensions relating to several aspects. Our approach does not suffer from such limitations by using the f-role mechanism, evolution diagram, and the role association adapted to the product-line concept.

6 Conclusion

In this paper, we have proposed an evolving variability model based on evolving roles. We have introduced a new association called "Role" that binds the variable classes to their roles we've called 'f-roles" .The latter are represented by classes that we have integrated in UML Meta Models. Variants are considered in the class diagram as roles may be played by variation points. The latter may play only a role among several or many roles simultaneously. New roles can be played and then, they will be added to role classes related to the variable class. Moreover, the role evolution represents the variability at the level of variation point. This evolving aspect of roles has also been integrated with related UML "Role" association that the model has been strengthened by another UML view: the state-machine. It provides to a developer a better visual representation of evolving variability. We have consolidated the model by a new diagram: the evolution diagram that includes a class view and a state-machine view according to the variable virtual partitions in a class diagram. The search for a simplified evolution diagram is our current work.

References

1. Seidl, C., Heidenreich, F., Aßmann, U.: Co-Evolution of Models and Feature Mapping in Software Product Lines. In: Proceeding of the 16th International Conference of the Sofware Poduct Line, Salvador-Brazil (2012)
2. Steimann, F.: A radical revision of uML's role concept. In: Evans, A., Caskurlu, B., Selic, B. (eds.) UML 2000. LNCS, vol. 1939, pp. 194–209. Springer, Heidelberg (2000)

3. Audibert, L.: UML 2 - Édition 2007, Institut Universitaire de Technologie de Villetaneuse , adresse auteur: Laurent.audibert[at]iutv.univ-paris13.fr (2007)
4. Jodłowski, A., Płodzień, J., Stemposz, E., Subieta, K.: Introducing Dynamic Object Roles into the UML Class Diagram. Varsovia University, Poland (2008)
5. Depke R., Engels G., Küster J. M.: On the Integration of Roles in the UML. Technical Report No. 214, University of Paderborn (2000)
6. Dahchour, M., Rayad, H., Lakhrissi, Y., Krioule, A.: Extension UML par les rôles.Département Informatique, INPT, Rabat, Maroc (2007)
7. Guido Boella, A., van der Torre, L., Verhagen, H.: Roles, an Interdisciplinary Perspective. University of Torino, Italy (2009)
8. Subietay, K., Lowskiz, A.J., HabelAx, P., PLodzie, J.: Conceptual modeling of business applications with dynamic object roles (2003)
9. Chernuchin, D., Dittrich, G.: Dependencies of Roles. Dortmund University (2006)
10. Zhu, H., Alkins, R.: Towards Role-Based Programming, Department of Computer Science. Nipissing University (2008)
11. Paulo, J., Almeida, A., Guizzardi, G.: Semantic Foundation for Role-Related Concepts in Enterprise Modeling. University of Espírito Santo, Brazil (2008)
12. Kozaki, K., Sunagawa, E., Kitamura, Y., Mizoguchi, R.: Role Representation Model Using OWL and SWRL ISIR. Osaka University, Japan (2007)
13. Djebar, Y., Kimour, M.T.: Un modèle pour la variabilité dans les LdP. In: Conférence Internationale de l'informatique appliquée -ICAI 2009-BBA, pp. 454–460 (2009)
14. UML -OMG,2007(a).: Unified Modeling Language,V2.1 (2007), http://www.OMG.org
15. Borba, P., Teixeira, L., Gheyi, R.: A theory of software product line refinement. In: Cavalcanti, A., Deharbe, D., Gaudel, M.-C., Woodcock, J. (eds.) ICTAC 2010. LNCS, vol. 6255, pp. 15–43. Springer, Heidelberg (2010)
16. Anquetil, N., Kulesza, U., Mitschke, R., Moreira, A., Royer, J.-C., Rummler, A., Sousa, A.: A Model-driven Traceability Framework for Software Product Lines. Software and Systems Modeling (2010)

An Approach for the Reuse of Learning Annotations Based on Ontology Techniques

Nadia Aloui and Faïez Gargouri

University of Sfax, ISIMS, MIRACL Laboratory - B.P. 242, 3021 Sfax, Tunisia
alouinadia@gmail.com, faiez.gargouri@isimsf.rnu.tn

Abstract. This article presents an approach for the reuse of learning annotations based on ontology alignment. We couple some Knowledge Management's con- cepts (as knowledge and collective memories) to the specific area of e-learning. Our approach of learning annotation reuse lies essentially on a similarity be- tween the two ontologies of context and annotation respectively to the subsys- tems of context and annotation of our ontology-based approach for reusing and learning through context-aware annotations memory.

Keywords: E-learning, Learning annotations, annotations memory, Ontology Alignment.

1 Introduction

In the field of e-learning, annotations are posted for a specific educational goal [7] as a trace of the activity of the reader, noticeable on a document as a mark placed in a specific purpose, and a specific location in which it cannot be dissociated. They can therefore contextualize the knowledge gained from learning objects. The actors then add gains in knowledge resulting from a learning process. Indeed, the annotations added to the learning objects involved in the readability of knowledge are not predictable [1]. Thus, sharing and reusing annotations allow achieving economic, computational, and educational opportunities as they are a real help in the drafting process, a reading support, and a content user evaluation. A memory of collective annotations for all actors in a CEHL is indispensable in order to capitalize and reuse learning's annotations. The CEHL actors benefit by sharing and reusing annotations. As a direct result of this annotation's knowledge capitalization, among others, the quality of learning will hugely improve [2], [1] and [3].

The adaptation of an application context can take many aspects, such us, behavior, content or presentation adaptation. In our approach, we focus on the adaptation of annotation memory to different learning contexts of different actors. Our objective is to provide these actors with learning adapted to their needs in different learning contexts. In addition, it is possible to reuse these different contexts, subsequently, in other contexts, and by the same actors or others. As a result, our approach will greatly increases the quality of learning by the annotations' added-value [1]. It seems therefore necessary to propose a top level ontology of context. This ontology will

A. Amine et al. (Eds.): *Modeling Approaches and Algorithms*, SCI 488, pp. 273–283.
DOI: 10.1007/978-3-319-00560-7_31 © Springer International Publishing Switzerland 2013

consider, on the one hand, all the contextual specificities of all the actors in the annotation memory and ensures the sharing of different learning contexts, on the other hand.

To realize all the cited objectives, we proposed OARLCAM (Ontology-based Approach for Reusing and Learning through Context-aware Annotations Memory) [4]. It is a new general architecture of the adaptive annotations memory with a detailed description of the functionalities offered by each module. OARLCAM is based on a context and annotation top level ontologies [5]. The former represents the learning context of the various actors of the annotations memory. The latter describes the semantics of the annotation to be reused, shared, and learned from knowledge included in these annotations according to a pedagogical objective.

In this paper, we present our approach of learning annotation reuse which lies essentially on a similarity between two ontologies of context and annotation respectively to the subsystems of context and annotation.

This paper is organized as follows. Section 2 gives a general overview of our approach OARLCAM. Section 3 presents the main contribution of this paper: ARLAOA, an approach for the reuse of learning annotations based on ontology alignment. Section 4 introduces our second contribution: similarities measurements.

2 Overview of Our Approach

Different models for describing annotations as activities where proposed, like [1], [2] [3], [6], [7], [8] and [9]. Our study of these approaches confirms that we can deduce their lack for the training of all CEHL actors based on annotations according to a given context.

In order to take into account this note, our annotations memory includes three subsystems : i) the contextualisation subsystem, containing the modules of context capture, context handling, context server, context presentation and a context top level ontology of training, ii) the learning subsystem, containing the modules of learning objects management, learning objects composer and follow-up of the training and iii) the annotation subsystem, containing the annotation module, the annotations' manager, the annotations' adapter, the annotation top level ontology, the annotations' presentation module and the annotations warehouse, for later reuses.

We also defined two ontologies: CTLO [5] (Top Level Ontology Context) and TLPOO (Top Level Pedagogical Objective Ontology).

CTLO is a generic and exhaustive context ontology which provides the context proprieties related to learning provided by our annotations memory. It is conceived to solve the limits and the insufficiencies of the existing context models. CTLO contains three facets: cognitive, semantic and contextual. On another side, the context top level ontology contains six facets.

TLPOO is defined to describe the learning objectives for the context-aware annotations memory [10], in which learning annotations are capitalized and reused for learning-based annotations. In this ontology, pedagogical objectives are closely correlated with the semantic of learning annotation. To design the TLPOO, we follow as fit CTLO the Noy [9] iterative method.

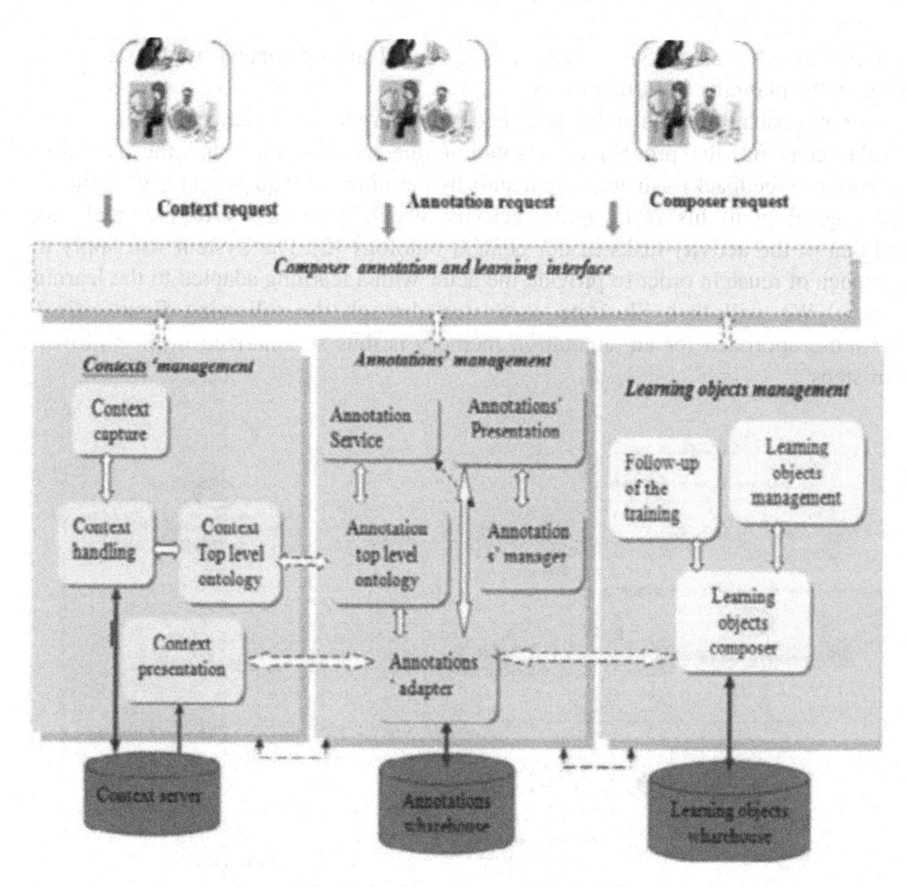

Fig. 1. Our approach OARLCAM

Our approach to reuse annotations and learning contexts is based on a mapping between the CTLO and the TLPOO to extract the appropriate annotations to a given learning objective.

3 ARLAOA: An Approach for the Reuse of Learning Annotations Based on Ontology Alignment

Our approach of learning annotation reuse lies essentially on a similarity between the two ontologies of context and annotation respectively to the subsystems of context and annotation. This has for objective to check the annotations which are adequate to the current learning context taking into consideration all the contextual properties which describe it. Our approach of reuse is concerned with the levels of reuse 1) the reuse of stored annotations to take advantage of the included knowledge, benefit from experiences feedback of different actors and consequently, improve the learning quality always according to a pedagogical objective and a current context; 2) the reuse of learning contexts stored in the context server within similar learning contexts by other actors 3) the reuse of the same learning context by the same actor through the planning of further users.

Our approach of reuse can be described by a four-step cycle namely: the search for similar contexts, the process of adaptation, the memorization and the evaluation (experience feedback). An actor shall start by defining the request that is to define his pedagogical goal, his pedagogical activity which corresponds to the goal facet and that of the activity-tasks of our context ontology [5]. Our system will apply the approach of reuse in order to provide the actor with a learning adapted to the learning context. We will then illustrate each step through the following figure (fig.2). Our reuse approach for an annotation memory is thus summarized in the following four steps.

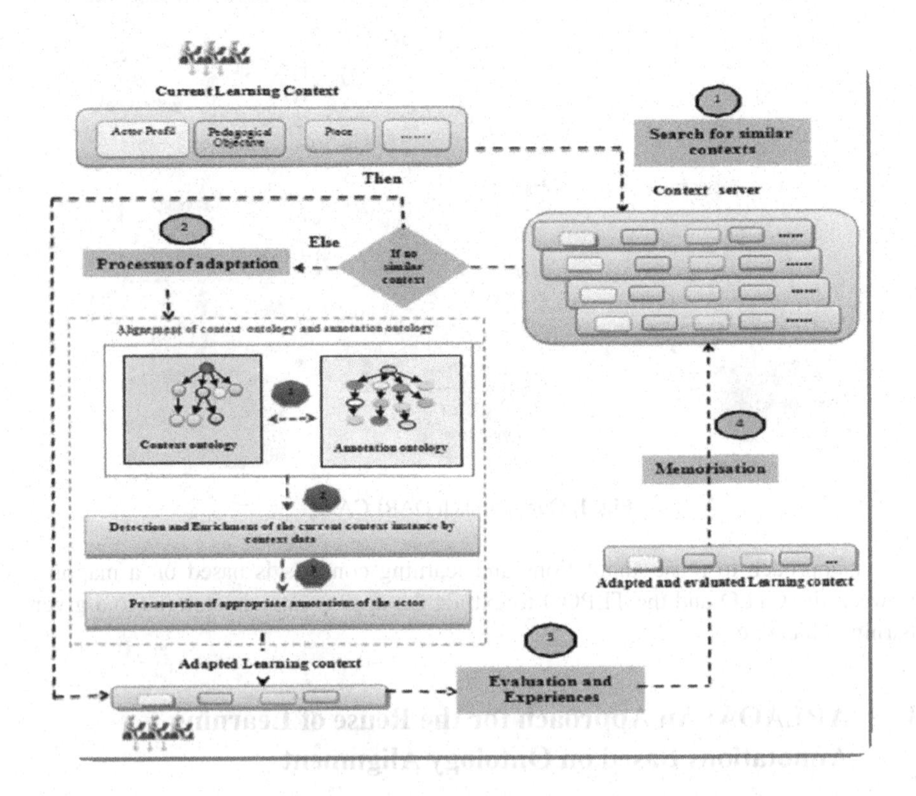

Fig. 2. The ARLAOA Approach

3.1 Searching a Similar Context (Step 1)

To meet the actor's needs for a learning adapted to his learning context X, we will start by searching, in the context server, for a learning context similar to X. If it is the case, we will enrich this context instance with the current computing contextual properties such as the date, the location, etc., and then we present it directly to our actor.

3.2 Adaptation (Step 2)

This step is primordial for adapting the learning annotations to the current context through the alignment of two ontologies of annotation and context. This alignment shall allow checking the annotations which correspond to the same properties of the current learning context (pedagogical goal, pedagogical activity…). We will adopt then an approach of measurement of semantic and structural similarities which are expressed by a better learning quality. Owing to this alignment, the checked out annotations will be enriched by the contextual data detected by the system context capture. Indeed, it is a learning based on context-aware annotation memory.

3.3 Memorization (Step 3)

Once we provided the actor with a learning adapted to his learning context as well as to his pedagogical goal, the learning-context instance is stored in the context server for further use within similar contexts by other actors or by the same actor.

3.4 Evaluation and Experience Feedback (Step 4)

The result/evaluation characteristic defines the level of pertinence of a learning session by our system. It is a criterion which shall be taken into account during the stage of reuse of previous learning contexts. In fact, the reuse lies, on one hand, on the learner's judgments concerning the pertinence of the quality of learning proposed also by other actors such as the time span which separates successive uses. During his learning, our actor then evaluates the pertinence of his current learning context.

Our reuse approach is actually based on interactions between different ontologies making explicit learning by context-aware annotation memory.

In consequence, this approach lies on a similarity between the annotation ontology and the context ontology in order to check the adequate annotations for a given learning context. We will, then, use our own algorithm of similarity measurement which shall be presented in the following section.

In the next section, we describe our method of similarity measurements between the context ontology and the annotations ontology, on one hand, and the various instances of context ontology for further reuses of the learning context on the other hand.

4 Similarities Measurements

Several similarity measurements are suggested in the alignment process [11]. The Terminology method compares the entity labels. It is divided into purely-syntactic methods and those using a lexic. A syntactic method performs the correspondence through the measurements of strings dissimilarity. Whereas the lexical approach performs the correspondence through the lexical relationships (synonymy, hyponymy, Etc.). The internal structure comparison method compares the entity internal structures (interval, value, attributes cardinality, etc.). The external structure comparison method compares the entity relationships with others. It is divided into entity comparison methods within their taxonomies and external structure comparison methods by taking into account the cycles. The semantic method compares the interpretations (or more exactly models) of entities.

In our learning approach, based on annotation memories, our interest is to find a semantic similarity between the context ontology and the annotation ontology in order to offer the adequate annotations for the current learning context of the current actor. This has for objective to provide an adapted learning to the current context. Therefore, it is indispensable to find semantic and structural similarities between the concepts explicating the learning context (learning field, pedagogical goal, pedagogical activity, type of activity, actor's profile…). In the case of further reuse, the measurement of similarities is performed between various instances of memorized learning contexts.

We calculate the global similarity with performing two successive steps. The first step allows calculating the semantic similarity. The second one allows calculating the structural similarity for each pair of concepts of both ontologies by exploiting the contiguity structure. We suggest then two functions in order to calculate respectively the structural similarity and the semantic similarity.

4.1 Measure of Semantic Similarities

We use the Jaro-Winkler functions as they give good results of similarity according to the comparative study of Cohne [12].

Moreover, recent works [13] and [14], for example, have shown the usefulness of WordNet in identifying synonymy between concepts of a given domain. We expect that our future experiments our algorithms are based on Wordnet, in addition to the measurement function.

The SimSEM function allows calculating the semantic similarities of the pairs of concepts C1 and C2 respectively of both ontologies O1 and O2 by adopting the following steps:

- For each concept C1 and C2, go through the concepts of O2 so that the category of C1 = C2,
- Calculate the semantic similarity according to the Jaro-Winkler [12] function,
- Store the semantic similarity (SimSEM) in the vector VSS.

4.2 Measure of Structural Similarities

The structural techniques exploit the ontology structure to be compared, often represented in graphics. The comparison of similarities between two concepts of both ontologies can be based on the position of the concepts in their hierarchies. These techniques are based on the following hypothesis: "(H) – if two entities of two ontologies are identical, their nearby entities will be the same in a certain way" [14] and [15].We suggest calculating the structural similarity between the entities of two ontologies. We are inspired from the works of Albosaly [13] and Zghal [14].

We calculate then the structural similarity by exploiting the semantic similarity (fig.3) of the pair of concepts to match as well as the contiguous structure by adopting the following steps:

- If the contiguity of a concept c1 noted V(c1) is similar to the contiguity of the concept c2 noted V(c2) then c1 and c2 are similar in a certain way,
- The position of V(c1) in relation to c1 shall be similar to V(c2) of c2, then we sugest to calculate the structural similarity of the pair (C1, V1) and (C2, V2) so as to get Simst (C1, V1) in order to determine the position of V(c1) in relation to C1 and respectively V(c2) in relation to C2.

```
Structural Similarity Algorithm
Function: Sim_ST
DONNEES :
    1) O1 et O2 : the two ontologies to align
    2) VSS : vector of semantic similarity
    3) II S : Similarity Weight of each category of concept
SORTIES : VSST : vector of structurel similarity
  BEGIN
      /* go to each concept of ontology 1 */
         For each (CO1 εO1) do
      /* go to each concept of ontology 2 */
         For each (CO2 εO2) do
      If C1.type==C2.type then
 /*Extract in VV1 vector neighboring concepts to C1 */
        VV1 =NeighboringCONCEPTS(C1)
  /* Extract in VV2 vector neighboring concepts to C2*/
        VV2 =NeighboringCONCEPTS(C2)
  /*Calculate the C1 structure of neighbourhood similarity */
        SimSV1= fonctSimSV1(C1,VV1)
 /* Calculate the C2 structure of neighbourhood similarity */
        SimSV2= fonctSimSV2(C2,VV2)
  /*calculate structurel similarity*/
        SimST = fonctSimST (SimS, SimSV1, SimSV2,VV1,VV2)
/* Add C1, C2 and SimST to VSST*/
        AJOUTER ((C1, C2, SimST), VSST)
        Return (VSST)
   END
```

Fig. 3. Structural similarity algorithm

Several functions measuring structural similarities based neighborhood calculating are proposed in the literature [16], [13]. The choice of a particular function depends closely on the results obtained. We are currently conducting a comprehensive study on the selection or improvement of any of these functions to suit our research. For the example in Figure 4, and to illustrate our algorithms, we use the measure of by exploiting the similarity measure neighborhood Wu & Palmer [17] (simWp) and the semantic similarity measure Jaro- Winkler (SimS) in the following manner:

Sim Str (c1, c2) = Σ SimS (VV1, VV2) * simWp (c1, c2) * (1 - | d1-d2 | ²) and d1 = simwp (c1,vv1(c1)) and d2 = simwp (c2,vrv2(c2)),

Example: We apply our structural method to make mapping between the two concepts reformulate et structure or design a course:

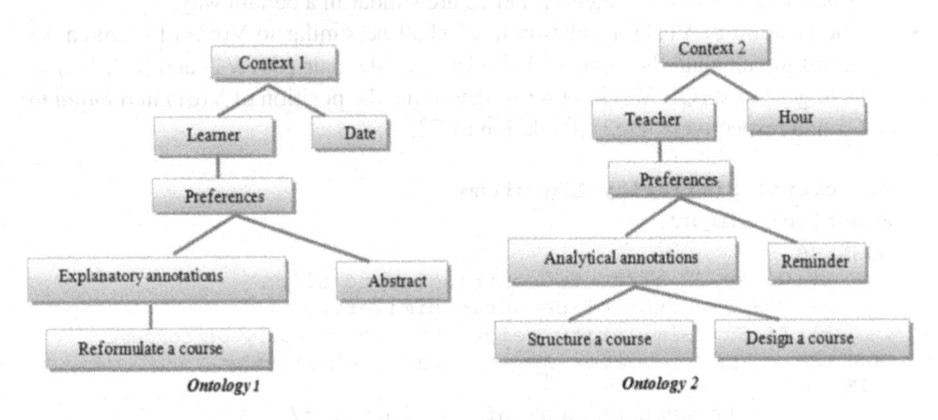

Fig. 4. Example of our measurement method

We consider that c1=reformulate a course and c2= structure or design a course.

We suppose that SimS (explanatory annotations, analytic annotations) are similair with SimS= 0.87;

Case-1: (c1, c2) = (reformulate a course, structure a course) = {(annotation explanatory, annotation analytical)} | V (c1, c2) | = 1;
SimWP (explanatory annotations, reformulate course) = 0.85 = (2 * 3/3 +4) = d1; SimWP (analytical annotations, structure a course) = d2 = 0.88;
simWP (reformulate a course, a course structure) = 0.9;

And finally Simstr (rephrase course, course structure) = 0.87 * 0.9 * (1 - | 0.85-0.88 | 2) = 0.78

Case-2: (c1, c2) = (reformulate a course, design course) = {(explanatory annotations, analytical annotations)} | VR (c1, c2) | = 1;
SimWP (explanatory annotations, reformulate course) = 0.85 (2 * 3/3 +4) = d1SimWP (analytical annotations, design a course) = 0.8 = d2;
simWP (reformulate a course, design course) = 0.7;

And finally Simstr (reformulate a course, design course) = 0.87 * 0.7 * (1 - | 0.85-0.8 | 2) = 0.6

As a conclusion, we can deduce from these calculations that the similarity between (reformulate a course, a course structure) is more similar than that between (reformulate a course, design a course).

4.3 Calculation of Global Similarity

We quantify the semantic similarity and the structural similarity in order to obtain the global similarity. Our aim is to align the ontology of the learning context and the ontology of annotation in order to provide the adequate learning for the current learning context. We seek a global similarity in order to optimize the learning quality as well as our annotation memory. The calculation of the similarity could be also used between the context of ontologies in the case of reusing the same learning context or searching for similar contexts in the context server, etc.

```
Algorithm: Similarities
INPUT:
1) O1 and O2: Ontology 1 and Ontology 2
2) VSS: Semantic vector of similarities
3) VSSt: structural vector of similarities
4) WSimSt: Weight of structural similarities
5) WSimS: Weight of semantic similarities
OUTPUT: VSG: global (Semantic and structural) vector of
Similarities
Begin
/* go to each concept of ontology 1 */ For each ( CO1 εO1) do
/* go to each concept of ontology 2 */ For each (CO2 εO2) do
If CO1.type==CO2.type then
/*Extract semantic similarities between CO1 and CO2 of
VSS*/
SimS=EXTRACTSIM (VSS, CO1, CO2)
/*Extract structural simiraties de CO1et CO2 of VSSt */
SimSt=EXTRActSIM (VSSt, CO1, CO2)
/*calculate global similarity*/ SimG = SimS + SimSt
/* Add CO1, CO2 and SimG in VSG*/ Add ((CO1, CO2, SimG), VSG)
Return (VSG)
END
```

Fig. 5. Global similarity algorithm

5 Conclusion

This paper represents a study to realize an adaptive annotation memory for context-aware training ARAMAC for the various actors in a CEHL. This memory can satisfy the need for training according to a given objective and a given context of all the actors in terms of utility, re-use, sharing and adaptability. ARAMAC [4] is based on ontology engineering.

The running mechanism of our framework articulates around a whole of modules. Each module allows a functional need well defined and is composed of Web services. This framework aims to facilitate its integration, its interoperability with other e-learning systems.

The top level context ontology (CTLO) [5] allows to formalize the proposed context data delivered by providers of context.

We developed the top level ontology of pedagogical objective (TLPOO)[9] in order to implement it in our annotations memory dedicated to all actors.

In this paper, we have also describing our process of reuse of two levels for the annotation as a first level and then for the learning contexts as a second level, that is to say the manner in which this knowledge will be exploited in order to generate automatically a learning relatively adapted to a given pedagogical goal in a current context, in addition to the reuse of these contexts later.

The reuse and capitalization in our approach are based on ontology alignment; we present thus our method of semantic, structural and global similarity measurements.

Several perspectives are possible, for this work. In Particular, We Aim to experiment and evaluate our method of calculation Similarities by adding improvements especially in terms of measurement functions that can give us better results.

References

1. Azouaou, F., Desmoulins, C.: Taking Teaching Context into Account for Semantic Annotation Patterns. In: EC-TEL 2006, pp. 543–548 (2006)
2. Bringay, S., Barry, C., Charlet, J.: Annotations: A Functionality to support Cooperation, Coordination and Awareness in the Electronic Medical Record. In: Hassanaly, P., Herrmann, T., Kunau, G., Zacklad, M. (eds.) Cooperative Systems Design, COOP 2006, Carry-le-Rouet, France, May 9-12, pp. 39–54 (2006)
3. Oouadah, A., Azouaou, F., Desmoulins, C.: 2ème conférence internationale sur l'informatique et ses applications. In: CIIA 2009, Modèles et Architecture d'une Mémoire d'Annotation context-aware pour l'Enseignant, Saida, Algérie, May 03-04 (2009)
4. Aloui, N., Gargouri, F.: Ontology-based Architecture for Reusing and Learning Through Context-aware Annotations Memory. In: ICCGIS, INRIA Confetrence, Lexumburg (2011)
5. Aloui, N., Gargouri, F.: Generic context ontology for a Learning Through Context-aware Annotations Memory. In: ACM ICICS, Oman (2012)
6. Euzenat, J., Shvaiko, P.: Ten Challenges for Ontology Matching. In: OTM Conferences (2), pp. 1164–1182 (2008)
7. Mille, D.: Thèse., in Département informatique, Université Joseph- Fourier, Modèles et outils logiciels pour l'annotation sémantique de documents pédagogiques: Grenoble, 173 pages (2005)
8. Mokeddem, H., Azouaou, F.: Cyrille Desmoulins. In: CIIA 2009: Ontologie de la Sémantique de l'Annotation Pédagogique de l'Apprenant (2009)
9. Moisuc, B.: Adaptation dans les systèmes d'information spatio-temporelle interactifs, INFORSID, Hammamet, Tunisie (2006)
10. Aloui, N., Gargouri, F.: A generic ontology of objective for learning through a context-aware annotations memory. In: IADIS e-learning Conference, Portugal (2012)
11. Shvaiko, P., Euzenat, J.: A survey of schema-based matching approaches. In: Spaccapietra, S. (ed.) Journal on Data Semantics IV. LNCS, vol. 3730, pp. 146–171. Springer, Heidelberg (2005)
12. Cohen, W., Ravikumar, R.P., Stephen, E.: A Comparison of String Distance Metrics for Name-Matching Tasks. American Association for Artificial Intelligence (2003)

13. Abolhassani, H., Hariri, B.B., Haeri, S.H.: On Ontology Alignment Experiments. Webology 3(3) (September 2006)
14. Zghal, S.: Thèse en informatique, Contributions à l'alignement d'ontologies OWL par agré- gation de similarités. Université de Tunis (October 23, 2010)
15. Fellah, A., Malki, M., Zahaf, A.: Alignement des ontologies: utilisation de WordNet et une nouvelle mesure structurelle. In: Conférence en Recherche d'Information et Applications, CORIA (2008)
16. Touzani, M.: Alignement des ontologies OWL-Lite. Master's thesis, University of Montreal
17. Wu, Z., Palmer, M.: Verb semantics and lexical selection. In: Proc. of the 32 nd Annual Meeting of Computational Linguistics, Las Cruces, pp. 133–138 (1994)

Very Large Workloads Based Approach to Efficiently Partition Data Warehouses

Gacem Amina[1] and Kamel Boukhalfa[2]

[1] ESI School, Algiers
a_gacem@esi.dz
[2] USTHB University, Algiers
kboukhalfa@usthb.dz

Abstract. Horizontal Partitioning (HP) is an optimization technique widely used to improve the physical design of data warehouses. However, the selection of a partitioning schema is an NP-complete problem. Thus, many approaches were proposed to resolve this problem. Nonetheless, the overwhelming majority of these works do not take into account the size of the workload which can be very large. Huge workload increases the time of HP selection algorithms and may deteriorate the quality of final solution. We propose, in this paper, a new approach based on classification and election to select an HP schema in the case of large-sized workloads. We conducted an experimental study on the ABP-1 benchmark to test the effectiveness and scalability of our approach.

Keywords: Horizontal Partitioning, Large Workloads, Classification.

1 Introduction

Data warehouses are at the heart of enterprises. They are interrogated simultaneously by several applications to help managers and analysts into their tasks. In large companies that store terabytes of data, it is not uncommon for a data warehouse to be subject to workloads consisting of multiple jobs performed by several clients beyond *thousands of queries* [1]. This stems from the growing need for analytical tasks. Large workloads can degrade the performance of the DBMS, and thus, slow applications and increase the response time to users, especially a decision makers, often demanding in time. To solve this problem, various optimization techniques have been proposed, including the Horizontal Partitioning (HP) [2]. Selecting a partitioning schema is known NP-complete problem [3]. Algorithms and heuristics abound about how to automatically set the best HP schema. Typically, these algorithms contain two phases: metadata and statistics collection and HP schema selection. Several works have addressed the HP selection problem [4,5,6,7,3,8,9,10]. The proposed algorithms were tested on workloads containing a small number of queries, although a lot of queries are continually submitted to data warehouses. Thus, it is crucial to test those algorithms on large workloads. However, a large-sized workload can increase dramatically the time needed to access the data warehouse and gather the meta-data. It

A. Amine et al. (Eds.): *Modeling Approaches and Algorithms*, SCI 488, pp. 285–294.
DOI: 10.1007/978-3-319-00560-7_32 © Springer International Publishing Switzerland 2013

lengthens also the evaluation of solutions produced by the selection algorithms because of the number of predicates that leads to an explosion of the search space. These effects will subsequently cause: (1) an extra time needed to analyze the queries for retrieving the statistics, (2) a delay in the search for a partitioning configuration because of the numerous possible solutions, (3) consumption of additional time to evaluate each partitioning configuration as the evaluation is performed on each query, and (4) a degradation in the quality of the final partitioning schema due to the exploration of various solutions. The administrator must make a compromise between time of selection algorithms and the quality of HP schema. Our main goal is to propose an approach supporting large workloads by reducing the execution time of selection algorithms without deteriorating the quality of the HP schema. We propose in this approach the addition of two phases to classical approaches: *classification* and *election*. Classification [11], very widespread in several issues, could be useful in this problem. Grouping similar queries into classes allows the use of a smaller set of classes instead of a large workload. The election generates a new query for each class that will be injected into the new workload. This paper is organized as follows: Section 2 presents the horizontal partitioning, then, the Section 3 discusses related works. Our HP selection approach is detailed in section 4. The experimental study is described in Section 5. Finally, we conclude the paper in Section 6.

2 Horizontal Partitioning

HP divides objects of the database (tables, views and indexes) into multiple horizontal rowsets called partitions [12]. A partition is defined by a set of predicates. Each row of a partitioned table belongs to a single partition. HP is a non-redundant optimization technique as it doesn't replicate data, but only divided tables into partitions depending on the used predicates [4].

There are two HP modes: *primary* and *derived partitioning.* The first one partitions a table using predicates defined on its attributes. The second one exploits references and partitions a table by using foreign keys from an already partitioned table.

Example 1. Given a data warehouse (DW) with a dimension table Customer and Sales fact table, if the DW administrator seeks distribution of customers in relation to their age, it may break the Customer table into three primary partitions as it follows:
- $Customer - Children = \sigma_{Age<18}(Customer)$
-$Customer - Adults = \sigma_{18 \leq Age<60}(Customer)$
- $Customer - Retirees = \sigma_{Age \geq 60}(Customer)$

The sales fact table is divided by reference into three referenced partitions as it follows:
- $Sales - Children = Sales \ltimes Customer - Children$
- $Sales - Adults = Sales \ltimes Customer - Adults$
-$Sales - Retirees = Sales \ltimes Customer - Retirees$

HP has been studied in several environments: *parallel* [13,6], *distributed* [4,14], *Grid* [5,15,9,7], *Cloud* [16]. It has been implemented in major DBMS: *Oracle*

[17], *SQL Server* [18] and *IBM DB2* [19]. Each DBMS has different modes of partitioning and provides a set of commands to create and manipulate partitions.

3 Related Works

The HP selection problem has been proven NP-Complete [20,3]. Several approaches were proposed to resolve this problem. They can be grouped into four categories: *Predicates* Based Approaches (PBA), *Affinity* Based Approaches (ABA), *Cost model* Based Approaches (CMBA) and *Classification* Based Approaches (CBA). In the first category, the selection starts by enumerating all the existing predicates in the workload, and subsequently generating a set of minterms [4,14]. A minterm is a conjunction of simple predicates. PBA approaches are simple to implement, but have an exponential complexity (if n is the number of predicates, 2^n minterms are generated). ABA approaches associate a partition to each group of predicate which haves a certain degree of affinity. The affinity between two predicates is defined as the sum of frequencies of queries referencing those predicates together. Two algorithms were proposed to generate the HP schema : *graph based algorithm* and *matrix based algorithm* [21]. The ABA approaches were proposed and applied for vertical partitioning [21] and has been adapted in horizontal partitioning by [22]. CMBA approaches are inspired from cost models used by the RDBMS optimizer to choose the most optimal execution plan. In these approaches, the HP problem is modeled as an optimization problem with an objective function that incorporates the execution cost of the workload. Then, a meta-heuristic like *Genetic Algorithm*, *Simulated Annealing* [3] explores the space of solutions. Some works consider a mathematical cost model [5,3], others consider the DBMS optimizer cost [10]. CBA approaches are based primarily on the K-means algorithm [23]. These approaches generates clusters of predicates extracted from the queries. Each cluster is used to generate a partition. The K-means algorithm has been used in classification problems in the context of distributed relational [9] and XML [7] data warehouses . The majority of works cited above have been tested on workloads containing between 20 and 60 queries. Moreover, no approach mentioned above does define its behavior when the workload is very large. As the number and complexity of decision-support queries grow continually, it is crucial to develop a *scalable approach* that can deal with the large-sized workload. We present in the next section our approach based on classification and election.

4 Our HP Selection Approach

HP selection problem has been widely studied in the literature. To reduce the complexity of the problems, existing approaches propose pruning attributes or dimension tables [24,25]. To the best of our knowledge, neither approach has addressed the problem of scaling when the workload is very large. Indeed, a large workload induces a couple of inconvenience: (1) the time taken to explore the solutions space becomes too long (2) the number of predicates considered by the

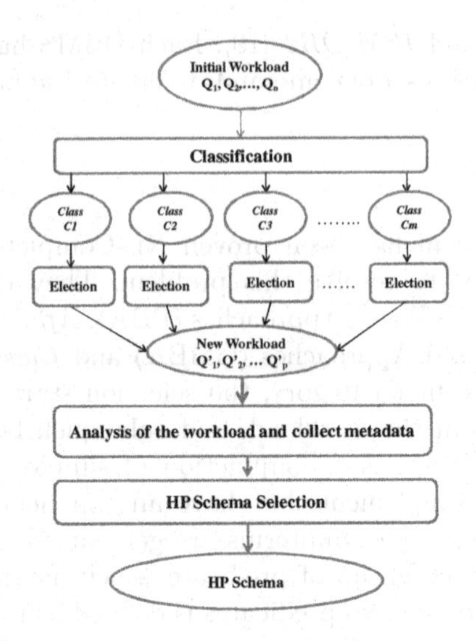

Fig. 1. An overview of our approach

partitioning process raises and extend the size of the search space. We propose, in this paper, a new approach that speeds HP selection without degrading the quality of final solution. Our approach is composed of two phases : (1) *classification* that classifies queries in order to reduce the size of workload and (2) *election* that chooses a query from each class to build a smaller workload. Figure 1 shows an overview of our approach.

4.1 Classification Phase

The goal of the classification phase is to group similar queries into the same class. It takes as input a workload consisting of N queries and outputs R class ($R \leq N$). We considered that two queries are related if they reference common tuples in the fact table. This principle is based on the idea that, if two queries references the same data, then the same HP schema could satisfy both queries. The ideal case is observed when the data referenced by the first query are included in the data referenced by the second. There are many clustering algorithms in the literature, we can mention: K-means [23], *DB-SCAN* [26], etc. We choose K-means algorithm in our work because it is simple to adapt to our problem. Hence, by using this algorithm, we can create classes from clusters and generate elected queries from centroids. Classification of queries in our approach involves four steps: (1) Extraction of predicates from the initial queries, (2) construction of the *query-predicate matrix*, (3) Computation of the number of classes and (4) Classification by K-means. The extraction of the selection predicates is done by examining the clause *where* for each query. A selection predicate has the following form: $A\ \theta\ Value$ where A is an attribute, $\theta \in \{<, >, \leq, \geq, =\}$ and $Value \in domain(A)$. From the set of predicates obtained in the previous step,

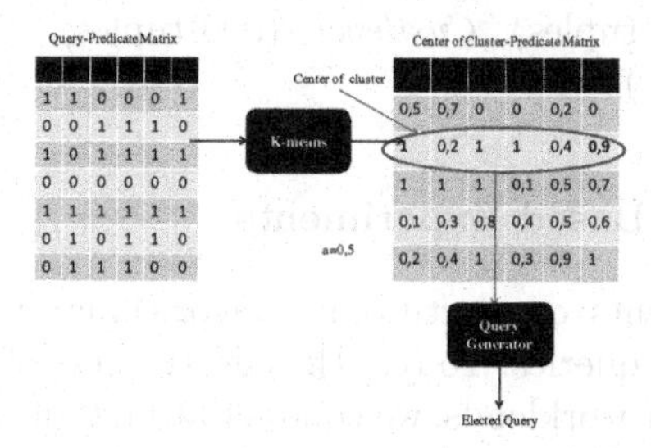

Fig. 2. A classification and election based on K-means

we construct a matrix M called *query-predicate matrix*. An element m_{ij} of M is defined by $m_{ij} = 1$ if the query Q_i use the predicate P_j, 0 otherwise. We use the similarity principle explained above to compute the suitable number of classes. The number of classes K corresponds to the number of inclusions found between queries. Hence, The K-means algorithm takes as input K and the *query-predicates matrix* and outputs K class of queries $\{C_1, C_2, \cdots, C_K\}$.

4.2 Election Phase

The K-means algorithm computes, in addition to classes, the *center* of each cluster. In our approach we use the center of each cluster to generate the elected query . A center is represented by a vector in which every element corresponds to a predicate. Each predicate is accompanied by a weight representing its importance in the construction of this center (Figure 2). The elected queries are constructed from the centroids. To choose which predicates should be integrated into the new query, we defined a threshold α that determines the minimal weight at which the predicate is incorporated into the elected query. This threshold can be set by the administrator or calculated experimentally. A predicate P_i is chosen to build an elected query if and only if $Weight(Pi) \geq \alpha$. For example, in Figure 4, if the parameter α has been set to 0.5, then only predicates p_1, p_3, p_4, p_6 are integrated into the elected query.

5 Experimental Study

We present in this section, an evaluation of our approach through an experimental study divided into three parts: (1) experiments based on the size of workloads (medium, large), (2)experiments based on the degree of heterogeneity (heterogeneous, homogeneous) and (3) Oracle validation. To select HP schema we use the genetic algorithm and cost model proposed in [3]. The parameters used in GA are: *initial population 50, selection rate 0.9, crossover rate 0.8, mutation rate 0.1, number of generations 500 and we set the maximum number of partitions at 1000.* We used the star schema generated by the APB-1 benchmark [27]. It consists of a fact table *Actvars* (24,786,000 tuples) and four dimensions tables, *Prodlevel* (9000 tuples), *Custlevel* (900 tuples), *Timelevel* (24 tuples) and *Chanlevel* (9 tuples).

5.1 Workload Size Based Experiments

In literature, experiments on partitioning algorithms are done on workloads which do not exceed 60 queries. To test the effectiveness of our approach and its ability to support larger workloads, we conduct at first an experimental study on a medium-sized workload (500 queries) and then, we consider a larger workload (1000 queries). We set the experimental value of α at 1 and 0.3. The results of experiments on 500 queries are shown in Figure 3. Our approach reduces the execution cost by 2 % for both values of α in comparison with the classical approach (approach without classification). To check the scalability of our approach, we generate a large-sized workload of 1000 queries. We measure the time needed to run the entire selection process (all steps of HP selection). This time is denoted T_{total} and is calculated as follows:

$T_{total} = T_{ld} + T_{an} + T_{cl} + T_{el} + T_{se}$ where:

T_{ld} is the time spent to load queries from source files, it does not depend on the used approach, so it has been ignored, T_{an} is the time needed to analyse metadata and compute statistics (it may vary according to the size of the workload), T_{cl} is the time required to classify queries, T_{el} is the time needed to perform the election and T_{se} computes the time of selection (genetic algorithms).

After running a workload of 1000 queries, we observe the performance of our approach. The results are shown in Figure 5. Our method improves dramatically the performance as it divides the entire time T_{total} by 10. Since the number of queries decreases, the time needed to perform analysis of DW and to explore the solutions space have been greatly reduced. For instance, the time spent to explore the solution space has been improved by 95% as the reduced workload only contains 615 queries. The Figure 4 shows the quality of partitioning schema generated for 1000 queries. Our approach improves by 20 % the quality of partitioning when compared to the classical approach when $\alpha = 1$. In addition, it produces an HP schema with similar quality to the one obtained by the classical approach when $\alpha = 0.3$. We conclude from these experiments that despite the

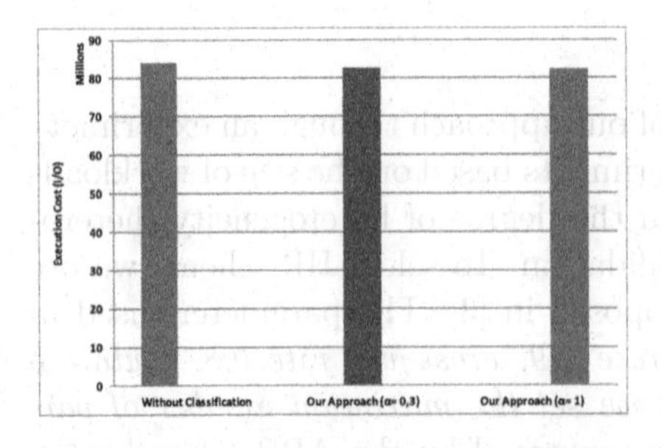

Fig. 3. Cost of 500 queries

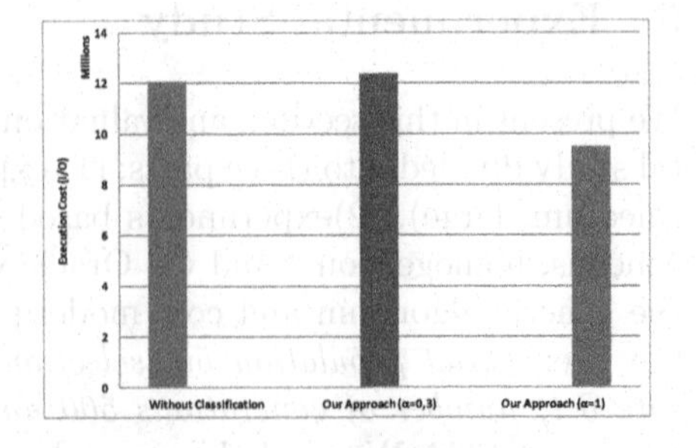

Fig. 4. Cost of 1000 queries

use of a reduced workload, our approach produces HP schema that have similar or better quality than the classical approach that uses the initial large-sized workload. Moreover, our approach improves the response time of HP selection. We can also deduce the threshold value α has an impact on the quality of the final solution.

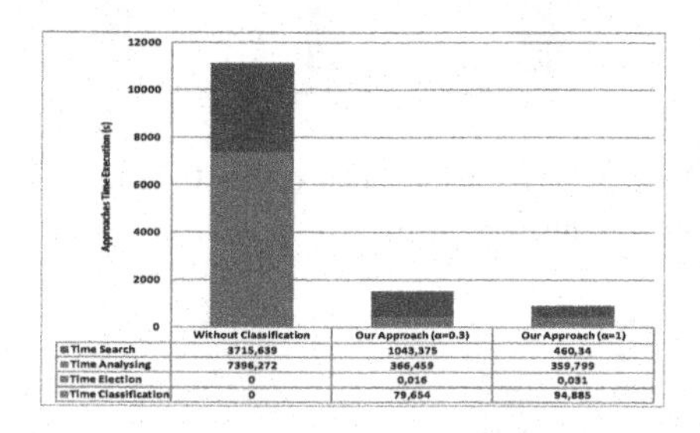

Fig. 5. Performance of our approach on 1000 queries

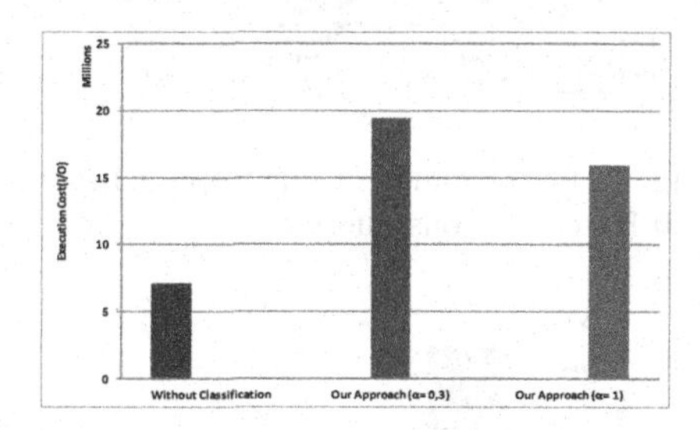

Fig. 6. cost of 730 Heterogeneous queries

5.2 Degree of Homogeneity Based Experiments

Workloads that were used in previous experiments were generated randomly. As the classification process and election are based on the similarities between queries, it is interesting to study the behavior of our approach towards heterogeneous and homogeneous workloads. Heterogeneous queries share only a few lines of data. In contrast, homogeneous queries share a large number of data lines. To conduct these tests, we design a module that generates workloads which can be homogeneous or heterogeneous.

Heterogeneous Workload: We generated 730 heterogeneous queries, containing predicates that differ from one query to another. Each predicate appears in at most three queries. Figure 7 represents the performance of our approach compared to the classical approach. The entire execution time is divided by two. However, we notice that there is not a significant decrease in the execution cost. Figure 6 shows that our approach does not improve the quality of HP schema.

Homogeneous Workload: We tested our approach on a workload of 1010 queries. To ensure the homogeneity of queries, we have ensured that each predicate is present in at least 200 queries. Figure 9 shows a significant decrease in the time spent to analyze queries and select the HP schema. Our approach made its best experimental results with the test on homogenous workload. We see in Figure 8 that our approach improves the cost of 52.33 % when $\alpha = 0.3$ and 70 % when $\alpha = 1$. These results are due to the homogeneity of queries which makes it easier to group queries into classes and elect for each class an appropriate query.

On the basis of experimental results, we can formulate a set of recommendations: (1) For medium loads (less than 500 requests), our approach significantly improves the query execution cost (20%) and divide by 10 the execution time of the approach. (2) When the workload is heterogeneous, our approach speeds

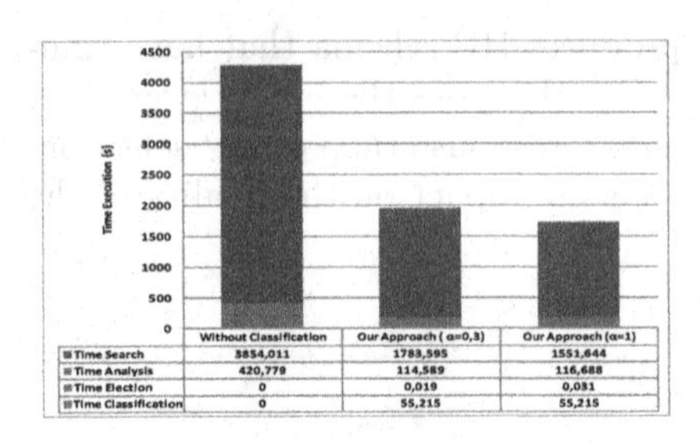

Fig. 7. Performance of our approach for 730 Heterogeneous queries

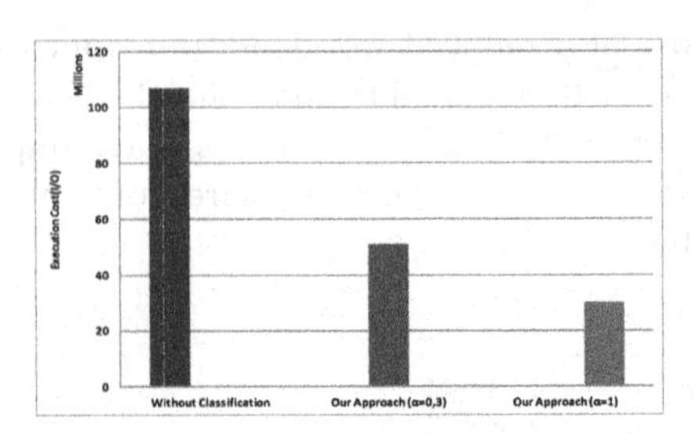

Fig. 8. Execution cost of 1010 homogeneous queries

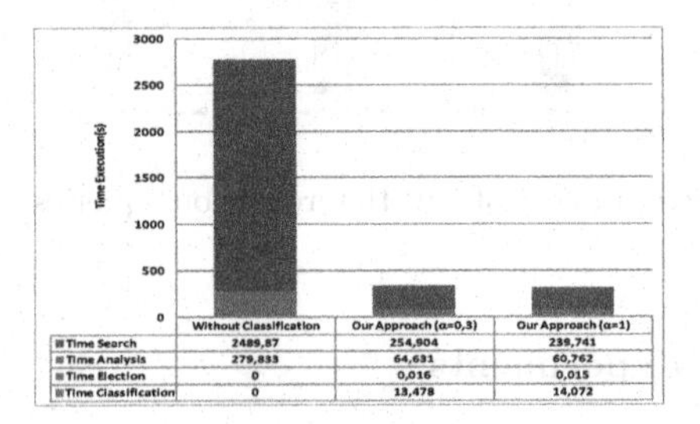

Fig. 9. Performance of our approach on 1010 homogenous query

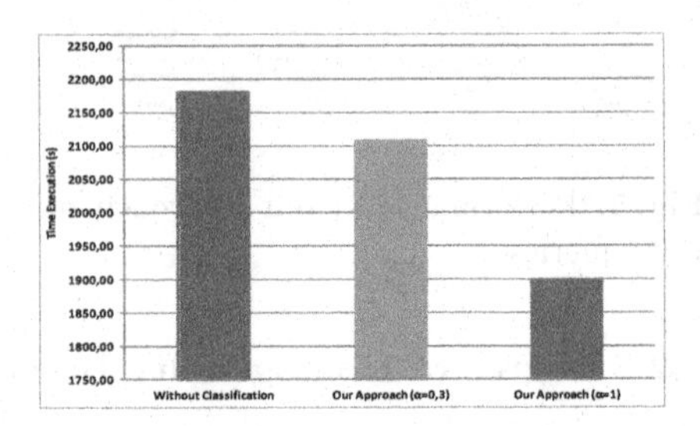

Fig. 10. Validation Oracle

the selection process but does not improve the quality of HP schema, in this case, the administrator should determine which is prior: the response time or the quality of HP schema. (3) when the workload is large and homogeneous, our approach provides a considerable gain in response time and significantly reduces the execution cost. We therefore strongly recommend our approach in the presence of a homogeneous workload.

5.3 Validation under Oracle 11g

In order to test the performance and effectiveness of our approach in real DBMS, we have implemented the HP schema obtained our approach on Oracle 11g and compare it to the one generated by the classical approach, we have then executed a workload of 1000 queries on the DBMS. The execution time of the initial workload is calculated for each generated HP schema. The results show that our approach, not only reduces the HP selection time, but also improves the quality of generated HP schema. We observe a gain of 3.4 % for $\alpha = 0.3$ and 13 % for $\alpha = 1$ in execution time of the workload(see Figure 10).

6 Conclusion

This paper addresses the physical design of DW when subjected to large workloads. It was dedicated to a non-redundant optimization technique used on physical design, the horizontal partitioning. We have seen that the size of the workload influences the execution time of the selection algorithms and the quality of the obtained partitioning schema. In fact, it is difficult to get a partitioning schema with a good quality in a reasonable time for such workloads. To solve this problem, we proposed an approach based on classification of queries in several classes followed by an election of a query for each class. The experimental study and validation in Oracle have shown that our approach gives better results compared to the classical one. It significantly reduces the time of HP schema selection process while generating a schemas with similar or better quality than those produced by the classical approach, especially in case of homogeneous workloads. Several studies are planned to improve this work. We can mention: (1) improvement of the classification and election processes by incorporating new criteria such as relationships between predicates, amount of data referenced by each query, etc., (2) conduct finest experiments to find the best values of the parameter α and (3) include other optimization techniques in our approach as indexes and materialized views.

References

1. Feinberg, D.: Database management systems. Technology trends, Gartner (2006)
2. Sanjay, A., Narasayya, V., Yang, B.: Integrating vertical and horizontal partitioning into automated physical database design. In: Proceedings of the ACM SIGMOD International Conference on Management of Data, pp. 359–370 (2004)
3. Bellatreche, L., Boukhalfa, K., Richard, P.: Data partitioning in data warehouses: Hardness study, heuristics and ORACLE validation. In: Song, I.-Y., Eder, J., Nguyen, T.M. (eds.) DaWaK 2008. LNCS, vol. 5182, pp. 87–96. Springer, Heidelberg (2008)
4. Ceri, S., Negri, M., Pelagatti, G.: Horizontal data partitioning in database design. In: Proceedings of the 1982 ACM SIGMOD International Conference on Management of Data (1982)
5. Bellatreche, L.: Utilisation des vues matérialisées, des index et de la fragmentation dans la conception logique et physique d'un entrepôt de données. Thèse de doctorat, Université de Clermont-Ferrand (2000)
6. Rao, J., Zhang, C., Megiddo, N., Lohman, G.: Automating physical database design in a parallel database. In: Proceedings of the 2002 ACM SIGMOD International Conference on Management of Data, pp. 558–569. ACM, New York (2002)
7. Cuzzocrea, A., Darmont, J., Mahboubi, H.: Fragmenting very large xml data warehouses via k-means clustering algorithm, 301–328 (2009)
8. Barr, M., Bellatreche, L.: A new approach based on ants for solving the problem of horizontal fragmentation in relational data warehouses. In: 2010 International Conference on Machine and Web Intelligence (ICMWI), pp. 411–415 (2010)
9. Karima, T., Abdellatif, A., Ounalli, H.: Data mining based fragmentation technique for distributed data warehouses environment using predicate construction technique. In: 2010 Sixth International Conference on Networked Computing and Advanced Information Management (NCM), pp. 63–68 (2010)
10. Rehme, R., Bruno, N.: Automated partitioning design in parallel database systems. In: Proceedings of the 2011 ACM SIGMOD International Conference on Management of Data, pp. 1137–1148. ACM, New York (2011)

11. MacQueen, J.B.: Some methods for classification and analysis of multivariate observations. In: Proceedings of 5th Berkeley Symposium on Mathematical Statistics and Probability, pp. 281–297. University of California Press (1967)
12. Bellatreche, L., Boukhalfa, K.: An evolutionary approach to schema partitioning selection in a data warehouse. In: Tjoa, A.M., Trujillo, J. (eds.) DaWaK 2005. LNCS, vol. 3589, pp. 115–125. Springer, Heidelberg (2005)
13. Valduriez, P.: Parallel database systems: open problems and new issues. Kluwer Academic Publishers, Hingham (1993)
14. Valduriez, P., Özsu, M.: Principles of Distributed Database Systems, 2nd edn. Prentice Hall, New Jersey (1999)
15. Fiolet, V., Toursel, B.: Intelligent database distribution on a grid using clustering. In: Szczepaniak, P.S., Kacprzyk, J., Niewiadomski, A. (eds.) AWIC 2005. LNCS (LNAI), vol. 3528, pp. 466–472. Springer, Heidelberg (2005)
16. Bajda-Pawlikowski, K., Abadi, D., Silberschatz, A., Paulson, E.: Efficient processing of data warehousing queries in a split execution environment. In: Proceedings of the 2011 ACM SIGMOD International Conference on Management of Data, pp. 1165–1176. ACM, New York (2011)
17. Baer, H., et al.: Oracle database vldb and partitioning guide 11g release 2. Technical report, Oracle, Inc, Oracle White Paper (2011)
18. Microsoft, C.: Sql server 2012 performance white paper. Technical report, Microsoft Corporation (2012)
19. Cain, M.: Table partitioning strategies db2. Technical report, IBM (2006)
20. Sacca, D., Wiederhold, G.: Database partitioning in a cluster of processors. In: Proceedings of the 9th International Conference on Very Large Data Bases, pp. 242–247. Morgan Kaufmann Publishers Inc., San Francisco (1983)
21. Navathe, S., Ceri, S., Wiederhold, G., Dou, J.: Vertical partitioning algorithms for database design. ACM Trans. Database Syst. 9(4), 680–710 (1984)
22. Bellatreche, L., Karlapalem, K., Simonet, A.: Horizontal class partitioning in object-oriented databases. In: Tjoa, A.M. (ed.) DEXA 1997. LNCS, vol. 1308, pp. 58–67. Springer, Heidelberg (1997)
23. Pham, D., Dimov, S., Nguyen, C.: An incremental k-means algorithm. Journal of Mechanical Engineering Science 7(218), 783–795 (2004)
24. Bellatreche, L., Boukhalfa, K., Richard, P., Woameno, K.Y.: Referential horizontal partitioning selection problem in data warehouses: Hardness study and selection algorithms. IJDWM 5(4), 1–23 (2009)
25. Bouchakri, R., Bellatreche, L., Boukhalfa, K.: Une sélection multiple des structures d'optimisation dirigée par la méthode de classification k-means. In: EDA, pp. 207–222 (2010)
26. Ester, M., Kriegel, H., Sander, J., Xu, X.: A density-based algorithm for discovering clusters in large spatial databases. Data Mining Knowlege Discovery KDD 2(2), 169–194 (1996)
27. OLAP-Council: Apb-1 benchmark. Technical report, OLAP Council (1998), http://www.olpacouncil.org/research/resrchly.htm

Evaluation of the Influence of Two-Level Clustering with BUB-Trees Indexing on the Optimization of Range Queries

Samer Housseno[1], Ana Simonet[2], and Michel Simonet[2]

TIMC[1] and AGIM[2] laboratories Université de Grenoble, France
{Samer.Housseno, Ana.Simonet, Michel.Simonet}@imag.fr

Abstract. A BUB-tree is an indexing structure based on B-trees and on a Z-order space filling curve, which transforms multidimensional data into a unique key, enabling the use of a mono-attribute index. We propose a two-level indexing structure relying on a partition of the data space into disjoint clusters. At the first level the clusters are indexed by a BUB-tree and at the second level the data of each cluster is itself indexed. Indexing the clusters provides an efficient query optimization because data filtering is performed at the cluster level, which reduces the data transferred from disk to memory. We compare the performance of our approach with single-level BUB-tree indexing on two types of queries: exact match queries and range queries, which play an important role in multidimensional databases, such as Data Warehouses or Geographic Information Systems. Our approach applies to any system supporting a partition of the attribute domains.

Keywords: Multidimensional index, UB-tree, B-tree, Space-Filling Curve, Range query optimization.

1 Introduction

The need to cluster multidimensional data is present in all applications where data must be selected according to conditions over several attributes. Clustering approaches based on certain criteria (e.g., similarity) partition the data space and lead to a possible physical organization of the data on disk in order to restrict the data of interest for any query based on the same criteria. The criteria allowing the data clustering can be based on the analysis of the data distribution at a given time, or on semantic criteria. Semantic criteria can be either defined in an ad-hoc manner by the database administrator (DBA) (e.g., in Oracle) or obtained by data analysis methods. It can also be automatically derived from the predicates defined on constraints the data must satisfy in order to belong to a class [1] or based on the selection predicates of predefined queries [2].

Data clustering favors range queries, where data is selected according to conditions on several attributes. The optimization of range queries has led to the emergence of a number of multidimensional indexes such as grid files, R-trees, R *-trees, whose

A. Amine et al. (Eds.): *Modeling Approaches and Algorithms*, SCI 488, pp. 295–304.
DOI: 10.1007/978-3-319-00560-7_33 © Springer International Publishing Switzerland 2013

performance heavily depends on the number of indexed attributes. For a particular type of index, when the number of dimensions exceeds some level, its performance degrades. For example, for an R*-tree the search time doubles when passing from 6 to 12 dimensions [3]. This phenomenon, known as "curse of dimensionality", limits the number of attributes that can be used with this kind of index.

One solution to limit the impact of the number of attributes on the performance of an index is to transform multidimensional indexing into one-dimensional indexing [4]. Space-Filling Curves (SFC) [5] and methods such as the Pyramid method [8] transform data from a high-dimensional space into a one-dimensional space, thus allowing the use of mono-attribute indexes such as B-trees, a standard indexing structure in most DBMS. UB-tree and its variant BUB-tree are indexes on multi-dimensional data structures based on the Z-order SFC and B+-tree. They have been recognized as particularly efficient for the evaluation of range queries in high-dimensional vector spaces [6]. It partitions the whole space into a set of disjunctive but consecutive clusters and offers a hierarchical representation of this space.

In order to better optimize range queries, we have proposed a new index, the 2L-Index, which can be used with all systems supporting a partitioning of the data space into clusters [7]. This index offers a two-level clustering: the first level indexes the partition (clusters) of the data space by a BUB-tree and the second level indexes the data itself. This structure limits data processing in main memory: a pretreatment is performed on the clusters of the first level and only the data of the selected clusters is loaded into the main memory.

The paper is organized as follows. We first present a short survey about the space filling curves used in the domain of multidimensional indexes, then we present our approach, 2L-Index, and the main Osiris concepts that are necessary to understand the partitioning approach of the data space. Finally a study of the influence of the two-level clustering on BUB-tree performance is presented, with its results.

2 Multidimensional Indexing

Mono-attribute indexing structures are well known and optimized. The need to index data on several attributes has become important with the emergence of multidimensional applications such as Data Warehouses, spatial databases, multimedia databases, computer graphics, Geographical Information Systems (GIS), and has motivated the adaptation of mono-attribute indexing in these contexts. The transformation of a multi-dimensional space into a one-dimensional space is an important and necessary step to use mono-attribute indexing data structures to index multi-dimensional data. The objective of this transformation is to generate for each tuple of data a unique identifier in a one-dimensional space, enabling the use of a mono-attribute indexing structure, e.g. a B-tree, to index the initial objects.

To generate such an identifier, different methods have been proposed. These methods can be classified into two categories: 1) Specific methods for each approach (Examples: Pyramid approach [8] and iDistance approach [9], and 2) Space Filling Curve (SFC) Methods [5]. Our indexing method is based on the Z-order SFC.

2.1 Space-Filling Curve

Space-Filling Curves (SFC) are methods to map a multi-dimensional space into a one-dimensional space. Such methods first appeared in 1890, when the mathematician G. Peano constructed a curve that maps the unit interval [0,1] to the unit square $[0,1]^2$ [10]. In 1891, Hilbert constructed a mapping of the whole space [11] and many curves have been proposed since that time [5].

Nowadays, the most commonly used SFC are the Peano curve (Z-order), the Hilbert order, the Gray order, the Scan order and the Sweep order. Each SFC has its own mapping function. A SFC visits the points of the multi-dimensional space one after another. The main difference between the curves is the choice of the next point to be visited. With a SFC, the multidimensional data space is linearized to a one-dimensional space by representing a multidimensional point by its position on the SFC. Consequently, the points are ordered, which permits to index them using a mono-attribute indexing data structure. The Z-order offers the best characteristics relatively to fairness[1], locality-preservation[2] and irregularity[3] [13-14]. These qualities, associated with the low cost of its algorithm to calculate the identifier of a tuple of data, make of the Z-order SFC a good choice to transform a multidimensional space into a one-dimensional space. The mapping of a data tuple (point) from a multi-dimensional space into a one-dimensional space is done by calculating its position on the Z-curve, which is called its Z-value. Based on a binary representation, the Z-value is assembled by cyclically taking a bit from each coordinate of a point and appending it to those taken previously (bit-interleaving).

2.2 UB-Trees and BUB-Trees

A UB-tree (Universal B-tree) [12] is a balanced multi-dimensional data structure based on the Z-order SFC and B^+-trees. In a UB-tree, Z-values (also called Z-addresses) are indexed in a B^+-tree. UB-trees offer a hierarchical representation of the space and they also partition the whole space into a set of disjunctive but consecutive clusters called Z-regions. Each Z-region containing the indexed data is inserted into one leaf node in the UB-tree. Z-regions represent clusters of points in the indexed space. A Z-region is bounded by two Z-addresses which are the lower and the upper Z-addresses inside it. On the other hand, the inner nodes contain super-Z-regions [12]. A super-Z-region bounds all the super-Z-regions in its subtrees. Contrary to UB-trees, a Bounding UB-tree (BUB-tree) does not index the empty Z-regions [15].

The algorithms for insertion, deletion and exact match queries are similar to those implemented in B^+-trees, except that the Z-address of the manipulated data must be computed before the execution of an algorithm. For range queries, the linear DRU

---

[1] A SFC is **fair** if it has similar behavior towards all dimensions.

[2] **Locality-preservation**: if two points are near to each other in the multidimensional space, then they will be near to each other in the one-dimensional space.

[3] **Irregularity** is measured for each dimension separately; it gives an indicator of how a SFC is far from the optimal order-preserving SFC. An optimal order-preserving SFC is one that sorts multi-dimensional points in ascending order for all dimensions.

algorithm is proposed in [6]. This algorithm is based on the intersection operation between the range query and the (super-)Z-regions. If a super-Z-region intersects the range query, so do its children. The complexity of this algorithm is linear according to Z-address bit-length; i.e., O(n log(|D|)), where n is the number of bits of a Z-address, and |D| is the number of dimensions.

3 Two-Level Clustering: 2L-Index

In our proposal, called 2L-Index (2L for two-Level), instead of directly indexing the data, first the clusters, named Eq-classes[4], are indexed (first indexing level) and then the data inside each Eq-class (second indexing level); BUB-trees are used for the indexing at both levels. This solution optimizes queries because a pretreatment can then be performed on the BUB-tree at the first level in order to determine the Eq-classes whose objects "concern" the query. Only the objects belonging to these clusters must be loaded into memory. Moreover, as we used a BUB-tree, only non-empty Eq-classes are present at the first level. We call Eq-tree the BUB-tree that indexes the Eq-classes. In an Eq-tree, the clusters/Eq-classes are stored in a Z-region (leaf node). The data of each cluster is indexed by a tree called D-tree (D for Data).

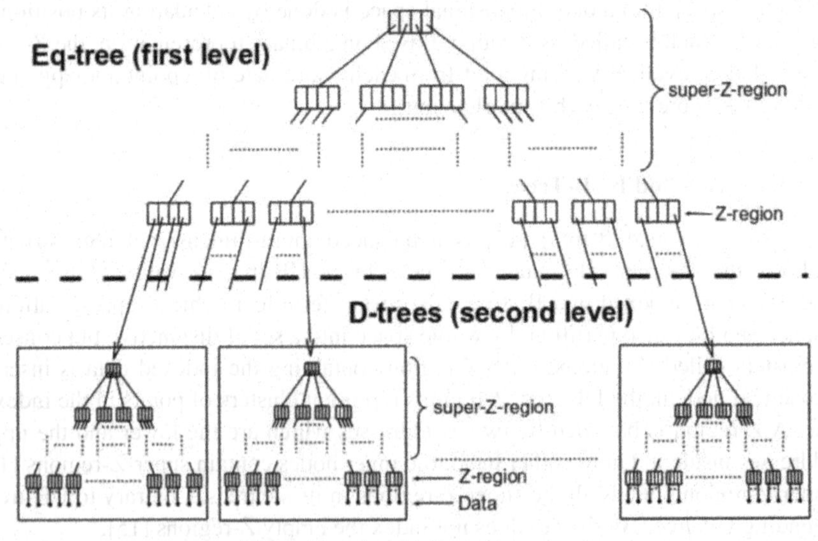

Fig. 1. Two-level indexing structure

There are as many D-trees as there are non-empty Eq-classes. In a D-tree the data is stored in a Z-region (leaf node). This structure is presented Fig. 1. Both types of trees have the same structure, with the following differences: 1) a leaf node of an Eq-tree contains a pointer to its D-tree; 2) the root of a D-tree contains the attributes

[4] Eq-classes are the elements of the semantic partitioning in the Osiris system, which has inspired our approach. Eq-classes group data tuples that are "equivalent" relatively to the Domain Predicates that define the views (subclasses) of a class [1].

subdomains and some measures about the cluster's data, e.g., its cardinality; and 3) A leaf node of a D-tree contains the cluster's data.

4 Evaluation

In order to evaluate the influence of two-level indexing Vs single level (classical) indexing, we have created a situation where only this parameter was modified. To do so, we have implemented a test bed where data is indexed by a BUB-tree and the same data is indexed through our two-level mechanism.

We have considered three parameters which were liable to influence the result: the number of tuples, the number of attributes and the number of subdomains.

4.1 Test Data

- 60 test bases have been indexed by both BUB-tree and 2L-Index. Each base corresponds to a given number of attributes (dimensions), a given number of SubDomains (SD) and to a given number of tuples. The values considered are:
  – Numbers of attributes: 2, 3 and 5.
  – Numbers of SubDomains (SD): 2, 3, 4, 5 and 6.
  – Number of tuples: 50 000, 200 000, 500 000 and 1,5 million. These numbers may vary depending on the generation procedure. The tuples have been generated randomly according to a uniform distribution.
  – Types of queries: Exact Match queries (Point queries) and Range queries.
- Each test has been repeated several times. The calculated response time is the average response time.
- The tests have been performed on a Lenovo Y512 with Intel Core2 Duo T8100, 2.10 GHz and 4 GB RAM.
- The datasets have been managed by JDBM, an Open Source transactional persistence engine for Java that offers B^+-tree management. We have implemented the management of BUB-trees and our 2L-Index in Java.
- The clusters (Eq-classes in the Osiris model) are defined by Domain Predicates[5] (DP) that lead to a partitioning of an attribute domain. For example, given an attribute X defined over the domain [0, 70], possible DP are: $X \leq 10$, $X \geq 20$, $X \geq 30$, $X \leq 40$, $X<50$, $X \geq 60$, which causes the partitioning of the domain of X into 7 SD:
  d_{11}=[0, 10], d_{12}]10, 20[, d_{13}=[20,30[,d_{14}=[30,40],d_{15}=]40, 50[, d_{16}=[50, 60[, d_{17}=[60,70]

  Similarly, for an attribute Y defined over the domain [0, 50], the DP $Y<10$, $20 \leq Y \leq 30$ and $Y>40$ would define the following partition for Y:
  d_{21}=[0,10[, d_{22}=[10,20[, d_{23}=[20,30], d_{24}=]30,40], d_{25}=]40,50]

  The product of the two partitions defines a partition of the two-dimension space into 7x5 Eq-classes. Fig. 2 illustrates this partition.

---

[5] Domain Predicates [1] are predicates of the form attr∈Domain. Examples of DP are: x=1 (x∈[1,1]), x<10 (x∈[0,10[), y∈{a,b,c}. Ozsu minterms [16] are a particular case of DP.

- The performance is evaluated by considering the gain percentage, i.e., the percentage of improvement of the response time of the 2L-Index against that of BUB-tree.

$$GainPercentage = 100 - \frac{responseTime\ 2LIndex \times 100}{responseTime\ BUB}$$

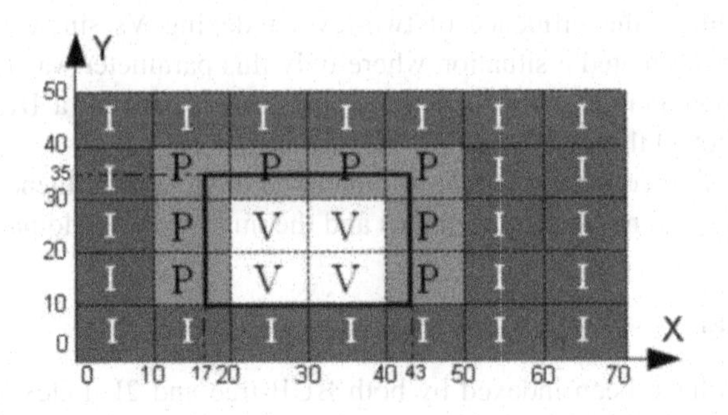

Fig. 2. Example of data space partitioning and range query: (X∈[17,43], Y∈[10,35[)

An **exact match query**, also called **point query**, e.g., (X=5, Y=24), defines a point in the data space, which is situated within one and only one Eq-class: (d_{11}, d_{23}) in this example. The evaluation of such a query consists in first identifying its Eq-class and then searching for it in the Eq-tree. As only the non-empty Eq-classes are represented in a BUB-tree, if the Eq-class is not present, this means that the searched object is not in the base and the query has an empty result; consequently there is no need to search for the object in the D-trees (second indexing level). If the Eq-class is present in the Eq-tree, then a search is performed in the corresponding Eq-tree, and only that one, which limits the search space for the query.

A **range query** defines a partitioning of the data space. This partitioning may intersect or subsume some Eq-classes. Fig. 2 illustrates the case of the query $17 \leq X \leq 43$ **and** $10 \leq Y < 35$

By definition of the Eq-tree, all the objects of the Eq-classes subsumed by the query are part of the answer, and there is no need for a search in the corresponding D-trees. Such Eq-classes are called **V**alid. Some Eq-classes are intersected by the query, they are called **P**otential, and their corresponding D-trees have to be checked for the query conditions. All the other Eq-classes are **I**nvalid, which means that none of their objects can satisfy the query. In the example of Fig. 2:

V: (d_{13}, d_{22}), (d_{13}, d_{23}), (d_{14}, d_{22}), (d_{14}, d_{23})
P: (d_{12}, d_{22}), (d_{12}, d_{23}), (d_{12}, d_{24}), (d_{13}, d_{25}), (d_{14}, d_{26}), (d_{15}, d_{22}), (d_{15}, d_{23}), (d_{15}, d_{24})
I: (d_{11}, *), (d_{12}, d_{21}), (d_{12}, d_{25}), (d_{13}, d_{21}), (d_{13}, d_{25}), (d_{14}, d_{21}), (d_{14}, d_{25}), (d_{15}, d_{21}), (d_{15}, d_{25}), (d_{16}, *), (d_{17}, *)

Only the tuples of the D-trees corresponding to the 8 **P**otential Eq-classes have to be checked for the query conditions. The tuples of the 23 **I**nvalid Eq-classes need not be examined as they cannot be part of the answer to the query, and the tuples of the 4 **V**alid Eq-classes are part of the answer, without individual checking.

4.2 Results

Some queries have an empty result. We have distinguished these cases, as the gain can be maximal in such situations.

In the other cases (non-empty result), for each of the tests, we obtain an elementary diagram like that of Fig. 3-left (Point query; 5 attr; 6 SD). We then have made a synthesis of the results involving the same number of attributes. Such a diagram is shown Fig 3-right (Point query; 5 attr; 2, 3 and 6 SD). The results of Fig 3-left correspond to the line of triangles in the synthetic diagram of Fig. 3-right

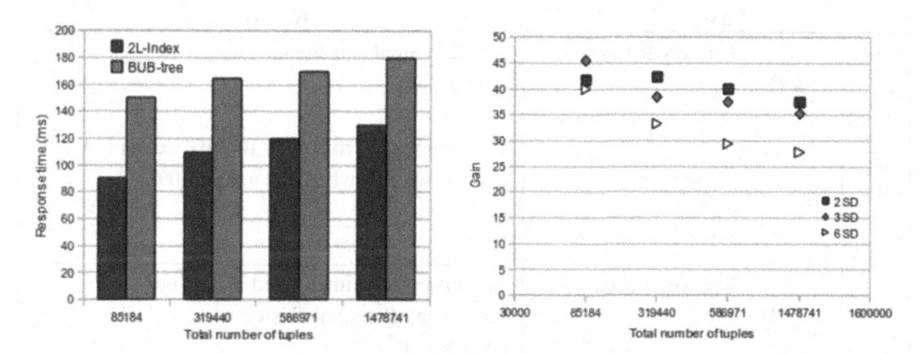

Fig. 3. Elementary (left) and synthetic (right) diagrams for Point queries with 5 attributes

There are 79 elementary diagrams and 22 synthetic diagrams.

Exact match queries (point queries)

The general conclusion for exact match queries is that when there are a very small number of Eq-classes (under 10) there is no benefit to perform two-level indexing. The gain can even be negative, due to the overhead of the first level indexing. In these situations, the benefit expected through the first level of indexing (select **the** Eq-class containing the answer) does not compensate the extra work to search the Eq-classes.

- In all the other cases the gain is effective: **from 22% to 40%**.
- In the case of queries with an **empty result**, we distinguish two situations:

1. The Eq-class that satisfies the conditions of the query is not present in the index. The gain is **above 84%.** The average response time of the 2L-Index has a stable value of about 20 ms, which is independent from the size of the base; it only depends on the size of the Eq-tree.
2. The cluster that satisfies the conditions of the query is present in the Eq-tree. The gain in this case is lower than in the previous case (**below 20%** in the case of two SD, **from 20% to 40%** otherwise). The explanation is that in both cases, the query searches in the first level (the Eq-tree), but in the first case, the query does not need to search in the second level (D-trees).

Range queries.
A range query divides the set of clusters into three disjoint subsets: 1) **V**: all their data satisfies the query (Valid Eq-classes); 2) **I**: no data satisfies the query (Invalid Eq-classes); and 3) **P**: some of their data may satisfy the query (Potential Eq-classes).

Based on these three sorts of clusters, we have identified eight possible situations, depending on whether the sets V, P and I are empty or not. We illustrate the most significant situations in Table 1.

Table 1. Testing results according to the situation of a range query within the data space

	Cases	*Results*
$V=\emptyset, P\neq\emptyset, I\neq\emptyset$	**Unitary P.I case** 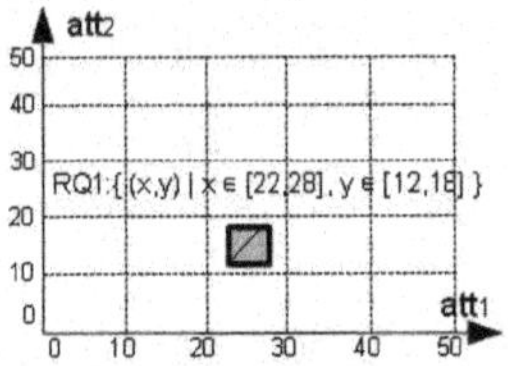 RQ1:{(x,y) | x ∈ [22,28], y ∈ [12,18] }	Only **P**otential and **I**nvalid clusters : Card(**P**) =1 The gain varies from **62% to 79%** except in the case of two SD, where the gain is 40%
$V\neq\emptyset, P=\emptyset, I\neq\emptyset$	**Unitary V.I case** 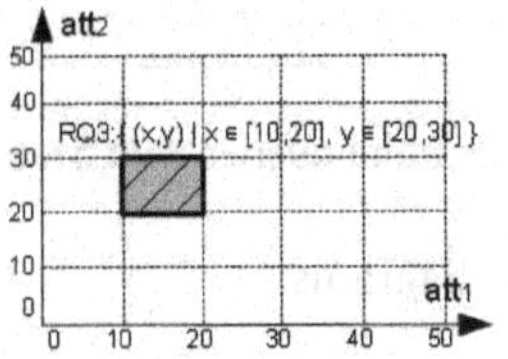 RQ3:{ (x,y) | x ∈ [10,20], y ∈ [20,30] }	All clusters are either **V**alid or **I**nvalid. There is no **P**otential cluster: Card(**V**) =1 The gain varies from **65% to 88%.**
	Multiple V.I case 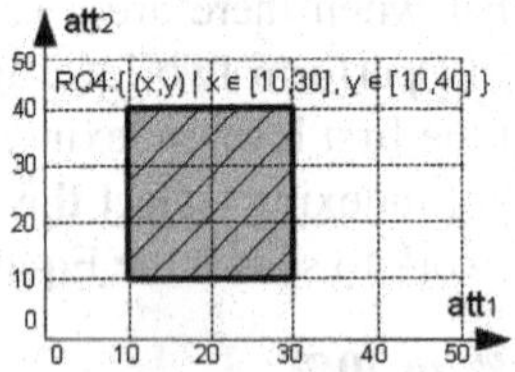 RQ4:{(x,y) | x ∈ [10,30], y ∈ [10,40] }	All clusters are either **V**alid or **I**nvalid. There is no **P**otential cluster : Card(**V**) > 1 The gain varies from **40% to 90%.** More clusters in the partitioning results in a higher improvement. The gain increases with the number of Eq-classes.
$V\neq\emptyset, P\neq\emptyset, I\neq\emptyset$	**V.P.I case** 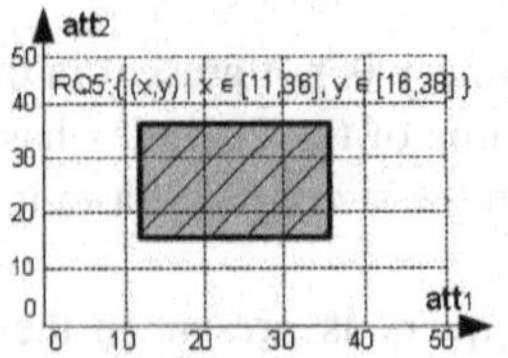 RQ5:{(x,y) | x ∈ [11,36], y ∈ [16,38] }	General case. There are **V**alid, **I**nvalid and **P**otential clusters. The gain varies from **62% to 79%.** With a small number of SD and a large data size, the gain is very small (5%) or even negative (-11%, case of two SD).

Fig. 4 illustrates the results of the evaluation of range queries with 5 attributes and 4 SD. In the general case (V.P.I.) on the left, the response time grows with the number of tuples but the growth variation is much more important with one-level indexing. In the case of two-level indexing, it is almost linear, which is confirmed on the other diagrams.

On the left diagram we have indicated the selectivity[6] of the query along with the number of tuples. Other diagrams show that the gain is more strongly related to the selectivity than to the size of the base.

In the case of range queries with an **empty result** (cf Fig. 4-right), the average response time of the 2L-Index is very small compared to the BUB-tree. The gain is **above 80%** and the average response time is almost constant (between 20 and 70 ms). The explanation is that in the 2L-Index only non-empty clusters are indexed. As these queries do not return data, the clusters that satisfy them are empty and are not present in the Eq-tree. Consequently the search stops at the primary level of the index (Eq-tree) and the search at the secondary level (D-tree) is not performed.

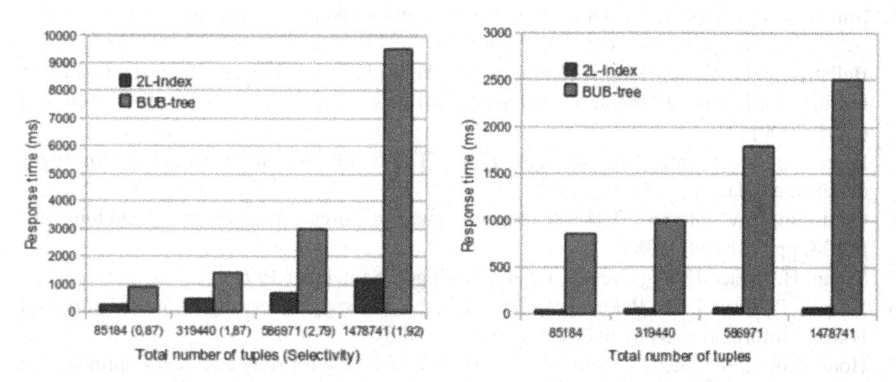

Fig. 4. Range queries with 5 attributes and 4 SD. General case (left), Empty result (right)

5 Conclusion

The indexing structure 2L-Index is based on a partition of the data space into clusters. Although the 2L-Index method has been developed in the framework of the Osiris model [1], which provides a native partition of the data space, it can be used in any system supporting a partitioning of attribute domains, e.g., commercial systems such as Oracle. It consists in indexing both clusters (the first level) and the data of each cluster (the second level) by BUB-trees. The BUB-tree structure is a multi-dimensional generalization of B-trees based on the Z-curve space filling curve.

To analyze the influence of indexing with two levels on the optimization of range queries and point queries, we have compared the performance of the BUB-tree (single-level clustering) with our approach 2L-Index. The evaluation results show that two-level indexing outperforms single level indexing. This is because a filtering is done at the first indexing level, which limits or even avoids the work at the second level. However, the benefit of 2L-Index degrades with a small number of clusters,

[6] The selectivity of a query is the percentage of the number of data tuples returned by the query relatively to the total number of data tuples of the database.

because of the overhead of the search in the first level index; we are then close to a situation of single-level indexing.

When no cluster at the first level contains data satisfying the query (empty response), there is no search at the second level, and the response time is very small and almost constant, whereas with BUB-tree indexing the search time grows with the size of the base. In this situation, the benefit of the filtering at the cluster level is maximal.

An adaptation of the 2L-Index to DW in order to optimize predefined queries is ongoing. Instead of BUB-trees, a bitmap index is used to index the tuples of the fact table at the second level. The first results are very promising.

References

1. Simonet, A., Simonet, M.: Objects with views and constraints: From databases to knowledge bases. In: OOIS, pp. 182–195 (1994)
2. Bellatreche, L., Karlapalem, K., Simonet, A.: Algorithms and Support for Horizontal Class Partitioning in Object-Oriented Databases. Distributed and Parallel Databases Journal 8(2) (April 2000)
3. Berchtold, S., Keim, D.A., Kriegel, H.-P.: The x-tree: An index structure for high-dimensional data, pp. 28–39 (1996)
4. Orenstein, J.A., Merrett, T.H.: A class of data structures for associative searching. In: PODS, pp. 181–190 (1984)
5. Sagan, H.: Space-Filling Curves, 1st edn. Springer (September 1994)
6. Skopal, T., Krátký, M., Pokorný, J., Snásel, V.: A new range query algorithm for universal B-trees. Information Systems 31(6), 489–511 (2006)
7. Housseno, S., Simonet, A., Simonet, M.: Ub-tree indexing for semantic query optimization of range queries. International Journal of Computer, Information and Mechatronic Engineering 35, 177–184 (2009)
8. Berchtold, S., Böhm, C., Kriegel, H.P.: The Pyramid-technique: Towards breaking the curse of dimensionality. In: SIGMOD Conference, pp. 142–153 (1998)
9. Jagadish, H.V., Rui Zhang, B.C.O.: Distance techniques. In: Shekhar, S., Xiong, H. (eds.) Encyclopedia of GIS, pp. 469–471. Springer (2008)
10. Peano, G.: Sur une courbe qui remplit toute une aire plaine. Mathematishe Annalen 36, 157–160 (1890)
11. Hilbert, D.: Ueber die stetige Abbildung einer Line auf ein Flächenstück. Mathematische Annalen 38, 459–460 (1891)
12. Bayer, R.: The universal B-Tree for multi-dimensional Indexing: General Concepts. In: Masuda, T., Tsukamoto, M., Masunaga, Y. (eds.) WWCA 1997. LNCS, vol. 1274, pp. 198–209. Springer, Heidelberg (1997)
13. Mokbel, M.F., Aref, W.G., Kamel, I.: Analysis of multidimensional space-filling curves. GeoInformatica 7, 179–209 (2003)
14. Mokbel, M.F., Aref, W.G.: Irregularity in high-dimensional space-filling curves. Distributed and Parallel Databases 29(3), 217–238 (2011)
15. Ramsak, F.: The BUB-Tree. In: Proceedings of 28 rd VLDB 2002, Hong Kong, China (2002)
16. Ozsu, M.T., Valduriez, P.: Principles of Distributed DataBase Systems. Prentice-Hall, Inc., Englewood Cliffs (1991)

Identification of Terrestrial Vegetation by MSG-SEVIRI Radiometer and Follow-Up of Its Temporal Evolution

Naima Benkahla[1] and Abdelatif Hassini[2]

[1] Laboratory of Analysis and Application of Radiation, University of USTOMB, El M'nouer. B.P.1505 Oran, Algeria
benkahla_naima@yahoo.fr

[2] Institute of Maintenance and Industrial Security, University of Oran Es-Senia, Algeria
hassini.abdelatif@univ-oran.dz

Abstract. The main focus of this work is to propose a method to depict one vegetation index from free clouds MSG2-SEVIRI (METEOSAT Second Generation 2 – Spining Enhanced in Visible and Infrared Imager) data. The proposed method uses the multi spectral high satellite data frequency (MSG2-SEVIRI image acquisition every 15 minutes), for rapid identification of Normalized Difference Vegetation Index (NDVI), and follow-up of its temporal evolution, to produce an approved monthly scale vegetation chart for a half terrestrial disk scanned by SEVIRI radiometer. To validate our method, we have compared each obtained result with other performed sensor embided on polar SPOT satellite called SPOT Vegetation.

Keywords: index of vegetation, temporal synthesis, NDVI, SEVIRI, satellite sensor.

1 Introduction

Vegetation covers two thirds of the surface of the continents with 24% of forests, 15% of meadows and tundra, 15% of savannas and 11% of cultures. These biomasses are essential with the wellbeing of humanity; they provide the bases of the life on Earth through their ecological functions, by controlling the climate and the resources water, while being used as habitats with the animal and vegetable biodiversity like by providing food products essential with the man. It is of primary importance to preserve them. An effective management of the environment and agricultural resources require a reliable and precise collection of information on the state of the forests and the cultures, but also, a permanent monitoring of their evolution.

The interaction radiation-matter is one of the bases of the interpretation of the images of spectral remote sensing. The behavior of the objects is a way privileged to analyze and interpret the remote sensed images, because it reweighs on the general laws of physics. Various types of surfaces such as water, bar soil or vegetation, differently reflect the radiations in various radiometric channels. The reflected radiations (radiances) according to the wavelength are called thematic spectral signature. The healthy vegetation absorbs most of the visible incidental light (emitted by the sun) and reflects most of the light near infra-red.

A. Amine et al. (Eds.): *Modeling Approaches and Algorithms*, SCI 488, pp. 305–314.
DOI: 10.1007/978-3-319-00560-7_34 © Springer International Publishing Switzerland 2013

The spectral signature of the house plants is very characteristic. The chlorophyl of a plant under development absorbs the visible light, especially in red, to use it in photosynthesis, while the light of the infra-red close relation is reflected very effectively because the plant does not have any utility of it. Fig. 1 shows some spectral signature concerning vegetation, nude soil and water. The vegetation strongly reflects in the range of the 0.7-0.9 μm, while it reflects little in the area of the 0.6-0.7 μm. The spectral signature of the vegetation being very characteristic, the distinction between bare soil and covered soil generally does not pose any problem. The difference between reflecting in the visible one and the infra-red close relation can, as we already indicated, help to determine the photosynthesis and the growth of the plants.

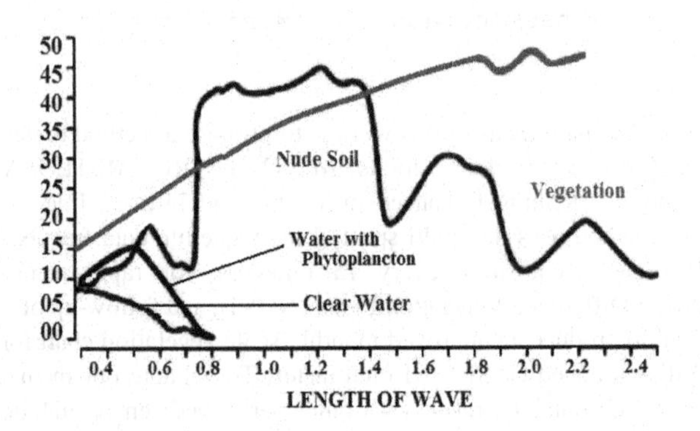

Fig. 1. Graph of some spectral signatures: water, soil and Vegetation

The index NDVI (Normalized Difference Vegetation Index) is an indicator of the presence of vegetation; it is calculated generally as a ratio composed by tow spectral fields, NIR for near infrared measurement and RED for visible centered in red measurement. More retails of NDVI product are explained in this work.

2 Materials and Methods

2.1 SEVIRI Sensor

The Spinning Enhanced in Visible and Infrared Imager (SEVIRI) instrument is based on a compact three-mirror telescope and scan assembly. The 42 detectors of the twelve channels are accommodated in the telescope's focal plane in two areas, one at 20°C for solar channels (centered at wavelengths around 0.6, 0.8, 1.6 μm and about 0.75 μm for High Resolution Visible (HResVis) channel. The thermal infrared detectors (centered at wavelengths around 3.9, 6.2, 7.3, 8.7, 9.7, 10.48, 12.0 and 13.4 μm) are passively cooled down (85°K or 95°K) to optimize their performance. The compact design allows the insertion of a small black body for full-pupil calibration. The response by every detector to the target's radiation is converted into an electronic signal by means of pre-amplifiers and a main detection unit. The amplification can be adjusted to the needs a various stages of the signal processing. The full image processing from raw counts to level 1.5 images is performed by the Image Processing Facility (IMPF) branch of EUMETSAT [1] [2]. SEVIRI sensor generate every 15 minutes a full scan of a half terrestrial disk centered in Africa. Each scan comprise 11 channels covering spectral bands from visible to the thermal infra-red, with a space resolution of 3km at sub satellite point, and a panchromatic channel with high space resolution (1km). Characteristics of the SEVIRI channels are described in Table 1.

Table 1. Some characteristics of SEVIRI- MSG2 channels

Chan-nels	*Spectral Bands*	*characteristics*			*Spatial Resolution*
		λmin	λ	λmax	
1	VIS0.6	0.56	0.635	0.71	3 km
2	VIS0.8	0.74	0.81	088	3 km
3	PIR1.6	1.50	1.64	1.78	3 km
4	IR3.9	3.48	3.90	4.36	3 km
5	WV6.2	5.35	6.25	7.15	3 km
6	WV7.3	6.85	7.35	7.85	3 km
7	IR8.7	8.30	8.70	9.10	3 km
8	IR9.7	9.38	9.66	9.94	3 km
9	IR10.8	9.80	10.80	11.80	3 km
10	IR12.0	11.00	12.00	13.00	3 km
11	IR13.4	12.40	13.40	13.40	3 km
12	HRV	Large band 0.4 - 1.1			1 km

The first use of the METEOSAT images is obviously weather forecasting. However other uses are easily possible, thanks to the low cost of installation of a station of reception and to the policy of EUMETSAT which authorizes the reception and the free diffusion of the images in time quasi-reality for scientific purposes.

2.2 Reception of SEVIRI-MSG2 Data

SEVIRI data are transmitted from MSG2 to EUROBIRD-9 viaThe EUMETSAT (European organization for the exploitation of Meteorological Satellites) DVB Service , a complete DVB DVB (Digital Video Broadcasting) system is installed in our Laboratory and comprise a satellite receiving dish to be mounted outside, an LNB (Low Noise Block) which converts the 11GHz signal down to the 1GHz region and amplifies it to overcome cable loss, good satellite cable terminated with connectors to connect the LNB to DVB card and the DVB card itself which fits into one of the PCI slots inside a Personal Computer. Note that a 5-volt PCI slot is required. For Meteosat-8 and Meteosat-9 Eumetsat recommend that to have a separate PC dedicated to data capture and file sharing and that it should be at least a 2 GHz Pentium IV system or equivalent. For data capture, we have currently using a Pentium IV (3 GHz) machine as Receiver PC. We used in our case Windows XP system. The length of the cable from LNB to PC is about 20 m. Fig. 1 shows synoptic of the acquisition system that we have installed[3][4]. To receive MSG data from EUMETSAT via EUROBIRD9 satellite relay, we have used special software, DVB card will communicate with TECHNISAT software when it installed. It turns the DVB card into a channel through which files are received from Eumetsat and saved on PC. All compressed acquired data are opened by the software xrit2pic and saved under the PGM format (Portable Gray map 16 bits) in order to keep the coding 10 bit of the image. Fig. 2 shows a flowchart of all parts of our receiving station.

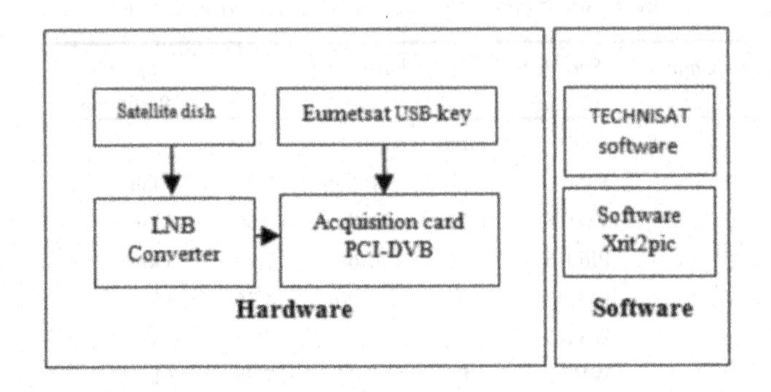

Fig. 2. Global synoptic of MSG acquisition system

2.3 Calibration of the Visible and Near Infra-red Channels

The calibration technique used in this research, consist to converts the raw numerical accounts into physical parameter measurement, the calibration algorithm of channels 1, 2,3 and 12 from METEOSAT MSG1 and MSG2 images is given by equation 1. MSG data as produced by EUMETSAT contain rectified SEVIRI images in a 10 bit digital format. The images are not only geolocated and transformed to a GEOS projection, they are also representing a fixed radiometric scale. This scale is provided via two linear scaling parameters in the image header (cal_slope and cal_offset). In the case of the solar channels they refer to the vicarious calibration and in the case of the thermal channels they state a pure scaling law for the radiances obtained from the blackbody calibration. From here, we can reproduce the spectral radiance for each spectral band by equation 2.

$$REFL\,(i) = 100 * R\,(i) \,/\, TOARAD\,(i\,) \,/\, cos(TETA) \tag{1}$$

Where
i: Channel number
REFL: Reflectance [%]
R: the spectral radiation or radiance mW m^2 sr^{-1} $(cm^{-1})^{-1}$].

$$R = CAL_{offset} + CAL_{slope}.\,Count \tag{2}$$

Count: binary value of the pixel (account of pixel, between 0 and 1023).
CAL_slope and CAL_offset [mW m^2 sr^{-1} $(cm^{-1})^{-1}$] are removed coefficients of calibration from the header part of MSG-SEVIRI images.
TOARAD: solar constant at the top of the atmosphere [mW m^2 sr^{-1} $(cm^{-1})^{-1}$].

TETA: solar angle of zenith (calculated starting from the date, of time, the lat., the lon.); for the twilight condition (TETA > 80°) TETA is placed at 80° to avoid index I of the problems of channel I (1 = VIS0.6; 2 = VIS0.8; 3 = NIR1.6; 12 = HRV)

$$\text{TOARAD (i=1, VIS0.6 channel)} = 20.76 / ESD^2$$
$$\text{TOARAD (i=2, VIS0.8 channel)} = 23.24 / ESD^2$$
$$\text{TOARAD (i=3, NIR1.6 channel)} = 19.85 / ESD^2$$
$$\text{TOARAD (i=12, HRV channel)} = 25.11 / ESD^2$$

ESD: outdistance ground-sun (in the astronomical units), which changes during the year according to the equation 3.

$$\text{ESD (JulianDay)} = 1 - 0.0167 \cos\left(2.\pi.\frac{(\text{JulianDay} - 3)}{365}\right) \tag{3}$$

In our laboratory (LAAR), data are initially received by TELLICAST software, follows the software either the XRIT2PIC, used automatically, to read and collect, after decompressing all segments of acquired image, and to save it as a full disk HRIT file data, and, then the raw images resulting are treated by other software's lake ENVI-RSI, or by MSG VIEWER which we are developed to extract some thematic products.

2.4 NDVI Data Calculation

Several methods have been performed for the detection of vegetation, some use a single channel, and others use a combination of channels. We chose the most used in such studies and also the most effective is: calculate the vegetation indices NDVI. To value the evolution of vegetation in the studied zones, we have calculated the NDVI measured during ten day in month of acquirement, it is calculated from the reflectance of the channels Vis06 and Vis08 (equation 4).

$$NDVI = \frac{\rho_{NIR} - \rho_R}{\rho_{NIR} + \rho_R} \tag{4}$$

Where, ρ_{NIR}, is the reflectance of the visible channel 0.8μm and ρ_R, is the reflectance of the visible channel 0.6μm. The real difference between the light of the sun reflected in the red part of the specter (channel 1), where vegetation is the absorbing energy for the photosynthesis and the energy reflected in the near infrared (the channel 2), give a qualitative measure for the activity of photosynthesis. Therefore from a set of the images of the MSG-SEVIRI sensor acquired by the station of the LAAR laboratory (Laboratory of analysis and application of the Radiance) from the 17 to July 27, 2011 to 12 :00.UTC, we have calculated a picture composite NDVI of 10 images (Fig.8), this operation has us permits to eliminate the clouds in order to construct a picture in clear sky. The NDVI index varies in the range of -1 .0 and +1.0 .the values bigger than 0.1 denote some degrees generally grow in the greenery and the intensity of vegetation. The values between 0 and 0.1 are generally features of the rocks and the naked soil, and the values less 0 indicate the clouds, snow, ices and water surfaces, respectively. The surfaces of vegetation have values of the NDVI spreading of 0.1 until 0.8 in the dense forest. Table.2 shows the values of NDVI for the different types of covers [5].

Table 2. The values of NDVI for the different types of covers

NDVI	*>0.7*	*>0.025*	*>0.002*	*>-0.046*	*>-0.257*
Type of cover	Dense vegetation	Dry naked soil	Clouds	Snow and ice	Water

3 Results and Discussions

3.1 Data Calibration Preprocessing

SEVIRI-MSG2 sensor product a high temporal resolution images (full terrestrial half disk scan every 15 minutes), this advantage is taken to obtain cloudless pixel temporal syntheses, which is focused to eliminate all pixels affected by the presence from clouds or shades of clouds and replace each them by free cloud one. In general, an image without clouds effect is comparable to an image of terrestrial Albedo, and a cloudiest regions in each acquired image in visible spectra, are looked as a higher brightness digital image count. The operation consist to blending four at least five consecutive image and replacing the higher brightness's in each point of resulting image by the lower one in the same pixel region. The resulted images which we obtained with minimum brightnes pixels are extracted from data sequence satellite and are representative of Albedo of each point of a terrestrial surface. Fig.4 shows result of radiometric calibration of the visible channel Vis 0.6 (central wavelength 0.6 μm), this operation consiststo convert every raw digital data from fig. 3 to Terrestrial Albedo images. The histogram of MSG image which is coded on 10 bits (1024 grays levels), is represented by a graphics propertied 1024 values in abscissas, and the number of elementary images in ordinates (Fig..3), but after the radiometer calibration of these images, this histogram will be represented as it is schematized on (Fig..4), possessing a values of reflectance from 0 to 100%.

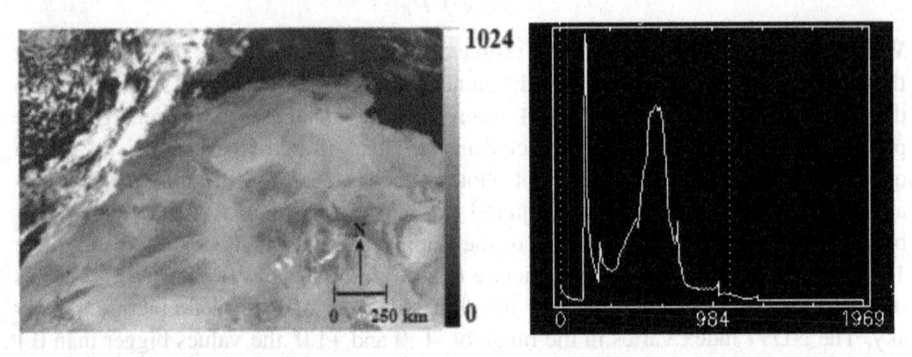

Fig. 3. Raw image of the visible channel VIS0.6μm received in October 17, 2012

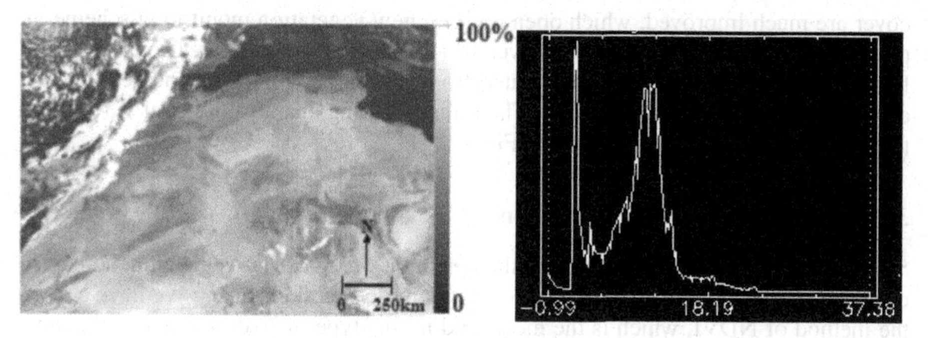

Fig. 4. Result of radiometric calibration of the visible channel VIS 0.6µm

3.2 Construction of a Image in Clear Sky

The initial goal of application of METEOSAT data is the survey of the atmosphere (meteorology and climatology). Actually, Scientist can used these data in a large field of application like, the oceanography and the terrestrial applications. indeed to be able to study terrestrial and oceanographically surface features us must create one image in clear sky, free of all clouds and it from sets of images has taken at the same hours laying several days, the creation of a images in clear sky rest therefore on the technique of the minimal brilliance recorded for a pixel given on a very determined period. In fact the pixel for which brilliance is strong are those forming the cloudy masses, while the pixels to weak brilliance are representative to the objects of the terrestrial surface, has know the sea and soils [6].

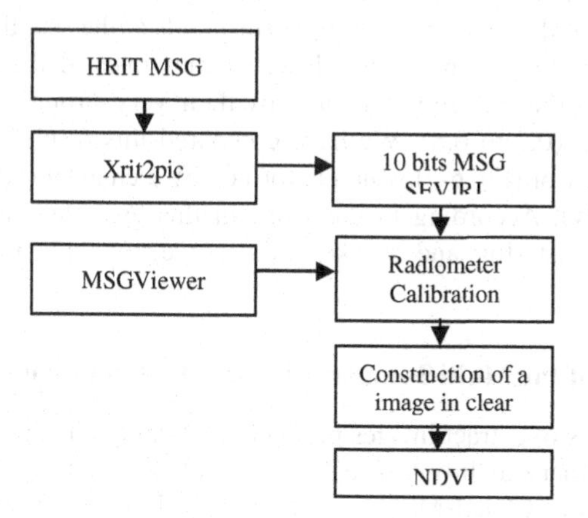

Fig. 5. A flowchart of NDVI data Processing product used

SEVIRI sensor detects radiation in 12 spectral bands among which two are specifically suited for vegetation studies. MSG data have the great advantage over data from the polar orbiting satellites that the frequency of observations is so much higher (15 min interval versus one time a day acquisitions), thus the chances to avoid cloud cover are much improved, which open up for a new vegetation monitoring scheme, in particular for the African continent over which MSG is located [7]. The Normalized Difference Vegetation Index is calculated from reflectance in the red and nearinfrared channels for each image acquisition. The full processing chain from Level 1.5 product to the final NDVI images is shown in Fig.5.

3.3 Vegetation Product Extraction

Several methods have been done for the detection of terrestrial vegetation, some use only one channel and others use the combination of several channels. We have chosen the method of NDVI, which is the most used in this type of studies and also the most efficient and a good indicator of vegetation.

A. THE CALCULATION OF THE VEGETATION INDICATION

The NDVI can be calculated from equation 5 by using equation 4, containing 0.6 µm SEVIRI-MSG2 channel (C1 visible channel centered in RED spectra) for the reflectance of the visible light, and 0.8µm SEVIRI-MSG2 channel (C2 near infrared NIR channel) for the infrared near reflectance. Fig. 5 shows a flowchart of all steps to extract vegetation product used in MSGViewer software.

$$NDVI = \frac{C2 - C1}{C2 + C1} \tag{5}$$

The primary NDVI image result is presented in fig.6-a. The classification of NDVI images permits to regroup the regions in homogeneous units of the NDVI and permits the identification of the big landscape units. To interpret data, we have referred to the spectral behavior of the objects, as it is translated by the NDVI images (Fig.6-b). The large part of North of the picture (Fig.6-b) corresponds at the sea, the index of vegetation is globally very low in this zone. However the zones of the plant cover in the north take out again the picture remarkably by their very strong values of indication of NDVI vegetation (0.2 to 0.8). We have compared this result by the composition RGB1of SEVIRI sensor (Fig.6-c) which is formed by a channel (NIR1.6, VIS0.8, and VIS0.6, respectively). According to the work in this area this enables RGB detect vegetation, snow, dust, fog and smoke [8] [9], we conclude that our results are satisfying.

3.4 Consistent of the Multi Temporal Vegetation Evolution

The analysis consists to extract images product, containing NDVI of different dates of the year (from February to July). The NDVI is calculated from visible and infrared channel images from SEVIRI-MSG2 sensor scenes. Resulted NDVI images product for these dates is represented in fig.7. The zones of middle cover of vegetation appear in green dark and in clear-green for the dense vegetation. NDVI images in April and May appears as having some more elevated values compared to the other dates (especially in the East of Algeria). Vegetation in this period is sufficiently covering and present a state of growth more developed (Fig.7-a and Fig.7-b). In July, the evolution of the cultures modifies the values of reflectance. This month presents the period of the maturation of the plant species of where lowers what decreases the values of the vegetation indications of chlorophyll activity (Fig.7-c). September is a period of the plugging. It is for this reason that the corresponding picture has values of weaker NDVI than the previous (Fig.7-d). This difference is not substantial in some zones (especially the drills of the conifers) that keep their verdures. NDVI always is constant enough all along the year for these zones.

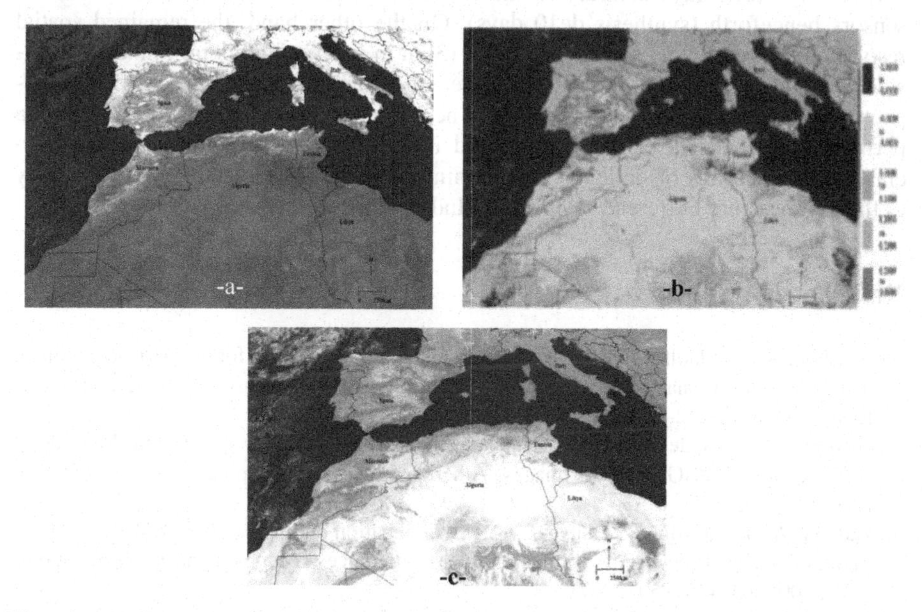

Fig. 6. -a- Chanel NDVI of 17 to 27 July 2011, -b- Result of classification of the NDVI picture.-c- Composition of R3G2B1 (NIR1.6, VIS0.8, and VIS0.6, respectively) 17 to 27 July 2011

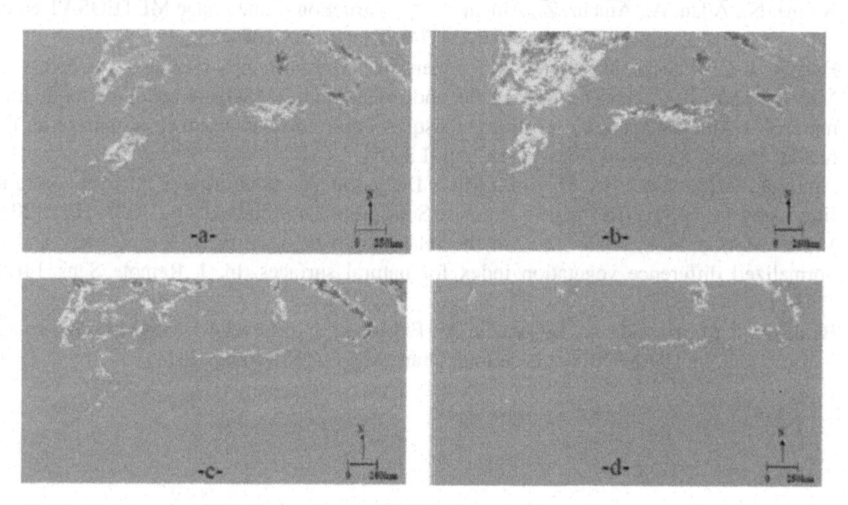

Fig. 7. Multi temporal NDVI results: -a-NDVI image taken in the month of April 2011, -b- NDVI image taken in the month of May 2011, -c- NDVI image taken in the month of July 2011,-d- NDVI image taken in the month of September 2011

4 Conclusion

In this work, we showed the possibility to get from SEVIRI- MSG2 sensor the temporal syntheses of indicators as for example NDVI vegetation index. We can think that the MSG2 satellite data will permit to achieve syntheses of good quality on periods of 4 to 5 days contrary to the AVHRR-NOAA or VEGETATION- SPOT sensors henceforth (synthesis de10 days). On the other hand, the remained spatial resolution limited to low spatial resolution (3km in sub satellite point), unless to consider methods of fusion of the data with the panchromatic channel to 1km. This perspective is the object of other works. It is necessary to underline that the MSG data permit to get the data of temperature and indication of vegetation, what drives to elaborate indications of deficiency combining these two parameters, simultaneously as the Temperature Difference Vegetation Index.

References

1. Fensholt, R., Sandholt, I., Stisen, S., Tucker, C.: Analyzing NDVI for the African continent using the geostationary meteosat second generation SEVIRI sensor. Remote Sensing of Environment, pp 212–229 (2006)
2. Hassini, A., Benabdelouahed, F., Benabadji, N., Belbachir, A.H.: Active Fire Monitoring with Level 1.5 MSG Satellite Images. American Journal of Applied Sciences, 157–166 (2009)
3. Hassini, A., Benabadji, N., Belbachir, A.H.: Acquisition and processing of MSG level 1.5 images data. In: Proceedings of International Conference on Physics and its Applications, ICPA 2007, pp. 145–150 (2007)
4. Hassini, A., Benabadji, N., Belbachir, A.H.: Reception of the APT weather satellite images. AMSE J., Advances B, France, 4, pp: 25–43 (2005)
5. Naima, N., Adan, A., Ameur, Z., Ameur, S.: Construction d'une image METEOSAT en ciel clair, Université Moulod Mammeri (UMMTO), Tizi ouzou (2009)
6. Bergès, J.-C., Lacaze, B., Smiej, M.F.: Vers un suivi en temps réel de la sècheresse au Maroc à partir des données Meteosat Seconde Génération (MSG). Actes du colloque international « Aménagement du territoire et risques environnementaux » en hommage au Professeur Hassan Benhalima, Fès 15–16 (April 2005)
7. Amraoui, DaCamara, C.C., Pereira, J.M.C.: Detection and monitoring of African vegetation fires using MSG-SEVIRI imagery. Remote Sensing of Environment 114, 1038–1052 (2010)
8. Van De Gried, A.A., Owe, M.: On the relationship between thermal emissivity and the normalized difference vegetation index for natural surfaces. In. J. Remote Sens. 14(21), 1119–1137 (1993)
9. Benkada, R.M., Hassini, A., Benabadji, N.: Exploitation of Level 1.5 MSG-SEVIRI Images Sensor Data. In: CIAM 2011 Proceeding, Oran, Algeria (November 2011)

Automatic Detection of Contours of Circular Geologic Structures on Active Remote Sensing Images Using the Gradient Vector Flow Active Contour

Djelloul Mokadem and Abdelmalek Amine

GeCoDe Laboratory, Tahar Moulay University of SAIDA
{dmokadem,amine_abd1}@yahoo.fr

Abstract. We try in this work to solve the problem of automatic detection of contours of circular geologic structures of the Adrar Tikertine (feuille de Tin-felki) on radar remote sensing images. The utility of these structures is irrefutable, particularly in mineral prospecting and geological cartography. To reach this goal, we use an active contour model called Gradient Vector Flow (GVF). With the difference to traditional approaches, the Gradient Vector Flow (GVF) concept includes simultaneously two operations: detection and edge link of contour points. The last one was always considered as a very complicated task in traditional approaches and must be done separately from detection of contour points. In fact, the strong point of the GVF active contours is the definition of new external force able to attract the deformable contour to concave regions, generally not attained with traditional active contours called "snakes".

Keywords: automatic detection, circular geologic structures, remote sensing, geological cartography, Gradient Vector Flow, active contours.

1 Introduction

In this work, we treat the problem of the extraction of the contours of circular geologic structures on active remote sensing images. We suggest applying Gradient Vector flow introduced by Chenyang Xu and J. Lee Prince [1] on SAR remote sensing images ERS-1 in order to de delimit the circular geologic structures in these images. Among the existing active contour models, the GVF model answers well this problem considering its properties.

The detection of contours in the images is a problem which motivated many research orientations without ever giving perfect solutions, because of the properties of remote sensing images. The first approaches were founded on the detection of zones of strong gradient. However, several problems prevent an efficient detection: contours are weakly contrasted sometimes; the presence of noise involves a high rate of false alarm. The approach of active contours or snakes introduced originally by Kass and al. in 1987 [2] made it possible to solve some of these problems, resting on the paradigm that a method which provides various possible answers and which depend on the choice of certain terms is better than a method bringing only one answer. This method

A. Amine et al. (Eds.): *Modeling Approaches and Algorithms*, SCI 488, pp. 315–327.
DOI: 10.1007/978-3-319-00560-7_35 © Springer International Publishing Switzerland 2013

met much success because of its capacity to integrate the two traditional stages of contours detection in only one stage (extraction and linking). In the second section, we will show the importance and the utility of the circular geological structures for the geological cartography and the mineral prospecting. In the third section, we study the Gradient Vector Flow deformable model while presenting its ascending traditional active contour called "snake". A sequence of experiments will be applied to synthetic images then on SAR remote sensing images ERS-1 to confirm the choice of GVF model. Finally, a conclusion on work carried out will enclose this paper.

2 State of the Art

The use of remote sensing technique constitutes a useful complement for the traditional methods of geological cartography by novel brought information, especially the structural order, which they contain. The forms of circular geological structures also called “cycléaments” by David and Perthuisot [4], are in close connection with morphology, geology and the observations resulting from other techniques of investigations such as geochemistry, or geophysics. [3]

The relation between these structures and morphology is defined by the study of the topographic funds of varied scales and the air photographs. This analysis permits to explain the landscape of certain structures. The circular granites of Taourirt and sub-circulars of Central Polycyclic Hoggar offer a good example of this type of morphology. These granites are beautiful solid masses with sub-circular contours and concentric structure to which we associated minerals like the fluorite, iron, the lead and other types of mineralization. Moreover, many of these forms of structures could correspond to phenomena of sub-surface; granitic cupola for example and whose detection is of a great interest in mining research. The contribution of the radar images taken by planes or by space shuttles appeared particularly useful, especially in the desert areas. Indeed, as the radar wave penetrates dry sand on a 5 m thickness (case of band C of the ERS-1 satellite), it reveals under a sand cover of deflation an hydrographic fossil network or another geological outcrop.

In the continuation of this paper, we will present a state of the art on the traditional deformable models (snakes) [2] and derived model Gradient Vector Flow [1] to isolate the most appropriate detector to our problem which consists in delimiting circular geological structures on radar remote sensing images.

3 The GVF Deformable Model GVF and Its Application to Remote Sensing Images

3.1 Classic Deformable Models

A deformable model or a deformable contour is defined as a discrete, continuous or continuous mathematic object by pieces (a material point, a curve, a surface, an area, a volume or an unspecified association of these five elements) of which the position, the orientation, the shape, topology and the composition are possibly variable [5].

These variations are regularized, i.e. the model comprises privileged states, or that it presents a natural tendency to be opposed to any change. The process of detection of contours results then from a compromise between these constraints of regularization and an objective related to continuous information in the image.

Mathematically, deformable contours are represented by parameterized curves (curves or surfaces), *x(s)*, where *s* is a curvature parameter. Its form is typically given by a variational formula whose general form is as follows:

$$\xi = \xi_{\text{int}} + \xi_{ext} = \int_S \lambda[\alpha |x'(s)|^2 + \beta |x''(s)|^2)] + E_{ext}(x(s))ds \tag{1}$$

with $$E_{ext}(x(s)) = E_{externe}(C) = -\int_a^b |\nabla(G_\sigma(x,y) * I(x,y))|^2 \partial s \tag{2}$$

where a and b are the ends of the curve C , ∇I is the gradient operator applied to the image $I(x,y)$ *and* G_σ is the Gaussian operator with standard deviation σ applied to the same image . To find *x(s)* which minimizes the functional calculus of energy: The functional calculus can be seen like a representation of the energy of the curve. The final form of the curve corresponds at least to this energy. The first term ξ_{int} represent a priori knowledge on the model such as its material properties (elasticity α and rigidity β). The second term ξ_{ext} is usually derived from the image data and its minimum is reached when deformable contour is on the borders of the targeted object [6].

3.2 GVF Deformable Contours

The flow of gradient vectors was introduced by Xu and Prince [7] to surmount the problem of the limited operating range of the conventional external forces to be used. This problem is due to a local definition of the force and the absence of a propagation mechanism of this information [8]. To solve this problem, and for all the forces derived from the gradient of a scalar field, these authors propose to generate a field of vectors which propagates the information of the gradient. The field is called GVF (Gradient Vector Flow). The GVF of a scalar field *f* is defined as the field of vectors $V= [u,v]$ which minimizes the functional calculus of energy (Equation (1)).

The approach developed by Xu and Prince consists in using an equilibrium equation of forces like a starting point for the design of deformable model GVF. The GVF field is equivalent to a static external force $F^{(g)}{}_{ext} = v(x\ ,\ y)$. To obtain a dynamic equation corresponding to the snake, the latter replaced the potential force - ∇E_{ext} by $v(x,y)$ and they obtained the following equation:

$$x_t(s,t) = \alpha x''(s,t) - \beta x''''(s,t) + v \tag{3}$$

They call the parametric curve which makes it possible to solve the equation above, the snake Flow of the gradient vector. They solve it by an iterative process after discretization exactly like they did in traditional active contours (snakes). Although the final configuration of a GVF active contour satisfies the equilibrium equation of forces (equation (4)), in general, the latter, does not represent the Euler equations of minimization of energy problem (equation (1)).

$$F_{\text{int}} + F_{ext}^{p} = 0 \tag{4}$$

This is due to the fact that *v(x, y)* will not generally be an irrotational field. The loss of this property of optimality, however, is well compensated by the performance improved of GVF active contour [7].

3.2.1 Edge Map

They define an edge map *f(x, y)* like a derivative of the image *I(x, y)*.This map characterize the zones of interest in the image by the great values:

$$f(x, y) = -E_{ext}(x, y) \tag{5}$$

Where E_{ext} *(x, y)* is external energy defined in the equations (2). It comes whereas ∇f point towards these interesting areas [9].

The edge map has three significant properties [7]:

1. The gradient vectors of the edge map ∇f point towards the objects borders, and are normal to contours;

2. The amplitude of the gradient vectors is significant in the immediate vicinity of contours [14];

3. In the homogeneous areas where *I(x, y)* is almost constant, ∇f is almost null.

The effect of the properties quoted above on the behavior of the traditional snake when the gradient of the edge map is used as an external force is as follows:

- The effect of the first property is commonplace; the snake initialized closer to the edge will converge towards a stable configuration in the immediate vicinity of this edge;
- As for the second property, the attraction area will be in general much reduced;
- For the third property, the homogeneous areas will not have any external forces.

The two last properties are undesirable in certain cases where the image strongly noised or is slightly contrasted. The GVF approach aims to extend the influence zone of the gradients by a numerical process of diffusion. What makes it possible the gradient vectors to point towards areas with concavities [6].

3.2.2 Gradient Vector Flow

Xu and Prince [1] introduced this new type of gradient to try to solve certain problems of traditional active contours: those dependent on initialization and their difficulty to follow contours with nonconvex forms (concave). A new external force, obtained by isotropic diffusion of external flow, was thus proposed thus making it possible to increase the attraction of the zones of strong gradient and to attract the points of the snake to previously inaccessible parts (nonconvex). The authors proceed by preliminary standardization then by isotropic diffusion of the external field of force in order to smooth the noise and to diffuse the strong values of external flow. This authorizes initializations inside the objects; the diffusion of the external field ensures the dilation of active contour towards the borders of the object.

The new definition of external energy is based on the preliminary calculation of the module of the field of gradient vectors $\|\nabla I\|$ derived from the image *I*. External energy is expressed through the field of gradient vector flow: $V = [u(x, y), v(x ,y)]^T$ who plays the role of external force.

This field must minimize the energy defined by the following equation:

$$\xi = \iint \mu.(u_x^2 + u_y^2 + v_x^2 + v_y^2) + |\nabla f|^2 |V - \nabla f|^2 \, dxdy \tag{6}$$

When the variations of the amplitude of the gradients ∇f are strong, the second term dominates the integral and its minimum is reached for $V = \nabla f$ and external energy is then equivalent to f. Conversely, if these variations are weak, the second term is insignificant and the minimization of ξ come down to minimize the sum of the squares of the space derivative u_x, u_y, v_x *et* v_y of the components of *V*. The coefficient μ balances the respective influence of the two terms and must be regulated according to the quantity of noise present in the image: μ is bigger as the noise is significant [12].

This methodolgy proves efficiency since it makes possible GVF-snakes to be fixed on the borders with concavity for which attraction anymore is not masked. Let us note that the field vector which minimizes energy ξ is neither entirely irrotational nor entirely solenoidal [5]. After calculus of variations, they note that *V* satisfied the following equations of Euler:

$$\mu\nabla^2 u - (u - f_x)(f_x^2 + f_y^2) = 0 \tag{7}$$

$$\mu\nabla^2 v - (v - f_y)(f_x^2 + f_y^2) = 0 \tag{8}$$

Where ∇^2 is laplacian operator (2nd derivative), and f_x, f_y indicate the derivative compared to *x* and to *y* respectively. They note that for the homogeneous areas (where *I (x, y)* is almost constant), the second term in each equation is close to zero since the gradient of *f* is almost null. Consequently, in such area, *u* and *v* are given each one by the equation of Laplace, and resulting field GVF is interpolated starting from the area of the borders, reflecting a competition among the contours vectors and a larger capture range[13], which makes it possible GVF vectors to move towards the the concavities of borders. Numercial plan for implementation of the GVF model is detailed in [6].

3.3 Implementation of GVF Deformable Model

For the calculation of the GVF on images in grayscale, we must calculate the edge map functions. The latter generally arises in two forms:

$$f^1(x,y) = |\nabla I(x,y)| \quad (9) \quad \text{or} \quad f^2 = |\nabla[G_\sigma(x,y) * I(x,y)]| \tag{10}$$

f^2 is recommended if the noise in the image is significant. The Gaussian filters have several advantages compared to the traditional balanced filters [10].

A mathematical explanation of the use of smoothing by Gaussian was formulated by CANNY [11]. The latter showed that the derivative first of a Gaussian filter provides a value close to the optimal value for the criterion by which he proposed to define a good edge detector.

The stages to follow to apply GVF model to binary images and level scale images are enumerated and ordered in the following diagram (cf. figure 1).

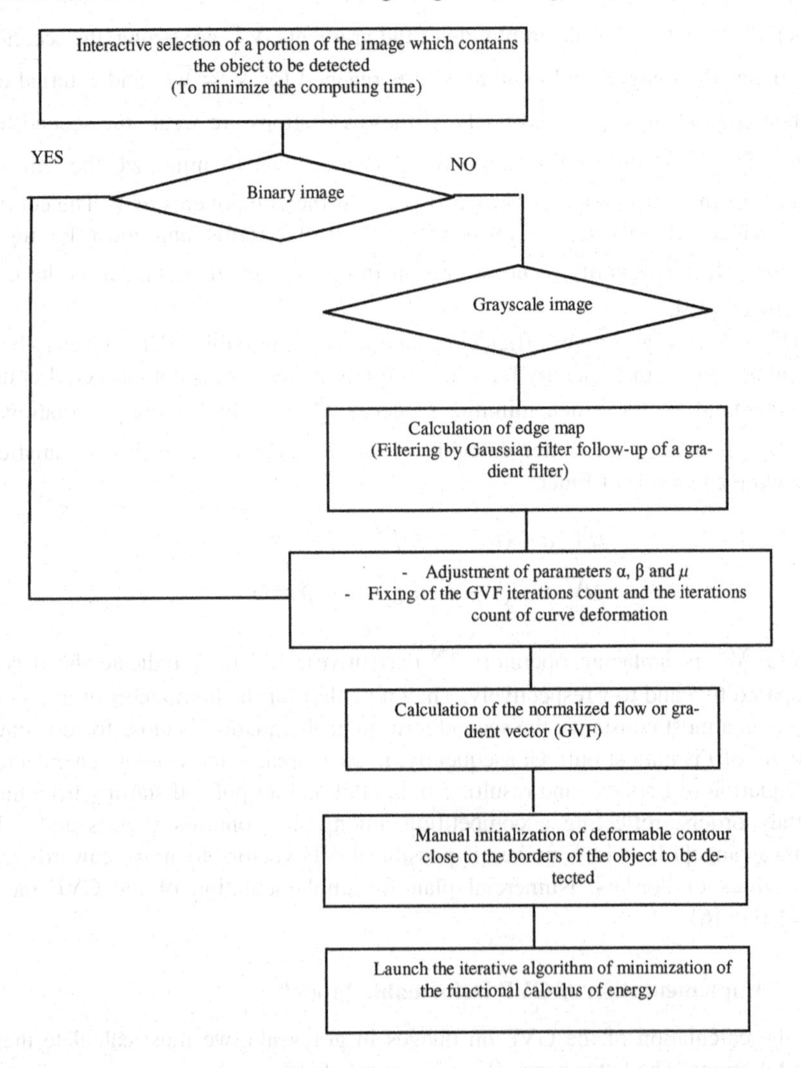

Fig. 1. Stages of application of GVF model on binary and grayscale images

3.3.1 Validation of GVF Deformable Model

The use of synthetic images is a very interesting tool to validate any method of image processing. This is why; we considered it useful to validate the GVF deformable model on test synthetic images to evaluate the performance of this model and to show its various properties.

In the beginning, we treated a 2D noisy synthetic image of size 220x203 pixels and representing a circular structure with coarse borders (cf. figure 2). Figure 3 shows the result of the application of the Gaussian filter on the image, while figure 6 illustrates the field of the gradient vector.

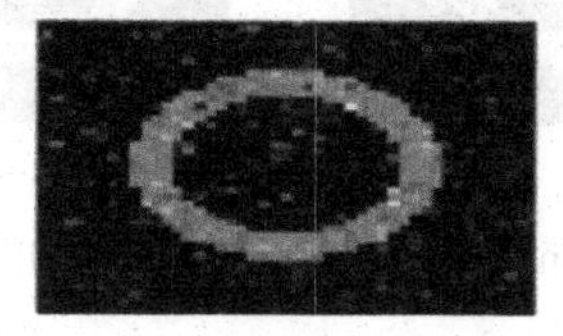

Fig. 2. Noisy synthetic image representing a circular structure with coarse borders

The choice of the parameters α (elasticity), β (rigidity) of the initial curve (initial contour) and the coefficient μ must care of the regularity of the shape of the target and the quantity of noise. The target to be detected in this image has a regular circular shape. Consequently, the initial curve must be much more elastic than rigid (α=0.05, β=0.01). The presence of noise on the image requires a higher value for the coefficient μ (μ = 0.2) knowing that μ usually takes value 0.1 for images very slightly noised where the targets are clearly contrasted.

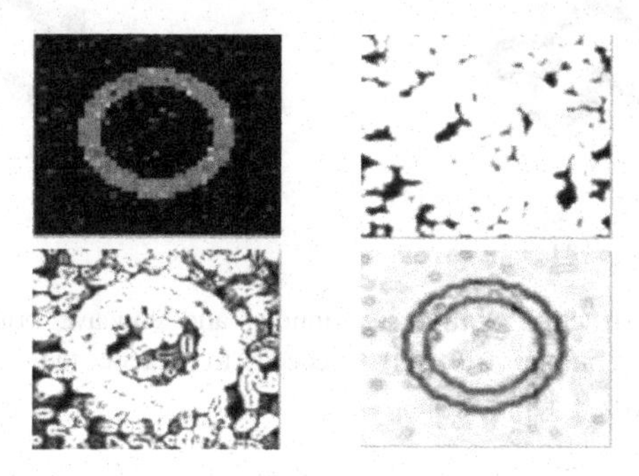

Fig. 3. Application of the Gaussian noise filter, gradient of the Gaussian noise filter, and gradient filter

Fig. 4. Field of normalized gradient vector with α=0.05, β=0.01 and μ = 0.2

Figure 5 illustrates the process of detection of the circular structure. In the figure (5.a) we present the initial curve at the moment t=0, this curve surrounds the object. The evolution is carried out thus towards the interior of the curve. The result is reached after 200 iterations. This result is presented in the figure 5.b. We notice as well as the delimitation of the circular structure from outside was carried out successfully.

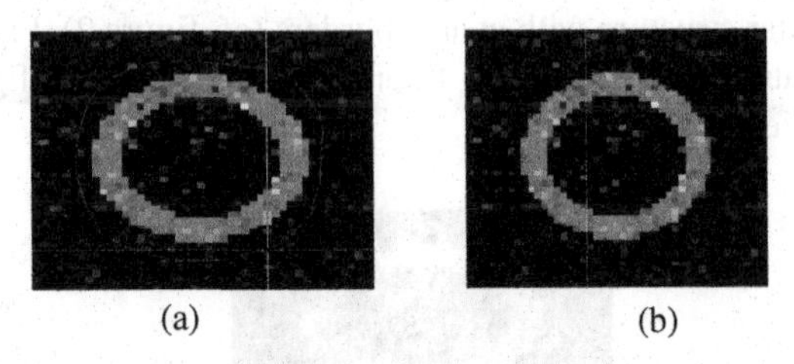

(a) (b)

Fig. 5. a) Initialization of contour b) final contour (in red) of the object to be detected

We then, treated a noisy 2D synthetic image representing a structure with discontinuous and concave borders (cf. figure 6.a). The figure 6.b shows the result of the application of the GVF active contours model on this image. We can note that the detection of the structure by the GVF is convincing.

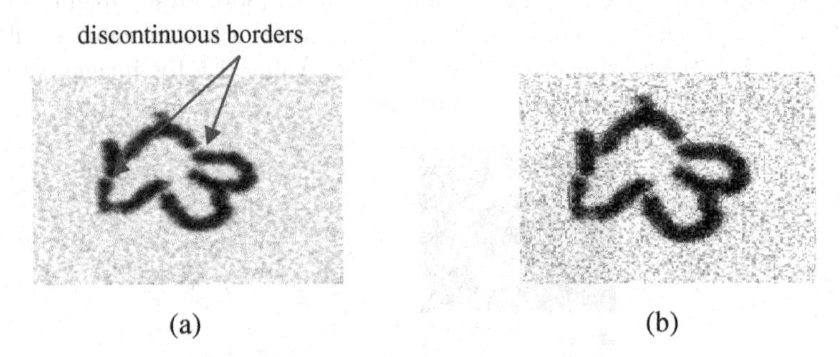

(a) (b)

Fig. 6. a) A closed structure with discontinuous and concave borders (147x141 pixels) b) deformable GVF contour (in red) succeed to detect the borders of the target. $\alpha = 0.5$, $\beta = 0.00$, $\mu = 0.1$

Finally, we treated a 2D noisy synthetic image representing an object in circular shape including two other objects (cf. figure 7.a). The figure 7.b shows the result of the application of GVF model on this image.We notice as well as deformable contour has detected only the outside borders of the circular shaped object. This property is very interesting because it makes it possible to avoid over-segmentation of the target in the image which is very appropriate for our problem.

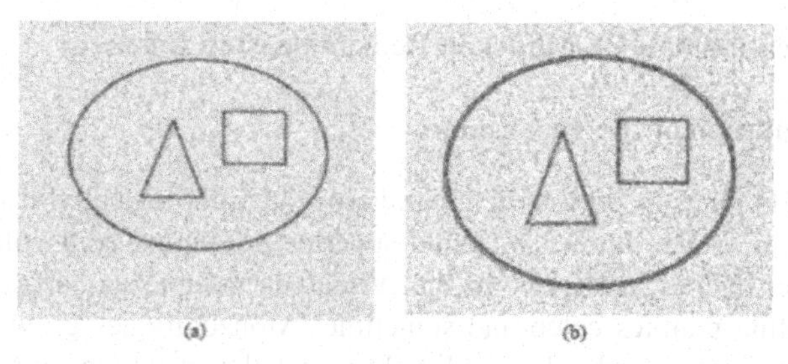

Fig. 7. a) Circular shaped object including two other objects (350x350 pixels) b) Deformable GVF contour (in red) detects only the outside borders of the objects what makes it possible to avoid over-segmentation of the image. $\alpha = 0.5$, $\beta = 0.00$, $\mu = 0.1$

Figure 8 shows the result of the application of the traditional edge detector of Canny on the ERS-1 image.

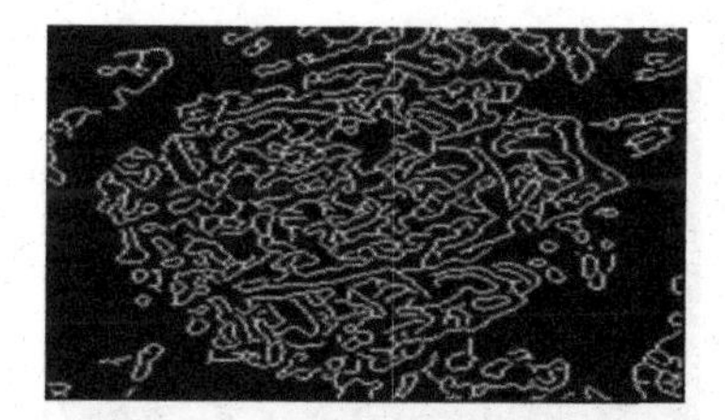

Fig. 8. Application of the CANNY detector on the ERS-1 image

By comparing this result with that of the application of GVF deformable model in the figure 13, we can see that the contribution of deformable contours is very clear. Indeed, we notice that the application of the CANNY detector to the circumscribed granite of TINKERTINE gives disconnected contours which are not realistic and do not reflect the reality of circularity and closing of the structure. The closing of contours is a very fastidious task for the traditional detectors because it generates a lot of errors which are difficult to repair. The linking of two distinct contours in is an example. Quite to the contrary the active contours models take care of the structure of contours, so the points of an active contour are connected, which removes the task of the linking.

3.3.2 Discussion

The results obtained on a diversity of synthetic images show that GVF model even succeed to detect contours of the irregular shaped structures with coarse and discontinuous borders. This model penetrates easily in the concave areas.

The geological structures on the remote sensing images have very often same characteristics. These latters are not enough homogeneous and present "holes" inside the global structure. This deformable contour model is thus appropriate for our problem, since it avoid over-segmentation of the target.

3.4 Application of GVF Model on Remote Sensing Images

3.4.1 Presentation of the Test Zone

The test area belongs to polycyclic central Hoggar. Lithologically, it is characterized by volcanogenic series known as sitted superior presenting acid volcanite rocks of rhyolites type, dacites, tuffs etc. In the Afedafeda complex, quartzic granodiorites, diorites, alkaline granites calco and sometimes Monzonitiques grow up. Geological formations of circular and sub circular shapes of the panafrican plutonic unit offer intrusive granite with circumscribed contour of type Taourirts. This last grow up in the Western Northern part of the region of Tin Felki (Adrar Tikertine).

We have used a radar image of 12.5m resolution provided by the European satellite ERS-1. It represents the circumscribed Granite of TINKERTINE of the region of TinFelki in central Hoggar. Its dimensions are of 400x450 pixels (cf. figure 9).

Fig. 9. Radar image ERS-1: Circumscribed granite of TIKERTINE of the region of TINFELKI., size: 400x450 pixels. Resolution: 12.5 m.

3.4.2 Experiments on ERS-1 Radar Image

The application of the deformable contours model GVF on the radar image ERS-1 led to the following results. Figure 10 illustrates the field of the normalized Gradient vector GVF of the ERS-1 image. It is the most essential stage; it generates the field of external forces with which the internal force of deformable contour must be regularized to minimize at maximum the total energy. In other words, contour must be fixed (according to its physical parameters of curve α and rigidity β) most perfectly possible on the zones where GVF field is maximum.

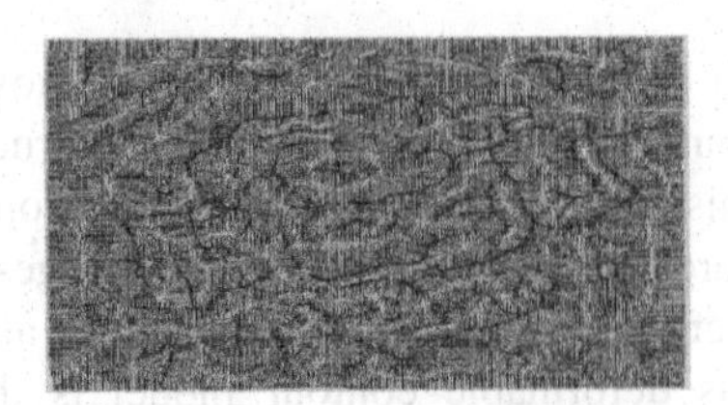

Fig. 10. Normalized GVF field (image ERS-1)

Figure 11 shows the initial curve at the moment t=0 including the object from outside. The evolution is carried out thus towards the interior of the curve. Figure 12 illustrates the process of detection of the circular structure. The result is reached after 200 iterations. It is presented in the figure 13. We can note that the delimitation of the circular structure from outside is satisfactory.

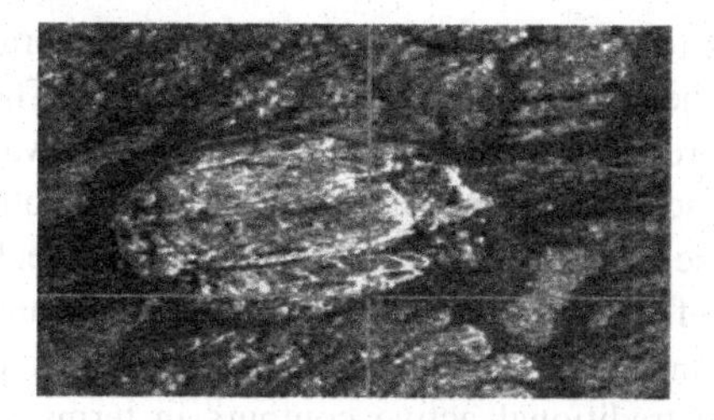

Fig. 11. Initialization of contour (image ERS-1)

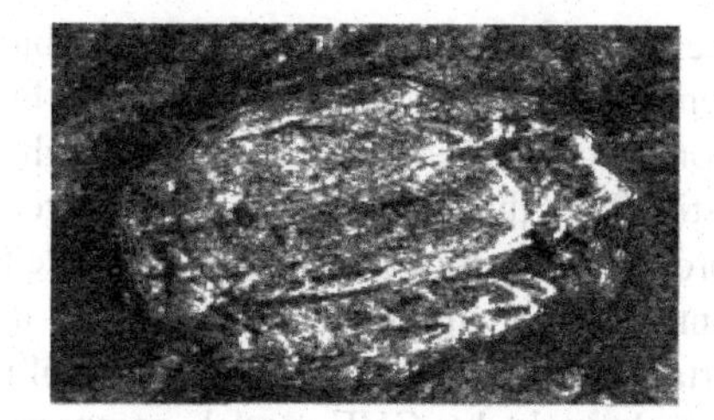

Fig. 12. Progression of deformable contour (image ERS-1)

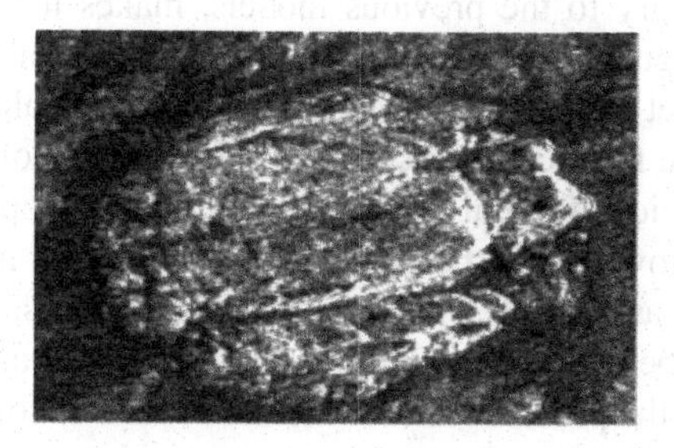

Fig. 13. Convergence of the deformable polygon towards the borders of the circular structure of the circumscribed Granite of TINKERTINE at iteration 200

3.4.3 Discussion

In the desert scene of the south of Algeria (Polycyclic Central Hoggar) the sandy cover hide a significant part of the studied target. This same sandy cover tends to present the target like a fragmented or discontinuous structure in optical mode.

This problem is not also embarrassing for the radar images ERS-1. The radar imagery is most adapted to detect the geological structures, though, many times, these structures have very irregular forms. The radar waves penetrate the ground until approximately 5 meters and are therefore, the best placed to characterize accurately the geometry of the objects.

The developed application cannot be a real time application since the images to be treated are generally big sized and cover broad scenes of the earth's crust. The calculation of the Flow of the gradient vectors is "greedy» in video memory and computing time. It is calculated by sweeping the entire image pixel by pixel. However, with a configuration Intel Pentium B960 2.2 GHZ with 4 Gb of RAM and 64 Mb of dedicated video memory, the calculation of the GVF did not exceed 7 seconds on test images of small sizes (220 X 203 pixels, 147x141 and 350 X 350 pixels).

4 Conclusion

The particularity of our goal and the objects to be delimited led us to adopt the Gradient Vector Flow model, introduced by Chenyang Xu and J Lee Prince [1]. The geometric and geodesic models, though more recent, are ignored. This choice was justified by the fact that the circular geological structures have often discontinuous borders, heterogeneous areas even inside these structures with sometimes a very weak contrast and a significant noise. So the use of these models proves unbeneficial, and certainly generates over-segmentation of the target, which is disagreeable. The advantages brought by GVF model, compared to traditional active contours in terms of convergence towards concave areas and lower sensitivity to the noise are proven. This model makes it possible to detect only one object at the same time.

This property, contrary to the previous models, makes it possible to avoid over-segmentation of the target even if it appears parcelled out. Moreover, it makes it possible to delimit the target from outside and gives a single result. However, the choice of this model is not free from criticisms. The implementation of active contours raises two difficulties: the choice of initial contour and the gradient operator which will push or pull active contour towards the borders of the structure by marrying all the details of the structure as well as possible. As for initial curve, it must be initialized closer as possible to contour to be detected to limit the iteration count and especially to ensure a good convergence of the process towards the most realistic solution.

References

1. Xu, C., Prince, J.L.: Gradient Vector Flow: A New External Force for Snakes. In: IEEE Proc. Conf. on Comp. Vis. Patt. Recog, CVPR 1997 (1997)
2. Kass, M., Witkin, A., Terzopoulos, D.: Snakes: Active Contour Models. International Journal of Computer Vision, 321–331 (1988)
3. Rouhana, R.: Détection des linéaments dans les images de RADARSAT par un réseau neuronique cellulaire. Master, Ecole technologique supérieure, Université du Quebec (January 1998)
4. David, E., Perthuisot, V.: Cycléaments de la province d'Alicante (Espagne). Photo-interprétation (1985)
5. Angella, F.: Modèles déformables et systèmes particulaires : application à l'extraction de structures arborescentes en analyse d'images. Thèse de Doctorat, Ecole doctorale de physiques, université de Bordeaux I. 9 (January 2001)
6. Xu, C.: Deformable models with application to human cerebral cortex reconstruction from magnetic resonance images. Ph.D. Thesis, Baltimore, Mayland (January 1999)
7. Xu, C., Prince, J.L.: Snakes, Shapes, and Gradient Vector Flow. IEEE Transactions on Image Processing 7(3) (March 1998)

8. Esteban, C.H., Schmitt, F.: Une approche par modèle déformable pour la reconstruction 3D de haute qualité d'objets photographies. Thèse de Doctorat, Ecole Nationale Supérieure des Télécommunications, France (2004)
9. Ikhlef, R.: (2005), http://www.tsi.enst.fr/tsi/enseignement/ressources/mti/gvf/gvf5.htm
10. Leger, C.: VISIONIQUE. Cours tutorial, école supérieure des procèdes électroniques et optiques (2002)
11. Canny, J.: A computational approach to edge detection. IEEE Transactions on Pattern Analysis and Machine Intelligence PAMI 8(6), 679–697 (1986)
12. Rana, P.K.: Prediction of Sea Ice Edge in the Antarctic Using GVF Snake Model. Journal geological society of India 78, 99–108 (2011)
13. Liu, L., Bovik, A.: Active contours with neighborhood-extending and noise-smoothing gradient vector flow external force. EURASIP Journal on Image and Video Processing (2012)
14. Schaerer, J.: Segmentation et suivi de structures par modèle déformable élastique non-linéaire. Application à l'analyse automatisée de séquences d'IRM cardiaques. thèse de doctorat, L'Institut National des Sciences Appliquées de Lyon, France (2008)

NALD: Nucleic Acids and Ligands Database

Abdelkrim Rachedi[1,2,*] and Khuphukile Madida[1]

[1] Wits-Bioinformatics Unit, School of Molecular and Cell Biology,
University of the Witwatersrand, Private Bag 3, Wits 2050, South Africa
[2] Current address: Department of Biology, Faculty of Sciences & Technology,
University Dr Tahar Moulay, Saida, Algeria
rachedi@bioinformer.co.uk

Abstract. The nucleic acids and ligands database (NALD) is concerned with the identification of ligands (drugs) that bind nucleic acids (NA) and provide users with sets of specific information in relation to the binding existing between both molecules. NALD thus annotates nucleic acids in complexes with ligands in terms of detailed binding interactions, binding motifs where binding occurs, binding properties, binding modes & classes and links to diseases may be in association with the ligands. These were calculated from entries of NA/Ligand complexes from the protein data bank (PDB) and also extracted by both automatic and manual means from scientific literature sources such as the PubMed web site (PMID) and publications (in hardcopy form). NALD provides online access to these types of information while it focuses on ligands that bind nucleic acids with implications on diseases of high prevalence in Africa and in particular in Algeria and the Southern African Development Community (SADC) region such as HIV/AIDS, cancer, hepatitis, malaria and tuberculosis. The database offers data integration in the form of links to the PDB, DrugBank, and other resources such as UniProt and PubMed databases. In addition and for those ligands classified as known drugs, drug information extracted from the DrugBank. NALD can be accessed freely from **http://www.bioinformaticstools.org/nald**.

Keywords: Bioinformatics, Database, Data mining, Data integration, Nucleic acids, Binding motifs, Ligands, Drugs, Diseases.

1 Introduction

Nucleic acid molecules are biologically very important found in all living organisms. They contain genetic material that must be synthesized and reproduced with high fidelity to ensure its proper function. Failure to do so conditions of compromised health and disease arise and for this reason nucleic acids are points of interest for therapeutic-drug targets during specific binding events. The protein data bank (PDB) [1] is an international repository of a large number of 3-D structures for macromolecules including protein, nucleic acids and their complexes. Many of the compounds binding nucleic acids known in general as ligands are considered drugs either designed as therapeutics

* Corresponding author.

A. Amine et al. (Eds.): *Modeling Approaches and Algorithms*, SCI 488, pp. 329–336.
DOI: 10.1007/978-3-319-00560-7_36 © Springer International Publishing Switzerland 2013

or as additives for structural and functional investigation. This project deals with the development of an online database, Nucleic Acids and Ligands Database (NALD), which annotates the nucleic acid (NA) in complexes with ligands (drugs) in terms of detailed binding interactions, NA motifs where binding occurs, binding properties, binding modes and classes and links to diseases. Focus will be centered on ligands that bind nucleic acids with implications on diseases of high prevalence in Algeria and the Southern African Development Community (SADC) region such as HIV/AIDS, cancer, hepatitis, malaria and tuberculosis. The database provide data integration in links to the PDB, PDBeChem [2], whenever possible to Drugbank [3] and other resources such as UniProt [4] and PubMed [5] databases.

2 Methods

The database consists of two models of information storage; MySQL Relational Module (MySQL tables) [6] and Fla-Files modules. The relational module, created in MySQL database platform, is where tables, governed by a database schema, are

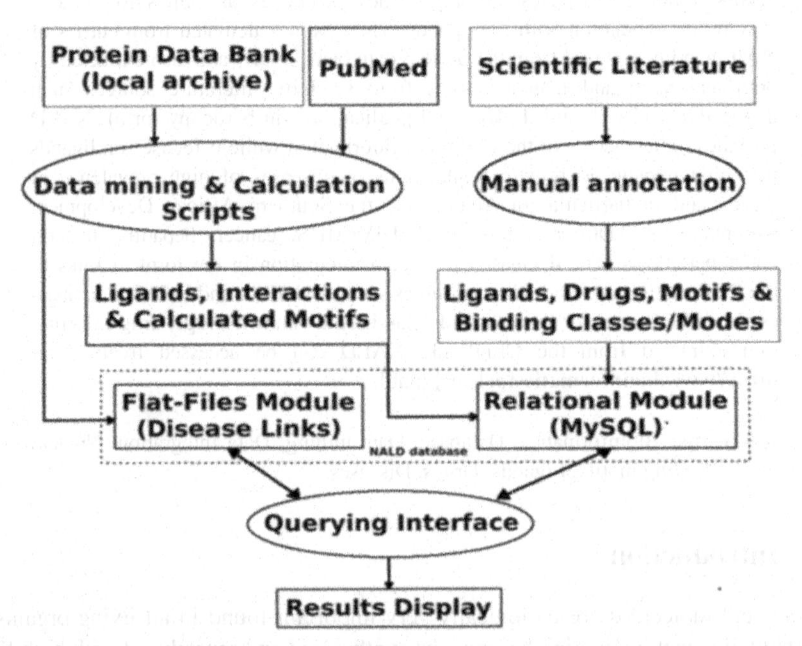

Fig. 1. Diagram description of the NALD database components. PHP scripts automatically mine data from the PubMed (accessed via internet) and a local archive of the PDB. Another set of PHP scripts calculates binding interactions and motifs. Mined data (disease links) from PubMed are stored in flat files (Flat-Files module) and the rest of data are stored as MySQL tables. Detailed empirical information gathered from scientific literature (hardcopy publications) was used to annotate (manual annotation) binding motifs, binding modes and properties. Annotation data is stored in MySQL tables. The Querying Interface module (the application's main web-page) uses another set of PHP scripts that analyze users input queries and fetch/display results.

used to store related information and the linking of the table components is via unique identifier columns or primary keys. The second module, flat-file structure, stores information in plain text files, Fig.1.

Data is mined from a local archive of the PDB (contains mainly PDB entries of nucleic acids complexes) and PubMed using PHP [7] scripts. These conduct calculation of NA/Ligands binding interaction and calculated binding motifs. Scientific literature (in hardcopy publications) is used for manual extraction of empirical detailed information about NA/Ligands binding motifs, binding modes and properties. The data are then stored into a MySQL based tables. Information about disease links to ligands and which is extracted automatically from PubMed is stored in flat-files format.

The tables present in MySQL were created using PHP scripts that acquire data and simultaneously create and load information directly into the MySQL tables. The tables were populated by data mined from the Sequence, Structure and Function Server (SSFS) website [8] which is a locally developed online system that uses a weekly updated local archive version of the PDB data. The tables are created with a hierarchical structural component from the PDBid table cascading down to the classes of binding, see next. The information pertaining to calculated binding motifs and ligand interactions with nucleic acids are based on exclusively complexes of nucleic acids (DNA, RNA and hybrids) with ligands as found in the PDB. Data related to binding motifs, binding modes/classes and properties have been manually mined from scientific literature (hardcopy), e.g. from Clark et.al., 1996 [9] and Bunkenborg et.al., 2002 [10], and then loaded in MySQL tables. Disease links data have been mined from the PubMed database and are stored in flat-files. The Querying Interface module represented by the NALD's main web-page uses another set of PHP scripts that analyze users input queries, fetch and display results.

3 Results

The database is updated on regular basis and currently contains 1970 NA/Ligands complexes and 1235 unique ligands, Fig.2-C. Annotated classes of NA–Ligands binding covers Intercalation, Modification, Addition and Cavity fitting classes in addition to binding modes of ligands to the DNA's Minor Groove, Major Groove and both Minor/Major Grooves. Annotation included binding motifs, binding sites and properties (Fig.2-B). Customized 3D graphics module, based on Jmol [11], is used to illustrate the binding of ligands to their NA targets, Fig.4-D.

3.1 Querying the NALD Database

NALD database offers two ways for data retrieval; "Find Binding details & Disease links:" and "Find Ligand Binding Motifs and Classes:", Fig.2-A. Suggestions on how to use the database are provided from the main web interface under the internal link "Directions", Fig.2-B.

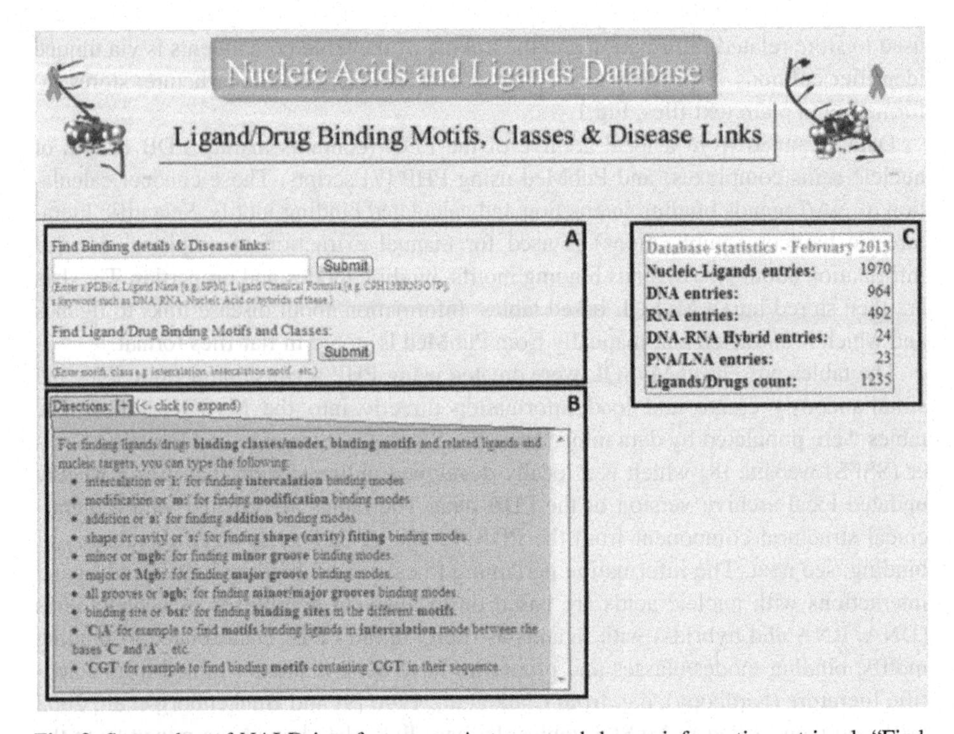

Fig. 2. Screenshot of NALD interface page. A. two search boxes information retrieval; "Find Binding details & Disease links:" and "Find Ligand Binding Motifs and Classes:". B. Internal link "Directions" for examples on how to find important information. C. Monthly updated statistics of the database content with URL links that provide direct browsability and exploration of the database based on relevant data entries.

3.2 Find Binding Details and Disease Links

This search box allows for both general and specific querying and browsing of the database. Keywords used in the search can be general like 'DNA', 'RNA', 'hybrid' or specific like ligand names such as 'SPM' (3 letter code for the drug known as SPERMIN), full names such as 'Spermin' and chemical formula like 'C10H26N4' and analogues.

NALD displays a results list, Fig.3, which gives a summary list of known 3D structures in the PDB database for nucleic acid molecules (first column), bound ligands (second column) and other useful information such as the title of the published molecules, method used to solve the structures and Resolution at which the structures were solved.

Exploration of the ligand binding details with the NA molecule, calculated binding motif and disease links in addition to other useful information can be generated by clicking on ligand ids, second column, Fig.4. Further information about each PDB entry can be displayed by clicking on the first column which retrieves the entry's summary from the SSFS system.

Your searchterm, "dna", returned (1176) hit(s).

PDBid	Ligands	Header	Title	Method	Resolution (Å)
100d	SPM	DNA-RNA HYBRID	CRYSTAL STRUCTURE OF THE HIGHLY DISTORTED CHIMERIC DECAMER R(C)D(CGGCGCCG)R(G)-SPERMINE COMPLEX-SPERMINE BINDING TO PHOSPHATE ONLY AND MINOR GROOVE TERTIARY BASE-PAIRING	X-RAY DIFFRACTION	1.90
101d	CBR, NT, MG	DNA	REFINEMENT OF NETROPSIN BOUND TO DNA: BIAS AND FEEDBACK IN ELECTRON DENSITY MAP INTERPRETATION	X-RAY DIFFRACTION	2.25
102d	TNT	DNA	SEQUENCE-DEPENDENT DRUG BINDING TO THE MINOR GROOVE OF DNA: THE CRYSTAL STRUCTURE OF THE DNA DODECAMER D(CGCAAATTTGCG) 2 COMPLEXED WITH PROPAMIDINE	X-RAY DIFFRACTION	2.20
106d	MCY	DNA	SOLUTION STRUCTURES OF THE I-MOTIF TETRAMERS OF D(TCC), D(5MCCT) AND D(T5MCC). NOVEL NOE CONNECTIONS BETWEEN AMINO PROTONS AND SUGAR PROTONS	SOLUTION NMR	N/A
107d	DUO	DNA	SOLUTION STRUCTURE OF THE COVALENT DUOCARMYCIN A-DNA DUPLEX COMPLEX	SOLUTION NMR	N/A
108d	TOT	DNA	THE SOLUTION STRUCTURE OF A DNA COMPLEX WITH THE FLUORESCENT BIS INTERCALATOR TOTO DETERMINED BY NMR SPECTROSCOPY	SOLUTION NMR	N/A
109d	IBB, MG	DNA	VARIABILITY IN DNA MINOR GROOVE WIDTH RECOGNISED BY LIGAND BINDING: THE CRYSTAL STRUCTURE OF A BIS-BENZIMIDAZOLE COMPOUND BOUND TO THE DNA DUPLEX D(CGCGAATTCGCG)2	X-RAY DIFFRACTION	2.00
10mh	5CM, 5NC, SAH	TRANSFERASE/DNA	TERNARY STRUCTURE OF HHAI METHYLTRANSFERASE WITH ADOHCY AND HEMIMETHYLATED DNA CONTAINING 5,6-DIHYDRO-5-AZACYTOSINE AT THE TARGET	X-RAY DIFFRACTION	2.55
110d	DM1	DNA	ANTHRACYCLINE-DNA INTERACTIONS AT UNFAVOURABLE BASE BASE- PAIR TRIPLET-BINDING SITES: STRUCTURES OF D(CGGCCG) /DAUNOMYCIN AND D(TGGCCA)/ADRIAMYCIN COMPL	X-RAY DIFFRACTION	1.90
115d	BRU	DNA	ORDERED WATER STRUCTURE IN AN A-DNA OCTAMER AT 1.7 ANGSTROMS RESOLUTION	X-RAY DIFFRACTION	1.70
119d	MG	DNA	CRYSTAL AND MOLECULAR STRUCTURE OF D(CGTAGATCTACG) AT 2.25	X-RAY	2.25

Fig. 3. Screenshot of the results page for "dna" keyword in the NALD

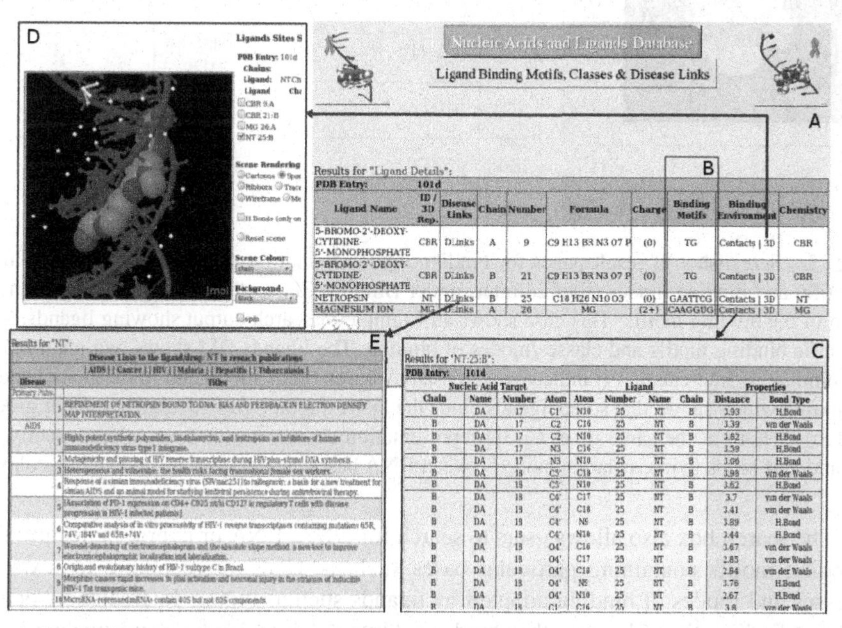

Fig. 4. Screenshot of result pages for ligands in the PDB entry 101d; DNA/Ligands complex. A. overall list of ligands in the 104d. B. Calculated binding NA sequence motifs for each ligand. C. Binding details with bond distances and possible types of bonds between the ligand Netropsin (NT) and DNA. D Jmol 3D representation of the DNA/NT binding. E. Disease links in relation with the ligand NT pointing to PubMed abstracts.

3.3 Find Binding Motifs and Classes

NALD offers two types of binding motifs, the calculated binding motifs seen above (Fig.4-B) and empiric binding motifs reported in scientific literature [8,9] with detailed annotation of the motifs, binding sites, modes, classes and properties (Fig.5-A). Classes of NA–Ligands binding covers Intercalation, Modification, Addition and Cavity fitting in addition to binding modes of ligands to the DNA's Minor Groove, Major Groove and both Minor/Major Grooves.

Detailed annotation is associated with each binding motif which includes the NA sequence, binding mode position(s) and types of NA bases involved in the binding, Fig.5-B. This is also reflected in 3D representation, Fig.5-C.

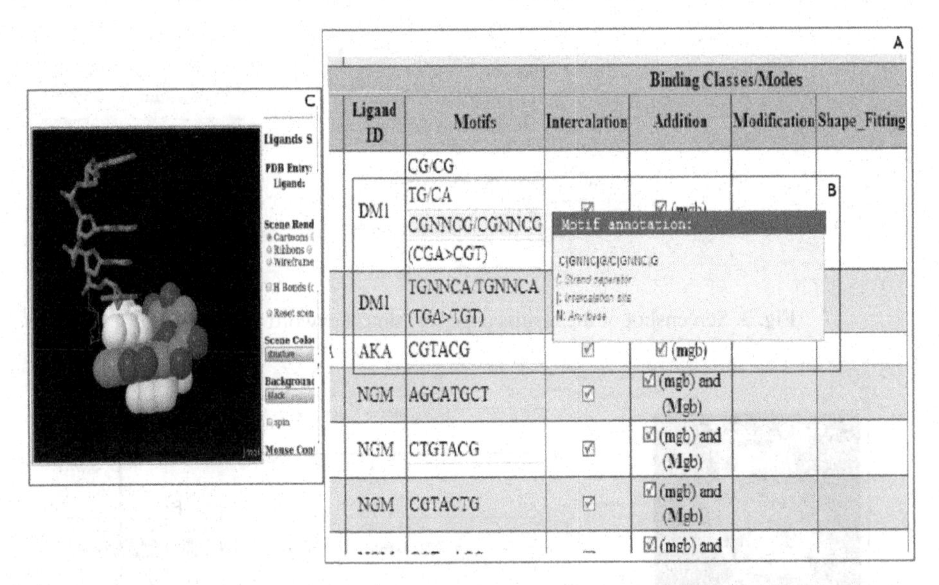

Fig. 5. Screenshot of result pages for binding modes & motifs: A. Binding motifs (column Motifs) and binding modes (four columns under Binding Classes/Modes). B. Detailed annotation of the binding motifs. This case shows an example of search output showing ligands, their specific binding motifs and classes/modes of binding. The ligands DM shows two main modes of binding 1. Intercalation (| indicates which DNA bases binds the ligand in the motif) and 2. Addition which shows that some ligands binds the DNA's minor groove (mgb) or major groove (Mgb) or both at the same time. C. 3D representation shows, in this case, where exactly the ligand DM binds in intercalation mode between DNA bases shown in white and yellow colours.

The search box also allows users to search for DNA, RNA or hybrid molecules that binding motifs containing particular bases such as 'CGC' or searching for existing classes and modes of binding adopted by ligands such as typing 'Intercalation' or just 'I:' for finding those ligands that binds in “Intercalation modes” and nucleic acids motifs binding them, and modification ('M:') for finding those ligands that causes “modification” when biding nucleic acids.

4 Conclusions

Nucleic acids are of great biological importance in all organisms and the inability to maintain integrity or tampering with nucleic acids leads to disease. Small chemical molecules or ligands bind nucleic acids in various ways and thus therapeutic strategies are designed from the the identification and study of binding details of these ligands including motifs bound to and modes of binding.

As described above, the NALD database is a system that provides specialized and integrated data through the implementation of a warehouse database type composed of in-house databases using both relational and flat-files database platforms. To enhance providing integrated data and fast recall service to users, the system implements direct URL links to source databases like the PDB and PubMed and generic content and links to sources like the Drugbank.

The NALD database, through the numbers of features summarized in points below, has the potential to be instrumental in helping with studies and processes involving the identification of potential nucleic acids targeted drugs and novel design of new drugs in the fight against diseases currently thriving in Africa and in particular Algeria and the SADC region including the HIV/AIDS pandemic, cancer and other conditions such as tuberculosis (TB), malaria and hepatitis.

1. NALD provides a point of acquisition for categorized information about nucleic acid binding ligand categories including binding motifs, types and classes of the ligands and drugs.
2. The information contained and level of annotations provided in NALD paves the way to drug design strategies and identifying potential in various ligands and drugs.
3. NALD supplies links to medically applicable information related to the ligands centered on current diseases facing large populations of Sub-Saharan Africa.
4. NALD facilitates cross-referencing with other currently larger databases such as the PDB, DrugBank, UniProt and PubMed and provides a user friendly environment for fast integrated data retrieval from a single system.

Access to Database

NALD can be accessed freely from http://www.bioinformaticstools.org/nald

Funding

This work was supported by the Faculty of Science, University of the Witwatersrand, Johannesburg, South Africa. Currently, NALD is supported by the Faculty of Sciences and Technology, University Dr. Tahar Moulay, Saida, Algeria.

References

1. Berman, H.M., Westbrook, J., Freng, Z., Gilliland, B.T.N., Weissig, H., Shindyalov, I.N., Bourne, P.E.: The Protein Data Bank. Nucleic Acids Res. 28, 235–242 (2000)
2. Dimitropoulos, D., Ionides, J., Henrick, K.: Using PDBeChem to Search the PDB Ligand Dictionary. Current Protocols in Bioinformatics; Chapter 14:Unit14.3, 14.3.1–14.3.3 (2006)
3. Knox, C., Law, V., Jewison, T., Liu, P., Ly, S., Frolkis, A., Pon, A., Banco, K., Mak, C., Neveu, V., Djoumbou, Y., Eisner, R., Guo, A.C., Wishart, D.S.: DrugBank 3.0: a comprehensive resource for 'omics' research on drugs. Nucleic Acids Res. 39(Database issue), D1035–D1041 (2011)
4. Suzek, B.E., Huang, H., McGarvey, P., Mazumder, R., Wu, C.H.: UniRef: Comprehensive and Non-Redundant UniProt Reference Clusters. Bioinformatics 23, 1282–1288 (2007)
5. PMID, `http://www.ncbi.nlm.nih.gov/PubMed`
6. MySQL: The MySQL RDBMS, `http://www.mysql.com`
7. PHP, `http://www.php.net/`
8. SSFS, `http://www.bioinformaticstools.org/ssfs`
9. Clark, G.R., Squire, C.J., Gray, E.J., Leupin, W., Neidle, S.: Designer DNA-binding drugs: the crystal structure of a meta-hydroxy analogue of Hoechst 33258 bound to d(CGCGAATTCGCG)2. Nucleic Acids Res. 24(24), 4882–4889 (1996)
10. Bunkenborg, J., Behrens, C., Jacobsen, J.P.: NMR Characterization of the DNA Binding Properties of a Novel Hoechst 33258 Analogue Peptide Building Block. Bioconjugate Chem. 13(5), 927–936 (2002)
11. Hanson, R.M.: Jmol – a paradigm shift in crystallographic visuali-zation. Journal of Applied Crystallography 43(5), 1250–1260 (2010) Jmol: an open-source Java viewer for chemical structures in 3D, `http://www.jmol.org/`

Maximality-Based Labeled Transition Systems Normal Form

Adel Benamira[1,2] and Djamel-Eddine Saïdouni[1]

[1] MISC Laboratory, Constantine 2 University, 25000 Constantine, Algeria
[2] Computer Science Dept., University of 08 May 45, 24000 Guelma, Algeria
{benamira,saidouni}@misc-umc.org

Abstract. This paper proposes an algorithm (functional method) for reducing Maximality-based Labeled Transition Systems (MLTS) modulo a maximality bisimulation relation. For this purpose, we define a partial order relation on MLTS states according to a given maximality bisimulation relation. We prove that a reduced MLTS is unique. In other word, it provides a normal form.

Keywords: Formal concurrency semantics, Maximality semantics, Maximality-based labeled transition systems, Bisimulation relation, Complete partial order.

1 Introduction

Action refinement has been deeply studied for characterising true concurrency semantics. For this purpose, several authors have proposed new semantics, and in the same context new equivalence relations proven to be preserved under action refinement and supporting action duration (See [6] for survey). The ST-semantics is one of the most studied of these propositions, originally defined in [8] over Petri nets, in which semantics, non atomic actions are split into start and end sub actions. The ST-semantics has been applied in the literature to process algebras [1,9].

The interleaving ST-bisimulation (ST-bisimulation in short) without silent moves has been defined on Petri nets [8], and further on prime event structures [7]. In [13], an alternative definition of the ST-bisimulation has been proposed for prime event structures with silent moves; the main point of this definition is that it does not require any-more to split actions as previously, the partial order relation among events being used instead for determining the set of maximal events in each configuration. The same idea has been used in [5] for defining the maximality preserving bisimulation on labeled P/T nets, and with the hypothesis that all visible actions are non atomics, the maximality preserving bisimulation coincides with the ST-bisimulation.

For implementing the ST-Bisimulation relation, in [3] an algorithm for a particular process algebra has been proposed. This approach consists in verifying the ST-bisimulation relation between process algebra terms; actions are split

A. Amine et al. (Eds.): *Modeling Approaches and Algorithms*, SCI 488, pp. 337–346.
DOI: 10.1007/978-3-319-00560-7_37 © Springer International Publishing Switzerland 2013

into start and end sub actions. Then, the proposed solution is ad hoc to the considered process algebra. Dealing with non dependability between concurrency semantic model and specification models, another concurrency semantic model, named maximality-based labeled transition system, has been defined in the literature and used for expressing the semantics of process algebras and P/T Petri net with the hypothesis that actions are not necessary atomic [4,11,12], i.e. actions are abstractions of finite processes and elapse in time. The main interest of maximality-based labeled transition system model is that it can be implemented and used in verification without splinting actions into starts and ends sub actions. In this paper, given a maximality-based labeled transition system *mlts* and a maximality bisimulation relation $\mathfrak{R}$ on *mlts* nodes, we propose an algorithm like a recursive function (functional implementation) for reducing *mlts* w.r.t $\mathfrak{R}$.

Let us take the example in [11]. Consider the Petri net of Fig.1.(a). By applying the approach of [12], the corresponding maximality-based labeled transition system of this Petri net is given by Fig.1.(b).

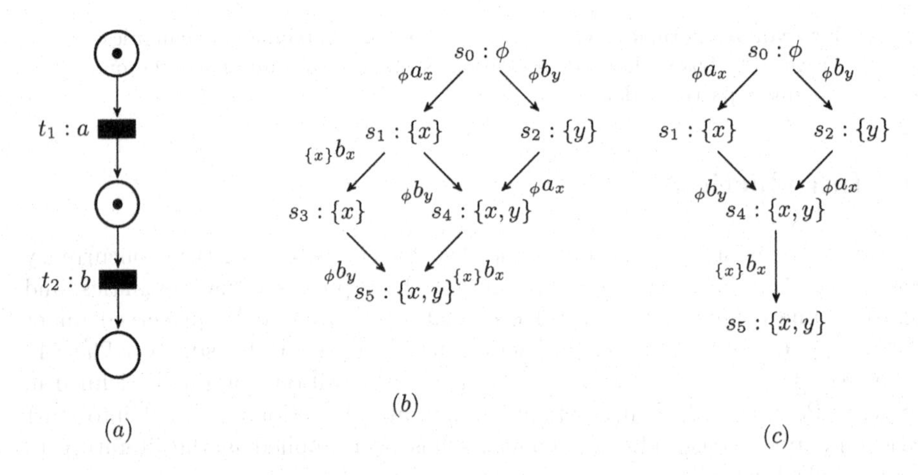

Fig. 1. Operational semantics of a Petri net in terms of MLTS

At first, recall that a maximality-based labeled transition system is given by a graph labeled on both states and transitions. Each state is labeled by a set of event names. Each event name identifies the start of execution of an action (eventually under execution) which occurred before this state. This action is said to be potentially under execution in this state. A transition between two states s_i and s_j is labeled by a 3-uple (M, a, x) (denoted $_{M}a_x$) where x is the event name identifying the start of execution of the action a and M denotes the set of event names representing some causes of the action a. Elements of M belong to state s. Occurrence of this transition terminates actions identified by M, thus, the set of event names corresponding to state s_j is that of s_i from which we subtract the set M and add the event name x. Formal definition of a maximality-based labeled transition system will be given in Sect.2.1.

In the initial state (state s_0) of the maximality-based labeled transition system of Fig.1.(b), no action is running, from where the association of the empty set with this state. From state s_0, actions a and b can start their execution independently, their starts are respectively identified by event names x and y. a and b can be launched in any order. The set $\{x\}$ (resp. $\{y\}$) in state s_1 (resp. s_2) stipulates that the action a (resp. b) are potentially under execution in this state. The set $\{x, y\}$ in s_4 shows that actions a and b can be executed simultaneously.

Note that when the system is in state s_1, while the action a has not been terminated yet, the only evolution concerns the start of b. However, when a terminates, we can start the action b caused by a or the action b which is independent from the end of a. Resulting states are respectively s_3 and s_4. We can observe that from state s_3, the start of b is always possible. However, the same ending constraint of a is imposed for the execution of b at the level of state s_4. Note that causal dependence between execution of b across from the action a is captured by the consumption of the produced token coming from the transition t_1 during the firing of t_2 in the Petri net.

Notice that from state s_1, transitions leading respectively to states s_3 and s_4 are due to the firing of the same transition t_2. In the first firing, the token of the initial marking is used whereas in the second firing, the used token is that produced by the firing of t_1. On the other hand, as noted above, the derivation by b leading to state s_4 is not conditioned by the end of the action a, while the derivation leading to state s_3 is conditioned by the end of a.

By maximality bisimulation relation, we can omit the derivations $s_1 \longrightarrow s_3 \longrightarrow s_5$ in the maximality-based labeled transition system of Fig.1.(b). In other words, the maximality-based labeled transition system of Fig.1.(c) is the reduced system modulo maximality bisimulation relation which preserve action refinement.

The paper is organized as follows. In Sect.2, we give the definition of a maximality-based labeled transition system and maximality bisimulation relation. In Sect.3, we define a partial order relation on a maximality bisimulation relation by witch an algorithm for reducting of maximality-based labeled transition systems modulo maximality bisimulation relation is described as a functional implementation. This paper is ended by some conclusions of this work. Proofs can be found in [2].

2 Preliminaries [4,10]

2.1 Maximality-Based Labeled Transition Systems

Definition 1. *Let $\mathcal{M}$ be a countable set of event names, a maximality-based labeled transition system of support $\mathcal{M}$ is a tuple $(\Omega, \lambda, \mu, \xi, \psi)$ with:*

- $\Omega = (S, T, \alpha, \beta, s_0)$ *is a transition system such that:*
  - S *is the set of states in which the system can be found, this set can be finite or infinite.*

- *T is the set of transitions indicating state switch that the system can achieve, this set can be finite or infinite.*
- *α and β are two applications of T in S such that for all transition t we have: $\alpha(t)$ is the origin of the transition and $\beta(t)$ its goal.*
- *s_0 is the initial state of the transition system Ω.*

– *(Ω, λ) is a transition system labeled by the function λ on an alphabet Act called support of (Ω, λ). In the other word $\lambda : T \to Act$.*
– *$\psi : S \to 2^{\mathcal{M}}$ is a function which associates to each state the finite set of maximal event names present in this state.*
– *$\mu : T \to 2^{\mathcal{M}}$ is a function which associates to each transition the finite set of event names corresponding to actions that have already begun their execution and the end of their executions enables this transition.*
– *$\xi : T \to \mathcal{M}$ is a function which associates to each transition the event name identifying its occurrence.*

such that $\psi(s_0) = \phi$ and for all transition t, $\mu(t) \subseteq \psi(\alpha(t))$, $\xi(t) \notin \psi(\alpha(t)) - \mu(t)$ and $\psi(\beta(t)) = (\psi(\alpha(t)) - \mu(t)) \cup \xi(t)$

Note 1. In what follows, we use the following assumptions:

– In this present paper we suppose the uniqueness of event name.
– Let $mlts = (\Omega, \lambda, \mu, \xi, \psi)$ a maximality-based labeled transition system such that $\Omega = \langle S, T, \alpha, \beta, s_0 \rangle$. $t \in T$ is a transition for which $\alpha(t) = s$, $\beta(t) = s'$, $\lambda(t) = a$, $\mu(t) = E$ and $\xi(t) = x$. The transition t will be noted $s \xrightarrow{E a_x} s'$.
– Let $f : E \to F$ be a function (bijection) such that domain $Dom(f) = E$ and codomain $Cod(f) = F$, and let D (respectively C) be a subset of E (respectively of F). Restrictions of f with respect to its domain and codomain are defined by:
  - $f\lceil D = \{(x, y) \in f | x \in D\}$
  - $f\lfloor C = \{(x, y) \in f | y \in C\}$

– $\mathfrak{F} \subseteq 2^{\mathcal{M} \times \mathcal{M}}$ is the set of all bijective functions between subsets of $\mathcal{M}$.
– Id_A is the identity function on elements of a set A.
– For $s \in S$ and $Y \subseteq S$: $T_{a_x}[s] = \{s' | (s,_M a_x, s') \in T\}$ and $T_{a_x}[Y] = \cup\{T_{a_x}[s] | s \in Y\}$.
– The set of Maximality-based labeled transition systems is noted $\mathfrak{Mlts}$.

2.2 Maximality Bisimulation Relation

Definition 2. *Let $mlts_1 = (\Omega_1, \lambda_1, \mu_1, \xi_1, \psi_1)$ and $mlts_2 = (\Omega_2, \lambda_2, \mu_2, \xi_2, \psi_2)$ be two maximality-based labeled transition systems such that $\Omega_1 = \langle S_1, T_1, \alpha_1, \beta_1, s_0^1 \rangle$ and $\Omega_2 = \langle S_2, T_2, \alpha_2, \beta_2, s_0^2 \rangle$. $mlts_1$ and $mlts_2$ are said to be* maximally bisimilar, *noted $mlts_1 \approx_m mlts_2$, if there is a relation $\mathfrak{R} \subseteq S_1 \times S_2 \times \mathfrak{F}$ with*

1. *$(s_0^1, s_0^2, \emptyset) \in \mathfrak{R}$. Initial states of $mlts_1$ and $mlts_2$ are related by the relation. Since the sets of maximal events in initial states are empty, the function relating these two sets is empty.*

2. *If* $(s_1, s_2, f) \in \mathfrak{R}$ *then*
   (a) $Dom(f) \subseteq \psi(s_1)$ *and* $Cod(f) \subseteq \psi(s_2)$.
   (b) *If* $s_1 \xrightarrow{{}_E a_x} s_1'$ *then there is* $s_2 \xrightarrow{{}_F a_y} s_2'$ *such that*
      i. $\forall (u, v) \in f$, *if* $u \notin E$ *then* $v \notin F$
      ii. $(s_1', s_2', f') \in \mathfrak{R}$ *with* $f' = (f\lceil(\psi(s_1') - \{x\}))\lfloor(\psi(s_2') - \{y\}) \cup \{(x, y)\}$
   (c) *If* $s_2 \xrightarrow{{}_F a_y} s_2'$ *then there is* $s_1 \xrightarrow{{}_E a_x} s_1'$ *such that*
      i. $\forall (u, v) \in f$, *if* $v \notin F$ *then* $u \notin E$
      ii. $(s_1', s_2', f') \in \mathfrak{R}$ *with* $f' = (f\lceil(\psi(s_1') - \{x\}))\lfloor(\psi(s_2') - \{y\}) \cup \{(x, y)\}$.

3 Partial Order on a Maximality Bisimulation Relation

In this section, we assume a given $mlts = (\Omega, \lambda, \mu, \xi, \psi)$ to be a maximality-based labeled transition system such that $\Omega = \langle S, T, \alpha, \beta, s_0 \rangle$ and $\mathfrak{R}$ a maximality bisimulation relation on $mlts$ states. We define two partial order relations. The first relation is over a set of states of mlts. The second relation is over the set of maximality-based labeled transition systems. This last relation will be used for computing the normal form of a maximality-based labeled transition system. We prove that both relations are complete partial orders. These partial order relations will be used to define a recursive function[1] of reduction of maximality-based labeled transition systems modulo maximality bisimulation relation. The reduced maximality-based labeled transition system constitutes its normal form.

3.1 Partial Order Over a Set of States

Definition 3. *Let* $(s, s', f) \in \mathfrak{R}$, $s \leq s'$ *if and only if:*

$$\forall x \in \psi(s) : \exists y \in \psi(s') \text{ and } (x, y) \in f.$$

Proposition 1. *Given* $(s, s', f) \in \mathfrak{R}$:

1. *We have* $s \leq s'$ *or* $s' \leq s$.
2. *The relation* $\leq$ *is a partial order.*
3. $(S, \leq)$ *is a Complete Partial Order (CPO).*

Example 1. In $mlts_1$ of Fig.2, we have $(s_2, s_3, f_1) \in \mathfrak{R}$, $(s_4, s_5, f_2) \in \mathfrak{R}$ and $(s_5, s_6, f_3) \in \mathfrak{R}$ such that $f_1 = \{(z, y)\}$, $f_2 = \{(z, y), (t, u)\}$ and $f_3 = \{(u, v), (x, x)\}$. In other words, we have $s_2 \leq s_3$, $s_4 \leq s_5$ and $s_6 \leq s_5$.

In this CPO, the chains which have the same least upper bound forms a partition on S. This partition is formally defined by Definition.4.

Definition 4. *Let* $D = (S, \leq)$ *be a CPO, we can define a partition* Υ_D *as follow:* $\Upsilon_D = \{B_i \subseteq S|$ *for any* $X \subseteq B_i$, *if* $\exists Y \subseteq S$ *with* $\bigsqcup X = \bigsqcup Y$ *then* $Y \subseteq B_i\}$. *Such that* X *and* Y *be two chains over* D.

Example 2. Given $mlts_1$ of Fig.2, we have $\Upsilon_D = \{\{s_0\}, \{s_1\}, \{s_2, s_3\}, \{s_4, s_5, s_6\}\}$, its stems from the fact that $\sqcup\{s_0\} = s_0$, $\sqcup\{s_1\} = s_1$, $\sqcup\{s_2, s_3\} = s_3$ and $\sqcup\{s_4, s_5\} = \sqcup\{s_5, s_6\} = s_5$.

[1] It is straightforward to propose an imperative algorithm

3.2 Normal Form of a Maximality-Based Labeled Transition System

Definition 5. *Let* $mlts = (\Omega, \lambda, \mu, \xi, \psi)$ *and* $mlts' = (\Omega', \lambda, \mu', \xi', \psi')$ *be two maximality-based labeled transition systems such that* $\Omega = \langle S, T, \alpha, \beta, s_0 \rangle$ *and* $\Omega' = \langle S', T', \alpha', \beta', s'_0 \rangle$, *we define a relation* $\leq$ *on* $\mathfrak{Mlts} \times \mathfrak{Mlts}$ *as follows:* $mlts \leq mlts'$ *if and only if :* $\forall s \in S . \exists s' \in S'$ *such that* $s \leq s'$.

Proposition 2. $(\mathfrak{Mlts}, \leq)$ *is a complete partial order.*

Consider the example in Fig.2, we have in $mlts_1$, $(s_2, s_3, f_1) \in \mathfrak{R}$, $(s_4, s_5, f_2) \in \mathfrak{R}$ and $(s_5, s_6, f_3) \in \mathfrak{R}$ such that $f_1 = \{(z, y)\}$, $f_2 = \{(z, y), (t, u)\}$ and $f_3 = \{(u, v), (x, x)\}$. In the other words, we have $s_2 \leq s_3$, $s_4 \leq s_5$ and $s_6 \leq s_5$. We can deduce easily that $mlts_1 \leq mlts_2 \leq mlts_3 \leq mlts_4$, so $Y = \{mlts_1, mlts_2, mlts_3, mlts_4\}$ is a chain with $\sqcup Y = mlts_4$. From this example, we can remark that $mlts_2$ is the $mlts_1$ after suppression the state s_6, because $s_6 \leq s_5$. From the fact that $s_2 \leq s_3$, we obtain $mlts_3$ from $mlts_2$ by suppression the state s_2. Since $s_4 \leq s_5$, we obtain $mlts_4$ by suppression the state s_4 of $mlts_3$.

In other words, we can obtain $mlts_4$ (seen as a normal form) from $mlts_1$ by the suppression of the states s_6, s_2 and s_4. We signal here, that the suppression of the states is not in the arbitrary order (see Definition.9). Also, the suppression order is not unique (Property.4 of Proposition.4). The reader may remark that the graph obtained from $mlts_1$ by suppression of s_4 is not a maximality-based labeled transition system ($\psi(s_5)$ is not respected).

To have a normal form, we propose a recursive function Γ (Definition.11) which is a continuous function on CPO $(\mathfrak{Mlts}, \leq)$. The function Γ is constructed by the following definitions.

Definition 6. *Let* s_1 *and* s_2 *be two states of mlts with* $s_1 \leq_f s_2$. *The state* s_1 *is an eliminated state if and only:* $\exists s'_1, s'_2 \in S : s'_1 =_{f'} s'_2$ *such that* s'_1 *predecessor of* s_1 *and* s'_2 *predecessor of* s'_2.

Definition 7. *Let* $(s, s', f) \in \mathfrak{R}$, *we define a substitution function as follows:*

- $\sigma_{\phi,mlts} = \iota$ *(identity substitution),*
- $\sigma_{f,mlts} = [z/x][z/y]\sigma_{f-\{(x,y)\},mlts[z/x][z/y]}$ *such that* z *doesn't appear in mlts (new event name).*

Proposition 3. $\mathfrak{R}' = (\mathfrak{R} - \{(s, s', f)\} \cup \{(s, s', f\sigma_{f,mlts})\}$ *is a maximality bisimulation relation on* $mlts\sigma_{f,mlts}$.

Definition 8. *Let* $s_1, s_2 \in S$ *with* $(s_1, s_2, f) \in \mathfrak{R}$, *and let* $mlts' = (mlts)\sigma_{f,mlts} = (\Omega, \lambda, \mu', \xi', \psi')$ *such that* $\mu' = (\mu)\sigma_{f,mlts}$, $\xi' = (\xi)\sigma_{f,mlts}$ *and* $\psi' = (\psi)\sigma_{f,mlts}$. *Let* $\Gamma : S \times S \times \mathfrak{Mlts} \to \mathfrak{Mlts}$ *be a function such that* $Dom(\Gamma) = \{(s, s', mlts)$ *such that* $(s, s', f') \in \mathfrak{R}$ *and verifies Definition.6 for mlts*$\}$. $\Gamma_{s_1,s_2}(mlts) = (\Omega', \lambda, \mu'', \xi'', \psi'')$ *is a maximality-based labeled transitions system in which* s_1 *is removed such that* $\Omega' = \langle S', T', \alpha', \beta', s_0 \rangle$ *with:*

1. $S' = S - \{s_1\}$,
2. $T' = T - \{In(s_1) \cup Out(s_1)\} \cup Set\_Out(s_1, s_2) \cup Set\_In(s_1, s_2)$,

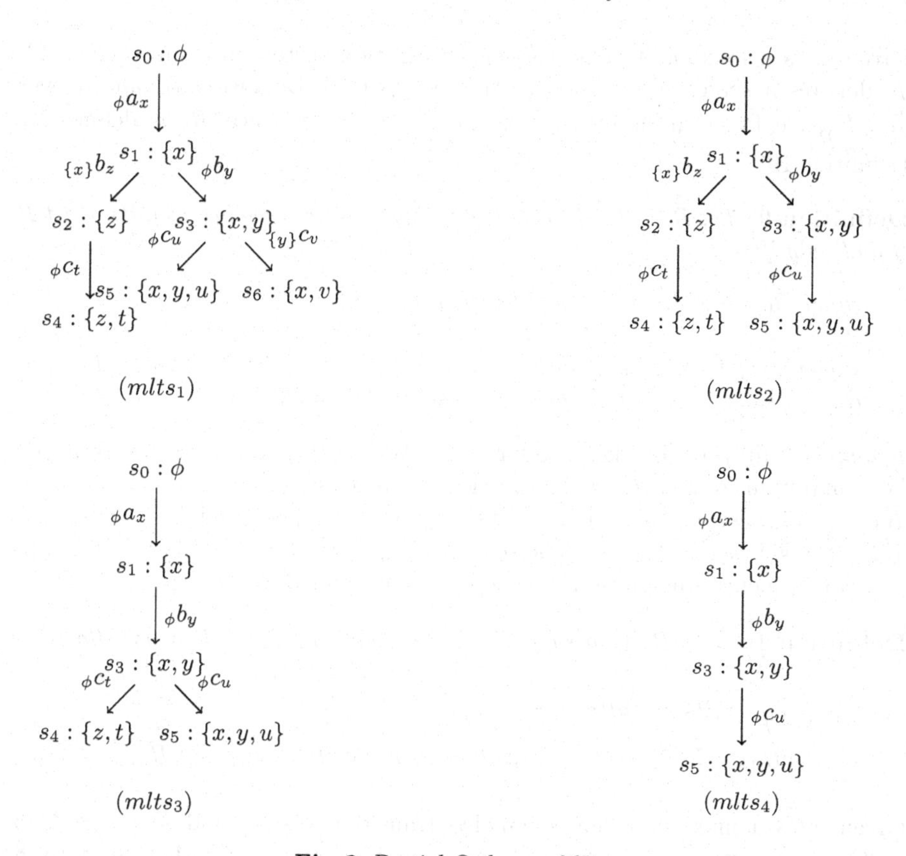

Fig. 2. Partial Order on $\mathfrak{Mlts}$

3. $\mu'' = \mu' - (\{((s_i, s_1), X_i)\} \cup \{((s_1, s_j), X_j)\})$,
4. $\xi'' = \xi' - (\{((s_i, s_1), x_i)\} \cup \{((s_1, s_j), x_j)\})$,
5. $\psi'' = \psi' - (s_1, X)$, $\alpha' = \alpha\lceil T'$ *and* $\beta' = \beta\lceil T'$.

with: $In(s) = \{t|\forall t \in T : \beta(t) = s\}$, $Out(s) = \{t|\forall t \in T : \alpha(t) = s\}$, $set\_In(s_1, s_2) = \{(s_{J},_M a_x, s_2)|\forall(s_j,_N a_x, s_1) \in T \wedge M = \psi(s_j) \setminus \psi(s_2)\}$ *and* $set\_Out(s_1, s_2) = \{(s_2,_M a_x, s_j)|\forall(s_1,_N a_x, s_j) \in T \wedge M = \psi(s_2) \setminus \psi(s_j)\}$.

Example 3. In Fig.2, we can affirm that: $\Gamma_{s_6,s_5}(mlts_1) = mlts_2$, $\Gamma_{s_2,s_3}(mlts_2) = mlts_3$ and $\Gamma_{s_4,s_5}(mlts_3) = mlts_4$.

Proposition 4. *Let mlts be a maximality-based labeled transition system:*

1. *For any* s_1 *and* s_2 *two states of mlts, if* $s_1 \leq_f s_2$ *then* $mlts \leq \Gamma_{s_1,s_2}(mlts)$,
2. Γ *is monotone : for any* s_1 *and* s_2 *two states of mlts, if* $s_1 \leq_f s_2$ *and* $mlts \leq mlts'$ *then* $\Gamma_{s_1,s_2}(mlts) \leq \Gamma_{s_1,s_2}(mlts')$,
3. Γ *is continuous :* $\sqcup\{\Gamma_{s_1,s_2}(mlts)|mlts \in Y\} = \Gamma_{s_1,s_2}(\sqcup Y)$,
4. *For any states* s_1, s_2, s_3 *and* s_4 *of mlts, if* $\Gamma_{s_3,s_4}(mlts)$ *and* $\Gamma_{s_1,s_2}(mlts)$ *are both defined then* $\Gamma_{s_1,s_2} \circ \Gamma_{s_3,s_4}(mlts) \cong \Gamma_{s_3,s_4} \circ \Gamma_{s_1,s_2}(mlts)$.

Given *mlts* a maximality-based labeled transition system and subset $R \subseteq \mathfrak{R}$. $\overrightarrow{R}$ denotes a set of suppression sequences w.r.t R (Definition.9) ranged over $\overrightarrow{R}_i$. $\Gamma_{\overrightarrow{R}_i}$: a function suppression a states w.r.t a sequence $\overrightarrow{R}_i$ is defined by Definition.10.

Definition 9. *Let $R \subseteq \mathfrak{R}$, $\overrightarrow{R}_i$ is called a suppression sequence of mlts w.r.t R if and only if:*

- *either $\overrightarrow{R_0} = \epsilon$ if $R_0 = \phi$ (empty sequence),*
- *$\overrightarrow{R_n} = (s_n, s'_n).\overrightarrow{R_{n-1}}$ such that $(s_n, s'_n, f_n) \in R$ and $\overrightarrow{R_{n-1}}$ is a suppression sequence of $\Gamma_{s_n,s'_n}(mlts)$ under a maximality bisimulation relation $R_{n-1} = R_n - \{(s_n, s'_n, f_n)\}$ (with the hypothesis that $R = R_n$).*

Example 4. In $mlts_1$ of Fig.2, we have $R = \{(s_2, s_3, f_1); (s_4, s_5, f_2); (s_6, s_5, f_3)\}$. We can obtain from R, three suppressions sequences:
$\overrightarrow{R}_1 = (s_2, s_3).(s_4, s_5).(s_6, s_5)$, $\overrightarrow{R}_2 = (s_2, s_3).(s_6, s_5).(s_4, s_5)$ and
$\overrightarrow{R}_3 = (s_6, s_5).(s_2, s_3).(s_4, s_5)$. The sequence $(s_4, s_5).(s_2, s_3).(s_6, s_5)$ is not a suppression sequence, therefore, Definition.6 is not verified for (s_4, s_5).

Definition 10. *Let $\overrightarrow{R}_i$ be a suppression sequence, we define $\Gamma_{\overrightarrow{R_i}}$ as follows:*

- *either $\Gamma_{\overrightarrow{R_i}}(mlts) = mlts$ if $\overrightarrow{R}_i = \epsilon$,*
- *$\Gamma_{\overrightarrow{R_n}}(mlts) = \Gamma_{\overrightarrow{R_{n-1}}} \circ \Gamma_{(s_n,s'_n)}(mlts)$ such that $\overrightarrow{R_n} = (s_n, s'_n).\overrightarrow{R_{n-1}}$.*

Given *mlts* a maximality-based labeled transition system and $D = (S, \leq)$ a CPO over $\mathfrak{R}$, we can have from Υ_D a suppression sequence with respect to $\mathfrak{R}$ (Proposition.5). We define an operator Γ which eliminates a set of states w.r.t $\mathfrak{R}$.

Proposition 5. *Let $mlts = (\Omega, \lambda, \mu, \xi, \psi)$ be a maximality-based labeled transition system such that $\Omega = \langle S, T, \alpha, \beta, s_0\rangle$, and $D = (S, \leq_f)$ a CPO. Given $\Upsilon_D = \{Y_i | i = 0..n\}$ with $Y_0 = \{s_0\}$ and $\exists a \in Act : (T_{a_{f(x)}}[Y_i] = Y_j) \wedge i < j$.*

1. *We can obtain, over $\Gamma_{\overrightarrow{Y_i}}$, the suppression sequence $\overrightarrow{Y_{i+1}} = (s_1, \sqcup Y_{i+1})...(s_j, \sqcup Y_{i+1})...(s_n, \sqcup Y_{i+1})$ with $s_j \in Y_{i+1}$ and $s_j \neq \sqcup Y_{i+1}$,*
2. *The composition in $\overrightarrow{Y_{i+1}}$ is commutative.*

Example 5. Given $\Upsilon_D = \{\{s_0\}, \{s_1\}, \{s_2, s_3\}, \{s_4, s_5, s_6\}\}$ of $mlts_1$ of Fig.2, we have $\Gamma_{\Upsilon_D}(mlts_1) = \Gamma_{(s_4,s_5).(s_6,s_5)} \circ \Gamma_{(s_2,s_3)}(mlts_1)$

Definition 11. *Let $mlts = (\Omega, \lambda, \mu, \xi, \psi)$ be a maximality-based labeled transition system such that $\Omega = \langle S, T, \alpha, \beta, s_0\rangle$, and $D = (S, \leq)$ a CPO. Given $\Upsilon_D = \{Y_i | i = 0..n\}$ with $Y_0 = \{s_0\}$ and $\exists a \in Act : (T_{a_{f(x)}}[Y_i] = Y_j) \wedge i < j$. We can define the function Γ as follows: $\Gamma(mlts) = \Gamma_{\overrightarrow{Y_n}} \circ .. \circ \Gamma_{\overrightarrow{Y_1}}(mlts)$.*

Proposition 6. *Let mlts be a maximality-based labeled transition system: $mlts \leq \Gamma(mlts)$ and $\Gamma(mlts)$ is unique.*

Definition 12. *Given mlts* $= (\Omega, \lambda, \mu, \xi, \psi)$ *a maximality-based labeled transition system such that* $\Omega = \langle S, T, \alpha, \beta, s_0 \rangle$*, mlts is in normal form if and only if: for any maximality bisimulation relation* $\mathfrak{R}$ *and for any* $s, s' \in S$*:* $((s, s', f) \in \mathfrak{R} \Rightarrow s = s')$.

Theorem 1. *Given mlts a maximality-based labeled transition system,* $\Gamma(mlts)$ *is the normal form of mlts.*

4 Conclusions

This paper proposes a functional method for reducing maximality-based labeled transition systems. The choice of this model is motivated by its independence from any concurrency specification model. For this purpose, we define a complete partial order on both the set of maximality-based labeled transition systems and the set of maximality-based labeled transition system states. These relations allow us to define an algorithm reducing a maximality-based labeled transition system according to maximality bisimulation relation. As a perspective of this work, it remain the definition of an algorithm computing a maximality bisimulation relation for any maximality-based labeled transition system and by the way analysing realistic systems specified in Petri nets and processes algebras using results of [11,12].

References

1. Aceto, L., Hennessy, M.: Adding action refinement to a finite process algebra. In: Leach Albert, J., Monien, B., Rodríguez-Artalejo, M. (eds.) ICALP 1991. LNCS, vol. 510, pp. 506–519. Springer, Heidelberg (1991)
2. Benamira, A., Saidouni, D.E.: Maximality-based labeled transition systems normal form (extended version). Research report, MISC Laboratory, Constantine 2 University, 25000 Constantine Algeria (2013)
3. Bravetti, M., Gorrieri, R.: Deciding and axiomatizing weak st bisimulation for a process algebra with recursion and action refinement. ACM Trans. Comput. Log. 3(4), 465–520 (2002)
4. Courtiat, J.P., Saïdouni, D.E.: Relating maximality-based semantics to action refinement in process algebras. In: Hogrefe, D., Leue, S. (eds.) IFIP TC6/WG6.1, 7th Int. Conf. on Formal Description Techniques (FORTE 1994), pp. 293–308. Chapman & Hall (1995)
5. Devillers, R.R.: Maximality preserving bisimulation. Theor. Comput. Sci. 102(1), 165–183 (1992)
6. van Glabbeek, R.: Comparative Concurrency Semantics and Refinement of Actions. Ph.D. thesis, Free University, Amsterdam (1990), Second edition available as CWI tract 109, CWI, Amsterdam (1996), `http://theory.stanford.edu/~rvg/thesis.html`
7. van Glabbeek, R.: The refinement theorem for ST-bisimulation semantics. In: Proceedings IFIP TC2 Working Conference on Programming Concepts and Methods, Sea of Gallilee, Israel, pp. 27–52. North-Holland (1990)

8. van Glabbeek, R.: Petri net models for algebraic theories of concurrency (extended abstract). In: de Bakker, J.W., Nijman, A.J., Treleaven, P.C. (eds.) PARLE 1987. LNCS, vol. 259, pp. 224–242. Springer, Heidelberg (1987)
9. Hennessy, M.: Concurrent testing of processes (extended abstract). In: Cleaveland, W.R. (ed.) CONCUR 1992. LNCS, vol. 630, pp. 94–107. Springer, Heidelberg (1992)
10. Saïdouni, D.E.: Sémantique de Maximalité: Application au Raffinement d'Actions en LOTOS. Ph.D. thesis, LAAS-CNRS, 7 av. du Colonel Roche, 31077 Toulouse Cedex France (1996)
11. Saïdouni, D.E., Belala, N., Bouneb, M.: Aggregation of transitions in marking graph generation based on maximality semantics for Petri nets. In: VECoS 2008, July 2-3. eWiC Series, The British Computer Society (BCS). University of Leeds, UK (2008) ISSN: 1477-9358
12. Saïdouni, D.E., Belala, N., Bouneb, M.: Maximality-based structural operational semantics for Petri nets. In: Proceedings of Intelligent Systems and Automation(CISA 2009), Tunisia, vol. 1107, pp. 269–274. American Institute of Physics (2009) ISBN: 978-0-7354-0642-1
13. Vogler, W.: Bisimulation and action refinement. In: Jantzen, M., Choffrut, C. (eds.) STACS 1991. LNCS, vol. 480, pp. 309–321. Springer, Heidelberg (1991)

Automatic Generation of SPL Structurally Valid Products Using Graph Transformations Approach

Khaled Khalfaoui[1], Allaoua Chaoui[2], Cherif Foudil[3], and Elhillali Kerkouche[1]

[1] Department of Computer Science, University of Jijel, Jijel, Algeria
[2] Department of Computer Science and its Application, University Constantine 2, Algeria
[3] Department of Computer Science, University of Biskra, Biskra, Algeria
{khalfaoui_kh,foud_cherif,elhillalik}@yahoo.fr,
a_chaoui2001@yahoo.com

Abstract. A Software Product Line is a set of software products that share a number of core properties but also differ in others. Differences and commonalities between products are typically described in terms of features. A software product line is usually modeled with a feature diagram, describing the set of features and specifying the constraints and relationships between these features. Each product is defined as a set of features. In this area of research, a key challenge is to ensure correctness and safety of these products. There is an increasing need for automatic tools that can support feature diagram analysis, particularly with a large number of features that modern software systems may have. In this paper, we propose using model transformations an automatic framework to generate all valid products according to the feature diagram. We first introduce the basic idea behind our approach. Then, we present the used graph grammar to perform automatically this task using the AToM^3 tool. Finally, we show the feasibility of our proposal by means of running example.

Keywords: Software Product Lines, Feature Diagram, Variability Modelling, Valid Products, Graph Transformations.

1 Introduction

Software product line engineering is an approach for developing families of software systems. The main advantage over traditional approaches is that all products can be developed and maintained together. Feature models [1] are widely used in domain engineering to capture common and variant features among the different variants. One of the major problems is the generation of valid products. In research area, we are interested by model transformations approach [2].

Model transformations are a very useful in the validation, manipulation and processing of diagrams. They are performed by executing graph grammars [3]. A graph grammar is composed of rules. Each one has their left and right hand sides (LHS and RHS). Rules are compared with an input graph called host graph. If a matching is found between the LHS of a rule and a sub-graph in the host graph, then

A. Amine et al. (Eds.): *Modeling Approaches and Algorithms*, SCI 488, pp. 347–356.
DOI: 10.1007/978-3-319-00560-7_38 © Springer International Publishing Switzerland 2013

the rule can be applied and the matching subgraph of the host graph is replaced by the RHS of the rule. Each rule has also application conditions that must be satisfied, as well as actions to be performed when the rule is executed. A graph rewriting system iteratively applies rules of grammar in the host graph, until no rules are applicable.

In this work, we propose an automatic framework based on this technique to produce valid products of an SPL. This paper is organized as follows: In section 2, we discuss some related works. Section 3 provides the background of our approach. We recall some basic notions about FD diagrams and we give an overview of graph transformations technique. In Section 4, we present our proposal. We first introduce the idea behind the generation of valid products. Then, we present the used graph grammar to perform automatically this task. We illustrate our framework through some examples in section 5. Finally, section 6 concludes the paper and gives some perspectives of this work.

2 Related work

Research in the field of SPL is becoming increasingly important, particularly through its ability to increase software reuse. But, the success of this approach is conditioned by the correctness of final products. Over the past few years, a great variety of validation techniques had been proposed.

Mannion in [4] proposed the adoption of propositional formula for formally representing software product lines. Sun et al. in [5] proposed a formalization of FMs using Z and the use of Alloy Analyzer for the automated support of the analyses of FMs. Wang et al. in [6] have proposed an approach to modeling and verifying feature diagrams using Semantic Web OWL ontologies. The specialists in the field asserts that three of the most promising proposals for the automated analysis of feature models are based on the mapping of feature models into Constraint Satisfaction Problem (CSP) solvers [7], propositional SAT is fiability problem(SAT) solvers [8] and Binary Decision Diagrams (BDD) [9]. For more details see [10], where Benavides presented a comprehensive literature review on the most important techniques and tools.

Nowadays, graph transformations are widely used for modelling and analysis of complex systems in the area of software engineering [11]. In [12], the authors have proposed a tool that formally transforms dynamic behaviours of systems expressed using UML Statechart and collaboration diagrams into their equivalent colored Petri nets models. In [13], it has proposed an approach to extract and integrate the parallel changes made to Object-Oriented formal specifications in a collaborative development environment. Zambon et al. in [14] have proposed an approach for the verification of software written in imperative programming languages. They used an explicit representation of program states as graphs, and they specified the computational engine as graph transformations.

In previous work [15], we have proposed an automatic approach for behavioral analysis of SPL products. The current paper concerns the structural aspect.

3 Background

Feature Modelling: Research on feature modeling has received much attention in software product line engineering community. Feature-Oriented Domain Analysis (FODA) [1] is the most the most popular. The success of this approach resides in the introduction of feature models, which contain a graphical tree-like notation that shows the hierarchical organization of features. In the tree, nodes represent features; edges describe feature relations. Fig.1 depicts a feature model of a mobile phone SPL.

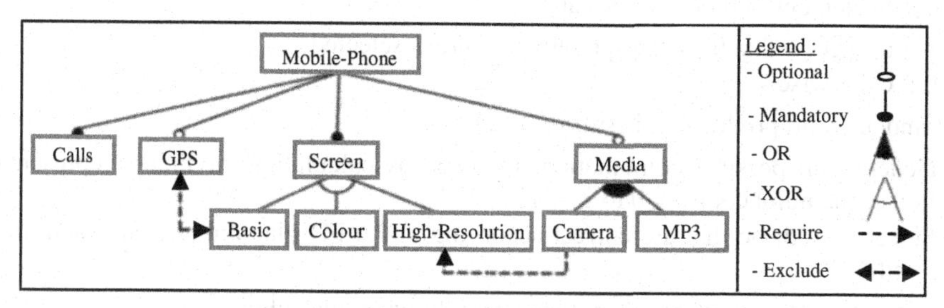

Fig. 1. Feature diagram of a mobile phone product line

Relationships between a parent feature and its child features (or subfeatures) are:

- Mandatory: If the father is selected, the child must be selected.
- Optional: If the father is selected, the child may be selected but not necessarily.
- XOR: If the father is selected, exactly one feature of the children must be selected.
- OR: If the father is selected, at last one of the OR-children must be selected.

In addition, cross-tree constraints are allowed. The most common are:

- Require: The selection of source feature implies the selection of the destination.
- Exclude: both features cannot be part of the same product.

Model Transformations: recently, model transformation techniques have become more focused. They are successfully applied in several domains. The translation is performed by executing graph grammars which are a generalization of Chomsky grammars for graphs [2]. A graph transformation rule is a special pair of pattern graphs called left hand side (LHS) and right hand side (RHS). In the rewriting process, rules are evaluated against an input graph, called the host graph. If a matching is found between the LHS of a rule and a subgraph of the host graph, then the rule can be applied. When a rule is applied, the matching subgraph of the host graph is replaced by the RHS of the rule. Rules can have applicability conditions, as well as actions to be performed when the rule is applied. Generally, rules are ordered according to a priority assigned by the user and are checked from the higher priority to the lower priority. After a rule matching and subsequent application, the graph rewriting system starts again the search. The graph grammar execution ends when no more matching rules are found. $AToM^3$ [16] is a visual tool for meta-modeling and model transformations. Its meta-layer allows a high-level description of models using the Entity-Relationship (ER) formalism.

4 A Graph Transformation Approach for Generating Structural Valid Configurations from an FD Diagram

Computation principle: Consider the general case, an FD with n features. Each product is defined by the selected features. So, it can be represented by a binary array:

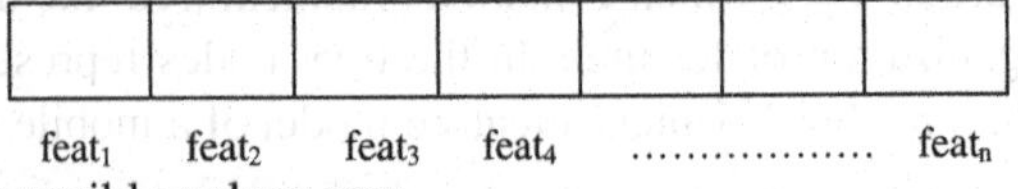

For each element, possible values are:

- 1 : to specify that the corresponding feature is selected.
- 0 : Otherwise.

To find all valid products, it is sufficient to:

- Generate all possible combinations obtained by calculating the binary representation of the numbers from 0 to $2^n - 1$.
- Select only those that are correct according to the FD diagram and edit them in a text file.

To automate this treatment, we propose the following algorithm:

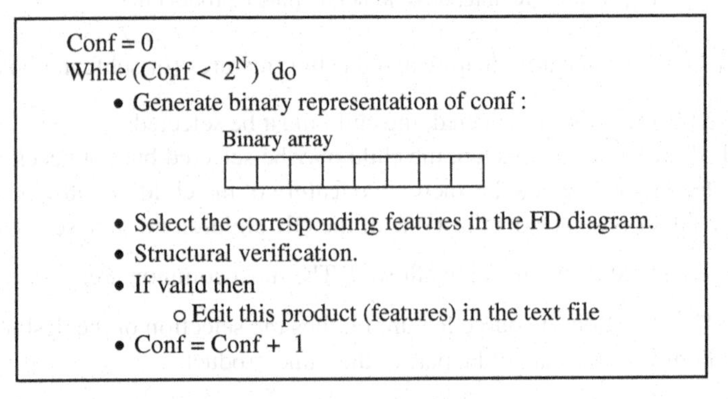

Fig. 2. Computation principal

To perform automatically all these computations, we have opted for the use of graph transformations technique. To facilitate the processing, we prefer to deal with a richer diagram called D-Tree model. It is the original FD diagram enriched with some interesting information. In the following, we first present the used meta-models. Then, we introduce the proposed graph grammar.

1. Meta-Modelling:

FD meta-model: It is composed of:

- Feature Entity: Each feature has two attributes:
  - Its identifier Name.
  - A boolean attribute called isRoot used to identify the root feature.
- OR, XOR, Mandatory, Optional, Include and Exclude Relationships: all of these relationships don't have attributes, but differ in their graphical appearance.

D-Tree meta-model: It consists on:

- Node Entity: It has three attributes:
  - o Name: to identify the node.
  - o The same, a boolean called isRoot.
  - o Index: to specify its index in the binary array.
- GenericLink Relationship: This association represents links between nodes in the D-Tree model. Graphically, they have the same appearance. To differentiate them, we add an attribute called RelationType.

2. The proposed Graph Grammar: we proceed in two steps:

Step1: The purpose of this step is to translate the FD diagram into a D-Tree model and calculate the number of possible configurations. Treatment begins with the creation of the D-Tree nodes. Each time the rewriting system locates an FD feature and associates it to a new D-Tree node. The attributes Name and isRoot are copied with the same values. In the same time, its index is adjusted. Then, we move on the creation of the D-Tree links. At this level, each FD relationship is transformed into a D-Tree GenericLink. The attribute RelationType is set according to the FD relationship type. Finally, we remove the FD features. To do this, we use:

- A temporary attribute for FD features called Translated: to indicate whether this feature has been previously treated or not.
- A global variable called Cpt-Features: it is used for numbering the D-Tree nodes.
- A global variable called Number-Confs (initialized to 1) to count the number of possible configurations. It will be used in the next step.

To perform this treatment, we propose eight rules (Fig.3).

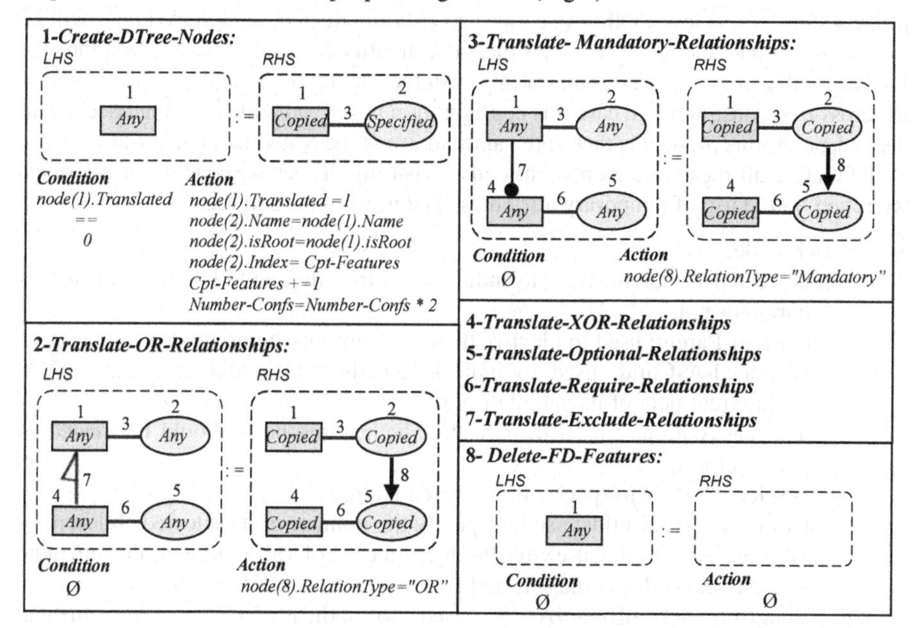

Fig. 3. FD to D-Tree rules

Step2: Now we pass looking for correct configurations using the obtained D-Tree model. Rules that allow realizing this step are grouped in two parts. Rules of the first part ($Part_1$) are used to generate all possible combinations. For each configuration, those of the second part ($Part_2$) allow checking its correctness. In case of validity, it will be edited in the text file.

$Part_1$: As presented in Fig.2, this treatment is performed by calculating the binary representation of the numbers (Conf) comprised between 1 and Number-Confs. This is an iterative processing where each time we have to:

- Reinitialize all temporary attributes: nodes and links.
- Increment the value of Conf by one and generate its binary representation. The result is copied into an array in order to facilitate its exploitation.
- Mapping this configuration in the D-Tree model to check its validity. We go through the binary array and for each element:
  - o If its value is 1: the node numbered with the same index is considered as selected. So, its attribute isSelected is set True.
  - o Otherwise: It set False.
- Performing the test stop:
  - o If Conf < Number-Confs: The verification is restarted. In this case, the first rule of the second part (Part2) will be activated.
  - o Otherwise: This process stops. No rule is applicable.

Part $_2$: Here, the purpose is to analyze only the configuration in question by checking all dependences defined in the FD diagram. First, we deal only parental relationships. For each node, if selected, we have to check the validity by examining its children. In case of anomaly, the current configuration is considered as invalid and the verification process stops at this level. Otherwise, once all children treated, the selected ones will be treated as parents and we start again the same treatment. So, we have to explore the D-Tree model from top to bottom, starting with the root up to the leaves. Then, we treat the cross-tree constraints. In case this configuration is still valid after having checked all dependencies, this product is therefore valid. So, it will be edited in the text file.

To realize all these treatments while also ensuring proper sequencing of the rules, we relied on the use of temporary attributes. The most important are:

- For each node:
  - o isTreated-AsParent: used to indicate whether this node has been treated as parent or not.
  - o Current-Parent: used to identify the node currently being treated as parent.
  - o isTreated-AsChild: used to indicate whether this child has been visited during treatment of its parent or not.
  - o ToTreat-AsParent: used to indicate whether this node should be treated as parent or not.
  - o isSelected: used to specify the selected features.
  - o Count-SelectedChilds: used to specify the number of its selected children.
  - o isEditedTextFile-InCaseValidProduct: in case of valid product, this attribute is used when editing the selected features in the text file.
  - o StateCopiedFromBinaryArray: used to indicate whether its attribute isSelected has been updated using the binary array or not.

- For each GenericLink association :
  - isChecked: It is used to indicate whether a Require or Exclude constraint has already been verified or not.

In addition, we use the following global variables:

  - Conf: Integer initialized to 0.
  - Current-Product: The binary array.
  - Continue: Boolean used to restart (or Stop) the verification of the new generated configurations.
  - is-StillValid: is used to indicate whether the current configuration is still valid or not.

To carry out this process, we propose the following rules:

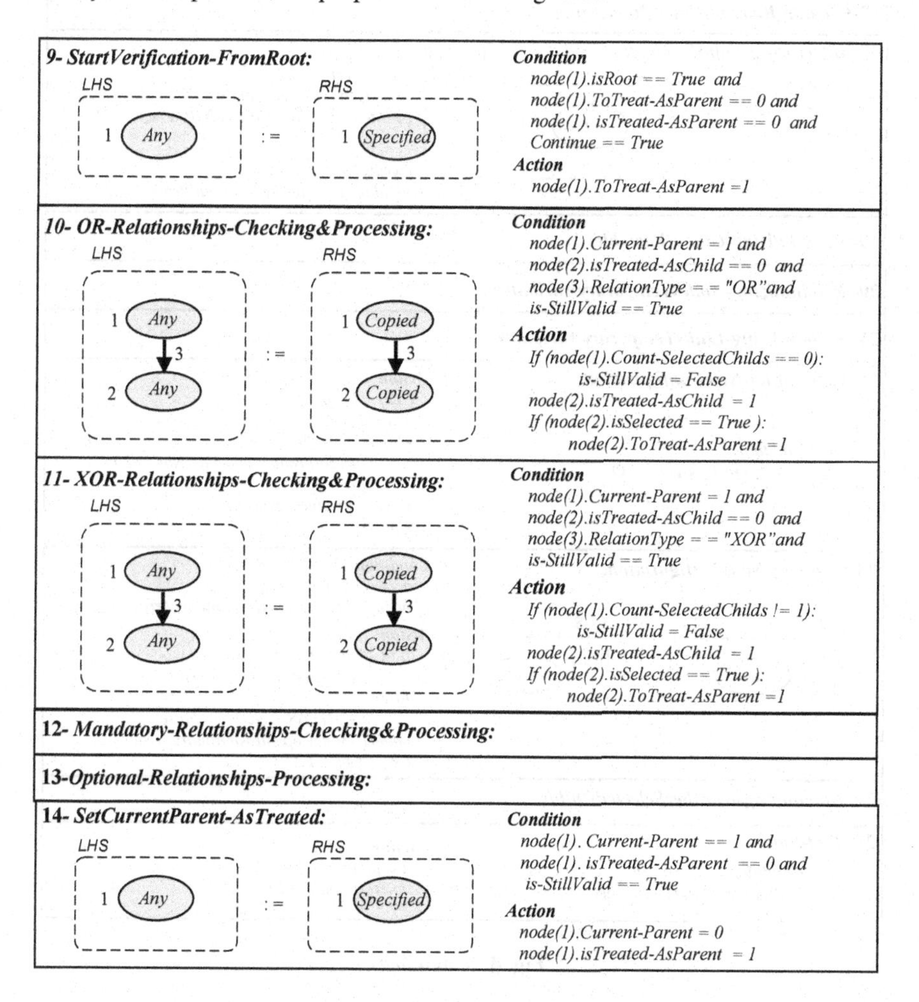

Fig. 4. Generating valid products rules

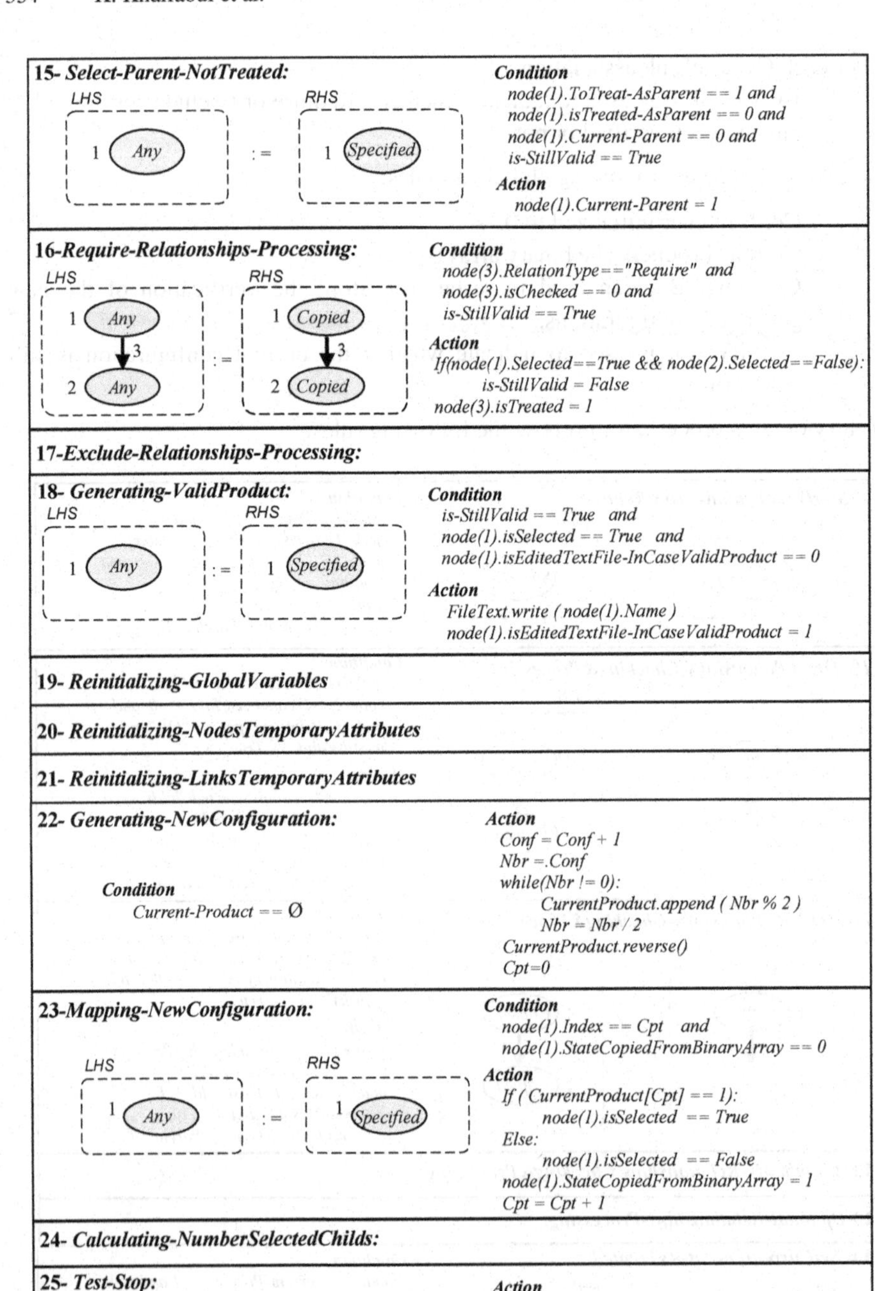
15- Select-Parent-NotTreated:
LHS
RHS
1 Any
:=
1 Specified
Condition
node(1).ToTreat-AsParent == 1 and
node(1).isTreated-AsParent == 0 and
node(1).Current-Parent == 0 and
is-StillValid == True
Action
node(1).Current-Parent = 1
16-Require-Relationships-Processing:
LHS
RHS
1 Any
3
2 Any
:=
1 Copied
3
2 Copied
Condition
node(3).RelationType=="Require" and
node(3).isChecked == 0 and
is-StillValid == True
Action
If(node(1).Selected==True && node(2).Selected==False):
is-StillValid = False
node(3).isTreated = 1
17-Exclude-Relationships-Processing:
18- Generating-ValidProduct:
LHS
RHS
1 Any
:=
1 Specified
Condition
is-StillValid == True and
node(1).isSelected == True and
node(1).isEditedTextFile-InCaseValidProduct == 0
Action
FileText.write (node(1).Name)
node(1).isEditedTextFile-InCaseValidProduct = 1
19- Reinitializing-GlobalVariables
20- Reinitializing-NodesTemporaryAttributes
21- Reinitializing-LinksTemporaryAttributes
22- Generating-NewConfiguration:
Condition
Current-Product == Ø
Action
Conf = Conf + 1
Nbr =.Conf
while(Nbr != 0):
CurrentProduct.append (Nbr % 2)
Nbr = Nbr / 2
CurrentProduct.reverse()
Cpt=0
23-Mapping-NewConfiguration:
LHS
RHS
1 Any
:=
1 Specified
Condition
node(1).Index == Cpt and
node(1).StateCopiedFromBinaryArray == 0
Action
If (CurrentProduct[Cpt] == 1):
node(1).isSelected == True
Else:
node(1).isSelected == False
node(1).StateCopiedFromBinaryArray = 1
Cpt = Cpt + 1
24- Calculating-NumberSelectedChilds:
25- Test-Stop:
Condition
Conf <= (Number-Confs – 1)
Action
Continue == True
is-StillValid == True

Fig. 4. (*continued*)

5 Illustrative Example

To illustrate our framework, let us consider the Mobile Phone example presented previously in Section 3. As input, we have to create the FD model using the visual environment generated by AToM^3 as presented in Fig.5.

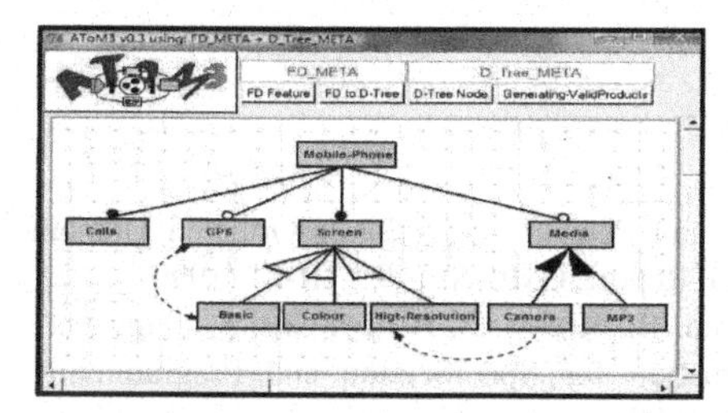

Fig. 5. The input FD diagram

Fig. 6. The generated D-Tree model

- By executing rules of the first step, we obtain the D-Tree model (Fig.6).
- To generate automatically all valid products, we have to execute rules of the second step on the D-Tree model. The generated text file is presented in Fig.7.

- Mobile-Phone, Calls, Screen, Basic
- Mobile-Phone, Calls, Screen, Basic, Media, MP3
- Mobile-Phone, Calls, Screen, Colour
- Mobile-Phone, Calls, Screen, Colour, Media, MP3
- Mobile-Phone, Calls, Screen, Colour, GPS
- Mobile-Phone, Calls, Screen, Colour, GPS, Media, MP3
- Mobile-Phone, Calls, Screen, Higt-Resolution
- Mobile-Phone, Calls, Screen, Higt-Resolution, Media, MP3
- Mobile-Phone, Calls, Screen, Higt-Resolution, Media, Camera
- Mobile-Phone, Calls, Screen, Higt-Resolution, Media, Camera, MP3
- Mobile-Phone, Calls, Screen, Higt-Resolution, GPS
- Mobile-Phone, Calls, Screen, Higt-Resolution, GPS, Media, MP3
- Mobile-Phone, Calls, Screen, Higt-Resolution, GPS, Media, Camera
- Mobile-Phone, Calls, Screen, Higt-Resolution, GPS, Media, Camera, MP3

Fig. 7. The generated text file

6 Conclusion

Software product line engineering is about producing a set of related products that share more commonalities than variabilities. Feature models are widely used in domain engineering to capture common and variant features among the different variants. Each legal product is defined as a combination of features that respects all dependencies defined in the FD model. With a large number of features, the generation of all valid products is extremely difficult.

To resolve this problem, we have proposed a novel approach based on graph transformations. To facilitate the processing, we proceed with two steps. The first one translates the FD diagram into a decorated tree noted D-Tree. The purpose is to obtain a richer model equipped with some interesting information. The second step generates all valid products by dealing this D-Tree model. The obtained result is edited in a text file. We have proposed an iterative processing where rules are grouped in two parts:

- Rules of the first part are used to generate all possible configurations.
- For each configuration, those of the second part allow checking its correctness. In case of validity, it will be edited in the text file.

In our future work, we plan to develop an integrated environment based on the current tool that supports the analysis of the extended feature diagrams. This requires the treatment of more complex relationships.

References

1. Kang, K., Cohen, S., Hess, J., Novak, W. and Peterson, S.: Feature-Oriented Domain Analysis (FODA) Feasibility Study. Technical Report, CMU/SEI- 90-TR-21 (1990)
2. Andries, M., Engels, G., Habel, A., Hoffmann, B., Kreowski, H.J., Kuske, S., Pump, D., Schürr, A., Taentzer, G.: Graph transformation for Specification and Programming. Science of Computer Programming 34, 1–54 (1999)
3. Rozenberg, G.: Handbook of Graph Grammars and Computing by Graph Transformation, vol. 1. World Scientific, Singapore (1999)
4. Mannion, M.: Using First-Order Logic for Product Line Model Validation. In: Chastek, G.J. (ed.) SPLC 2, pp. 176–187. Springer, London (2002)
5. Sun, J., Zhang, H., Li, Y., Wang, H.: Formal Semantics and Verification for Feature Modeling. In: Proceedings of the ICECS 2005, pp. 303–312 (2005)
6. Wang, H., Li, Y., Sun, J., Zhang, H., Pan, J.: Verifying Feature Models Using OWL. In: Web Semantics: Science Services and Agents on the World Wide Web, vol. 5, pp. 117–129 (2007)
7. Benavides, D., Trinidad, P., Ruiz-Cortés, A.: Automated Reasoning on Feature Models. In: Pastor, Ó., Falcão e Cunha, J. (eds.) CAiSE 2005. LNCS, vol. 3520, pp. 491–503. Springer, Heidelberg (2005)
8. Bagheri, E., Noia, T.D., Gasevic, D., Ragone, A.: Formalizing Interactive Staged Feature Model Configuration. Journal of Software 24, 375–400 (2012)
9. Mendonca, M., Wasowski, A., Czarnecki, K., Cowan, D.: Efficient Compilation Techniques for Large Scale Feature Models. In: Proceedings of GPCE 2008, pp. 13–22. ACM Press, USA (2008)
10. Benavides, D., Segura, S., Ruiz-Cortés, A.: Automated Analysis of Feature Models 20 Years Later: A Literature Review. Journal of Information Systems 35, 615–636 (2010)
11. Ehrig, H., Engels, G., Kreowski, H.-J., Rozenberg, G. (eds.): ICGT 2012. LNCS, vol. 7562. Springer, Heidelberg (2012)
12. Kerkouche, E., Chaoui, A., Bourennane, E.B., Labbani, O.: On the Use of Graph Transformation in the Modeling and Verification of Dynamic Behavior in UML Models. JSW 5, 1279–1291 (2010)
13. Taibi, F.: Automatic Extraction and Integration of Changes in Shared Software Specifications. IJSEIA 6, 29–45 (2012)
14. Zambon, E., Rensink, A.: using Graph Transformations and Graph Abstractions for Software Verification. ICGT-DS, EC of the EASST 38 (2011)
15. Khalfaoui, K., Chaoui, A., Foudil, C., Kerkouche, E.: Formal Specification of Software Product Lines: A Graph Transformation Based Approach. JSW 7, 2518–2532 (2012)
16. de Lara, J., Vangheluwe, H.: AToM: A Tool for Multi-formalism and Meta-modelling. In: Kutsche, R.-D., Weber, H. (eds.) FASE 2002. LNCS, vol. 2306, pp. 174–188. Springer, Heidelberg (2002)

Modeling On-the-Spot Learning: Storage, Landmarks Weighting Heuristic and Annotation Algorithm

Shivendra Tiwari and Saroj Kaushik

Dept. of Computer Science and Engg., IIT Delhi, New Delhi, India 110016
{shivendra,saroj}@cse.iitd.ac.in

Abstract. Huge information is intrinsically associated with certain places in the globe such as historical, geographical, cultural and architectural specialties. The next generation systems require access of the site specific information where the user is roaming at the moment. The on-the-spot learning (OTSL) is a system that allows the users to learn about the location, landmarks, regions where he/she is walking through. In this paper, we have proposed an OTSL model that includes the storage, retrieval and the landmark weighting heuristic. Apart from learning about the individual landmarks, we have proposed two ways of storing the spatial learning objects. *First*, use the administrative hierarchy of the region to fetch the information. This can be easily done by the reverse-geocoding operation without actually storing the physical hierarchy. *Second*, spatial chunking, creates the region based on the groups of landmarks in order to define a learning region. A *hybrid* solution has also been considered to achieve the advantages of both the region based methods. We use a weighting model to select the correct landmarks in the basic model. We extend the core model to include other factors such as speed, direction, side of the road etc. A prototype has been implemented to show the feasibility of the proposed model.

Keywords: On-the-Spot Learning (OTSL), Location Based Tour Guide, Landmark Based Learning, Region of Interest (ROI), Point of Interest (POI).

1 Introduction

There are a lot of cities which are full of history, it is easy to verify that learning about the particulars and seeing a picture of a historic site is different to the real world experience by physically visiting them. This reflection leads us to the idea to extend the capabilities of conventional navigation systems to the location based learning systems, also called on-the-spot learning (OTSL) systems. Walking through the historic sites allows the tourists to gain a much deeper insight and gives a different experience than looking at a picture of the same building. The learners can see the site from different angles, they can touch the stone, and they can experience the site with all the senses. It is easy to verify from personal insights that we remember facts about places we have visited much better than about places we have only read about. The core objectives and activities of a tourist visit of a city are learning about the city, about certain periods, about culture, lifestyle, events, architecture etc [1]. The OTSL

A. Amine et al. (Eds.): *Modeling Approaches and Algorithms*, SCI 488, pp. 357–366.
DOI: 10.1007/978-3-319-00560-7_39 © Springer International Publishing Switzerland 2013

is based on three major geographical properties such as Points (i.e. landmarks, historical locations, monuments etc), Lines (i.e. along the route, roads, rivers information) and finally the Area (polygonal area made of Points and Lines). Landmarks are cognitively salient, prominent features in the environment. Landmarks play a central role in human spatial cognition and learning. Most of the geographical learning is centered at the landmarks or the geographical regions having natural or synthetic specialties [2].

Landmarks are defined as prominent features in the environment that are unique or contrast with their neighborhood and as natural, built or culturally shaped features that stand out from their environment. The landmarks tend to exhibit significant qualities of form (e.g. size, inflow of people and visual attributes of the facade); significant qualities of visibility and significant meaningfulness of the building etc. These characteristics are found in landmarks irrespective of geographic scale, the same qualities being important for both local and global comparisons. The current tourist systems use lowest level of the Tourist POI details for the tour guides. Storing Tourist POI data and using it with the same granularity does not fit well in the next generation OTSL systems. There is a need for the systems that support on-demand information granularity, from the lowest level tourist POIs to the high level country information. The system should be able to reduce or expand the information as desired depending on the user interest, the level of details required and other constraints i.e. speed, travel direction etc. Use of region based objects (known as ROI) can fulfill granularity requirements. The *ROI* is a region where moving objects pause or wait in order to complete activities, i.e. restaurants, museums, parks, places of work, and so on [3]. The ROI includes all the geographical regions right from a small monument to the postal regions, city boundaries, states, and countries. The ROI is defined as a polygonal geographic area that is of interest for multiple purposes [4]. *Spatial Chunking* is a process of logical grouping the spatial elements [5]. Landmarks located along a route can be used to chunk certain parts of the route [6]. Such landmarks are considered point-like if they are located at a specific spot along the route, and are only functionally relevant for this spot. We want to use the spatial chunking for erecting the ROIs in order to serve the OTSL requirements.

Apart from the introduction in this section, the related work has been discussed in section 2. The section 3 describes the OTSL modelling in terms of selecting the landmarks using the weighted heuristics, and storage methods of the ROIs. The OTSL model has been improved further in section 4 by using the dynamic attributes of the user such as travel direction, speed etc. We have implemented the system using the web databases and the LBS infrastructure in the mobile environment and studied some of the attributes in section 5.

2 Related Work

Landmarks are used for the story telling and location based historical references in real world. We are trying to exploit the landmarks and the associated information for the OTSL applications. Richter et al [7] studies the selection of landmarks from the perspective of the route providing the spatial and structural context of the selection process. Klippel et al [5] exploit various relationships between landmarks and a route towards generating route instructions of varying granularity. Their hierarchy of granularities starts at elementary turn-by-turn instructions. From here they form chunks, either by the structure of the route, or by landmarks available along the route. The literature suggests that landmarks improve the quality of route directions in terms of their cognitive ergonomics. However, there is no universally accepted mechanism for including landmarks in route directions. Landmark identification and selection is an important part in the OTSL systems. So far, three main approaches to identify landmarks from spatial datasets are known in the literature.

The first approach proposed by Raubal et al [8], and Nothegger et al [9], constructs a set of evaluation functions for the visual and semantic salience of building facades. These functions measure differences of individual properties, such as for visual salience [10]. The second approach, by Kolbe et al [11], computes the salience of building facades using information theoretic measures. This approach determines the peculiarity or degree of surprise in visual characteristics of facades. The entropy of the visual appearance of a facade is 0 if its probability is 1, and becomes larger if the probability of its occurrence in this neighborhood is smaller. The facade with the largest entropy in a neighborhood is identified as a desired landmark. The third approach, by Elias et al [12], aims to identify interesting buildings; in this case, facades play only one contributing factor. Compared to the previous methods, this approach does identify salient buildings, but does not provide a measure for relative or absolute ranking in a neighborhood. Instead, it aims to identify landmarks by detecting outliers [2]. Tomko et al. [13] suggested a hierarchical ranking of streets. The salience of streets was determined by their centrality, arguing that centrality correlates with prominence.

The focus mobile tourist guides lies mainly on the accurate delivery of additional content to the exhibits or points of interest [14, 15, 24]. In the museum scenarios, the user is guided by the concept of the museum director who installs the exhibits in the real world. In the case of the mobile tourist guides the user is often guided by the incidental regional proximity of landmark. In these systems the drawback is that the information is usually stored in proprietary formats and is therefore hardly reusable [16]. There are several static multimedia content based tour guide systems [17, 18]; however they lack of dynamic content delivery. None of the tour guides presented in survey by Kenteris et al [18] on the dynamic guidance based on the location.

3 On-the-Spot Leaning (OTSL) Model

In this section we define a model for the OTSL. The landmarks category, reviews, popularity information about features in geographic environments is typically provided by the content providers. The tourist websites, online encyclopedias and the internet discussion forums are the main sources of the details of the landmarks such as monuments, historical regions etc. Our core of the OTSL Model has been described in the following subsections with the key components.

3.1 Selecting the Landmarks Using Weighting Heuristics

Our approach constructs a set of evaluation functions for a set of saliences of the landmarks. These functions measure differences of individual properties, such as for visual salience, the size or the form factor of the facade, from average properties in the local neighborhood. While visual and semantic salience describes the global properties, structural salience describes the properties of the landmark by their location in the structure of the environment. As a result, the overall salience of a learning object s is computed (using eq (1)) from weighted components of descript-ability salience s_d, visual salience s_v, semantic salience s_s and structural salience s_t, category salience s_c [2]. The visual salience of the landmarks can be calculated offline based on the road network and the landmarks positions. The descript-ability salience s_d is determined based on the content availability in the identified sources. The landmarks with lesser details are given lower priority. The other salience values are determined using the method in [2].

$$s = s_d \cdot w_d + s_v \cdot w_v + s_s \cdot w_s + s_t \cdot w_t + s_c \cdot w_c \tag{1}$$

Here, weights w_d, w_v, w_s, $w_t > 0$ and $\sum w = 1$. A local maximum filter can identify the most salient facades in a neighborhood by identifying a candidate set of landmarks to enrich the route directions. The popularity of the location is a very important factor on the location visits and the desire to gain the information about it. The popularity factor can be taken from the yellow pages [2] or can be computed by the visitors count methods [25]. The assumption is that the popularity is an attribute of the landmarks in the database. The overall-salience s' in the context of a route is then computed by multiplying s with popularity salience s_p:

$$s' = s \cdot s_p \tag{2}$$

Although the popularity can be added as a weighted factor in eq (1); however, we have used it as a multiplication factor to differentiate the popular landmarks clearly. Thus, the popular landmarks have higher overall-salience value.

3.2 Storage and Retrieval: Granulize the Information

The storage structure for the landmark is required in such a way that the information can be granulized as required at runtime. But achieving granularity entirely based on the run-time, expansive computation can create the performance issues. There are three approaches for erecting the ROIs – *First*, the administrative hierarchy based ROI – postal region, city, county etc. *Second*, the landmark chunking based ROI – a set of landmarks with the common properties are grouped to create a region. *Finally*, hybrid ROI – based on the strength of first two solutions, the lower level ROI uses landmark chunking; however the higher level ROI consists of the admin based regions. The different ROI storage approaches can be used in the different variety of applications.

Administrative Hierarchy Based ROIs Approach: The administrative ROIs are based on the local government decided geographical divisions. Information granularity feature allows the user to choose the specific level of details about the region in question. A huge continent level region can be further deepened down at the country, state, city, post regions, or the landmarks and to the street level. Based on the user's current location, the ROI of the different levels can be fetched and served accordingly. The process of getting the administration hierarchy details against the geo-coordinates is known as the reverse-geocoding. The reverse-geocoding efficiently returns the nearby landmarks, the parent city and other administrative names of the underlying location. We have implemented a prototype using the open-source technology to show the proof-of-concept in the later sections. The reverse-geocoding reveals the names of the cities, states and the countries of the landmarks.

Landmark Chunking Based ROIs Approach: In this approach, the actual data is stored at the lowest level landmark database. The landmark database also keeps the summarized form of the landmark description known as Façade Content (FC). The higher level ROIs keep only the references of the landmark's façade records. They keep only a set of highly weighted landmarks and lower level ROIs. The façade content of ROIs are defined as below:

$$\mathrm{ROI}_{FC} = \{L : L \in \Gamma, Salience\ (L) \geq \Delta\} \tag{3}$$

Here, Γ is a set of landmarks, Salience(L) is the salience value of an individual landmark L calculated using eqs (1) and (2), Δ is a threshold value for the minimum criteria of having weights to get included in the ROI. Fig 1 shows the landmark based contextual hierarchy. A group of chunkable landmarks are collected together in the logical manner to create the ROIs. The ROIs can also be created offline for the pre-identified road networks and associated landmark chunks.

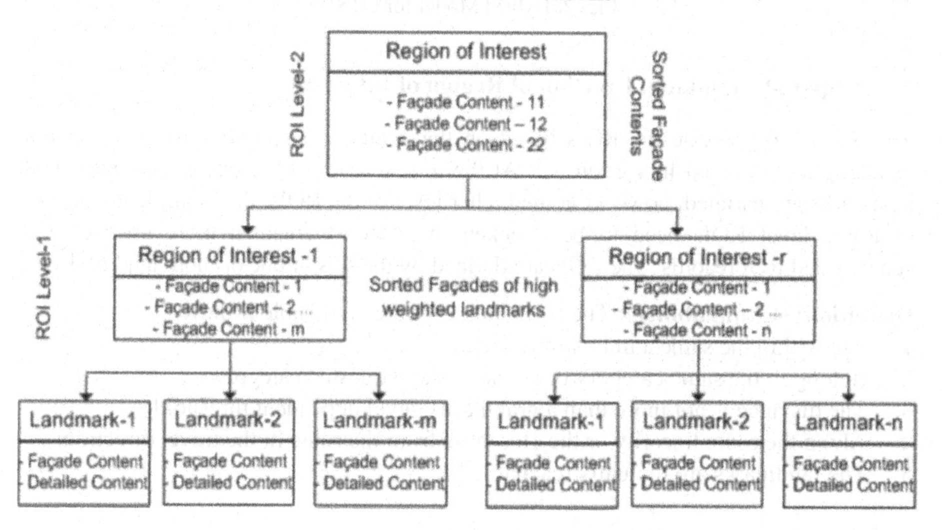

Fig. 1. Landmark Based Contextual Hierarchy

Hybrid Model: The chunking works well at the lower level ROIs where the number of landmarks is small. As the hierarchy goes up, the number of landmarks increase in the chunk, so the more expensive operations have to be carried out because of the bigger database records,. Also, at the higher level in the hierarchy, probably the user is willing to learn the high level details of the city or the country. In such cases the admin hierarchy based ROIs are more useful. Fig 2 shows architecture of hybrid model where the lower level records in the hierarchy have the detailed landmarks and the higher level ROIs are represented in terms of admin names. Hence, hybrid solution is useful in a real system implementation – use the landmark based chunking at the lower level and the admin based ROIs at the higher level in the hierarchy. It balance the level of details and the performance of the overall system.

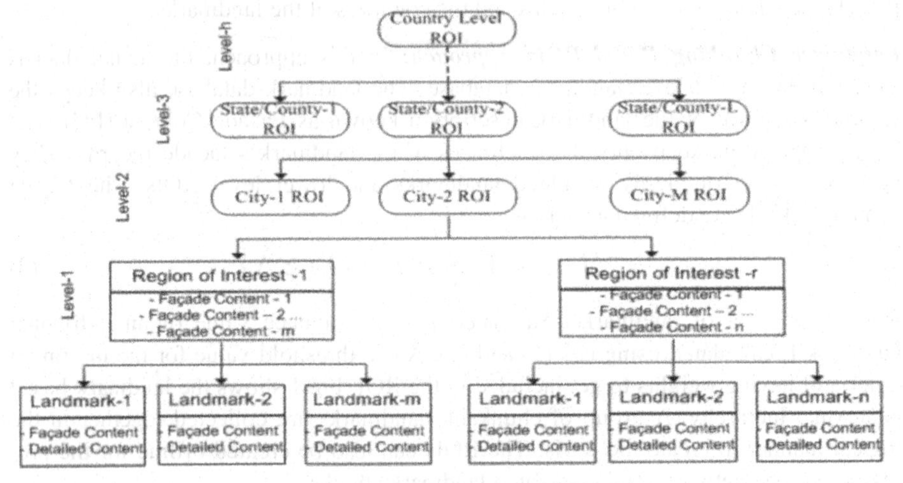

Fig. 2. Hybrid Model for OTSL

3.3 Spatial Chunking: Erection of Region of Interests

The idea is to generate the ROIs based on the available landmarks using the spatial chunking as discussed in section 3.2. At the lowest level of the details the individual landmarks are narrated; however at the higher level of the ROIs, the multiple landmarks or lower level ROIs need to be chunked in order to create a more abstract and summarized ROI records. The ROIs are defined by the sets of one or more landmarks.

Definition: (Chunkability): The two landmarks are chunkable if they:

1. Are within the same administrative region.
2. Belong to the same category or is one of the requested categories.
3. The distance is not more than a granularity-level dependent threshold.
4. Along the route linearity or the closeness from the route in the travel direction.
5. The landmarks are mutually reachable.

3.4 Annotation Algorithm for OTSL

In this section we describe about the systematic steps for the learning content finding and delivery in the OTSL. The first step is to determine the user's direction and speed based on the user's movement. Based on the user direction, the learning region is identified; however the speed is used to determine the level of details about the region. If the routing service is available to integrate with the OTSL system, then the content can be prepared in advance. A person on the highway would most likely wants to know about an overall city rather than the specific things. The higher level ROI contain the façade content grouped together at the runtime based on the chunkability of the landmarks. The summarized content for the individual landmarks is prepared offline to save the performance. OTSL annotation algorithm for annotating the learning content with landmarks is given below.

Algorithm. OTSL Annotation Algorithm

1. Generate a route.
   a. Given graph G = (V,E), find shortest path from origin o ∊ V to destination d ∊ V, using any standard shortest path algorithm.
   b. Determine the speed direction and travel mode from the user's movement and positing methods.
2. Find a set of landmarks L' ⊆ L that lie on the travel direction and along the way.
3. The chunkable landmarks are grouped together to form an ROI.
4. Determine the ROI and the granularity requirement based on the dynamic attributes in step (3).
5. For each ROI chunk,
   a. Fetch the learning from the landmarks/lower level ROIs.
   b. For each landmark in ROI
      i. Generate the narration such as "You are near <landmark name> at your <left/right/surrounding>. The <ROI name> includes the <landmark count> landmarks i.e. <L1, L2, L3,...Ln>".
      ii. Deliver the narration the underlying landmark

4 Extensions in OTSL Model

It is possible to design a range of further refinements on the content quality, while still relying only on widely available data sources, the road network and a geocoded directory service as discussed below.

Refining Landmarks Weights: One of the simplest extensions to the OTSL model is to repeat the landmark suitability weighting process, described in the section 3.1, for different user contexts. For example, the question of what is a good landmark from the learning point of view can depend on the mode of travelling, user interest, mood etc. Features in a car driver's focus of attention, such as traffic lights, crosswalks or bridges, are the places to avoid landmark narrations for drivers as it may cause a distraction. However, the rear seat users may use the narrations throughout the whole journey irrespective of the routing situations. Conversely, a pedestrian's concentration is generally directed towards other urban features, due to their lower travelling speed and also their separate travelling space so they can be narrated about the landmark content in more detail.

Adjusting Landmark Instance Weights: Two specific refinements are identified here [2]. First, the *Side of the Road* – prominent features that are easily visible from the direction of travel for a route are more salient to a learning system as the user can visualize the landmarks better. Also, the rear seat users should be able to mention the side they are sitting in the car or a tourist vehicle. The side of the street matters from the convenience point of view as the users driving in the same side may decide to stop by to visit the landmark. Hence, features on that side should be ranked more highly by a landmark weighting process. Second, *Multiple Landmarks* – it is possible that several landmarks of the same category occur on a particular route. In such cases, the landmark chunks should be formed in order to create a grouped ROI. The ROI creation again considers the chunkability of the landmarks.

Absence of the Landmarks: In some situations landmarks may not be available where needed. However, since any feature in the dataset has at least some salience, it may also be desirable to also apply some minimum suitability threshold such that no landmark is ever selected with a suitability weighting below it. Also, in such scenarios, the granularity level can be reduced and hence the higher level ROI can be delivered. In these cases the administrative hierarchy solutions are very useful.

Landmark's Shape Considerations: Generally, the landmarks have all been assumed to be exactly the points. However, at the level of granularity of human navigation, some spatial objects may be better represented with spatial extents of the interesting regions. Such shapes can be simply adopted as standalone Region of Interests (ROI); however the remaining algorithm does not change.

5 Prototype Implementation of OTSL Model

The prototype has been implemented using J2ME's Connected Limited Device Configuration (CLDC) platform with the Sun's Wireless Toolkit (WTK) 2.5.2 [19]. We have used Java's JSR-179 methods to determine the mobile device location. Once the location (latitude and longitude) is known, it is required to determine the semantic information against the query point (i.e. nearby landmarks names, street, city, state and country names) called as ROI's linguistic labels using reverse-geocoding operation of OpenStreetMap [20]. The reverse-geocoding returns the landmarks name, street name, city, state, country information in the desired language. These names are used further to find the learning content from the web-database. In our implementation, we have used the Wikipedia as the web database for ROI learning content source. However, the other web-databases such as internet news, discussion forums, landmark reviews etc can also be used as the data sources. The java based MER-C's Wikipedia editing APIs [21] have been used to search the region names from the Wikipedia's rich documents. The region data contains the detailed description in terms of its history, attractions, conception, culture etc depending on the specialty of the ROI. The region data format is based on the Wiki Markup Language [22] that includes text description along with the image names that can be accessed separately with a specific HTTP transaction. ROI content retrieved from the web database in the wiki markup format is parsed using WikiSense's parsing APIs

[23]. The parsing module generates structured object that can be directly used in the presentation layer.

Fig 3 (a) shows a simple performance comparison of the different approaches taking a route from The Embarcadero to Golden Gate Bridge in San Francisco. The Landmark Chunking (ROI-LC) based approach for representing ROI is more accurate as it contains all the important landmarks. However, the Administrative Hierarchy (ROI-AH) based approach is more efficient, but it contains only a high level detail about the underlying region. Fig 3(b) shows comparison between the individual POI and the ROI data modeling in terms of average content accuracy percentage..

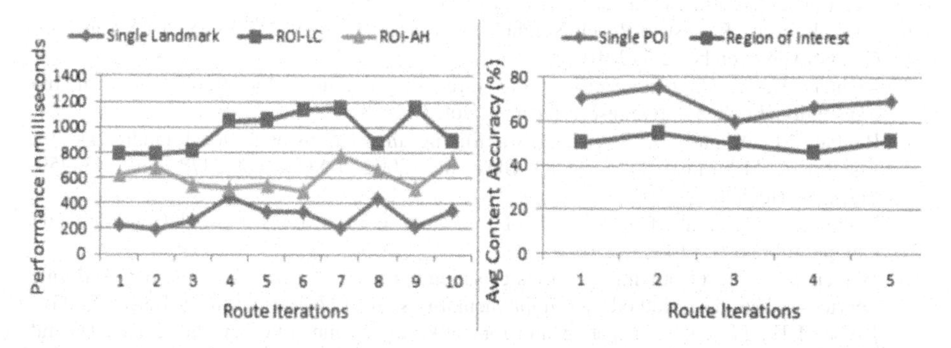

Fig. 3. (a) Performance comparison between different Approaches. (b) Content coverage accuracy between the single POI and ROI approach.

6 Conclusion and Future Work

The paper presents a conceptual model and related algorithm for on-the-spot learning (OTSL) with landmarks. The key features of this model are two fold – storage and retrieval of the learning content and the weighting system. The weighting system relies on commonly available data about categories of landmarks, and also considers some of the detailed instance-level data about the visual characteristics and facades of landmarks. Thus, the process of annotating the learning content with landmarks has two phases: an offline process of landmark identification from available categories; and a run-time landmark selection process based on route-specific or the person specific factors. We have implemented a prototype that exploits the open source navigation tools and the freely available web-databases. The system is easily pluggable with the existing navigation infrastructure that support site and region oriented on-demand information granularity of the interesting locations. We have tested the concept with the prototype for the text and images; however a rich learning system can be implemented by using some other multimedia region data sources. The Text-to-Speech conversion of the region description and narrating the description to the user in the form of stories will be a great extension of this prototype. The performance and the accuracy of the system needs to be checked on a variety of the devices and limited computing environments.

References

1. Tiwari, S., et al.: Fusion of Navigation Technology and E-Learning Systems for On-the-spot Learning. In: Proc. of ICWCA 2012, Kuala Lumpur, Malaysia, October 8-10 (2012)
2. Duckham, M., et al.: Including Landmarks in Routing Instructions. Journal of Location Based Services 4(1), 28–52 (2010)
3. Uddin, M., et al.: Finding Regions of Interest from Trajectory Data. In: MDM 2011 (2011)
4. Alvares, L.O., et al.: A model for enriching trajectories with semantic geographical information. In: Proc. of ACM-GIS 2007 (2007)
5. Klippel, A., et al.: Urban granularities: a data structure for cognitively ergonomic route directions. GeoInformatica 13(2), 223–247 (2009)
6. Shao, J., et al.: Easiest-to-Reach Neighbor Search. In: Proc. of ACM GIS 2010, San Jose, CA, USA, November 2-5 (2010)
7. Richter, K.-F.: Context-specific route directions, Berlin. Monograph series of the transregional collaborative research center, vol. 3 (2008)
8. Raubal, M., Winter, S.: Enriching wayfinding instructions with local landmarks. In: Egenhofer, M., Mark, D.M. (eds.) GIScience 2002. LNCS, vol. 2478, pp. 243–259. Springer, Heidelberg (2002)
9. Nothegger, C., et al.: Computation of the salience of features. Spatial Cognition and Computation 4(2), 113–136 (2004)
10. Winter, S., et al.: Focalizing measures of salience for way finding. In: Map-based Mobile Services – Theories, Methods and Implementations, pp. 127–142. Springer, Berlin (2005)
11. Kolbe, T.H.: Fuga ngernavigation und routenplanung in innensta dten und geba uden mit videos und panoramen. In: GI-technologien fur verkehr und logistik, pp. 337–356. IFGI (2002)
12. Elias, B.: Extracting landmarks with data mining methods. In: Kuhn, W., Worboys, M.F., Timpf, S. (eds.) COSIT 2003. LNCS, vol. 2825, pp. 375–389. Springer, Heidelberg (2003)
13. Tomko, M., et al.: Experiential hierarchies of streets. Computers, Environment and Urban Systems 32(1), 41–52 (2008)
14. Tiwari, S., et al.: Extracting Region of Interest (ROI) Details using LBS Infrastructure and Web-databases. In: Proc. of 13th International Conference on Mobile Data Management (MDM 2012). IEEE Xplore Digital Library, Bangalore (2012)
15. Cheverst, K., et al.: Developing a context-aware electronic tourist guide: Some issues and experiences. In: Proc. of the CHI 2000, Amsterdam, Netherlands (2000)
16. Schneider, G., et al.: A Location Based E-Learning System. In: Proceedings. of WEBIST (2007)
17. Abowd, G.D., et al.: Cyberguide: A mobile context-aware tour guide. Mobicom (1996)
18. Kenteris, M., et al.: Electronic mobile guides: a survey. Pers Ubiquit. Comput. (2011)
19. Sun Java Wireless Toolkit for CLDC (J2ME WTK) Overview, `http://java.sun.com/products/sjwtoolkit/overview.html` (last accessed on May 24, 2012)
20. OpenStreetMap (2011), `http://nominatim.openstreetmap.org/reverse?lat=37.809648&&lon=-122.477303&zoom=3` (last accessed on Februar 24, 2012)
21. MER-C: Wiki-java – a lightweight Java wiki bot framework (2011), `http://code.google.com/p/wiki-java/` (last accessed on May 24, 2012)
22. WikiFormatting – The Track Project, `http://trac.edgewall.org/wiki/WikiFormatting` (last accessed on May 24, 2012)
23. Wikixmlj: A Java API to parse Wikipedia XML dumps, `http://code.google.com/p/wikixmlj/` (last accessed on May 24, 2012)
24. Tiwari, S., Kaushik, S., Jagwani, P., Tiwari, S.: A Survey on LBS: System Architecture, Trends and Broad Research Areas. In: Kikuchi, S., Madaan, A., Sachdeva, S., Bhalla, S. (eds.) DNIS 2011. LNCS, vol. 7108, pp. 223–241. Springer, Heidelberg (2011)
25. Khetarpaul, S., et al.: Mining GPS Data to Determine Interesting Locations. In: Proc. of IIWeb 2011, Hyderabad, India, March 28 (2011)

Towards an Integrated Specification and Analysis of Functional and Temporal Properties:
Part I: Functional Aspect Verification

Mokdad Arous and Djamel-Eddine Saïdouni

MISC Laboratory, Mentouri University, Constantine, 25000, Algeria

Abstract. *Maximality-based Labeled Stochastic Transition Systems* (MLSTS) was presented [6, 11] as a new semantic model for characterizing the functional and performance properties of concurrent systems, under the assumption of arbitrarily distributed (i.e. non-Markovian) durations of actions. The MLSTS models can be automatically generated from S-LOTOS specifications according to the (true concurrency) maximality semantics [6]. The main advantage is to pruning the state graph without loss of information w.r.t. ST-semantic models [11]. As a first work on MLSTS, we focus in this paper on in the verification of functional properties of systems, using a variant of model-checking technique.

Keywords: CTL, Formal Verification, Maximality Semantics, Model-Checking, Semantic Models, Labeled Transition Systems.

1 Introduction

Since the use of concurrent and distributed systems has become more and more important in industry, the analysis, prediction and evaluation (of both qualitative and quantitative aspects) of their behavior have become mandatory. However, in concurrent systems multiple active agents interact together and with the environment, and this leads to phenomena like uncertainty, non-determinism and randomness in the global behavior of the system, and exhibit complex functional and temporal behavior that are often complex to predict. For instance, in transmission systems, concurrency and transmission errors in the traffic flow produced randomly, lead to various communication delays. Therefore, there is a need for adequate stochastic timing models to well specify and verify such (stochastic) behaviors.

Actually, two main approaches have been adopted for expressing random time properties of stochastic systems. In the first one, e.g. [1, 9, 13], the specification of the durations of actions is limited to exponential distributions, hence, it takes advantage from the memoryless property of exponential distributions, which yields analytically tractable models in the form of Continuous Time Markov Chains [2, 7]. Such models accord with the interleaving semantics [1, 9], therefore actions are considered as atomic (see Fig. 1-a), and the parallel execution of two actions is assumed to be equivalent to their interleaving execution.

A. Amine et al. (Eds.): *Modeling Approaches and Algorithms*, SCI 488, pp. 367–377.
DOI: 10.1007/978-3-319-00560-7_40 © Springer International Publishing Switzerland 2013

For responding to the limitation in expressiveness of the exponential distribution laws and the interleaving semantics imposed by the first approach, the second approach adopts general probability distributions to specify action durations, and it refers to the true concurrency semantics [10]. This last one allows to escape the action atomicity hypothesis imposed by the interleaving semantics. Thus, the system behaviors are not anymore represented like totally ordered sequences, but adequately like partial order ones. In fact, existing non-Markovian approaches, e.g. [10, 14, 15, 16, 17], introduce an explicit representation of the start and end events for every running actions (for example, see Fig. 1-b, where a^+ and a^- represent respectively the events of start and termination of execution of the action *a*). This allows considering a specific true concurrency semantics called ST-Semantics. In the ST-semantic models, the progression of a delay is represented as a combination of two events: the delay starting and termination. However, the price to pay is the generation of models which suffer from the state space explosion problem, due to the splitting of running actions into start and end events.

For our approach, we refer to an appropriate true concurrency semantics, namely Maximality semantics [3, 4]. Its principle consists in using the dependence relations between actions occurrences and by associating to every state of the system the set of actions which are potentially in execution (see Fig. 1-c).

Our *integrated* approach *based on MLSTS* aims at handling true concurrency notions and arbitrarily distributed durations for specifying functional and performance properties, without suffering from the state space explosion problem inherent to the splitting of actions. We interest in this paper in the concurrency logic verification, based on our MLSTS model, of functional aspects of concurrent systems. Along this paper, we assume that the reader is familiar with Labeled Transition Systems, Model-Checking, formal description technique LOTOS [8] and the temporal logic CTL.

The paper is scheduled as follows: In section 2, the main principles and formal definition of the MLSTS models are presented. In section 3, we present our approach for functional aspect verification based on MLSTS models by adopting a CTL model checking variant. Finally, section 4 concludes the paper and opens some perspectives.

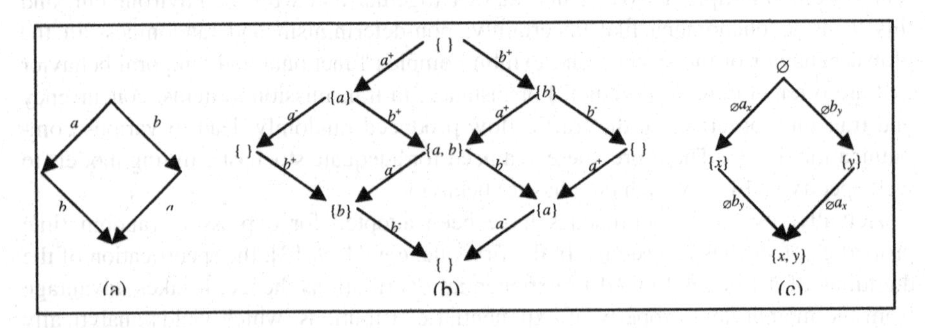

Fig. 1. Behavior of two parallel actions according to the interleaving semantics (a), the ST-semantics (b) and the Maximality semantics (c)

2 Maximality-Based Labeled Stochastic Transition Systems

The aim of our approach is to combine functional and performance modeling and analysis, and allow any kind of probability distribution function for specifying action durations. From a modeling views point, such an approach showed interest in easing the modeling stages. The main interest of our maximality based approach is to make possible to deal with true concurrency with less prone to state space explosion problems [4, 6].

2.1 Informal Presentation of MLSTS

Within the semantic model MLSTS, each transition only represents the start of an action execution. Since actions are not considered as atomic, the concurrent execution of multiple actions can be represented, and distinguishing between sequential and parallel executions is possible.

The running actions are represented at the states level, and each instance of running actions is called a maximal event and is identified by a distinct name. In fact, each state of the system is featured by a unique configuration. The configuration of a state s is denoted ${}_{M}[E]$, such that M is the set of maximal events and E is the behavior expression in the state s. Every transition defined from s is labeled by ${}_{C}(a, f)_{x}$ whenever a is an action that can be activated from E iff. the maximal events of the subset $C \subseteq M$ are terminated. Further C is called the causality set of the transition, and x is the name identifying the start event of the new execution of a. The event identification is required to avoid confusion since several instances of running actions can have the same action name.

Fig. 2 illustrate by a simple example that MLSTS models allow distinguishing between concurrent and sequential executions of actions. Consider tow actions a and b (whose durations follow probability distribution functions f and g respectively), and the following two behavior expressions:

E= (*a*,*f*); *stop* |[]| (*b*,*g*); *stop*
F= (*a*,*f*); (*b*,*g*); *stop* [] (*b*,*g*); (*a*,*f*); *stop*.

Such that E executes a in parallel with b, and F executes either a followed by b or b followed by a.

Initially, no action has yet been executed, then the set of maximal events is empty, and the initial configurations associated with E and F are, respectively, ${}_{\emptyset}[E]$ and ${}_{\emptyset}[F]$. From configuration ${}_{\emptyset}[E]$ the actions a and b can be executed independently (according to the semantic of the parallel composition operator |[]|), and the both execution paths lead to the configuration ${}_{\{x\}}[stop]$ |[]| ${}_{\{y\}}[stop]$, where x and y are the event names identifying the starting of a and b respectively. However, from configuration ${}_{\emptyset}[F]$, depending on the action which happens first (for example the action a), and according to the semantic of the prefix operator (;) expressing the sequentiality in execution, the start of the other action (i.e. the action b) is constrained by the causality dependence against the event identifying the potential execution of the first one (i.e. the action a). This results in the following execution path:

$$\varnothing[F] \xrightarrow{\varnothing(a,f)_x} {}_{\{x\}}[(b,g); stop] \xrightarrow{\{x\}(b,g)_y} {}_{\{y\}}[stop]$$

In a symmetric scenario, if the action b happens first, the behavior can be represented by the following execution path:

$$\varnothing[F] \xrightarrow{\varnothing(b,g)_y} {}_{\{y\}}[(a,f); stop] \xrightarrow{\{y\}(a,f)_x} {}_{\{x\}}[stop]$$

The MLSTSs representing the behaviors of E and F, obtained by applying the maximality semantics, are represented in **Fig. 2**. It is clear that from **Fig. 2**-a and **Fig. 2**-b the two behaviors of E and F are not equivalent.

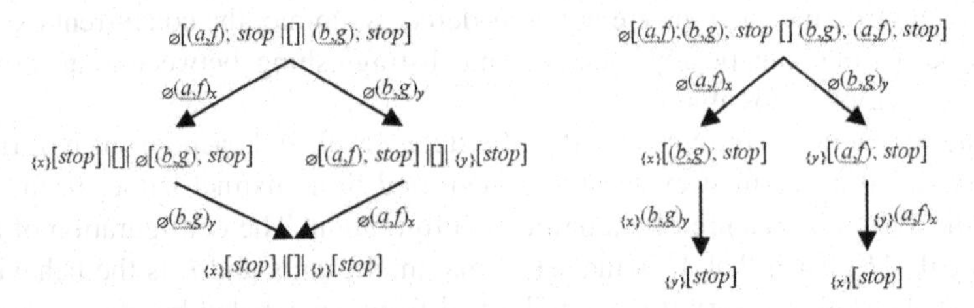

Fig. 2. MLSTSs representing behaviors of concurrent and sequential executions

2.2 Formal Definition of MLSTS

An MLSTS is defined as follows:

Definition 1. Maximality-based Labeled Stochastic Transition System (MLSTS).
Let $\mathcal{M}$ be a countable set of event names.
An MLSTS is a structure $(\Omega, A, DF, L, \mu, \xi, \psi)$ with:

- $\Omega = (S, s_0, T, \alpha, \beta)$: is a Transition System s.t. S is the countable set of states for the system, at least including the initial state s_0; T is the countable set of transitions specifying the states changes; α and β are two functions: $T \rightarrow S$ mapping every transition with its source $\alpha(t)$ and its target $\beta(t)$.
- A : is a (finite) set of actions.
- DF : is a finite set of probability distribution functions ($\Re \rightarrow [0, 1]$).
- $L: T \rightarrow (A \times DF)$: is a function which associates to each transition a pair composed of an action and a probability distribution function specifying the action duration.
- $\psi : S \rightarrow 2^{\mathcal{M}}_{fn}$: is a function which associates to each state the finite set of maximal event names in this state.
- $\mu : T \rightarrow 2^{\mathcal{M}}_{fn}$: is a function which associates to each transition the finite set of maximal event names of actions that have started their execution so that their terminations allow the start of this transition (i.e. the direct causes of the transition).

- $\xi : T \rightarrow \mathcal{M}$: is a function which associates to each transition an event name identifying its occurrence, such that for any transition $t \in T$:

$\mu(t) \subseteq \psi(\alpha(t))$
$\xi(t) \notin \psi(\alpha(t)) - \mu(t)$
$\psi(\beta(t)) = (\psi(\alpha(t)) - \mu(t)) \cup \{\xi(t)\}$ ♦

MLSTS is intended for modeling both qualitative (functional) and quantitative (stochastic temporal) behavior. Moreover, it is able to deal, in a true concurrency semantics, with any kind of probability distribution instead of restricting only to the exponential distributions. We also defined a Stochastic Process Algebras (SPA), called S-LOTOS [12], as a language to describe MLSTS. From S-LOTOS specifications, the underlying MLSTS models can be generated automatically, according to the maximality semantics [6].

3 Functional Aspect Verification Based on MLSTS

From an MLSTS of a given system, one can derive two semantic models [12]: a functional one enhancing true concurrency behaviors, and a performance one allowing quantitative evaluation. The functional model is obtained by abstracting the quantitative information related to the various durations of actions, whereas the performance model is obtained by abstracting the functional information. We showed in [12] that the performance model of an MLSTS is a Generalized Semi-Markov Process (GSMP). An interest for S-LOTOS is that it can be considered as a high level formalism for GSMP, and further analyses over the GSMP structures can yield performances evaluations. In this section, we focus in the functional aspect verification based on MLSTS models, using a variant of model checking approach resumed in Fig. 3.

In this approach (based on models), the system to be verified is firstly specified by means of the S-LOTOS language [12]. Next, the specification will be translated in an operational way towards the underlying MLSTS model [6, 11]. The expected properties of good behaviors of the system are written in CTL (Computation Tree Logic) [18], and they are finally verified by means of model-checker.

In spite of temporal logics facilitation of specification of system properties to be verified [5], model checking approach is limited by the state space explosion problem, particularly when the specification model underlying semantics is the ST-semantic one. A priori, the maximality transition relation appears to be more complicated than that of interleaving semantics or ST-semantic, because supplementary information is associated to states and transitions. However, this information can allow more reductions without loss of (qualitative and quantitative) information w.r.t. the ST-semantic models [11]. To benefit from the expression power of the MLSTS, in this paper, we present, as a first step in the analysis of this model, the application of the model checking technique to verify behavioral properties.

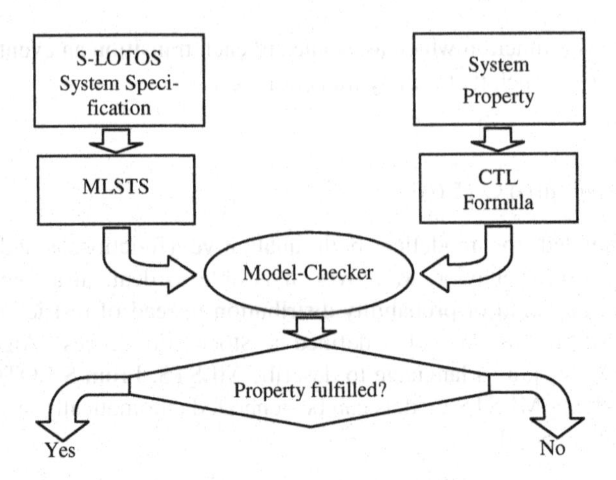

Fig. 3. Model checking based verification approach

3.1 CTL Model-Checking

Qualitative properties of concurrent systems refer to their behaviors, that is to say the sequences of states or actions generated during their executions. Temporal logics are well suited to specify these properties; because they allow to obtain abstract and modular specifications of systems (i.e. specifications are independent from any implementation and easily changeable).

Many temporal logics were defined and studied in the literature. In our approach, we adopt the Computation Tree Logic (CTL) as language to define the properties to be verified, mainly because the evaluation of CTL formulae is more efficient. In fact, classical algorithms are polynomial in the size of the model and also in the size of the formula [18, 19].

CTL is a branching time temporal logic widely used in the model-checking verification techniques. CTL contains the usual logical operators ($\neg$, $\vee$, $\wedge$, $\Rightarrow$ and $\Leftrightarrow$), and the usual temporal operators: X (Next: The X operator in "X ϕ" means that ϕ has to hold at the next state), F (Finally: The F operator in "F ϕ" means that ϕ eventually has to hold somewhere on the subsequent path), G (Globally: The G operator in "G ϕ" means that ϕ has to hold on the entire subsequent path), and U (Until: The U operator in "ϕ U φ" means that ϕ has to hold at least until at some position φ holds), whom have to be at once preceded by one of the path quantifiers: A (along All paths : The A operator in "A ω" means that ω has to hold on all paths starting from the current state.), and E (there Exists one path: The E operator in "E ω" means that there exists at least one path starting from the current state where ω holds).

In CTL, the operators must always be grouped in two: one path quantifier followed by a temporal operator. For example, "AG p" is satisfied in a state if for all paths from this state, p is always true. Thus, we can distinguish eight basic operators: AX, EX, AG, EG, AF, EF, AU and EU. Using these operators, CTL can express formally behavioral properties of concurrent systems, including principally safety and liveness properties.

Definition 2. The syntax of CTL formulae is as follows:

$\varphi ::= true \mid false \mid p \mid (\neg\, \varphi) \mid (\varphi \wedge \varphi) \mid (\varphi \vee \varphi) \mid (\varphi \Rightarrow \varphi) \mid (\varphi \Leftrightarrow \varphi) \mid$
$AX\, \varphi \mid EX\, \varphi \mid AF\, \varphi \mid EF\, \varphi \mid AG\, \varphi \mid EG\, \varphi \mid A\, [\varphi\, U\, \varphi] \mid E\, [\varphi\, U\, \varphi]$

Where $p \in AP$ is an atomic proposition. ♦

Obviously, several model-checkers were developed in the literature. The global state graph of the system can be generally viewed as a finite Kripke structure, wherein each state is labeled with a set of atomic propositions true in that state, and an efficient algorithm is given to determine whether a structure is a model of a particular formula.

3.2 MLSTS Based Qualitative Verification

In MLSTS models, the information included in the states represents the actions that are potentially in execution. For this fact, one can express belonging properties such as mutual exclusion in a more natural way, as well as from new properties that concern actions and their concurrent execution. The expression of these properties does not require the use of a new logic or the introduction of a new operators since one can use CTL temporal logic and consider actions in states as being atomic propositions. However, what changes is the intuition behind formulae. For example, the formula "EF $(a \wedge b)$", where a and b are names of actions, means that there is at least a path which leads to a state where parallel execution of a and b can take place. In a similar way, one can explain intuitively all the formulae of the CTL logic that may be checked using MLSTS model as follows:

- "$a \wedge b$" in a state s means that a and b can be executed in parallel in the state s.
- "$\neg\, a$" in a state s means that the execution of a cannot take place in the state s.
- "EX a" in a state s_0 means that there is at least a path (s_0, s_1, …) where a will be able to comply in the state s_1.
- "AX a" in a state s_0 means that for any path (s_0, s_1, …), a may be executed at state s_1.
- "E [a U b]" in a state s_0 means that there is a path (s_0, …, s_k, …), where b will be able to comply in the state s_k and a will be able to comply in every state of this path that precedes the state s_k.
- "A [a U b]" in a state s_0 means that for any path (s_0, …, s_k, …), there is a state s_k in this path where b may be comply and a may be comply in every preceding state.

Therefore, to express behavioral properties, it is not necessary to use logical formulae indicating the state of evolution of a process, we only would reason directly about actions. By proceeding so, properties will be easier to express and their meaning seems more natural. One can express new properties such as specifying actions incompatibility, we may express that a and b are incompatible by "AG$\neg$ $(a \wedge b)$" which means that the actions a and b will never be able to be executed concurrently. If one takes the example of a simple system of concurrent readers and writers of a shared variable, the mutual exclusion between readers and writers can be simply expressed by "AG $\neg$(*write* $\wedge$ *read*)", where *write* is the action of writing on the variable by a writer process, and *read* is the action of reading the variable by a reader process.

Moreover, in the MLSTS structure, every action is associated with an event name that allows distinguishing between multiple executions of the same action at any state (auto-concurrency). Considering this point will allow us to reason about the number of parallel execution of an action at any state, in other words, we may verify the degree of the auto-concurrency in a system. For this aim, one will have the form of proposition "*a*:*n*" (where *n* is a positive natural number) to express the fact that there is *n* parallel execution of the action *a*. Hence, one can express new properties, for instance, "AG $\neg$(*write*:2)" expresses that along all possible executions, two writing actions may not be in execution simultaneously.

3.3 Model Checking Algorithm

In this section, we adapt the standard CTL model-checking algorithm, e.g. [18], to our study for properties verification on MLSTS models by considering actions at MLSTS states as atomic propositions.

Let us suppose that one has an MLSTS model $M = (\Omega, A, DF, L, \mu, \xi, \psi)$ and a CTL formula φ. The purpose is to compute (recursively) states *s* of *M* satisfying φ, we call the set of these states $Sat(\varphi)$. Finally, the (model of the) system satisfies the desired property, denoted by $M \models \varphi$, if the initial state $s_0 \in Sat(\varphi)$. The algorithm presented in **Fig. 4** treat only the formulae of the form $(\varphi' \wedge \varphi'')$, $(\neg\varphi)$, EX φ, E[φ' U φ''], A[φ' U φ''] which are sufficient. It may be noted that the other logical and temporal operators are implicit and defined as much as the following abbreviations:

$\varphi' \vee \varphi'' \equiv \neg(\neg\varphi' \wedge \neg\varphi'')$	$\mathrm{EF}\ \varphi \equiv \mathrm{E}\ [\mathrm{true}\ \mathrm{U}\ \varphi]$
$\varphi' \Rightarrow \varphi'' \equiv \neg\varphi' \vee \varphi''$	$\mathrm{AF}\ \varphi \equiv \mathrm{A}\ [\mathrm{true}\ \mathrm{U}\ \varphi]$
$\varphi' \Leftrightarrow \varphi'' \equiv (\varphi' \Rightarrow \varphi'') \wedge (\varphi'' \Rightarrow \varphi')$	$\mathrm{EG}\ \varphi \equiv \neg\,\mathrm{AF}\ \neg\varphi$
$\mathrm{AX}\ \varphi \equiv \neg\,\mathrm{EX}\ \neg\varphi$	$\mathrm{AG}\ \varphi \equiv \neg\,\mathrm{EF}\ \neg\varphi$

We only consider the case in which φ = A[φ' U φ''] here, since all the other cases are either straightforward or similar. For this case, one will need information about successor states of *s* as well as on the state *s* itself, because A[φ' U φ'']= $\varphi'' \vee (\varphi' \wedge$ AX A[φ' U φ'']). Initially, A[φ' U φ''] is added to all the states already labeled by φ''. Then, A[φ' U φ''] will be propagated and added to any state labeled by φ' having all successor labeled by A[φ' U φ'']. In the same way one may argues for E[φ' U φ'']. We also take into account the case where φ has the form (*a*:*n*), in this case, a state *s* satisfies φ if it is labeled by *n* event names representing the execution of the action *a*. we use the function *act* which has in input an event name *x* and a state *s*, and returns the action to which is associated the event name *x* in this state.

In the worst case, this algorithm version require $O(length(\varphi) * card(S)^2 * card(T))$ running time. Therefore, this CTL model-checker has a linear temporal complexity according to the length of formula to be verified and quadratic complexity according to the size of the structure *M*.

```
Function
Input :
  A CTL formula φ,
  An MLSTS structure M=(Ω, A, DF, L, μ, ξ, ψ),
with S is the set of states for the system, T is the
set of transitions, and ψ is the function that asso-
ciates every state with a finite set of maximal
event names in the state.
Output : The set Sat(φ) of all states satisfying
           formula φ.
Begin
   Initialize Sat(φ) by ∅
   Case φ of :
      φ = a (atomic proposition representing that
             the action a is in execution) :
         for all s ∈ S do
           if ∃ x ∈ ψ(s) / act(x, s) = a
              add s to Sat(φ)
           endif
         endfor
      φ = φ' ∧ φ" :
         Compute Sat(φ')
         Compute Sat(φ")
         for all s ∈ S do
           if s ∈ Sat(φ') and s ∈ Sat(φ") then
              add s to Sat(φ)
           endif
         endfor
      φ = ¬φ':
         Compute Sat(φ')
         for all s ∈ S do
           if s ∉ Sat(φ') then
              add s to Sat(φ)
           endif
         endfor
      φ = EX φ':
         Compute Sat(φ')
         for all s ∈ S do
           if ∃ successor s' of s / s' ∈ Sat(φ') then
              add s to Sat(φ)
           endif
         endfor
      φ = A [φ' U φ"]:
         Compute Sat(φ')
         Compute Sat(φ")
         for all s ∈ S do
           if s ∈ Sat(φ") then
              add s to Sat(φ)
           endif
         endfor
         for j = 1 to Card(S) do
           for all s ∈ S do
              if s ∈ Sat(φ') and if ∀ successor s' of
                     s / s' ∈ Sat(φ) then
                add s to Sat(φ)
              endif
           endfor
         endfor
      φ = E [φ' U φ"]:
         Compute Sat(φ')
         Compute Sat(φ")
         for all s ∈ S do
           if s ∈ Sat(φ") then
              add s to Sat(φ)
           endif
         endfor
         for j = 1 to Card(S) do
           for all s ∈ S do
              if s ∈ Sat(φ') and if ∃ successor s' of
                     s / s' ∈ Sat(φ) then
                add s to Sat(φ)
              endif
           endfor
         endfor
      φ = (a:n),
         for all s ∈ S do
           if ∃ x1, x2,... xn, / x1, x2,... xn ∈ ψ(s) and
                act(xi, s)=a, 1≤i≤n, then
                add s to Sat(φ);
           endif
         endfor
   endcase
   return (Sat(φ))
end
```

Fig. 4. MLSTS based Model-checking algorithm

4 Conclusion

Considering functional correctness and performance evaluation in a common framework is desirable. Our integrated approach is based on an algebraic formalism for specifying functional and performance properties of concurrent systems, which lies in the class of Stochastic Process Algebras. The main advantage is the capability of specifying and modeling, in the same approach, stochastic concurrent systems under the assumption of generally distributed durations of actions, and reducing, relatively, the state space models, w.r.t. standard true concurrency ST-semantic models.

In this paper, we are interested in the functional verification part, we present the possibility of applying the model-checking approach on our MLSTS model for verifying functional behavior and concurrency properties. This contribution was mainly led by the presence of information about actions potentially in execution in states.

As perspective of this work, we can intend, on one hand, to apply our approach for the study of concrete applications of, for instance, communication protocols domain. Moreover, the final goal of our work is to give a unifying framework for functional verification and performance evaluation. We plan to complete this work by improving (existing) tools to investigate the problems of performance verification.

References

[1] Clark, A., Gilmore, S., Hillston, J., Tribastone, M.: Stochastic Process Algebras. In: Bernardo, M., Hillston, J. (eds.) SFM 2007. LNCS, vol. 4486, pp. 132–179. Springer, Heidelberg (2007)

[2] Haverkort, B.R.: Markovian models for performance and dependability evaluation. In: Brinksma, E., Hermanns, H., Katoen, J.-P. (eds.) FMPA 2000. LNCS, vol. 2090, pp. 38–83. Springer, Heidelberg (2001)

[3] Saïdouni, D.E., Courtiat, J.P.: Prise en compte des durées d'action dans les algèbres de processus par l'utilisation de la sémantique de maximalité, In. In: Proceedings of CFIP 2003, Hermes, France (2003)

[4] Saidouni, D.E.: Sémantique de maximalité: Application au raffinement d'actions en LOTOS, PhD Thesis, LAAS-CNRS, 7 av. du Colonel Roche, 31077. Toulouse Cedex France (1996)

[5] Lamport, L.: What good is temporal logic? In: Manson, R.E.A. (ed.) Information Processing. IFIP, vol. 83, pp. 657–668. Elsevier Science Publishers B.V., North Holland (1983)

[6] Arous, M., Ilié, J.M., Saidouni, D.E.: A Compact Semantic Model for Characterization of Stochastic Temporal Properties of Concurrent Systems. IJCSI International Journal of Computer Science Issues 96(1) (November 2012)

[7] Stewart, W.J.: Performance modelling and markov chains. In: Bernardo, M., Hillston, J. (eds.) SFM 2007. LNCS, vol. 4486, pp. 1–33. Springer, Heidelberg (2007)

[8] Bolognesi, T., Brinksma, E.: Introduction to the ISO specification language LOTOS. Computer Networks and ISDN Systems 14, 25–59 (1987)

[9] Hillston, J.: Process Algebras for Quantitative Analysis. In: Proceedings of the 20th Annual IEEE Symposium on Logic in Computer Science (LICS 2005), pp. 239–248 (2005)

[10] Katoen, J.-P., D'Argenio, P.R.: General distributions in process algebra. In: Brinksma, E., Hermanns, H., Katoen, J.-P. (eds.) FMPA 2000. LNCS, vol. 2090, pp. 375–429. Springer, Heidelberg (2001)

[11] Arous, M., Saidouni, D.E., Ilié, J.M.: Addressing State Space Explosion Problem in Performance Evaluation Using Maximality-based Labeled Stochastic Transition Systems. In: IPCSIT, vol. 54, pp. 41–48. IACSIT Press, Singapore (2012); Proceedings of the 2nd International Conference on Computer and Software Modeling (ICCSM 2012), Cochin, India

[12] Arous, M., Saidouni, D.E., Ilié, J.M.: Maximality Semantics based Stochastic Process Algebra for Performance Evaluation. In: 1st IEEE International Conference on Communications, Computing and Con-trol Applications (CCCA 2011), Hammamet, Tunisia, March 3-5 (2011)

[13] Bernardo, M.: Theory and Application of Extended Markovian Process Algebra. Ph.D. Thesis, University of Bologna, Italy (1999)
[14] Bravetti, M., Gorrieri, R.: The theory of interactive generalized semi-Markov processes. Theoretical Computer Science 282, 5–32 (2002)
[15] Bravetti, M., Bernardo, M., Gorrieri, R.: Towards performance evaluation with general distributions in process algebras. In: Sangiorgi, D., de Simone, R. (eds.) CONCUR 1998. LNCS, vol. 1466, pp. 405–422. Springer, Heidelberg (1998)
[16] Bravetti, M.: Specification and Analysis of Stochastic Real-time Systems. PhD thesis, Università di Bologna, Padova, Venezia (2002)
[17] D'Argenio, P.R., Katoen, J.-P.: A theory of stochastic systems. Part II: Process algebra. Information and Computation 203, 39–74 (2005)
[18] Clarke, E.M., Emerson, E.A., Sistla, A.P.: Automatic Verification of Finite-State Concurrent Systems Using Temporal Logic Specifications. ACM Transactions on Programming Languages and Systems 8(2), 244–263 (1986)
[19] Schnoebelen, P.: The Complexity of Temporal Logic Model Checking. In: Advances in Modal Logic, pp. 393–436 (2002)

Overhead Control in DP-Fair Work Conserving Real-Time Multiprocessor Scheduling*

Muhamad Naeem Shehzad, Anne-Marie Déplanche,
Yvon Trinquet, and Richard Urunuela

LUNAM Université, Université de Nantes, IRCCyN UMR CNRS 6597
L'Institut de Recherche en Communications et Cybernétique de Nantes
Nantes, France
{Naeem.Shehzad,Anne-Marie.Deplanche,Yvon.Trinquet,
Richard.Urunuela}@irccyn.ec-nantes.fr

Abstract. In real-time multiprocessor scheduling, the optimal global scheduling algorithms are criticized for excessive overhead due to the frequent scheduling points, migrations and preemptions. *DP-Fair* model is an optimal scheduling which combines the notion of fluid scheduling (ideal fairness) with deadline partitioning. It has lower number of scheduling points as compare to *PFair* which is the first optimal scheduling algorithm proposed for real-time multiprocessor systems. *DP-Fair* model exists both for non-work conserving as well as work conserving cases. In [14,15], we used some simple heuristics which lower the overhead by reducing the number of migrations and preemptions in the non-work conserving context. In this article, we show that the very same heuristics can be envisaged in case of work conserving scheduling, and we evaluate their efficiencies for lowering the overhead.

1 Introduction

Real-time systems have become very significant in the recent times and are seen everywhere from mobile phones, computer tablets upto the automobiles and space systems. This new scenario of technology constantly requires for an increase in the processing power. Multiprocessing is one of the solutions which addresses to this problem. In real-time multiprocessor systems, the scheduling problem has got a lot of attention after emergence of multicore architectures. The two main branches of multiprocessor scheduling algorithms in real-time systems are: partitioned scheduling and global scheduling. Contrary to a partitioned scheduling [7], a single queue of ready tasks and a single scheduler for all the processors is there in global scheduling. Moreover, migration of tasks and their jobs from processor to processor is allowed. Migration increases the schedulability which consequently improves the resource utilization. An important advantage is that a number of global scheduling algorithms are found to be optimal.

* This work has been supported by the French Agence Nationale de la Recherche through the RESPECTED project (Contract ANR-2010- SEGI-002). See **http://anr-respected.laas.fr**

A. Amine et al. (Eds.): *Modeling Approaches and Algorithms*, SCI 488, pp. 379–388.
DOI: 10.1007/978-3-319-00560-7_41 © Springer International Publishing Switzerland 2013

Though theoretically optimal, the global scheduling techniques are questioned about their practicality. The optimality is achieved at the cost of a huge number of migrations, preemptions and scheduling points. Theoretically, the cost of migrations, preemptions and scheduling points is considered to be zero but practically their effect in the system cannot be neglected when these occur frequently. On some modern multicore architectures, the cost of migration is much lower than that in past but is still a non-zero value. Thus frequent migrations and preemptions in the system lead to an increase in worst-case execution times which may result in the missing of deadlines.

There are two classes of optimal global scheduling algorithms: *PFair* [4] and deadline partitioning fair which is shortly known as *DP-Fair* [10]. Both classes are based on the principle of fairness, but they differ on how much fairness is required. In the recent years, two new classes of optimal algorithms are proposed which either use low version of fairness or no fairness at all. *RUN* [13] proposed by Regnier et al. is based on a very weak version of proportional fairness and *U-EDF* proposed by Nelissen et al. [12] extends *EDF* to multiprocessor systems and releases fairness. Both concentrate on reducing the number of preemptions and migrations.

This article concentrates only on work conserving model. A scheduling algorithm is said to be work conserving, if and only if, it never leaves any processor idle when there exists at least one active task waiting for the execution in the system. The system is non-work conserving otherwise. There is no difference between work conserving and non-work conserving scheduling when total utilization factor is equal to the number of processors. The work conserving system has better response time than an otherwise system. Furthermore, some unnecessary preemptions are avoided which result in lower run time cost.

PFair (Proportionate Fair) was presented by Baruah et al. in 1996 [4]. It uses the concept of fairness which suggests that processor share of each task is proportional to its utilization factor at any instant. In *PFair*, the time is divided into small intervals of equal length, and an interval is called quantum. Scheduling of all the tasks is done at the start of each quantum. *PFair* has been extended to work conserving and this is known as Early Release Fair or *ERFair* [2]. In *ERFair*, a task T_i with utilization factor $T_i.u$, must execute more than $\lfloor t * T_i.u \rfloor$ units of time at any instant t. The optimal *PFair* algorithms like *PF*, *PD* and PD^2 [4] can be easily adapted to allow early releases. As an extension of *PFair*, *ERFair* schedule is considered impractical due to runtime overhead [16] resulting from frequent scheduling points, preemptions and migrations.

In 2010, Kim and Cho [9], proposed *PL*, an optimal work conserving algorithm. It is a laxity based algorithm and uses approximate proportional fairness at each task period. In *PL* scheduling, a sub-job is associated with a remaining execution time of each task. Scheduling of the sub-jobs are computed using principle of *LLF* and the tie breaking rules of PD^2 [4]. It gives lower number of scheduling decisions and preemptions than *PFair*.

DP-Fair combines the notion of fluid scheduling (ideal fairness) with deadline partitioning while still guarantees the optimality. All the *DP-Fair* strategies choose to subdivide time into slices where all the tasks have a common (nodal) deadline. These deadlines are called boundaries and are defined by the points of task release (that coincide with task deadlines for an implicit deadline taskset). These are abbreviated as b_0, b_1,...b_k etc. Distance between any two boundaries is also called a *node*. In *DP-Fair,* the fairness is required to achieve only at the boundaries leading to an interval based scheduling and reducing the number of scheduling points compared to that of *PFair.* Scheduling in each node comprises two steps. The computation of execution time units (called *nodal execution time*) for each task for that node and the decision for dispatching these time units among the processors. The execution time units are called nodal execution times because they are assigned only for that particular interval and can be different from the total execution time of the tasks. Nodal execution time units may be a discrete or a non-discrete value depending on the technique used for their computation.

Some of well known work conserving algorithms following the principle of *DP-Fair* are *NVNLF* based on *Extended TN-plane* [5] and *NNLF* based on *TR-plane* [8]. They differentiate in the way work conserving behavior is achieved : the first one adapts the computation of nodal execution times, while the second one extends the dispatching rules (more details in section 2).

Almost all of the work related to the overhead control is done using the case of non-work conserving case. This is due to the fact that work conserving has naturally much lower overhead due to the number of migrations and preemptions than that of non-work conserving. Aoun et al. [3] used the processor affinity technique to reduce the number of migrations in *PFair* scheduling. Megel et al. [11] proposed a linear programming formulation and a local scheduler technique. In [14,15], we proposed the use of some simple heuristics to control the preemptions and migrations while still keeping the optimality, for a class of non-work conserving global algorithms. The results showed some significant improvement in the overhead. Recently proposed *RUN* [13] and *U-EDF* [12] also concentrated on reducing the number of preemptions.

The motivation behind this article is to apply some overhead control techniques used in [14,15], for a work conserving algorithm. Although work conserving algorithms has naturally fewer number of preemptions and migrations, because a processor does not remain idle if there is an active task for execution, but idea is to further reduce the overhead. Techniques are important in the sense that they do not disturb the optimality of the algorithm and do not involve too much computations.

System Model. We consider that τ is a set of N synchronous periodic tasks T_i where $i = 1, 2...N$ to be scheduled on M identical processors. Each task T_i has a period $T_i.p$ equal to its relative deadline $T_i.d$ (implicit deadline), an execution time $T_i.e$ and a utilization factor $T_i.u$. (which is $T_i.e/T_i.p$) in range (0, 1] such that $\sum_{i=1}^{N}(T_i.u) \leq M$. Each task in such a system is invoked or released repeatedly in accordance with its period $T_i.p$. Each such invocation is called a job of the task. We assume that all the tasks are independent, i.e. they do not share any common resource and do not have any precedence with each other. The costs of migration, preemption and context switch are assumed to zero.

The organization of the rest of the paper is as follows. Section 2 briefly explains the *NVNLF*, a work conserving scheduling algorithm. Section 3 revises the heuristics proposed in [14]. The results of experimental study of those heuristics are discussed in section 4. Section 5 concludes the paper.

2 NVNLF

NVNLF [5] stands for "No Virtual Nodal Laxity diagonal First". It is a work conserving algorithm proposed by Funaoka et al. [5]. Any periodic taskset τ with utilization $U \leq M$ will be scheduled to meet all deadlines on M processors by *NVNLF* [5]. As for any *DP-Fair* algorithm, scheduling of a taskset in a node is divided into two parts, computation of nodal execution time units and dispatching of taskset to the processors. It is briefly discussed here. Further information can be found in [5].

2.1 Computation of Nodal Execution Time

In this technique, in addition to the "conventional" nodal execution time, the time units left unused in the node are distributed among the tasks at the start of each node. These units are distributed such that the resulting nodal execution time of any task is not more than the length of the node or its total remaining execution time. If $T_i.l^*$ is the initial value of computed nodal execution time in a node between b_k and b_{k+1}, it can be computed by following equation at b_k:

$$T_i.l^* = min\{T_i.ret, T_i.u * (b_{k+1} - b_k)\} \tag{1}$$

where $T_i.ret$ is the total remaining execution time. Suppose RU are unused remaining units then they can be calculated as:

$$RU = M * (b_{k+1} - b_k) + \sum_{i=1}^{N} T_i.l^* \tag{2}$$

If $T_i.a$ is the additional time units allocated to a task, then the final value of computed nodal execution time in a node is calculated as:

$$T_i.l = T_i.l^* + T_i.a \tag{3}$$

In the *NVNLF* [5] presented in the literature, the nodal execution time of a task may be non-integer and free units are arbitrarily distributed among the ready tasks. Our implementation differs slightly. So as to avoid the impracticality of non-integer values, we use *BFair* [18] method for computation of $T_i.l^*$. Moreover, the distribution of free units gives preference to the tasks with the lower value of remaining execution time is given preference for allocation of free units because it was found on the bases of an experimental study that it lowers the migrations and preemptions.

2.2 Scheduling Rules

Once the nodal execution times of the tasks are finalized, the tasks are dynamically dispatched to the processors as that in the non-work conserving *LNREF* technique. The main scheduling rules of *NVNLF* inside a node are:

- When a waiting task reaches a state of zero nodal laxity, it preempts an arbitrarily running task with non-zero nodal laxity and starts execution. This is also known as event C;
- When a task finishes its nodal execution time, it is preempted. This is also known as event B.

3 Overhead Control

Once the nodal execution times for the taskset are computed and the simple scheduling rules are established, one can see that there is still room to design complementary dispatching strategies so as to reduce the number of task preemptions and migrations. Our objective is to explore this space and to decrease the overhead due to migration and preemption. Our proposal has been guided by few observations and gives rise to the related heuristics.

The first one comes from the fact where it has been shown that the order in which the M tasks are selected for execution inside a node is not important, provided that they have non-zero nodal laxity [6]. Thus, this heuristic is concerned with the running task selection criterion. We also name it Preemption Control Heuristic (*PCH*).

The second heuristic results from the standard assumption of instantaneous preemptions and migrations. Then in theory it does not matter which processor is hosting a given task, but only which tasks are running at a given time. That is why most algorithms give no explicit description about how to assign tasks to processors. Thus, this heuristic deals with the task to processor assignment criterion. We also name it Migration Control Heuristic (*MCH*).

3.1 Preemption Control Heuristic (PCH)

At a primary scheduling point (a boundary instant), all the ready tasks have equal priority and atmost M of them can be chosen arbitrarily for execution. *PCH* attempts to control the preemptions at these scheduling points. According to this technique, the tasks executing on a processor just before the scheduling point are given priority to re-execute, provided they still are ready. By continuing such executions, some unnecessary preemptions are avoided.

The steps are explained in algorithm 1. The for loop checks for each task of the *ReadyList* at the start of a node. If any of them was running just before the end of the previous node, it will be given a priority to run again at the beginning of the node instead of having a preemption.

The computational complexity of *PCH* per node is $O(\max(M,N))$

Algorithm 1. Preemption Control Heuristic (*PCH*)

Suppose

- H_B is the list of running tasks. Maximum size of H_B is M
- *ReadyList* is the list containing unsorted ready tasks at t
- *size()* returns the size of the list
- *add*(X) adds the object X to the list
- *getFirst()* returns the first object of the list
- *isRunning()* returns true if the task was running at $t-\epsilon$ where ϵ is a number which is infinitely small

Inputs *ReadyList*
Outputs H_B
at $t = b_k$

```
1.    H_B.clear();
2.    for (T ϵ ReadyList )
3.          if (T.isRunning())
4.                H_B.add(T);
5.    end for
6.    while((H_B.size () < M ) && (ReadyList.size() > 0))
7.          T = ReadyList.getFirst();
8.          H_B.add(T);
9.    end while
```

3.2 Migration Control Heuristic (MCH)

This heuristic is meant to control the migrations. Usually, running tasks are assigned to the available processors without considering their previous histories. According to *MCH*, a task keeps the record of the processor on which it was executed last time and then an affinity relation exists between task and processor. *MCH* takes into account this relation and if possible, tries to assign a newly running task to the same processor on which it was scheduled the last time. This heuristic works at start of a *node,* after the decisions of next running tasks has been taken, as well as at secondary scheduling events (event B or event C occurrences inside a *node*). The computational complexity of *MCH* per node is $O(\mathrm{M})$. The algorithm 2 describes *MCH* in a simple way.

3.3 Hybrid = (Migration + Preemption) Control

When both *MCH* and *PCH* are used with *NVNLF,* we name the resulting algorithm as *hybrid* algorithm. It keeps the benefits of both *MCH* and *PCH* and tries to reduce both migrations and preemptions.

When a main scheduling point arrives, it is *PCH* which starts working at first. It generates a list of running tasks H_B and after that, *MCH* starts its operation. At a secondary scheduling point we use only *MCH*.

Algorithm 2. Migration Control Heuristic *(MCH)*

Suppose

- P is an object that represents a processor. T is an object that represents a task
- *getProc*() returns the processor on which the task was executed last time
- *setProc*(X) set X as the processor allocated to the task
- *isIdle*() returns true if the processor is free
- *searchIdleProc*() returns an idle processor

Inputs H_B

at $t = b_k$ *or* instant of *event C or* instant of *event B*

```
1.  for(T ∈ H_B )
2.      if (!T.isRunning() )
3.          P = T.getProc();
4.          if (!P.isIdle() )
5.              P =searchIdleProc();
6.      T.setProc( P );
7.  end for
```

4 Simulation Results

A series of simulation based experimental studies were performed to find the effect of using the overhead control heuristics with *NVNLF* (our adapted form). *PCH* and *MCH* were added with *NVNLF* to find their combined effects on the number of migrations and preemptions compared to *NVNLF* algorithm without them. *NVNLF* with both the heuristics is termed as *hybrid-WC*. We have expressed migrations and preemptions of *hybrid-WC* as a percentage of corresponding values of *NVNLF*.

The experimental data was created using a data generator which uses Roger Stafford's randfixedsum algorithm [17] at the heart. We used STORM [1] as the simulation tool. Since at $U{=}M$, there is no difference between work conserving and non-work conserving scheduling, the results are presented for $U{=}0.75M$ and $U{=}0.5M$. The simulation study has been completed for four different sets of task periods and the results are the average of the results of these four sets.

4.1 Overhead Control at U=0.75M, U=0.5M

The figures 1 (a) and 1 (b) present the migration control and preemption control of *Hybrid-WC* in reference to the *NVNLF* algorithm at $U{=}0.75M$ and $U{=}0.5M$ respectively. The horizontal axis represents the varying number of processors. Each point in the graph shows an average result of experiments conducted on 480 tasksets. It results in less preemptions and consequently in less migrations.

The figure 1 (a) shows that *hybrid-WC* has around 50% of migrations at $U{=}0.75M$ which further decreases to about 40% at $U{=}0.5M$. At lower value of total utilization factor, there are relatively more chances of a task re-execution on the same processor.

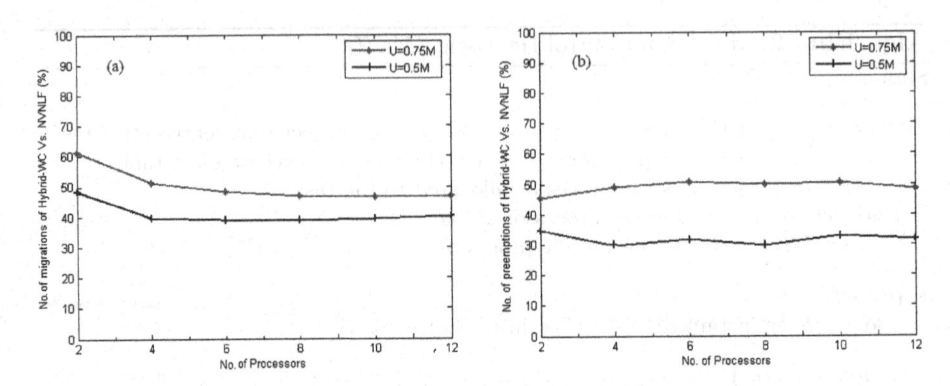

Fig. 1. Migration and Preemption control at variable number of processors

The preemption control of *hybrid-WC* has same behavior. It has around around 50% of migrations at *U=0.75M* which decreases to about 32% at *U=0.5M*.

4.2 Hybrid-WC Algorithm at Variable Task to Processor Ratio

Figures 2 (a) and 2 (b) present the migration control of *hybrid-WC* algorithm at variable task to processor ratio, while figures 2 (c) and 2 (d) present the

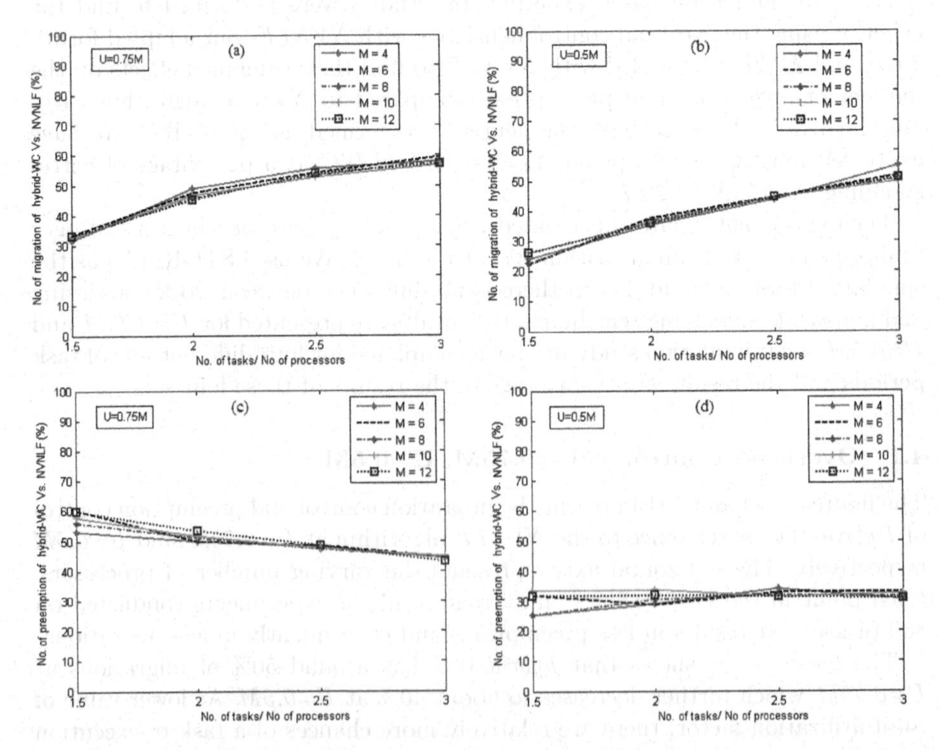

Fig. 2. Migration and preemption control at variable task to processor ratio

preemption control. Each point in the graph shows an average result of experiments conducted on 120 tasksets. The number of tasks N varies between 6 and 36 in variation with the number of processors.

The performance of migration control lowers with an increase in task to processor ratio both at U=*0.75M* and U=*0.5M*. This is because with more number of tasks on each processor, there are fewer chances of task re-execution on the same processor.

The performance of preemption control gets better with an increase in task to processor ratio at U=*0.75M*. At a relatively higher value of task to processor ratio and at relatively higher value of total utilization factor i.e. U=*0.75M*, more number of tasks continue upto the end of the node where the *PCH* is applied and it avoids the preemptions. However, at relatively lower utilization factor i.e. U=*0.5M*, the preemption control remains approximate same over different values of task to processor ratio.

5 Conclusion

In this article, some simple overhead control heuristics are used for an optimal work conserving *DP-Fair* algorithm known as *NVNLF*. The original version of algorithm *NVNLF* is modified to make it efficient and practically implementable. The simulation results have shown a significant reduction in the number of migrations and preemptions. The minimum reduction in the migration and preemptions is 40% and 50% respectively. The results are pretty consistent across different number of processors and different values of processor at different values of task to processor ratio. It will be quite interesting to find some dispatching techniques to decrease the number of events C inside a node to avoid preemptions.

References

1. http://storm.rts-software.org
2. Anderson, J., Srinivasan, A.: Early-release fair scheduling. In: The Proceedings of the 12th Euromicro Conference on Real-Time Systems Stockholm, Sweden, pp. 35–43 (2000)
3. Aoun, D.: Pfair scheduling improvement to reduce interprocessor migrations. In: The Proceedings of 16th International Conference on Real-Time and Network Systems, Rennes, France, pp. 131–138 (2008)
4. Baruah, S.K., Cohen, N.K., Varvel, D.A.: Proportionate progress: A notion of fairness in resource allocation. Algorithmica 15, 600–625 (1996)
5. Funaoka, K., Kato, S., Yamasaki, N.: Work-conserving optimal real-time scheduling on multiprocessors. In: The Proceedings of the 20th Euromicro Conference on Real-Time Systems. Barcelona, Spain, pp. 13–22 (2008)
6. Funk, S., Nadadur, V.: Lre-tl an optimal multiprocessing scheduling algorithm for sporadic task sets. In: The Proceedings of the 17th International Conferenece of Real-Time and Network Systems, Paris, pp. 159–168 (2009)

7. George, L., Goossens, J., Lupu, I., Courbin, P.: Multi-criteria evaluation of partitioning schemes for real-time systems. In: The Proceedings of the 15th IEEE International Conference on Emerging Techonologies and Factory Automation Bilbao, Spain, pp. 1–8 (2010)
8. Funakao, S.K.K., Yamasaki, N.: New abstraction for the optimal real time scheduling on multiprocessor. In: The Proceedings of the 2008 14th IEEE International Conference on Embedded and Real-Time Computing Systems and Applications Kaohisung, Taiwan, pp. 357–364 (2008)
9. Kim, H., Cho, Y.: A new fair scheduling algorithm for periodic tasks on multiprocessors. Inf. Process. Lett. 111(7), 301–309 (2011)
10. Levin, G., Funk, S., Sadowski, C., Pye, I., Brandt, S.: Dp-fair: A simple model for understanding optimal multiprocessor scheduling. In: The Proceedings of the 22nd Euromicro Conference on Real-Time Systems, pp. 3–13 (2010)
11. Megel, T., Sirdey, R., David, V.: Minimizing task preemptions and migrations in multiprocessor optimal real-time schedules. In: The Proceedings of the 31st IEEE Real-Time Systems Symposium, San Diego, USA, pp. 37–46 (2010)
12. Nelissen, G., Berten, V., Nélis, V., Goossens, J., Milojevic, D.: U-edf: An unfair but optimal multiprocessor scheduling algorithm for sporadic tasks. In: The Proceedings of 24th Euromicro Conference on Real-Time Systems, Pisa, Italy, pp. 13–23 (2012)
13. Regnier, P., Lima, G., Massa, E., Levin, G., Brandt, S.A.: Run: Optimal multiprocessor real-time scheduling via reduction to uniprocessor. In: The Proceedings of the 32nd IEEE Real-Time Systems Symposium, Vienna, Austria, pp. 104–115 (2011)
14. Shehzad, M.N., Déplanche, A.-M., Trinquet, Y., Urunuela, R.: Overhead control in real-time global scheduling. In: The Proceedings of 19th International Conference on Real-Time and Network Systems. Nantes, France, pp. 45–52 (2011)
15. Shehzad, M.N., Déplanche, A.-M., Trinquet, Y., Urunuela, R.: Efficiency evaluation of overhead control heuristics in dp-fair multiprocessor scheduling. In: The Proceedings of 17th IEEE International Conference on Emerging Technologies & Factory Automation, Krakow, Poland (September 2012)
16. Srinivasan, A., Anderson, J.: Optimal rate-based scheduling on multiprocessors. In: The Proceedings of the 34th ACM Symposium on Theory of Computing, pp. 189–198 (2001)
17. Stafford, R.: Random vectors with fixed sum (2006), `http://www.mathworks.com/matlabcentral/fileexchange/9700`
18. Zhu, D., Mosse, D., Melhem, R.: Multiple-resource periodic scheduling problem: How much fairness is necessary? In: The Proceedings of the 24th IEEE International Real-Time Systems Symposium, Cancun, Mexico, pp. 142–151 (2003)

Initializing the Tutor Model Using K-Means Algorithm

Safia Bendjebar and Yacine Lafifi

LabSTIC Laboratory, University of 8 May 1945 Guelma, BP 401 Guelma, Algeria
bendjebar_s@hotmail.fr, laf_yac@yahoo.fr

Abstract. This paper proposes an approach for the initialization and the construction of tutor's model in the e-learning systems. This actor has several roles and different tasks from a system to another. His main purpose is tracking and guiding students throughout their learning process. In their first interaction, the system has rather little information about its new tutors. The proposed approach serves to offer much information for each specific tutor based on the models of other similar tutors. The problem of initializing the tutor model can be resolved by assigning the tutor to certain group of tutors. Thus, a data mining algorithm, namely k-means is responsible for creating clusters based on the pre-entered information on tutors. Then, each new tutor is assigned to his closest cluster center. This model facilitates the assignment of tutors to learners for adapting the monitoring process.

Keywords: E-learning, Tutoring, Tutor model, Profile, Initialization, K-means.

1 Introduction

In the past few years, there has been an increasing focus on the use of e-Learning systems. The most of these systems provide a web-based learning, so that students can use the same online courses via the Internet [1], according to their profiles and their behaviors. It provides a set of tasks that are performed by actors, who so have different roles (teacher, tutor and learner) to improve the learning process of learners and supervise their interaction with the system.

There are not trainers who can present the content in an e-learning system but there are tutors who play a very important role to help and supervise all the activities of the learners. This actor applies his activity within an open and flexible distance learning environment. He assists students to communicate in an understandable form during their learning process. Furthermore, he helps to avoid or decrease the negative effects of a major problem in distance learning, which is the abandonment.

Tutors have many roles in distance learning systems such as: pedagogic, supervisor, psychologist, and so many others. The quality of monitoring helps to ensure the motivation of the learners.

In the most current learning systems, the learner is assigned to one and only one tutor [2]. This situation has problems when the tutor is not available during the most of the times to discuss the needs of the learner and to follow each one individually.

A. Amine et al. (Eds.): *Modeling Approaches and Algorithms*, SCI 488, pp. 389–398.
DOI: 10.1007/978-3-319-00560-7_42 © Springer International Publishing Switzerland 2013

The identification of the tutors' characteristics (roles, styles and methods of tutoring, behavior, etc.) can solve the problems mentioned above. These characteristics can be taken into account by tutors' model. This latter describes the state of their skills. It can be changed or updated depending on their interventions with the system.

Creating an adaptive model that meets tutors requirements can be challenging since tutors have different roles and different tutoring styles.

Tutor modeling can be defined as the process of gathering relevant information in order to infer the current cognitive state of the tutor, and to represent it so that can be accessible and useful to other actors. The process of the initialization of a tutor model has a great importance in the educational systems. When the tutor starts working with the learning system, the latter has no prior knowledge about his cognitive level or his behavior.

The aim of this research work is to initialize the tutor model, which can facilitate the assignment of tutors to learners for adapting the monitoring process and to get the profile that the tutor deserves. There is no research, which deals with the problem of the initialization of the tutor's model. However, there are several systems that have been developed to initialize the learner's model. Recently, a number of these systems have used data mining techniques such as statistics, clustering, classification, outlier detection, association rule mining, sequential pattern mining, text mining, or subgroup discovery [3]. The result of our study indicates that data mining techniques provided some tools for improving student's learning. It shows how useful data mining can be used in higher education, in particularly for modeling the characteristics of students [4], predicting their performance [5], analyzing their behavior [6], etc.

For this reason, we describe an approach to initialize the tutor model by using a clustering algorithm, which is k-means. The initialization of the tutor model is carried out dynamically for each tutor taking into account information from the other tutors while using the system. The k-means algorithm is used to refine the estimation about the new tutor's behavior. This is done by using the tutor's similarity with other tutors based on their roles, skills and performance.

The rest of this paper is organized as follows. Section 2 summaries the related works about the initialization of the user model. Section 3 presents the main problematic of this research work. In section 4, we discuss the description of the proposed approach. Section 5 presents the k-means algorithm. Finally, in section 6, we give a conclusion with a discussion about the usability of the initialization of the tutor's model, and the future works.

2 Related Works

After a thorough investigation in the related scientific literature, we mention that there is no implementation of the initialization of the tutor's model. However, there are different research works that focused on the instrumentation of the tutors' roles ([7], [8], [9], [10]). Therefore, there are a lot of researches that focused on the identification of efficient methods to initialize the student model. The main goal of these systems is to extract the basic components constituting the initial model using different techniques. A great number of educational systems initialize the models of new students by assuming that they know nothing, or that they have some standard prior knowledge about the domain being taught. In these systems, the student model should not only reflect his knowledge structure, but also his personalized characteristics. For obtaining a student model, which includes all his stereotypes information, data mining techniques have been applied.

Stereotypes are used in many platforms to initialize the student model, because they are constructed by hand before real users have interacted with the system and they are not updated until a human does it so explicitly [11].

Data mining techniques have been applied in education where a new research area has been recently created, which is Educational Data Mining (EDM). The latter uses many techniques such as: Decision Trees, Neural Networks, Naïve Bayes, K- Nearest neighbor, and many others. By using these methods, many kinds of knowledge can be discovered such as association rules, classifications and clustering [3].

The first use of stereotypes was introduced by Rich [12] in a book about a recommendation system. She developed a system called "Grundy", which can save explicit statements realized by the user himself and make inferences from a user's behavior. This system is useful to build and exploit user models. After that, stereotypes have been used in many other educational systems as means for initializing the student model ([13], [14], [15] [16]). An interesting utilization of stereotypes is proposed by Aïmeur et al. [13]. The authors developed CLARISSE, which is a machine learning tool for the initialization of the student model. CLARISSE generates an adaptive pre-test to classify new students.

Tsiriga and Virvou [17] have initialized the student model in a framework called ISM (Initializing Student Models) based on the k-nearest neighbor algorithm. The latter is a technique that can be applied to initialize any profile. However, this method ignores some personalized characteristics that will influence on the learning process such as cognitive ability.

Our approach for initializing the model of a new tutor shares some similarities with the approach and the method followed by ISM. In particular, ISM framework has been able to construct the students' models based on direct observations of their behavior. Predefined stereotypes are used to assign students to a certain category of students. Then, the model of a new student is initialized taking into account certain personal student characteristics. In our case, to define tutors' characteristics we have based on previous researches done on the LETline system [10]. In this system, the roles of the tutors are dynamic and depend on their skills. The number of the roles that can be assigned to tutors is fourteen.

From what were cited above, we propose an approach to initialize the tutor model by using a type of data mining techniques, which is k-means algorithm. In the next section, we will describe the reasons to model the tutor.

3 Context and Problematic

The goal of the tutor model is to provide learning systems with relevant information to be adapted according to the characteristics and the preferences of learners. There are so many reasons for modeling the tutors such as:

- Give a description as complete and faithful as possible about all aspects related to this user behavior.
- Obtain a unique profile of each tutor because each one has his own characteristics.
- Select the roles that are specific for the tutors' job.
- Accompany, supervise and regulate the activities of the learner, and manage the activity of tutoring.
- Show how the model is developed through time to make the better choice of a profile.
- Adapt the model according to the characteristics and the needs of the tutor.
- Assess the tutors' knowledge.
- According to his profile, we can define an external tutor to transmit his experience.

Finally, the most important uses of tutors' model are:

- to reduce the distance and the isolation,
- to visualize the status of each tutor,
- to take into account the specificities of each learner, and
- to increase learners' motivation and optimize their learning.

A central part in the proposed approach is the tutor modeling component, which is responsible for acquiring and representing the necessary information about each tutor. The tutor modeling performs two main functions:

1- Initialize the tutor model when a new tutor connects for the first time.
2- Update the tutor model based on the tutors' interactions with the system.

In this paper, we have based on the initialization of the tutor model. To build this model, it is necessary to answer the following research questions:

1) What information is including in the model?
2) How to get it?
3) What type of tutor is he?
4) What the tutor has already done? and finally,
5) How can we initialize the model?

4 The Proposed Approach

When the tutor starts working within the system, this latter has no prior information about his behavior. However, the system couldn't give any preferences for him. Therefore, the model of this actor must have an efficient way for inferring initial information about the tutor.

The proposed approach is taken into account by an educational system (Fig 1).

In this system, all the necessary information is collected to build the profile of this actor. For obtaining his tutorial profile and knowledge level, a pre-test is used. The tutor can choose one or more roles according to his skills to do a task.

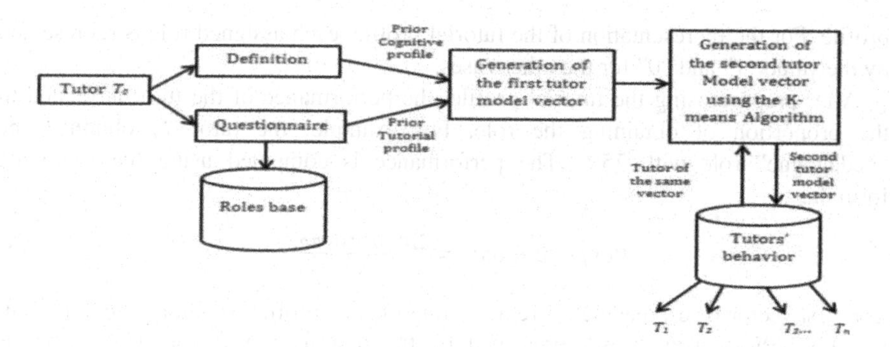

Fig. 1. Description of the proposed approach

a. The first vector (i.e. the first tutor):

The initialization of the tutor's model is based on his cognitive and tutorial profiles:

- ***Cognitive Profile*:** it depends on tutor's choices and skills. The tutor has as cognitive profile one value from the following {excellent, good, average, bad or very bad}.
- ***Tutorial Profile*:** The job of the tutor is related to his roles. During our research period, we noticed that the tutor has several nominations. From [10], the authors used fourteen roles. After testing the LETline system [19], we found that the number of roles was high. For this reason, we decided to reduce this number to eight roles (Fig. 2).

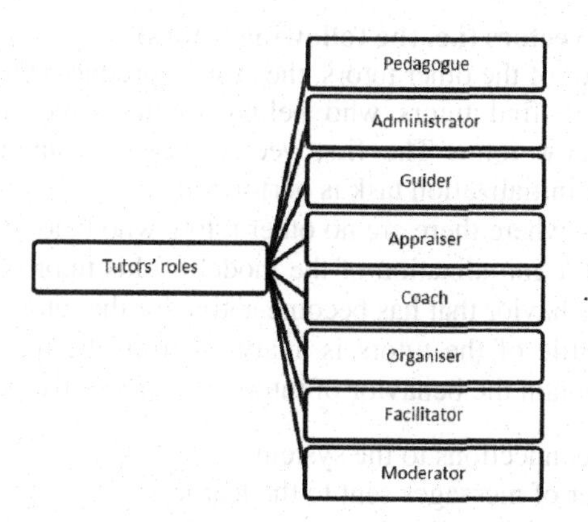

Fig. 2. The roles of the tutor used in our system

For initializing the tutorial profile of each tutor, we used a questionnaire that contains questions about the tasks of roles. From this questionnaire, if the responses to the questions related to a role X are more than 50% then the tutor can get the role X. If the responses are less than 50%, then the role X doesn't belong to the tutor's tutorial profile. For the representation of the tutorial profile, each assigned role is represented by the value "1" and "0" for the other cases.

After preprocessing the tutorial profile, the performance of the tutor is related to the proportion of obtaining the role. For example, the tutor T_1 obtained the "pedagogue" role with 75%. The performance is computed using the following formula:

$$Performance_R = \frac{\sum Proportion_R}{N} \tag{1}$$

The first vector is responsible for representing the two profiles (cognitive and tutorial) and his performance. It is represented by the first feature vector. This vector is defined as follows:

Id_T	Pedagogue	Administrator	Guider	Appraiser	Coach	Organizer	Facilitator	Moderator	Performance	Skills

For example:

Tut₁	0	1	1	0	0	0	1	1	65%	0.75

The vector represents the first tutor who has as roles: Administrator, Guider, Facilitator and Moderator. His total performance is 65%, and his skill is good.

b. The following vectors (i.e. the following tutors)

Based on the behavior of the other tutors, the system produces the second vector using k-means algorithm to find tutors who belong to the same cognitive and tutorial category with this new tutor. The first vector serves as an input to the k-means algorithm so that the initialization task is performed.

However, in cases where there are no other tutors who belong to the same category with the new tutor, the initialization of the model of this tutor is based on the default assumptions of the behavior that has become active for the tutor.

The behavior profile of the tutors is analyzed from the interactions tutor-learner and tutor-tutor. To obtain the behavior of tutors, we choose five key indicators:

a. The number of connections to the system.
b. The total number of messages sent to the learners.
c. The number of received messages.
d. The total number of demands (helps, support, and encouragement).
e. The total number of the accepted demands (helps, support, and encouragement).

For each indicator, we calculate the percentage of the number of activities performed by the tutor on the total number of activities performed by all the active tutors. An active tutor is a tutor who accessed the platform at least once.

For example, to calculate the score of the first indicator, we use the formula:

$$Connection = \frac{N\_C}{N\_C_t} \tag{2}$$

Where:

N_C: is the number of connections of the tutor who has already enrolled;
N_C_t: is the total number of connections of all active tutors.

To find the last behavior score of a tutor, we calculate the average of the scores of the five indicators (cited before) obtained by this tutor. Tutors are initially assigned to one of the five types of behavior depending on the interventions with the system. The set of rules used to extract the behavioral profile are the following:

- If the calculated percentage belongs to [0%, 20% [, the behavioral profile will be very isolated.
- If the calculated percentage belongs to [20%, 40% [, the behavioral profile will be isolated.
- If the calculated percentage belongs to [40%, 60% [, the behavioral profile is little dynamic.
- If the calculated percentage belongs to [60%, 80% [, the behavioral profile is dynamic.
- If the calculated percentage belongs to [80%, 100%], the behavioral profile is highly dynamic.

The proposed tutor model (Fig. 3) comprises two types of information. The first one has information that does not change during the learning process (Static features). While the second one includes those that can be changed during the learning process (Dynamic features) such as: the cognitive and the behavioral profile.

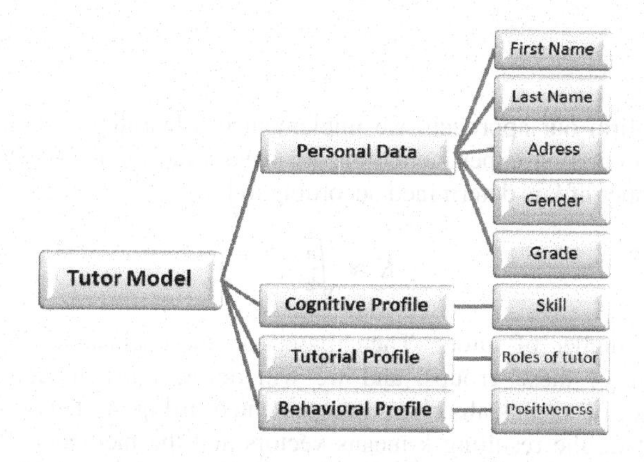

Fig. 3. Template of the tutor model

5 K-Means Algorithm

There is a large number of clustering algorithms in the literature. Its choice depends on the particular application. For our purposes, we need an algorithm that is capable of dealing with our real data. K-means is one of the simplest unsupervised learning algorithms that solved the well-known clustering problem.

This algorithm is used to classify or to group objects into K groups. K is a positive integer number, which is fixed a priori. The main idea is to define k centroids, one for each cluster, using the Euclidean distance metric. These centroids should be placed in a cunning way because different locations cause different results. Then, it is taken to be the new center value.

In order to better clarify the clustering algorithmic process, the pseudo-code of k-means algorithm is the following:

```
Program Centroid-Tutor
Input: a dataset T that represents all the tutors,
Output: K clusters
Begin
 Initialize cluster centroids (randomly);
 While not convergent
   For each object o in T do
       Find the cluster c whose centroid is most close to
       o;
     Allocate o to c;
   End
   For each cluster c do
       Recalculate the centroid of c based on the objects
       allocated to c;
   End
 End
End
```

In order to confirm our approach, we implemented a learning system that uses our propositions. In our first experimentation, we have fixed the number of tutors to 18 tutors. The number of k is determined according to [18]

$$K \approx \sqrt{\frac{n}{2}} \tag{3}$$

with n as the number of tutors. Then, based on these clusters, the new tutor is assigned to one of these groups, and we will define his behavioral profile. The implementation of k-means algorithm is represented in Fig. 4. This figure shows the initial tutors' data, the resulting k-means vectors and the members of means of the first tutor.

Actually, we are in the phase of the analysis of the obtained results.

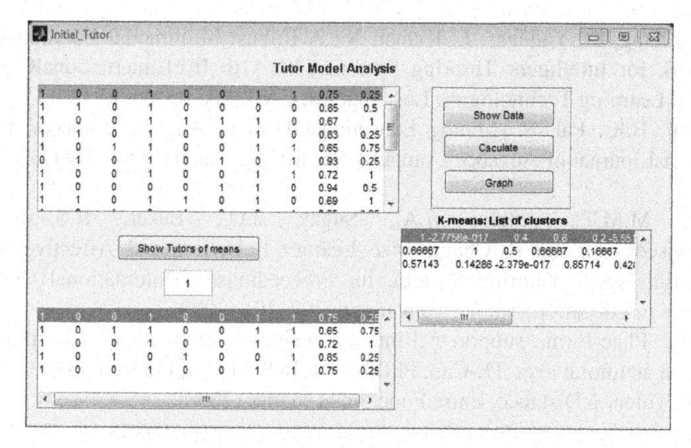

Fig.4. Snapshot of k-means to classify tutor

6 Conclusion and Future Works

In this paper, we have described an approach for improving the tutor model by taking into account its initialization process in educational systems. This initialization is performed for each tutor taking into account his personal data. Firstly, the system cannot define his behavioral profile. Based on his information, which is becoming as input to the k-means algorithm, we can generate a group of tutors constituting a cluster. This group is based on the similarities between the new tutor and the other tutors.

In this way, the initialization procedure is automatically updated each time the tutors interact with the system. We believe that this approach will produce good results since it uses well known techniques, already implemented in other similar scientific areas with quite promising reports.

Within the future plans of this research is the evaluation of the approach using real tutors to evaluate it and test its usefulness as an educational tool. Also, we plan to use a Bayesian approach to evaluate the effectiveness of the classification realized with k-means algorithm. This kind of evaluation needs a lot of time and a lot of tutors. Finally, we can apply other data mining algorithms to predict the tutors' performances and their last behaviors.

References

1. Sani, M.R., Mohammadian, N., Hoseini, M.: Ontological learner modeling. In: WCES 2012, ProcediaSocial and Behavioral Sciences, vol. 46, pp. 5238–5243 (2012)
2. Depoyer, C., De Lièvre, B., Peraya, D., Quintin, J.-J., Jailler, A.: Le tutorat en formation à distance. Perspectives en éducation & formation, DeBoeck (2011)
3. Klösgen, W., Zytkow, J.: Handbook of data mining and knowledge discovery, p. 1064. Oxford University Press, NewYork (2002)

4. Liu, X., Zhang, L., Yadegar, J., Kamat, N.: A Robust Multimodal Emotion Recognition Framework for Intelligent Tutoring Systems. In: 11th IEEEInternationalConference on Advanced Learning Technologies, Los Angeles, CA, USA, pp. 63–65 (2011)
5. Bharadwaj, B.K., Pal, S.: Mining Educational Data to Analyze Students' Performance. International Journal of Advance Computer Science and Applications (IJACSA) 2(6), 63–69 (2011)
6. Rodrigo, M.M.T., Anglo, E.A., Sugay, J.O., Baker, R.S.J.d.: Use of UnsupervisedClustering to Characterize Learner Behaviors and Affective States while Using anIntelligent Tutoring System. In: Proceedings of International Conference on Computers in Education, Taipei, Taiwan, pp. 392–401 (2008)
7. Garrot, E.: Plate-forme support à l'interconnexion de communautés de pratique (ICP). application au tutorat avec TE-Cap. PhDThesis, INSA Lyon, France (2008)
8. Rodet, J.: Tuteur à Distance, Entre Fonction et Métier.(2008), http://jacques.rodet.free.fr/intjrci.pdf (accessed at: October 01, 2012)
9. Lekira, A.R.: Rendre compte des effets des interventions des tuteurs à travers uneApproche Orientée Indicateurs. PhDThesis, MaineUniversity, France (2012)
10. Lafifi, Y., Azzouz, K., Faci, H., Herkas, W.: Dynamic Management of Tutor'sRoles in an Online Learning System. International Journal of Learning Technology (IJLT) 5(2), 103–129 (2010)
11. Tsiriga, V., Virvou, M.: Dynamically initializing the student model in a web-basedlanguage tutor. In: Proceedings of the 2002 First International IEEE Symposium"Intelligent Systems", vol. 1, pp. 138–143. IEEE Computer Society Press, Piraeus Univ (2002)
12. Rich, E.: User modelling via stereotypes. Cognitive Science 3(4), 329–354 (1979)
13. Aïmeur, E., Brassard, G., Dufort, H., Gambs, S.: CLARISSE: A machine learning tool to initialize student models. In: Cerri, S.A., Gouardéres, G., Paraguaçu, F. (eds.) ITS 2002. LNCS, vol. 2363, pp. 718–728. Springer, Heidelberg (2002)
14. Bontcheva, K.: Adaptivity, adaptability, and reading behaviour: Some results from the evaluation of a dynamic hypertext system. In: De Bra, P., Brusilovsky, P., Conejo, R. (eds.) AH 2002. LNCS, vol. 2347, pp. 69–78. Springer, Heidelberg (2002)
15. Zhao, C., Sun, Z., Liu, Q., Shang, C., Shen, D.: Initializing Students Models for theOnline Network Educational Systems. In: Proceedings of the 35th Annual Conference on Frontiers in Education, FIE 2005, China, pp. 14–17 (2005)
16. Virvou, M., Alepis, E., Troussas, C.: Centroid-based clustering for student modelsin computer-based multiple language tutoring. In: Proceedings of the InternationalConference on Signal Processing and Multimedia Applications and InternationalConference on Wireless Information Networks and Systems (SIGMAP), Rome, Italy, pp. 198–203 (2012)
17. Tsiriga, V., Virvou, M.: Initializing the Student Model using Stereotypes and Machine learning. In: El Kamel, A., Mellouli, K., Borne, P. (eds.) Proceedings of the 2002 IEEE International Conference on System, Man and Cybernetics, pp. 404–409 (2002)
18. Mardia, K., Kent, J.T., Bibby, J.M.: Multivariate Analysis, p. 521. Academic Press, London (1979)
19. LETline system, http://www.labstic.net/letline/

Towards a Generic Reconfigurable Framework for Self-adaptation of Distributed Component-Based Application

Ouanes Aissaoui[1], Fadila Atil[1], and Abdelkrim Amirat[2]

[1] Department of Computer Science, Badji Mokhtar University, Annaba, Algeria
aissaoui.ouenes@gmail.com, atil_fadila@yahoo.fr
[2] Department of Computer Science, Mohamed Cherif Messaadia University, Souk-Ahras, Algeria
abdelkrim_amirat@yahoo.fr

Abstract. Software is moving towards evolutionary architectures that are able to easily accommodate changes and integrate new functionality. This is important in a wide range of applications, from plugin-based end user applications to critical applications with high availability requirements. This work presents a component based framework that allows introducing adaptability to the distributed component-based applications. The framework itself is reconfigurable and it based on the classical autonomic control loop Mape-k (Monitoring, Analysis, Planning, and Execution). The paper introduces a prototype framework implementation and its empirical evaluation that shows encouraging results.

Keywords: Dynamic adaptation, Distributed systems, component-based application, Autonomic computing.

1 Introduction

With the development of the networks and the increasingly significant delocalization of the enterprises, the implementation of distributed applications was essentially being a solution allowing the cooperation of the various constituent actors of the enterprise. For the purpose of reducing the construction costs of distributed applications increasingly sophisticated development tools were designed.

Nowadays, more and more of distributed applications run round the clock, seven days out of seven. The shutdown of this type of application to improve it, or quite simply to perform updating operations costs a significant sums of money. A solution to this problem is to provide mechanisms allowing the evolution or the modification of an application during its running without stopping it [1]. So, we speak about the dynamic reconfiguration of distributed applications which can be defined as the whole of the changes brought to a distributed application at runtime.

A. Amine et al. (Eds.): *Modeling Approaches and Algorithms*, SCI 488, pp. 399–408.
DOI: 10.1007/978-3-319-00560-7_43 © Springer International Publishing Switzerland 2013

In the critical systems the adaptation must take place at runtime and the application should not be entirely stopped. Unfortunately, such adaptation is not trivial; there are several conditions and constraints to be satisfied, and this leads to many problems to overcome.

The problems treated in this paper accost the domain of research around the dynamic adaptation of the computing systems and in particular, the distributed component-based systems. Generally, the existing approaches provide solutions for (1) reconfiguration in non-distributed systems [5] or (2) reconfiguration in distributed systems but not distributed reconfiguration [10] [11] which is composed of multiple distributed processes.

Our objective is to facilitate the addition of the dynamic adaptation capabilities to existing component-based applications by providing a solution of management of the distributed and coordinated dynamic adaptation. For that, we propose a component-based framework to add flexible monitoring and adaptation management concerns to a running component-based application. In the proposed framework, we separate the concerns involved in the classical autonomic control loop MAPE (Monitoring, Analysis, Planning, and Execution) [3] and implement those concerns as separate components. As we treat in our context the distributed applications, we integrated in our framework a mechanism to manage the distributed coordinated adaptations. These components are attached to each managed sub-system.

The remainder of the paper is organized as follows. Section 2 presents an overview of our solution for the distributed and dynamic reconfiguration. Section 3 details step by step the design of our framework following the autonomic computing MAPE-K phases. In Section 4, we give implementation details for a prototype of our framework. Finally, Section 5 concludes the paper.

2 Overview of Our Solution for the Distributed and Dynamic Reconfiguration

We consider an application as a self-adaptable if it's composed of an adaptation system on the one hand and of a set of functional components on the other hand. The adaptation system is responsible for the management of the application context (collection of data, analyzes…) and of its adaptation, whereas the components represent logic trade of the application (functional code). Such separation is also suggested in many works such as [10], [13], [15]. In our context the adaptation system is represented by the framework which we present in this paper. Figure 1 shows an overview of our solution. For reasons of clearness, only two sites are represented.

As we treat the distributed applications, we find at each site a sub-system (a set of application components) plus one instance of our framework which manages the sub-system. The negotiation of adaptation strategy and the execution coordination of an adaptation operation are done via special components integrated in the framework. This organization makes the architecture of our solution decentralized what avoids the problems of the centralized approaches [14].

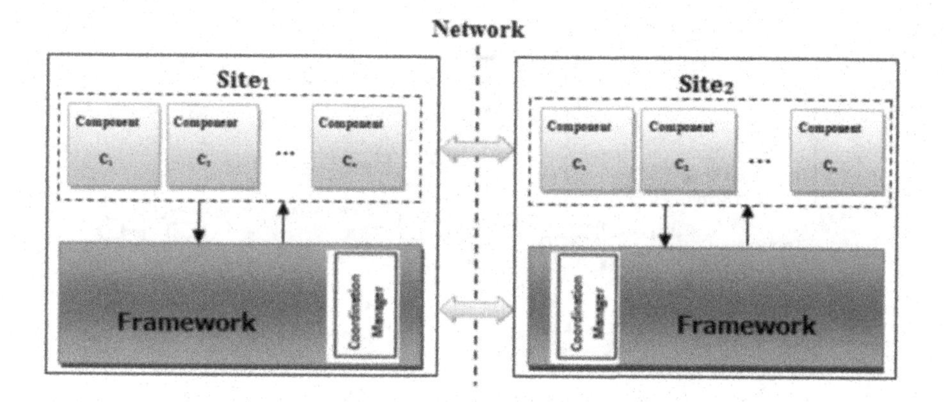

Fig. 1. Overview of our solution for the dynamic and distributed adaptation

3 Design of the Proposed Framework

For the definition of our framework, we consider a set of constraints which are: (i) independence of the existing component models, (ii) flexibility and extensibility of the framework, (iii) adaptability of the framework, and (iv) taking into account of the distributed nature of the software to make it adaptable.

Our framework is based on the classical autonomic control loop Mape-k (Monitoring, Analysis, Planning, and Execution) [3]. This loop is used in many works treating the dynamic adaptation [8] [11], [12]. The difference between these current research activities is in the implementation way of the Mape-k loop.

So, in our framework we separate the concerns involved in a classical autonomic control loop and implement those concerns as separate components. The monitoring, analysis and adaptations are carried out by this control loop. We have merged the two phases of analysis and planning and we have integrated them in the same component. A significant part of the coordination, negotiation and the checking (checking of the application structure and the behavior of its components) were externalized of the control loop. The Figure 2 shows an overview of our framework.

The coordinator coordinates the execution of the adaptation operations; the negotiator negotiates an adaptation strategy with its similar at the other sites. The checking component carries out the checking of the application structure as well as the checking of the behavior of its components following the running of an adaptation operation. This checking operation is carried out on the architecture description of the application. For that, the component <<translator>> (presented hereafter) forwards the changes carried out on the application to its architecture description each time that an adaptation operation is done. This is for assuring a causal connection between the architecture description of the application and the system in running.

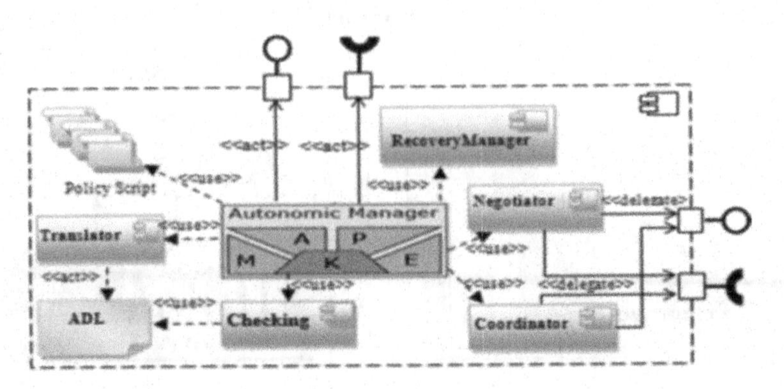

Fig. 2. Overview of the proposed framework

Knowledge Manager. It's a component used by the various components of the framework. It allows (1) to safeguard information on the knowledge base and (2) to provide information contained in the knowledge base to the other components according to their need.

Monitor. It's the first component in the chain which constitutes the control loop. It's composed of sub-component of the type <<CaptureContext>> and a set of sub-components of the type <<Sensor>>.The first supervises the application environment (e.g. memory, CPU, bandwidth…) whereas, the second supervises the various functional components of application (i.e., sub-system of the application managed by the framework) for that, it has a cardinality 1-*. These components (<<CaptureContext>> and <<Sensor>>) are responsible to (1) gather periodically the information of the controlled elements and (2) to pass them to the next object of the chain (the context manager).

ContextManager. It's the second object in the execution chain. It's a composite component in charge of the management of the running context. For that, it's composed of two components <<AcquisitionManager>> and <<Interpreter>> with a cardinality 1-1 for both. The first (1) gathers the information collected by the monitor and saves it in the knowledge base via the component <<KnowledgeManager>>, and (2) delegates the execution to the component <<Interpreter>>. This last, interprets data provided by the acquisition-manager. The received data are separately interpreted for each type of measurement in order to provide a significant contextual data. For example, a decreasing bandwidth event can be alone non representative. On the other hand, if it's repeated in time, it can indicate that the user moves away from an access point and thus being significant. So, the interpreter stores the values measured by event type. As the decision maker <<DecisionMaker>> is registered with events near the context manager, the detection of a suitable context change triggers the notification of the decision maker. In this case, the interpreter delegates the execution to the next component of the chain (DecisionMaker).

DecisionMaker. It's the third object in the execution chain of the control loop. It's responsible of making an adaptation decision and provides in exit the adaptation strategy to be applied. For that, it subscribes with events near one or more context managers. The decision maker (1) starts the interpretation of the adaptation policy (script) which is of type ECA "Event, Condition, Action". It is possible that several rules (i.e. several adaptation operations) so, several strategies are applicable at the same time. In this case, the decision maker must order these strategies according to their priorities, then (2) it initiates a negotiation operation of this strategy via the component <<Negotiator>> according to the given sequence. This negotiation is necessary since we speak here about the distributed adaptations. At the end of the negotiation, the decision of the negotiator is the notification of the participants of the negotiation failure or of its success with the strategy selected. Thereafter, the <<DecisionMaker>> (3) delegates the execution to the next object (Executor) in the chain.

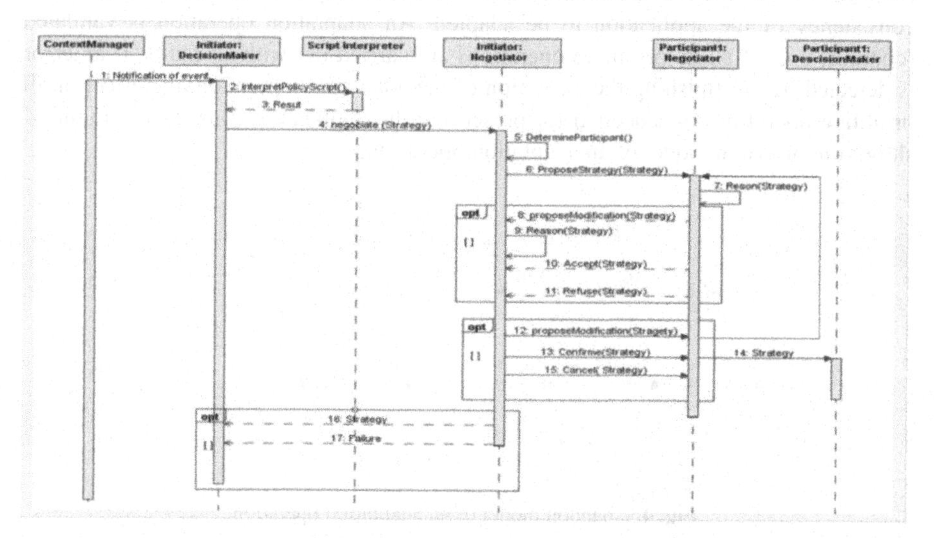

Fig. 3. Sequence diagram of negotiation between two adaptation managers

The diagram showed in figure 3 describes the sequence of messages for the negotiation of a strategy between an initiator and participants. For reasons of clearness, only one participant is represented.

The initiating decision maker chooses an adaptation strategy. Then, it asks its negotiator to negotiate the strategy which it chose. This negotiator proposes simultaneously to each participant the strategy that the decision maker chose. The negotiator of each participant receives the strategy and interprets its policy to reason on its applicability. It can then accepts, refuse or propose a modification of the strategy. Then, it answers the initiating negotiator. When this last receives all the answers, it thinks on the acceptances and/or the applicability of the modifications asked. When all the participants accept the strategy, the negotiation succeeds. Otherwise, it detects and solves the conflicts and it can then in its turn propose a modification of the strategy. The negotiation process is stopped if one negotiator refuses a strategy or if a stop condition is checked.

This condition is in connection to the authorized maximum time of negotiation or with the maximum number of negotiations cycles. If the negotiation succeeds, the initiating negotiator returns to the initiating decision maker the strategy resulting from the negotiation and sends to the negotiator of each participant the final strategy. At the reception of this strategy, the negotiator of the participant asks to this last (3) to adopt the strategy resulting from the negotiation and delegates the execution to the next object in the control loop <<Executor>>.

Executor. We adopted the transaction-based system technique [16] to make our adaptation operations transactional i.e. having the properties ACID (Atomicity, Consistency, Isolation, Durability) of transactions. So, we consider an adaptation operation as a set of primitive operations of adaptation.

The purpose of this decomposition is to facilitate the detection of errors during the running of these operations and much more their recovery what allows to preserve the consistency of the application to be adapted. An adaptation operation is validated (commit) only if all its primitive operations are carried out without faults. If an error is detected before finishing the execution of the adaptation operation, the effect of all primitive operations is cancelled for preserving the application consistency. Figure 4 shows our abandon model of an adaptation operation.

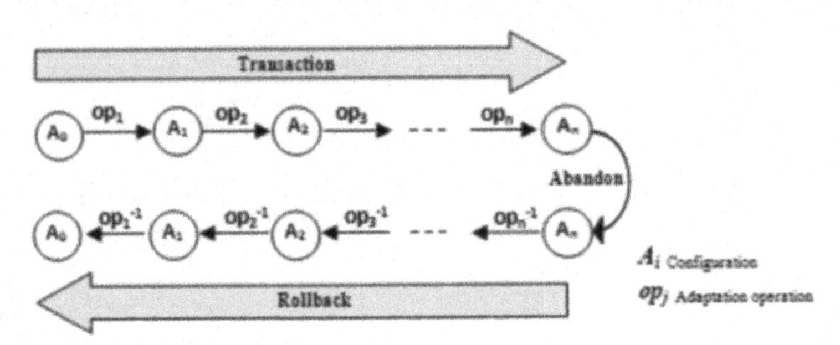

Fig. 4. Abandon model of an adaptation operation

According to our model the effect of the reconfiguration operation is cancelled by the running of the reverse action of each primitive operation done. We define the concept of opposite operation which is used to undo the effects of a reconfiguration operation and which is useful to ensure its atomicity. All the reconfiguration operations are not necessarily invertible. The operations of modification of the component properties are generally their own reverse. In addition, the opposite (or reciprocal) operation of a primitive operation is not necessarily a primitive operation but can be a composite operation.

Given a configuration A, by application of the composite operation $op^{-1} \circ op$ on A, we obtain normally $op^{-1} \circ op\ (A) = A$ according to the following diagram where $\overset{op}{\Rightarrow}$ indicates the reconfiguration by the operation op:

$$A \overset{op}{\Rightarrow} \acute{A} \overset{op^{-1}}{\Longrightarrow} A$$

For example, the reverse of the operation *removeComp* which allows removing a component is the operation *addComp* which allows adding a component.

In certain component model, certain operations are not invertible, like in the Fractal model where the operation *new* for which the opposite operation would correspond to an operation of destruction of component does not exist. For the particular case of the non reversible operations, the cancellation of reconfiguration requires a specific treatment in the event of abandonment of the transaction. Compensation operations can then be associated to these operations; moreover no guarantee on atomicity can then be given because the state of the system resulting from the abandonment cannot be completely identical to the state before the running of the reconfiguration.

Algorithm 1

```
1: Begin
2:   For all op_j ∈ startegy do
3:        RunOp(op_j) ;
4:        TranslateChanges(); // translate changes to the architectural description
5:        if not IsConsistentApplication () then
6:              SendMessageToCoordinator ("Adaptation failure");
7:              For all executed primitive operation op_j do   RecoveryManager.undo(op_j) ;
8:              end_for
9:        end_if
10:       else
11:            SendMessageToCoordinator ("ApplyNextOperation");
12:            response←coordinator.decisionCoordinator() ;
13:            if response != " ApplyNextAction "   then
14:                 SendMessageToCoordinator("Adaptation failure") ;
15:                 For all executed primitive operation op_j do RecoveryManager.undo(op_j);
16:                 end for
17:                          BREAK; // to exit the more external loop (for)
18:                      end_if
19:        end_else
20:  end_for
21:  if all operations in strategy are executed // if the adaptation is succeeds
22:        LogExecutedOps();
23:  end_if
24: End.
```

The <<Executor>> is the component responsible for the execution of the adaptation strategy suggested by the component <<DecisionMaker>> and of its control. For that, it (1) triggers the execution of each reconfiguration action in the strategy according to the order of their appearances.

We consider the adaptation of distributed application as a global adaptation process composed of distributed local adaptation processes. For that, a coordination component of the execution of an adaptation is necessary.

Following the running of each primitive adaptation operation, the <<Executor>> (2) calls the translation function of the component <<Translator>> for transferring the changes performed in the application in running to its architectural representation. After, it (3) carries out the checking of the consistency of the application structure and the checking of the validity of the behavior of its components via the component

<<Checking>>. If a constraint is violated, the <<Executor>> asks to the recovery component <<RecoveryManager>> to carry out the rollback for preserving the consistency of the application. In this case there, the component <<RecoveryManager>> undoes the effect of all the primitive adaptation operations which are already executed through the execution of their reverse operations as explained in the previous section. Moreover, this initiating executor notifies the coordinator of the failure of execution of the primitive adaptation operation in question. This last, deals with the notification of the other participants of this failure (the participants are the coordinators of the adaptation execution which are deployed at the other sites) so that they can undo the effect of the primitive operations already carried out at their level in order to preserve the global consistency of the application.

In the opposite case, i.e. if the <<Checking>> does not detect any error following the running of a primitive operation of adaptation, the <<Executor>> sends a message *"ApplyNextAction"* to the coordinator. This last awaits the reception of all the participants' messages. If one of them replies negatively (i.e. adaptation failure), the coordinator announces the failure of the execution of the adaptation operation. Otherwise, it indicates to the participants to carry out the next primitive adaptation operation and the process is still repeated. After the running and the validation of all the primitive operations of all adaptation operations in the strategy, the <<Executor>> (4) logs these executed operations in the journal of the application for a future use. The end of the execution of this operation determines the end of the control loop cycle. The running of the <<Executor>> is summarized by algorithm 1.

4 Implementation and Validation

In this section, we give details and technical choices made to implement an instance of our framework. We present also the result of the evaluation of this framework.

4.1 Background

For the implementation of the elements of our framework which we have presented in the section 3, we have used the component model ScriptCOM [9] which is an adaptable model extension of the model COM (Component Object Model) [2]. It's a component model which we have proposed in an earlier work. We have used this model because it allows the development of a scripting component as it's based on the use of the scripting languages. These languages allow the incremental programming, i.e. the possibility of running and developing simultaneously the scripts which represents in this context the components implementation. This adaptation is possible via a set of three controllers which are: the Interface controller, script controller and property controller. Moreover, this model benefited from contributions and advantages of the COM model since it's an extension of the latter. We have chosen this component model in order to make our framework itself adaptable.

4.2 Framework Implementation

The framework is implemented via the component model ScriptCOM as a set of non functional components that can be added, removed or modified at runtime. We have designed a set of predefined components that implement each one of the elements which we have described in Section 3. This is just one of possible implementations and particularly, this has been designed to provide self-adaptable capabilities to the framework.

4.2 Validation Plan

In order to validate our proposal, we have used the industrial model EJB [4] for the development of an application example which is an http server. We have chosen this model to prove that our framework is generic because it's implemented via the component model ScriptCOM and the adapted application is developed via another component model (EJB). We have chosen this application (HTTP Server) since it's used in the evaluation of many works [5-7]. Therefore, it represents a reference for us. This application of the type server does not interact directly with a user. However, the need for performances implies that it must be able to adapt to the characteristics of its host and the type of load which it undergoes. In this example, the significant context for the adaptation will be thus that of the material and software resources rather than the characteristics of the end-user. In order to improve the performances, we have integrated a mechanism to put in cache the content of files which it reads.

The objective of the validation in this paper is to test the adaptation mechanism influence on the application response time and the adaptation time. We have obtained encouraging results, where the influence on the response time is stable and that overhead time is about 15%. The adaptation time average is approximately 2 seconds. Of course, this figure is large compared to the response time of one request which is approximately 30ms. Notice, that this test is done via machines equipped with Intel(R) Core(TM) 2 Duo CPU T5670 @ 1.80GHz 1.79GHz and 1 GB of RAM.

5 Conclusion

We have presented a generic reconfigurable component-based framework for supporting the dynamic adaptation of distributed component-based applications. Our framework is based on the classical autonomic control loop Mape-k (Monitoring, Analysis, Planning, and Execution). It implements each phase of the autonomic control loop as a separate component, and allows multiple implementations on each phase, giving enough runtime flexibility to support evolving non functional requirements on the application. To the difference of the others frameworks, our framework is conceived to support the distributed adaptations. Moreover, it's independent of the component models and designed to minimize the cost and the time of the addition of capacities of self-adaptation to a large variety of system. A prototype of this framework has been implemented using an adaptable component model *ScriptCOM*. Moreover, an empirical evaluation of this prototype is done and it shows encouraging results.

Our future work focuses on improving the response time by the improvement of the negotiation and coordination algorithms.

References

1. Taylor, R.N., Medvidovic, N., et al.: Software Architecture: Foundations, Theory, and Practice, p. 736. John Wiley & Sons (2008)
2. Microsoft Corp., Component Object Model, `http://www.microsoft.com/COM`
3. IBM. An architectural blueprint for autonomic computing. Autonomic computing white-paper, 4th edn. (2006)
4. Matena, V., Hapner, M.: Enterprise Java Beans Specification v1.1 - Final Release. Sun Microsystems (May 1999)
5. David, P.C.: Développement de composants Fractal adaptatifs: Un langage dédié à l'aspect d'adaptation. PhD thesis, université de Nantes, France (2005)
6. Léger, M.: Fiabilité des Reconfigurations Dynamiques dans les Architectures à Composant. PhD thesis, Ecole Nationale Supérieure des Mines de Paris (2009)
7. Dormoy, J., Kouchnarenko, O., Lanoix, A.: Using Temporal Logic for Dynamic Reconfigurations of Components. In: FACS, 7th Int. Ws. on Formal Aspects of Component Software, Portugal (2010)
8. Ruz, C., Baude, F., Sauvan, B.: Flexible adaptation loop for component-based soa applications. In: ICAS 2011, The Seventh International Conference on Autonomic and Autonomous Systems, pp. 29–36 (May 2011)
9. Aissaoui, O., Atil, F.: ScriptCOM an Extension of COM for the Dynamic Adaptation. In: Proc. of 2nd IEEE International Conference on Information Technology and e-Services, Tunisia, pp. 646–651 (2012)
10. Garlan, D., Cheng, S.W., Huang, A.C., Schmerl, B., Steenkiste, P.: Rainbow: Architecture-based self-adaptation with reusable infrastructure. IEEE Computer 37(10), 46–54 (2004)
11. Maurel, Y., Diaconescu, A., Lalanda, P.: Ceylon: A service-oriented framework for building autonomic managers. In: Seventh IEEE International Conference and Workshops on Engineering of Autonomic and Autonomous Systems (EASe), pp. 3–11 (2010)
12. Gauvrit, G., Daubert, E., Andr, F.: Safdis: A framework to bring self-adaptability to service-based distributed applications. In: SEAA 2010, Proceedings of the 36th EUROMICRO Conference on, Software Engineering and Advanced Applications, pp. 211–218. IEEE Computer Society (2010)
13. Baresi, L., Guinea, S.: A3: Self-Adaptation Capabilities through Groups and Coordination. In: ISEC 2011, Kerala, India (2011)
14. Tan, C., Mills, K.: Performance characterization of decentralized algorithms for replica selection in distributed object systems. In: WOSP, pp. 257–262. ACM (2005)
15. Zouari, M., Segarra, M.T., André, F.: A Framework for Distributed Management of Dynamic Self-adaptation in Heterogeneous Environments. In: IEEE International Conference on Computer and Information Technology, pp. 265–272 (2010)
16. Gray, J., Reuter, A.: Transaction Processing: Concepts and Techniques. Morgan Kaufmann Publishers Inc., San Francisco (1992)

Dynamic Bayesian Networks in Dynamic Reliability and Proposition of a Generic Method for Dynamic Reliability Estimation

Fatma Zohra Zahra[1], Saliha Khouas-Oukid[2], and Yasmina Assoul-Semmar[3]

[1] High School of computer science, Algiers, Algeria
f_zahra@esi.dz
[2] Computer Science Department, Saad Dahlab University, Blida, Algeria
oukhouas@univ-blida.dz
[3] Department of Aeronautics, Saad Dahlab University, Blida, Algeria
assoulyasmina@yahoo.fr

Abstract. In this paper, we review briefly the different works published in the field of Dynamic Bayesian Network (DBN) reliability analyses and estimation, and we propose to use DBNs as a tool of knowledge extraction for constructing DBN models modeling the reliability of systems. This is doing, by exploiting the data of (tests or experiences feedback) taken from the history of the latter's. The built model is used for estimating the system reliability via the inference mechanism of DBNs. The proposed approach has been validated using known system examples taken from the literature.

Keywords: dependability analysis, dynamic Bayesian networks, structure learning, knowledge discovery, dynamic reliability estimation.

1 Introduction

The reliability of a system can be defined as the probability that the system can perform its intended function for a specified period under specified conditions [1]. Estimating the reliability of systems is a very important problem from the point of view of companies reputation, customer satisfaction and cost of the system model that can be directly linked to failures of the system, as it represents a real challenge for engineers because, current techniques for estimating reliability require a very high level of background in reliability analysis, as well as familiarity with the system to be studied [3].

Understanding the interaction of components of new complex systems, whether in the design (modeling) or deployment stage becomes a difficult problem that requires action by experts. Therefore, the old problems of reliability estimation develop into new challenges when addressed with respect to the modern complex systems and operation tasks. The main modeling features to assess the reliability of complex moderns systems [2], [4] are:

A. Amine et al. (Eds.): *Modeling Approaches and Algorithms*, SCI 488, pp. 409–418.
DOI: 10.1007/978-3-319-00560-7_44 © Springer International Publishing Switzerland 2013

- The representation and modeling of the system (the complexity and the size of the system, the temporal aspect, the dependence between events such as failures and the nature of the components (Multi-State)).
- Quantification of the model system (the integration of qualitative information with quantitative knowledge at different abstraction levels).
- The representation, the propagation and the quantification of uncertainty in the behavior of the system (the uncertainty in the estimation of parameters).

The modeling of these characteristics poses many problems, to solve these problems, the BNs have been proposed as an alternative to the traditional approaches of reliability analysis and estimation [1]. This method of modeling is not the solution of all the problems, but it seems to be very suitable in the context of the complex systems [1-2].

BNs [5] can be described briefly as Directed Acyclic Graphs (DAGs) which define a factorization of a joint probability distribution over the variables that are represented by the nodes of the DAGs, where the factorization is given by the directed links of the DAGs. The DBNs are at the same time, an extension of the BNs in which the temporal evolution of the variables is represented. There is tow approaches of integration of temporal dimension in the BNs, the first one is called generically the time-sliced approach [6-7] and the second approach is called generi-cally the event-based approach [6], [8].

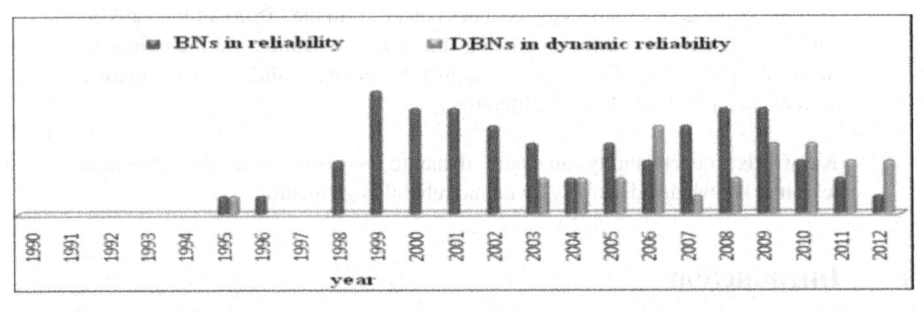

Fig. 1. Trend of publications in the area of reliability estimation based BNs and DBNs

The researchers have focused initially on the standard BNs. However, these latter's fail to capture the dynamic behavior of complex system components, where not only the combination of failing components reflects the state of the system, but also the sequence in which these components fail. DBNs can models this aspect in intuitive way. Consequently, they have been proposed for estimating the dynamic reliability [3], [6], [10-16].

This paper reviews the different works published in the field of DBNs for reliability analyses and introduces an approach that uses the methods of structure and parameters learning of DBNs from data to automatic building of DBN model representing a specific system. Sections 2 and 3 present a review and discussion on the use of DBNs for estimating and analyzing systems reliability, the proposed approach and its validation will be presented respectively, in sections 4 and 5, and we will finish with a conclusion.

2 Overview on the Use of Dynamic Bayesian Networks for Estimating and Analyzing the System's Reliability

The Fig. 2 shows that 31% of the topics treated in the literature relating to the application of the BNs in the field of the system's reliability estimation relate to the use of the DBNs for modeling the reliability of the dynamic systems. The statistical study is made on approximately 100 papers, which represent the principal contributions published in the journals: RESS (Reliability Engineering and System Safety), IEEE (Transactions One Reliability), and QREI (Quality and Reliability International Engineering). This study also shows that over 20 years period (from 1990 until 2012), the researchers started to be interested in the dynamic models of BNs and their application in the estimation of the dynamic reliability at the beginning of the ten last years (from 2003). The interest is continued until now (end of year 2012, see the Fig. 1).

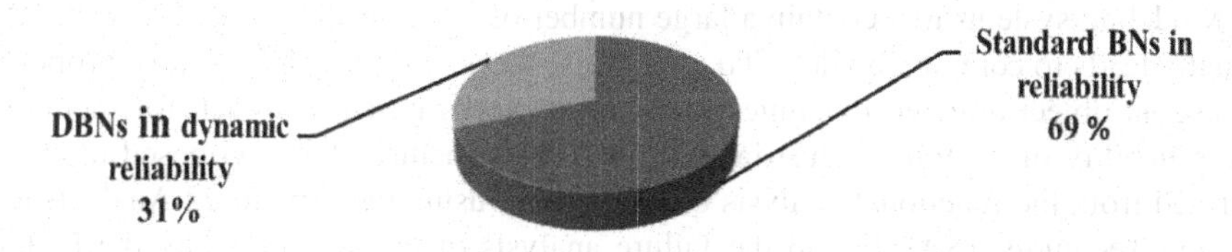

Fig. 2. Distribution of the references on the topics: DBNs in dynamic reliability and BNs in reliability

Bobbio et al [9] have proposed the use of BNs to improve the dependability analysis, and this by converting the standard FT in an equivalent BN. The Fault Tree Analysis (FTA) is one of the most popular techniques among the engineers of reliability [1], [9].

The dynamic Fault trees (DFTs) are an improvement of the conventional FTs, used for modeling the dynamic reliability. They introduce four (04) new logic gates to capture complex patterns of dependency called dynamic logic gates. *Montani et al* [10] have developed the work presented above by using DBNs to model the dynamic logic gates of DFTs. The same researchers presented in [11], RADYBAN (Reliability Analysis with Dynamic Bayesian Networks), a software for automatic conversion of a DFT into a DBN. A set of inference algorithms of DBNs has been implemented in this software for estimating the reliability.

Portinale et al [12] present two important extensions of the DFTs called: probabilistic dependency gate (PDG) and repair boxes gates. In particular, the repair process is very important for reliability analysis. The modeling capabilities offered by DBNs have enabled the authors to model the new logic gates of the DFT, and to exploit the power of DBNs in reliability estimation and analysis. The conversion of the logic gates (PDG, Repair Boxes) in DBN model is implemented and integrated into the compilation process of RADYBAN, the software developed in [11].

Boudali and Dugan [6] describe a Discrete Time Bayesian Network (DTBN) formalism to model the reliability of dynamic systems. They used the event-based approach for the time representation in the BNs. In this formalism the DFT of a system is converted into an equivalent discrete time BN.

The work of *Boudali and Dugan* presented in [13] generalizes the DTBN framework developed in [6] by the same researchers. The network structure is the same, the difference is that the variables in the DTBN are discrete, whereas in the new formalism called Continuous Time Bayesian Network (CTBN) are continuous. Therefore, the conditional probabilities become a density functions probability and the joint probability distribution becomes a density function of the joint probability. The benefits of the CTBN framework compared at the DTBN framework are:

- Gain of memory conditional probabilities are expressed as parametric functions, there are no multidimensional tables.
- Appropriate solutions; exact analytical expression of system reliability is derived.

Weber et al [14] propose to use the DBNs to model the dependencies between components of complex industrial systems and their environment. This allows representing the influence of time and exogenous constraints on the degradation and failure modes of the system.

Modeling systems that contain a large number of components (variables) with BNs usually leads to complex models. To avoid this, *Weber and Jouffe* [15] have proposed to use an object-oriented dynamic Bayesian networks based approach for modeling the reliability of complex industrial systems. The structure of the proposed model is derived from the functional analysis of the system, using the Structured Analysis and Design Technique (SADT) and the failure analysis of the system, using the Failure Modes and their Effects Analysis (FMEA) method.

3 Discussion

The Continues Time Markov Chains (CTMC) are very suitable for modeling the reliability of dynamic systems, however, their main limitations is the combinatorial explosion of the number of states, implying a exponential running time. Hence, among the main advantages of using DBNs, whatever the approach used is to avoid this problem, in addition of the advantages of modeling possibilities offered by the graphical probabilistic models that are:

- Uncertainty about the local dependencies (probabilistic logic gates).
- Multi-states Variables (multiple modes and behaviors).
- Uncertainty in the model parameters.
- Dependences between components (e.g., represented by a common environment).

Without forgetting that the reliability analysis can be done in both directions: the diagnosis [17] and prognosis, thanks to inference algorithms. These benefits are common to all approaches presented above, although each approach has its individual advantages and disadvantage.

The works of *Bobbio et a*l [9], *Montani et al* [10] and *Portinale et al* [11] realized by the software RADYBAN allowed reliability experts unfamiliar with BN to exploit the power of DBNs in the reliability estimation and analysis, without given up the conventional approaches (FT, DFT). However the discretization step width influences the results of the reliability estimation. Therefore, the results obtained using the DBN are not exactly identical to the results obtained using the CTMC (the method usually adopted for analyzing the reliability of DFT), In fact both methods are not exactly equivalent. There is a trade-off between the approximation provided by the discretization step and the computational effort needed for the analysis: Smaller is the discretization step, more accurate are the results obtained, and close to the continuous case computation. The DBNs type used in these works is restricted to Markovian systems.

Using the events based approach for representing the time in DBNs in *Boudali and Dugan* [6] frameworks has avoided the unnecessary complexity of using the Time-Sliced approach, and has allowed to apply this formalism to all systems, not only to the Markovian systems. In this framework, the conversion of a DFT in DBN can be automated. The CPTs and the a priori probability tables are defined using failure probability distribution of the basic components and the type of logic gates which are endearing. There is no need for BN experts, the structure of the DBN and the a priori probability tables (CPTs). The all can be automatically derived from the DFT. Nevertheless, these works consider only non-repairable systems. The extension of this work for repairable systems remains a prospect.

The work of *Weber Jouffe* [15] presents a very interesting formalism for modeling the reliability of complex systems, however, an important set of methods for analyzing the reliability of systems such as (FMEA, SADT) is used to construct the DBN which requires the intervention of several experts in the reliability field.

4 Proposed Approach

In summary, we can say that the principle of approaches using DBNs for the reliability estimation and analysis is the same in all the different methods presented in the previous sections, the network variables (nodes) represent the system components and the logic gates that connect them, links (arcs) represent the dependency and temporal relationships between components, for the quantitative part, the conditional probability tables represent the degree of force of the relationships (links).

Current approaches use a special DBN for a specific system. The construction of the DBN is done by the experts [18]. The network construction is done directly in collaboration with domain experts and BN experts, conditional probability tables are given by the domain experts, either the network structure is given by the domain expert using a conventional method of reliability analysis, the method often used is the DFT analysis, the DFT is then converted into a DBN and can be enriched by the views of domain experts, the priori conditional probability tables are given by the experts.

We propose a generic method for analyzing and estimating the reliability of systems (static and dynamic) by learning and extracting the knowledge about a system under study in the form of a DBN from its test or historical data.

To realize the process of extraction and exploitation of knowledge necessary for modeling the reliability of a system as a DBN (see Fig. 3), we must pass through the following phases:

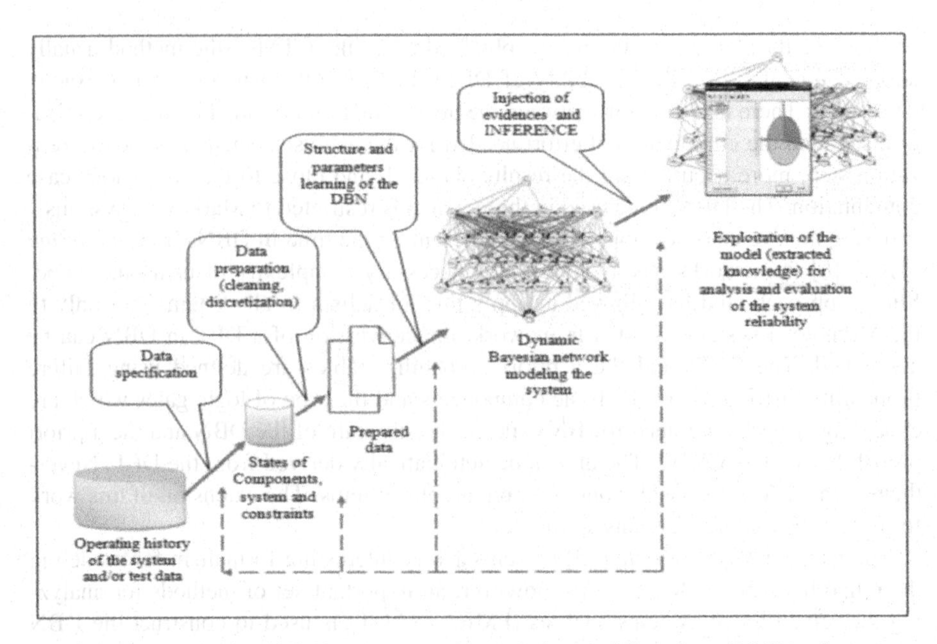

Fig. 3. Discovery and exploitation process of knowledge modeling the reliability of a system using DBNs

Phase 1 (Data Specification): The first step is to select the data that can meet our problem "modeling system reliability". The data collected throughout years of operation of a system representing the operational history of the latter are of different forms. Each organism stores these data in a different ways. These data essentially contain the operating parameters of components (such as temperature, vibration ... etc.), system failures encountered, their causes and duration as well as the maintenance done and problems found during preventive, programmed or corrective maintenance. Complex systems are equipped with a modern digital control and HMIs (Human Machine Interfaces) used to manage the system and know the states of its components, represented by their operating parameters through sensors. Therefore, all these information are automatically stored. We can use also the test data (in the case of a new system which it was not put yet in exploitation).

To model the dynamic reliability of a system, we need to know the state of the system and the states of its components at each time *t* for a period of operation *T*, the state of system and its components can be summarized as {working / failure}, as it can take other modalities, for example, a backup component may be in down, running or waiting state. We can add also variables that represent external constraints such environmental and operational factors.

Phase 2 (Data preparation): The second phase involves preparing the selected data specified in the previous phase so that they are easily exploited. Until now, there are no efficient algorithms for learning hybrid DBNs. Therefore, the continuous data must be discretized. Incomplete data are taken into account by the learning algorithms of DBNs [7]. Therefore, we will keep the partial observations of the system. This problem is addressed by using the appropriate learning algorithm.

Phase 3 (Model construction): The third phase consists of applying one of the methods of learning DBNs. Before starting the construction phase of the DBN model modeling the reliability of a system. We first had to choose between the event-based approach and the time-sliced based approach that incorporate temporal dimension in a BN. Each approach has advantages and disadvantages in the context of reliability modeling. The time-sliced based approach has been chosen based on two very important points:

- We can't model repairable systems using the event-based approach. Therefore we can't extend this work for risk analysis and maintenance planning.
- The inference mechanism of the event-based approach is not well developed, and most importantly, the learning mechanism of structure and parameters of the network is completely absent.

Nodes (variables) in the proposed DBN model represent the components of the studied system. The links (arcs) represent the dependency relationships between components and temporal relations representing the dynamism of the system. Nodes representing the exogenous constraints (environmental and operating conditions of the system) can be added. The arcs connecting the exogenous variables and the endogenous variables (system components) indicate accurately the impact (influence) of the environment and operating conditions on the system reliability.

We used in this work the 2TBN model of DBNs (Two-Slice Temporal Bayes Net) that is similar to the Markov model with independent dynamic variables. It means that in calculating the variables in step (i+1), step (i-1) is forgotten due to the Markov property, which says that the future state depends only on the present one. A 2TBN is defined to be a pair, $(B_0, B_\rightarrow)$ where B_0 is a BN which defines the prior P(X♭o♯), and $B_\rightarrow$ defines $P(X[t]|X[t-1])$ by means of a DAG [19].

For learning the DBN model representing a specified system, we assume that we have a database that contains N cases observed at T time sequences. Each sequence contains N_l cases Such as: $N = \sum_l N_l$.

The system to be modeled contains n_c components, and we also have, n_{ex} exogenous variables to represent. Consequently, we will have n variables (nodes). Where, $n = n_c + n_{ex}$. We add another variable which represents the state of the system in entirety, therefore the total number of the variables is: $n + 1$.

The initial structure B_0 (that of the first slice) is learned from the whole of data made up of the first case of each sequence. Then, we have N_0 cases to build the structure of the initial network, which is built independently of the transition network, and by using one of the algorithms of structure learning of standards Bayesian networks.

We have also, N Instances of transitions from which we can learn the structure of the transition network $B_\rightarrow$. The temporal relations are represented in this network. The pair $(B_0, B_\rightarrow)$ represents the DBN model.

To learn the structure of $B_\rightarrow$, we use a equivalence classes with ant colony optimization algorithm [20] by adding an extra constraint to the learning algorithm, i.e. the head of an arc must always be in the layer at time t. By following this, any BN structure learning algorithm that searches through the space of DAGs can be used and will produce a valid network.

The parameters of the model are estimated by following the method used in [21]. The main advantage of learning the DBN parameters from data in the context of system reliability is that we did not have to search what probability distribution (Exponential, Normal, Lognormal ... etc) that best represents the failure of the system components and determine the failure rate of each component. All conditional probability tables represent in implicit way the distribution of the system reliability.

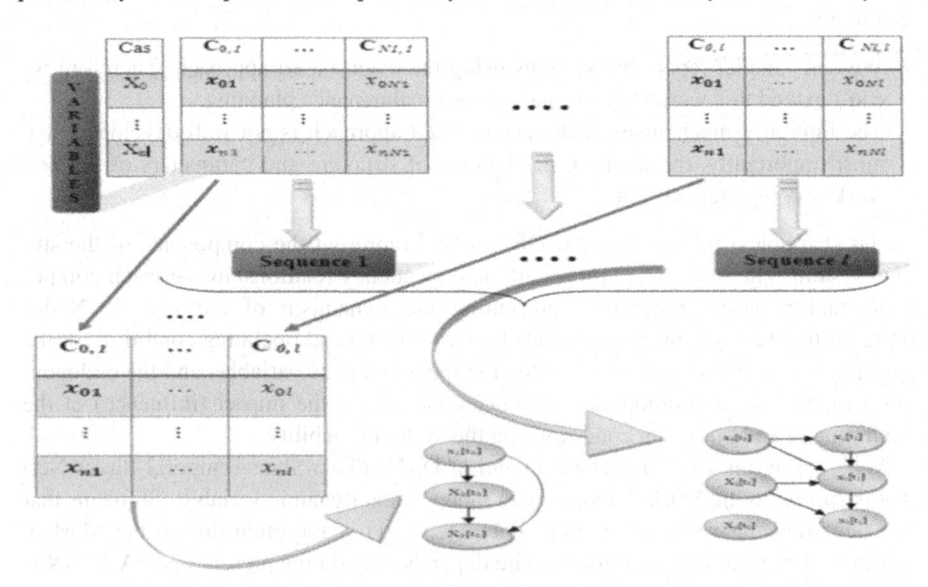

Fig. 4. Diagram representing the DBN model structure learning from a cases base

Phase 4 (Reliability estimation and analysis): The exploitation of the generated model for the analysis of reliability is made via the inference algorithms which makes it possible to make a diagnostic, predictive or inductive analysis of the modeled system. To assess the reliability of the system using the generated model, it suffices to calculate the probability of the variable *S* that represents the overall state of the system knowing that we observed wholes components or a part or even in the case of no observation.

5 Experimentation

To validate the proposed method, we have generated a historical data for system examples taken from the literature using their DFTs, inspired by the work of *Guillaume et al* [22]. The first example is taken from [11]. It represents an Active Heat Rejection System (AHRS). The second example is taken from [10] is part of a real system "a cardiac assist device", this part is the CPU which manages the other components. We have used the same DBN inference algorithm with the same parameters as in [11] and [10], using the INTEL PNL C++ libraries for DBN inference. The results obtained using our approach are very close and similar to the results obtained in [11] and [10] (see Fig. 5).

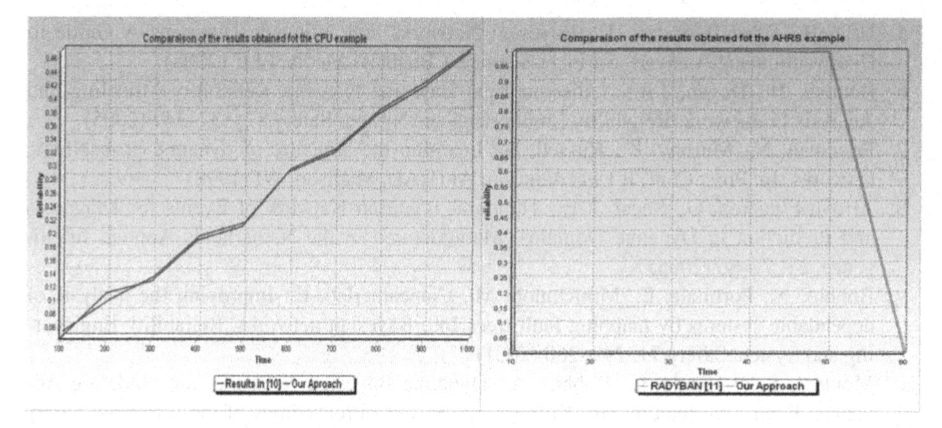

Fig. 5. Results obtained for AHRS and CPU examples

6 Conclusion

As any other building model, learning DBN models which model the system reliability from data is subject to a set of compromises, data must be left to govern the model. Data must be submitted to a pretreatment, several discretization techniques are applied to data. Several learning algorithms can be applied, and sometimes the model is post-processed by removing events that are impossible, hence there are number of degrees of freedom when the DBN model is constructed from data.

The main advantage is that the dependency relationships between the variables (the system components and the reliability factors) become apparent, and this reveals new system reliability knowledge. In addition, the strength of the dependence relations between variables are quantified so that a predictive model is established, it is based on solid statistical principles that any degree of domain knowledge can be applied. The main problem is the execution time of the learning procedure, because of the dimensionality of the state space is equal to the number of variables, the estimation of probabilities requires a large amounts of data to produce a reliable model which increases the execution time of the learning procedure.

References

1. Helge, L., Luigi, P.: Bayesian Networks in Reliability. Reliability Engineering and System Safety 92, 92–108 (2006)
2. Weber, P., Medina-Oliva, G., Simon, C., Iung, B.: Overview on Bayesian networks and applications for dependability, risk analysis and maintenance areas. In: Engineering Applications of Artificial Intelligence (2010)
3. Jose, O.D., Ramirez-Marquez, E.: A generic method for Estimating System Reliability using Bayesian networks. Reliability Engineering and System Safety 94, 542–550 (2009)
4. Zio, E.: Reliability Engineering: Old Problems and New Challenges. Reliability Engineering and System Safety 95, 125–141 (2009)

5. Uffe, B., Kjbrulff-Anders, L.: Bayesian Networks and Influence Diagrams A Guide to Construction and Analysis. Springer Science + Business Media, LLC (2008)
6. Boudali, H., Dugan, J.B.: A discrete-time Bayesian Network Reliability Modeling and Analysis Framework. Reliability Engineering and System Safety 87, 337–349 (2005)
7. Friedman, N., Murphy, K., Russell, S.: Learning the structure of dynamic probabilistic networks. In: Proc. Conf. It Uncertainty in AI (UAI), Madison, WI (1998)
8. Arroyo-Figueroa, G., Sucar, L.E.: Temporal Bayesian Network of Events for Diagnosis and Prediction in Dynamic Domains. Manufactured in the Netherlands Applied Intelligence. 23, 77–86 (2005)
9. Bobbio, A., Portinale, L., Minichinob, M., Ciancamerlab, E.: Improving the analysis of dependable systems by mapping fault trees Into Bayesian networks. Reliability Engineering and System Safety 71, 249–260 (2001)
10. Montani, S., Portinale, L., Bobbio, A.: Dynamic Bayesian Networks for Modeling Advanced Fault Tree features. In: Kolowrocki, K. (ed.) Proceedings of the European Safety and Reliability Conference (ESREL 2005). Balkema Publisher, Tri City (2005)
11. Montani, S., Portinale, L., Bobbio, A., Codetta-Raiteri, D.: RADYBAN: A Tool for Reliability Analysis of dynamic fault trees-through conversion Into Dynamic Bayesian networks. Reliability Engineering and System Safety 93, 922–932 (2008)
12. Portinale, L., Codette Raiteri, D., Montani, S.: Supporting Reliability engineers in Exploiting The Power of Dynamic Bayesian Networks. International Journal of Approximate Reasoning 51, 179–195 (2010)
13. Boudali, B., Dugan, B.J.: A Continuous-Time Bayesian Network Reliability Framework Modelin and Analysis. IEEE Transactions on Reliability 55, 86–87 (2006)
14. Weber, P., Munteanu, P., Jouffe, L.: Dynamic Bayesian Networks Modelling the dependability of degradations and exogenous Systems with Constraints. In: Proceedings of the 11th IFAC Symposium on Information Control Problems in Manufacturing (INCOM 2004), Salvador-Bahia, Brazil (2004)
15. Weber, P., Jouffe, L.: Complex System Reliability modeling With Dynamic Object Oriented Bayesian Networks (DOOBN). Reliability Engineering and System Safety 91, 149–162 (2006)
16. Hosseini, S.M.H., Takahashi, M.: Combining Static/Dynamic Fault Trees and Event Trees Using Bayesian Networks. In: Saglietti, F., Oster, N. (eds.) SAFECOMP 2007. LNCS, vol. 4680, pp. 93–99. Springer, Heidelberg (2007)
17. Lampe, M., Andrews, J.D.: Bayesian Belief Networks for System Fault Diagnostics. Quality and Reliability Engineering International 25, 409–426 (2009)
18. Sigurdsson, J.H., Wall, T., Quigley, J.L.: Bayesian Belief Net for managing expert Judgement and modeling reliability. Quality and Reliability Engineering International 17, 181–190 (2001)
19. Murphy, P.K.: Dynamic Bayesian Networks: Representation, Inference and Learning. PHD Thesis, University of California, Berkeley (2002)
20. Daly, R., Shen, Q.: Learning Bayesian network equivalence classes with ant colony optimization. Journal of Artificial Intelligence Research 35, 391–447 (2009)
21. Ghahramani, Z.: Learning Dynamic Bayesian Networks. In: Giles, C.L., Gori, M. (eds.) IIASS-EMFCSC-School 1997. LNCS (LNAI), vol. 1387, pp. 168–197. Springer, Heidelberg (1998)
22. Guillaume, M., Jean-Marc, R.: Modélisation algébrique des arbres de défaillance temporels. In: JD-MACS 2007 (2007)

Semantic Annotations and Context Reasoning to Enhance Knowledge Reuse in e-Learning

Souâad Boudebza[1], Lamia Berkani[1,2], Faiçal Azouaou[1], and Omar Nouali[3]

[1] Higher National School of Computer Science, ESI, Oued Smar, Algiers, Algeria
{s_boudebza,l_berkani,f_azouaou}@esi.dz
[2] Department of Computer Science, USTHB University, Bab-Ezzouar, Algiers, Algeria
l_berkani@hotmail.com
[3] Department of Research Computing, CERIST, Algiers, Algeria
onouali@cerist.dz

Abstract. We address in this paper the need of improving knowledge reusability within online Communities of Practice of E-learning (CoPEs). Our approach is based on contextual semantic annotations. An ontological-based contextual semantic annotation model is presented. The model serves as the basis for implementing a context aware annotation system called "CoPEAnnot". Ontological and rule-based context reasoning contribute to improving knowledge reuse by adapting CoPEAnnot's search results, navigation and recommendation. The proposal has been experimented within a community of learners.

Keywords: community of practice of e-learning, e-learning, knowledge reuse, semantic annotation, context, reasoning.

1 Introduction

With the large amount of pedagogical knowledge which is constantly growing among Communities of Practice of e-learning (CoPEs), the issue of knowledge reuse remains a serious problem [1]. Through the participation to the CoPE, e-learning actors create both, tacit and explicit knowledge. The main concern, however, lies in reusing tacit knowledge. Some research studies explore the use of semantic approaches for knowledge modeling and reuse. The works carried out in [2] [3] [4] rely on using ontological approaches to indexing resources. These approaches are useful to index and manage explicit knowledge, but are not suitable for eliciting tacit knowledge which requires externalization mechanisms. According to [5], semantic annotation approaches are more useful to modeling both tacit and explicit knowledge. These approaches have been shown their effectiveness in knowledge modeling disregarding how to reuse and reap benefit from that knowledge. In this regard, we consider that the preservation of knowledge context can be very useful at their reuse. The context of knowledge refers to parameters describing the situation in which this knowledge is modeled or reused. Few works have introduced the notion of context [5] [6], but without taking into consideration important aspects such as context reasoning.

A. Amine et al. (Eds.): *Modeling Approaches and Algorithms*, SCI 488, pp. 419–428.
DOI: 10.1007/978-3-319-00560-7_45 © Springer International Publishing Switzerland 2013

The object of this paper is to use both semantic annotations and context for modeling knowledge within CoPEs. The resulted model is used to improve knowledge reuse and sharing by benefiting from context reasoning.

The paper is organized as follows: In section 2, we describe the proposed context-based semantic annotation model. In section 3, context reasoning mechanisms are described. In section 4, the context-aware architecture of the proposed annotation system is presented. Section 5 discusses the implementation and experimentation of our proposal. Finally, section 6 contains concluding remarks and future work.

2 Context-Based Semantic Annotation Model

We propose in this section, a contextual annotation model of four dimensions (Fig. 1): Resource, Annotation, Controlled vocabulary and Context. Accordingly, the model represents the important aspects of annotation, which includes the description of the annotated resource, the representation of various elements of annotation and their links to the controlled vocabularies, as well as the description of members' context during the process of creation, evaluation or reuse of annotations. The model is implemented using ontology.

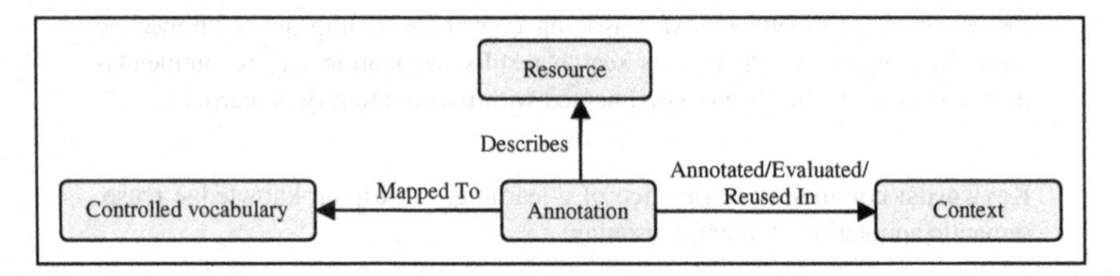

Fig. 1. General annotation model

2.1 Resource

This dimension represents the resource or the part of the annotated resource. Resources are heterogeneous and varied on their nature, form, size, etc. This dimension includes the following attributes (see Fig. 2): URL, title, authors, description, and type (e.g. course, exercise, presentation, etc.).

2.2 Annotation

This dimension represents the externalized knowledge which reflects personal knowledge and experiences of the annotator, and also those of recipients of annotation; those who reuse the annotation may express their judgments and feedback about the annotation via another annotation. This dimension is formalized based on the annotation models in [5] [7]. The conceptual model of annotation is presented in Fig. 2, two categories of annotation are distinguished: personal and shared.

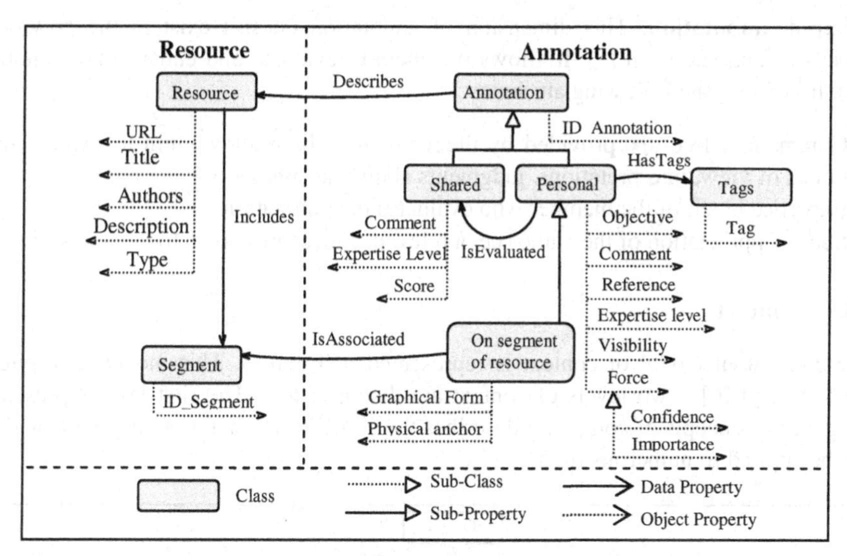

Fig. 2. Conceptual model of Resource and Annotation

"Personal" annotation. It is associated to the author of the annotation. In the case of annotation on the whole resource, the annotation has the following attributes:

- Tags: this is one or more keywords associated to the resource. It can better organize the resources, as well as it provides a simple and effective browsing technique.
- Objective: it represents the reason why the annotation is created. It serves to reuse the annotation, and it is associated with a controlled vocabulary.
- Comment: it contains free text, allowing the annotator to freely express his points of views, opinions and expertise about the annotated resource.
- Reference: it represents a link to another resource (e.g. reference book, citation, URL, etc.). It allows the annotator to argue or enrich his annotation.
- Expertise level: this attribute is important and people tend to trust an expert over a novice.
- Visibility: it refers to the access rights to annotation, we distinguish three types: *Private*, *Public*, and *Group*.
- Force: is the value that represents the annotation for the annotator, including "Importance" which describes the significance character of annotation relative to its creator and "Confidence" which means the assertion about the annotation.

In the case of annotation on a segment of the resource, the annotation includes also the following attributes:

- Graphical form: it represents the graphical aspect of annotation (highlighting, underlining, etc.). That is used to change the appearance of information to make it more visible [7].
- Physical anchor describing the annotated segment in the resource.

"Shared" annotation. This dimension of annotation doesn't exist in the previous models of annotation in [5]. It allows members to evaluate and enhance the annotation. It includes the following attributes:

- Comment: a free text provided by the recipient, which allows him to express his points of view, interpretations, judgments about the annotation.
- Expertise level: of the member who evaluates the annotation.
- Score: appreciation of the value (i.e. a relevance measure) given to the annotation.

2.3 Context

The conceptual model of context is represented in figure 3. This model is inspired from [5] and [6]. Two levels of context are distinguished. The first level represents the generic concepts of context, it describes the context of annotation in general and it can be applied to numerous fields.

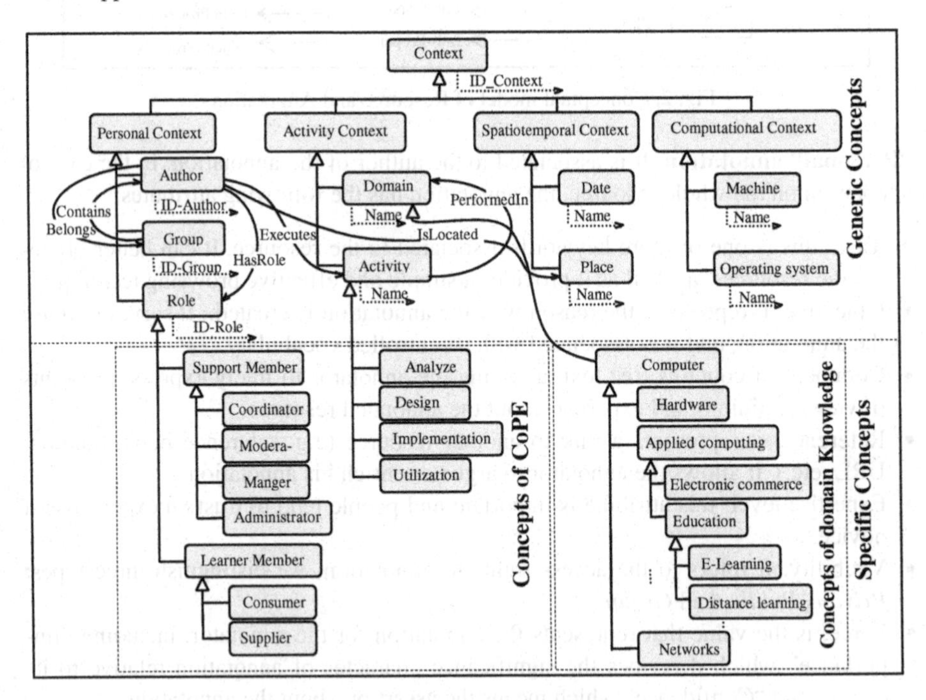

Fig. 3. Conceptual model of Context

The model of context is composed of four components:

- Personal Context: it includes the "Author" representing the member, the "Role" of member in the CoPE, and the "Group" to which the member belongs to.
- Activity Context: it includes the "Domain" which represents the knowledge domain (e.g. mathematics, physics, computer science, etc.), and the member's "Activity" in the CoPE.

- Spatiotemporal Context: it describes the "Date" and the "Place" in which the member creates, evaluates or reuses the annotation.
- Computing Context: it includes the "Operating system" installed on the host and the "Machine" on which turns the annotation tool.

The second level represents ontologies describing specific concepts of context. The ontology of CoPE [8] describes concepts related to CoPE. ACM Computer Classification System [9] is used to describe computer science domain.

2.4 Controlled Vocabulary

It represents ontologies associated to different elements of annotation like tags and attributes (e.g. graphical form, objective of annotation, etc.). We opt for the ontology proposed in [9] that including a rather comprehensive list of annotation graphical forms. As far as vocabulary associated to the objective of annotation, we reuse the ontology proposed in [10], describing learners' annotation objectives. Thereafter, other controlled vocabularies can be developed.

3 Context Reasoning

Formal approaches for context modeling such as ontology, offer many advantages, the foremost advantage is the reasoning capabilities. Context Reasoning aims to check consistency of the model as well as to infer new information about context and to derive high level of context. Indeed, the contextual information provided by the system, user, sensors, etc. leads to elementary data about context, whereas some contextual information are useful only if they are combined with other elementary or composite contexts. Context reasoning is used to support the knowledge reuse by providing annotations that best fit member's context. The reasoning tasks are grouped into two categories: ontological reasoning and rule based reasoning.

3.1 Ontological Reasoning

The OWL-DL language provides efficient reasoning, which makes it the ideal to represent the context ontology. The standard reasoning rules supported by this language includes: subClasseOf, subPropertyOf, TransitiveProperty, disjointWith, inverseOf, etc. They are used to infer the implicit context from the explicit context.

Fig. 4 shows some examples illustrating the use of ontological reasoning rules in our context ontology. For instance, "ActivityContext" is subclass of "Context" and "Activity" is subclass of "ActivityContext". Thus, "Activity" can be defined as subclass of "Context" using "subClassOf" rule. Furthermore, the concepts "Analyze" and "Design" are disjoint. The rule "disjointWith" can be used to infer a contradiction when the instance "ScenarioConception" is defined as instance of both classes at the same time. Also, "Belongs" is an inverse property of "Contains", the explicit context shows that "Author1" "Belongs" to "Group1", through the rule "inverseOf", a new context that "Group1" "Contains" "Author1" can be implicitly deduced.

Explicit context	Implicit context

<pre><owl:Class rdf:ID="ActivityContext">
   <rdfs:subClassOf>
      <owl:Class rdf:ID="Context"/>
   </rdfs:subClassOf>
</owl:Class>
<owl:Class rdf:ID="Activity">
   <rdfs:subClassOf>
      <owl:Class
rdf:ID="ActivityContext"/>
   </rdfs:subClassOf>
</owl:Class>
<owl:Class rdf:ID="Analyze">
   <owl:disjointWith>
      <owl:Class rdf:ID="Conception"/>
   </owl:disjointWith>
</owl:Class>
<owl:ObjectProperty rdf:ID="Belongs">
   <owl:inverseOf>
      <owl:ObjectProperty
rdf:ID="Contains"/>
   </owl:inverseOf>
   <rdfs:range rdf:resource="#Group"/>
   <rdfs:domain
rdf:resource="#Author"/>
</owl:ObjectProperty>
<Author rdf:ID="Author1">
   <Belongs rdf:resource="#Group1"/>
</Author></pre>
	<pre><owl:Class rdf:ID="Activity">
	   <rdfs:subClassOf>
	      <owl:Class rdf:ID="Context"/>
	   </rdfs:subClassOf>
	</owl:Class>
	<Conception
	rdf:ID="ScenarioConception">
	<Analyze rdf:ID="ScenarioConception">
	--- Error
	<Group rdf:ID="Group1">
	   <Contains
	rdf:resource="#Author1"/>
	</Group></pre>

Fig. 4. Ontological reasoning

3.2 Rule-Based Reasoning

Some contextual information cannot be easily inferred using ontological reasoning. Accordingly, we propose the use of predefined rules, considered as a flexible mechanism to infer other contextual information. Inference rules are described with Generic Rule Language specified by Jena API and based on first order logic.

Table 1. Context tuples

ID-Context	ID-Author	ID-Group	Role-Name	Activity-Name	Domain-Name
C1	Author1	Group1	Manager	Conception	E-learning
C2	Author2	Group2	Coordinator	Conception	E-learning
C3	Author3	Group1	Moderator	Conception	Distance Learning

The tuples in table 1 correspond to individuals of Context. The first tuple "C1" represents the current context of annotation. Context reasoning basis on the rule "R" (Fig. 5) infers a new context that the tuple of context "C1" has the same group and the same activity as the tuple "C3". More precisely, the rule R defines the relationship "SameGAc" between two instances of "context", when their authors belong to the same group and execute the same activity. This rule is based on relationships defined in the other inference rules ("Sameidc", "InC", "SamePerson", "SameGroup" and "SameActivity").

```
[R:(?c1 rdf:type prefix:Context)(?c2 rdf:type prefix:Context)
noValue(?c1 prefix:Sameidc ?c2)(?a1 rdf:type prefix:Author)
(?a2 rdf:type prefix:Author)(?a1 prefix:InC ?c1)(?a2 prefix:InC ?c2)
noValue(?a1 prefix:SamePerson ?a2)(?g1 rdf:type prefix:Group)
(?g2 rdf:type prefix:Group)(?a1 prefix:Belongs ?g1)(?a2 prefix:Belongs ?g2)
(?g1 prefix:SameGroup ?g2)(?ac1 rdf:type prefix:Activity)
(?ac2 rdf:type prefix:Activity)(?a1 prefix:Executes ?ac1)(?a2 prefix:Executes
?ac2)(?ac1 prefix:SameActivity ?ac2) -> (?c1 prefix:SameGAc ?c2)]
```

Fig. 5. Reasoning rule

4 Context-Aware Architecture

We propose, in this section, the context-aware architecture for our annotation system called CoPEAnnot. Many researchers have proposed several context-aware architectures and most of them are proposed in pervasive and mobile computing domain. The works in [6] [11] proposed architectures for context-aware annotation systems. The proposed architecture (Fig. 6) differs from the above architectures by the reasoning support it provides. It consists of two main components: context and annotation management, as proposed in [12]. The authors suggest that the body of application must be designed in isolation from contextual data.

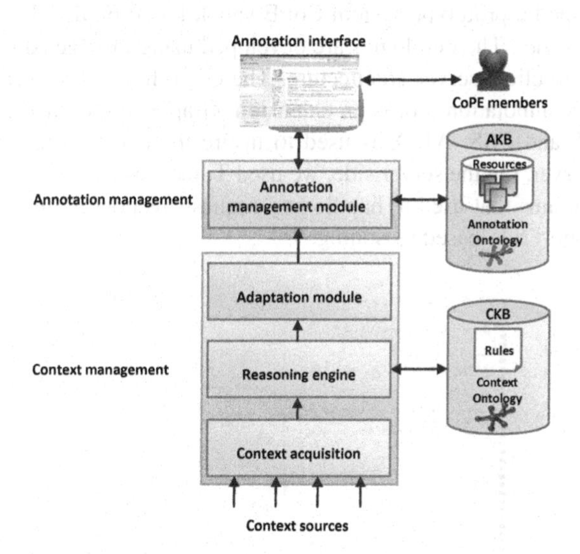

Fig. 6. CoPEAnnot Architecture

1. Context Management: it includes the following major steps:

– Context acquisition: is responsible for collecting contextual information from different sources (operating system, user model, physical sensors, etc.), for interpreting contextual information (transform them into more useful and meaningful information) and for their storage in accordance to the context ontology.
– Reasoning engine: is in charge of reasoning about contextual information acquired by the acquisition module. Based on ontological and rule-based reasoning,

the reasoning engine infers information about annotations' context which is semantically closest to the current context.

– Adaptation module: this module adapts the functionalities of context-aware system according to the contextual information provided by the reasoning engine.

2. Annotation management: it includes the following major steps:

– Annotation management module: is in charge to insert, store and update, research, navigation and recommendation of annotations. These last three features are adapted according to the current context of the annotation.

– Annotation interface: it represents the graphical interface that allows the exchange and interaction between the CoPE members and the annotation tool.

The context knowledge base (CKB) contains the context ontology and the inference rules. The annotation knowledge base (AKB) includes the resources and the ontology which defines the annotation model and their controlled vocabularies.

5 Implementation and Experimentation

We have developed a prototype system CoPEAnnot, based on the above architecture and annotation model. The ontologies are developed using Protégé editor. The system is implemented in client-server architecture. The client has as browser Mozilla Firefox, in which the annotation tool is an extension. Graphical interface was built using XUL, DOM, JS and CSS. AJAX is used to insure the communication between the client and the server. On the server side, we used Tomcat as a Servlet container. Servlet are java programs that used to handle http requests/responses. Jena frame work is also used to support rule-based reasoning.

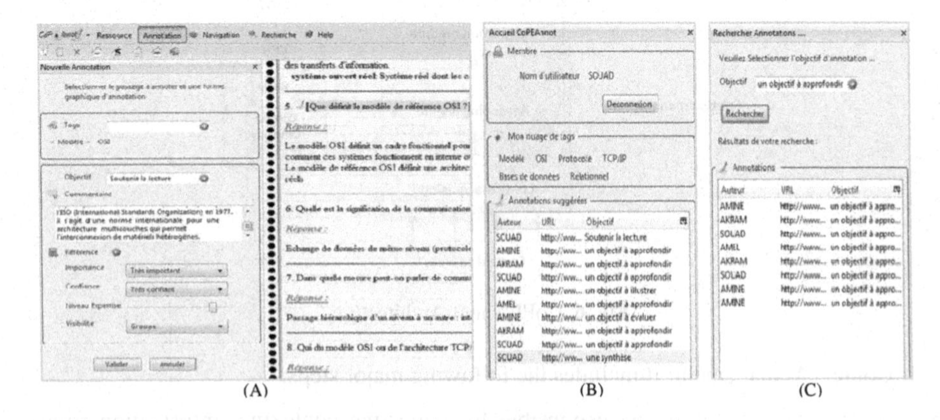

Fig. 7. Screenshots of CoPEAnnot tool

The annotation extension provides the following main features: CoPEAnnot (Home), resource, annotation, navigate, search and help. Home sidebar shows the tag cloud and the recommended annotations adapted to the current context of the member (part B Fig. 7). Members can annotate any type of resources. Graphical forms can be used to annotate on a part of html resources (part A Fig. 7). Thereafter, members can also edit, share, and evaluate an annotation. In addition to the standard features of navigation, the tag cloud enables faster discovering of knowledge. The tool provides also contextual semantic search based on controlled vocabularies (part C Fig. 7).

In order to validate CoPEAnnot, we consider the following experimental process:

1. Identify the key dimensions of evaluation: a questionnaire has been established based on the evaluation dimensions proposed in [13]: "utility", "usability" and "acceptability". Further, we evaluated the adaptation quality of the tool.

   - The utility represents the accordance between the features offered by the system and those expected by the user. We evaluate members' satisfaction about each feature of CoPEAnnot.
   - The usability indicates the ability to learn and use the system. We evaluate user-friendliness and simplicity of CoPEAnnot. Thus, we check if the members are become more familiar with annotations, tags and taxonomies.
   - The acceptability represents user's mental attitude towards the system. We measure members' satisfaction on using the tool and we see how often they want to use it.
   - The adaptation quality measurement, by assessing the appropriateness of tag cloud, recommended annotations, navigation and search results according to the current context of members.

2. Test Organization: this step involves gathering information from members of a community of learners in academic discipline in computer science. The questionnaire and the CoPEAnnot tool have been made available for the members.
3. Results Analysis: this step includes a statistical analyses of experimental data. Twenty four students (10 man and 13 women) were interviewed using the questionnaire. The average age of the respondents is 25 years. The results of this investigation were satisfactory and the learners have expressed a high level of satisfaction. The figure below presents the satisfaction rates of each dimension in the questionnaire.

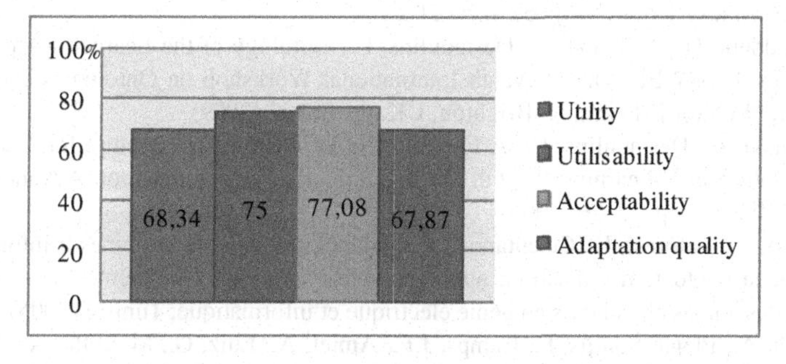

Fig. 8. Experimental results of CoPEAnnot

6 Conclusion

This paper recognizes the knowledge reuse problem within CoPEs and proposes a framework based on semantic annotation and context reasoning. A contextual semantic annotation model has been proposed to model the shared knowledge. Ontological and rule-based context reasoning mechanisms are used to adapt annotation system features according to the current members' context. The experimentation results of CoPEAnnot were very promising. Our future work will focus on improving the tool by extending reasoning capabilities on other elements of context like member's profile and developing controlled vocabularies. Semi-automatic annotation mechanisms can be further integrated to facilitate the annotation activity.

References

1. Berkani, L., Chikh, A.: A Process for Knowledge Reuse in Communities of Practice of E-learning. Social & Behavioural Sciences Journal (2010)
2. Benayache, A.: Construction d'une mémoire organisationnelle de formation et évaluation dans un contexte e-learning: le projet MEMORAe. PhD Thesis, Compiègne Univ. (2005)
3. Leblanc, A., Abel, M.H.: Knowledge Sharing via the EMEMORAe2.0 Platform. In: 6th International Conference on Intellectual Capital, Knowledge Management & Organisational Learning, Montreal, Canada, pp. 10–19 (2009)
4. Tifous, A., El Ghali, A., Dieng-Kuntz, R., Giboin, A., Evangelou, C., Vidou, G.: An ontology for supporting communities of practice. In: KCAP 2007, The Fourth International Conference on Knowledge Capture, Whistler, BC, Canada (2007)
5. Azouaou, F.: Modèles et outils d'annotations pour une mémoire personnelle de l'enseignant. PhD Thesis, Université Joseph Fourier - Grenoble I (2006)
6. Ouadah, A., Azouaou, F., Desmoulins, C.: Models and Architecture for a teacher's context-aware annotation memory. In: IADIS Int. Conf. e-Learning, pp. 215–223 (2009)
7. Mille, D.: Modèles et outils logiciels pour l'annotation sémantique de documents pédagogiques. PhD Thesis, Université Joseph Fourier - Grenoble I (2005)
8. Berkani, L., Chikh, A.: Towards an Ontology for Supporting Communities of Practice of E-Learning "CoPEs": A Conceptual Model. In: Cress, U., Dimitrova, V., Specht, M. (eds.) EC-TEL 2009. LNCS, vol. 5794, pp. 664–669. Springer, Heidelberg (2009)
9. ACM Computing Classification System (CCS), `http://www.acm.org/class/2012/`
10. Mokeddem, H., Azouaou, F., Desmoulins, C.: Ontology of the Learner's Annotation Objectives. In: SWEL 2009, Seventh International Workshop on Ontologies and Social Semantic Web for E-Learning, Brighton, UK, pp. 88–92 (2009)
11. Azouaou, F., Desmoulins, C.: A Flexible and Extensible Architecture for Context-Aware Annotation in E-Learning. In: 6th IEEE International Conference on Advanced Learning Technologies, pp. 22–26 (2006)
12. Chaari, T., Laforest, F., Celentano, A.: L'adaptation dans les systèmes d'information sensibles au contexte d'utilisation: approche et modèles. In : Cinquièmes journées scientifiques des jeunes chercheurs en génie électrique et informatique, Tunisie (2005)
13. Tricot, A., Plégat-Soutjis, F., Camps, J.F., Amiel, A., Lutz, G., Morcillo, A.: Utilité, utilisabilité, acceptabilité: interpréter les relations entre trois dimensions de l'évaluation des EIAH. In: Desmoulins, C., Marquet, P., Bouhineau, D. (eds.) Environnements Informatiques Pour l'apprentissage Humain, pp. 391–402. ATIEF/INRP, Paris (2003)

Change Impact Study by Bayesian Networks

Chahira Cherif and Mustapha Kamel Abdi

Département d'Informatique, Université d'Oran Es-sénia. B.P. 1524,
El M'Naouer, 31000 Oran, Algérie
cherifchahera@yahoo.fr, abdimk@yahoo.fr

Abstract. The study of change impact is a fundamental activity in software engineering because it can be used to plan changes, set them up and to predict or detect their effects on the system and try to reduce them. Various methods have been presented in the literature for this sector of maintenance. The objective of this project is to improve the maintenance of Object Oriented (OO) systems and to intervene more specifically in the task of analyzing and predicting the change impact. Among several models of representation, Bayesian Networks (BNs) constitute a particular quantitative approach that can integrate uncertainty in reasoning and offering explanations close to reality. Furthermore, with the BNs, it is also possible to use expert judgments to anticipate the predictions, about the change impact in our case. In this paper, we propose a probabilistic approach to determine the change impact in OO systems. This prediction is given in a form of probability.

Keywords: Change impact analysis, maintenance, probability model, Bayesian networks, bayesian inference.

1 Introduction

Over time the software has become more complex and their maintenance requires more time and expense, that is why the computer industry must develop more efficient software from a high quality in order to reduce maintenance costs which have become over time more expensive because of the great importance of this phase in the life cycle of software. The maintenance phase is the longest phase and it only ends with the end of Software life. In industry, the cost of software maintenance (compared to the total budget) has increased from 40 to 60% in the 80s, more than 75% or even 80% in early 2000. It is estimated that more than half of this maintenance is dedicated to the understanding of the program itself. The high cost of maintenance and especially the understanding of programs, incites researchers and industry to focus on this phase of the life cycle of a system in order to understand the factors that affect the cost. On the other hand, OO programming has been defined to model real-world entities, that is to say, the definition of objects corresponding to entities (contrary to procedural programming that defines a sequence of functions treatments intended to represent data). The changes effects in the system must be taken into consideration. Indeed, a small change can have extensive and unexpected

A. Amine et al. (Eds.): *Modeling Approaches and Algorithms*, SCI 488, pp. 429–438.
DOI: 10.1007/978-3-319-00560-7_46 © Springer International Publishing Switzerland 2013

effects on the remainder elements of the system. The danger in modification is the impact consequence for a given change. The modular design object, properly used, limit the effects related to changes. The purpose of this paper is to improve the maintenance of object systems and to intervene more specifically in the task of analyzing and predicting the change impact. Identifying the potential impact of a change, reduces the risk of costly and unpredictable changes. To do this, we try to give further explanation of the factors responsible for the actual change impact and its evolution. There are several models of representation and Bayesians Networks forming a particular quantitative approach which can integrate uncertainty in the reasoning (Naim al.2004) [1], providing explanations close to reality.

2 Change Impact Analysis

The impact analysis describes how to conduct, at an effective cost, complete analysis of a change in existing software. Maintainers must have an intimate knowledge of the structure and the content of the program. They use this knowledge to make impact analysis that identifies all the products and software systems affected by a request to change software and establish an estimate of the resources needed to accomplish the change. In addition, the risk of change is determined. Change request, called sometimes modification request (MR) and also called problem report (PR), must be analyzed and translated first in software terms. It is made after a change request is entered in the configuration management process.

2.1 The Objectives of Impact Analysis

- Determining the scope of a change to develop a plan and implement the work;
- Development of accurate estimates of needed resources to perform the work;
- The analysis of costs / benefits of the requested change;

The severity of a problem is often used to decide how and when a problem will be corrected. The maintainer then identifies the affected components. Several potential solutions are given and then a recommendation is made on the best course of action. Software designed with maintainability in mind greatly facilitate the process of impact analysis.

The change impact analysis is performed mainly during maintenance of the programs in order to evaluate the change impact on the rest of the system and reduce the risk of become caught up in costly expenses. This assessment includes the estimation of human, financial, time, effort and planning's required to accomplish the change.

2.2 Change Impact Analysis of OO Systems

The object-oriented concepts as encapsulation, inheritance or polymorphism makes systems more difficult to maintain, as well as identification of the parties affected by the changes [2]. Li and Offut [3] analyzes the propagation of changes in three different ways: encapsulation, inheritance and polymorphism. They detect the possible changes, then classify them in changes data members, methods or classes, and then propose a set of algorithms to determine the classes influenced by the changes. Finally, the following points illustrate why OO systems can be difficult to maintain [4]:

- Even if relatively easy to understand data structures and member functions of classes of objects, understanding the functionality and combined effects is not trivial.
- The complex relations between object classes make it difficult to anticipate and identify the propagation of change effects.
- Data dependencies, control dependencies, and dependencies of state behavior make difficult the preparation of test cases and test data generation to re-test the components affected by the changes.

3 Change Impact Analysis and Bayesian Networks

Tang et al. [5] used Bayesian networks to analyze the change impact at the architectural level. Indeed, in this article, the authors present the model AREL (Architecture Rational and Element Linkage) which is used to show the relationship between decisions to make and design elements that depend on these choices change architecture. Subsequently this model was transposed into BNs where the nodes are the elements of AREL and arcs are dependency relations between these nodes. BNs have been the subject of a case study that consists in an application of the banking sector.

Abdi et al. [6] were interested in the understanding of real factors that are responsible for the change impact and evolution. Metrics of design and implementation are studied to understand their effects on systems. The authors have proposed a probabilistic approach that uses the BNs to determine the change impact.

Mirarab [7] proposed an approach that uses two sources of information: dependency metrics and historical data. Metrics reflect the degree of connection between packages system expressed in number of calls between the classes of each package. These dependencies may correspond to method calls or uses of variables. For the detection of these dependencies, the authors use static techniques. The second type of information that is historical data represents data on co-changes in the past. Although this information does not present spread changes, the authors hypothesize that the elements changed at the same time in the past are likely to change in the same way in future versions. Once these data are collected, the second step consists to build BN to predict the change impact. BNs used in this approach are three types:

- BDM (Bayesien Dependency Model)
- BHM (Bayesian History Model)
- BDHM (Bayesian Dependency and History Model).

Zhou et al. [8] present a BN model that predicts the probability of co-changes between system entities. The information used as an entry point to networks are extracted from historical data and dependencies in the code, which are essentially the age of the change, its author, the frequency of change, the objective of change and other information used to predict the classes potentially affected by a change.

4 Reminder about Bayesian Networks

Bayesian Networks, which take their name from the work of Thomas Bayes in the eighteenth century "the probability of causes" are the result of research carried out in the 80s, Bayes' theorem is used to determine the probability of A given B:

$$P(A/B) = \frac{P(B/A)\,P(A)}{P(B)}$$

BNs are a modeling tool at the junction of probability theory and graph theory allowing to describe relations governing a set of random variables and perform probabilistic reasoning on them.

Definition

A Bayesian network is defined by:

- A directed graph without circuit (DAG) G = (V, E), where V is the set of nodes of G, and E the set of edges of G;
- A finite probability space (Ω, Z, p);
- A set of random variables associated with the nodes of the graph and defined on (Ω, Z, p), such as: $P(V_1, V_2, \ldots V_n) = {}_{1..n}\Pi_i = [p(V_i \mid C(V_i))]$

Where $C(V_i)$ is the set of causes (parents) of V_i in the graph G [1].

5 Case Study

Abdi et al. [6] proposed a probabilistic approach which uses BNs to determine the change impact. The main steps underlying the approach are:

1-Construction of the graph structure (Bayesian network) from practical knowledge (empirical studies)
2-Assignment of parameters (probability tables)
3-Bayesian Inference (algorithms, tools)
4-Results

5.1 Impact Network

According to Abdi et al. [6], the impact network as shown in figure 1. The objective of this work is the understanding of the real factors that are responsible for change impact and its evolution. Five types of metrics influence the change impact. Some of these metrics are considered design metrics (AMMIC and OMMIC) because it can be used at the design stage, while others are implementation metrics (MPC CBOU, CBONA) requiring presence of the source code.

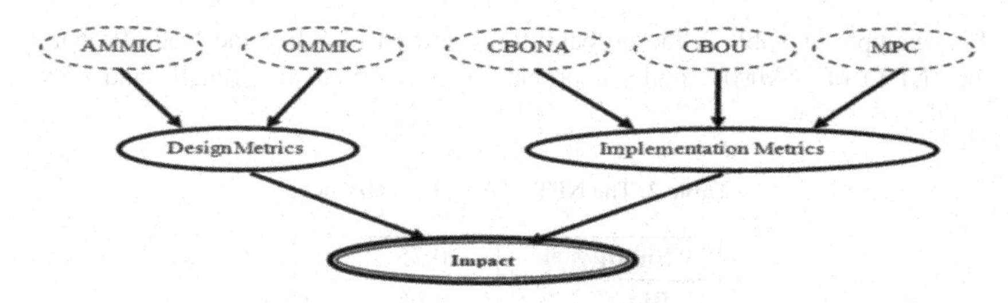

Fig. 1. The change impact Network

The sectioned metrics are defined in following table Tab1:

Table 1. The metrics used in the impact network

Metric	Definition
CBOU	Coupling Between Object: number of classes with which a class is coupled
CBONA	CBO No Ancestors: CBO without considering the ancestor classes.
MPC	Message Passing Coupling: number of messages sent by a a class in the direction of the other classes of the system
OMMIC	Others Method–Method Import Coupling: number of classes (other than super-classes and sub-classes) with which a class has a method-type interaction and coupling method type IC (Import Coupling).
AMMIC	Ancestors Method–Method Import Coupling: number of parent classes with which a class has a method- method type of interaction and a coupling type IC (Import Coupling).

5.2 Parameters Assignment

Nodes in the BN are broken down into two categories: input variables and intermediate variables. The probabilities of the input nodes are directly determined from the measurements of these variables given by a testing system. In our case, it is BOAP (Toolbox for Analysis Programs) developed at the Computer Research Institute of Montreal (CRIM) (Alikacem and Snoussi, 2002) [9]. It is a set of integrated software tools, which allows an expert to quickly evaluate the quality of software (conceptual weaknesses or structural, too complex instructions, etc.). BOAP system (version 1.1.0) contains a total of 394 classes. The metrics considered in this work are extracted from this system.

Entry Nodes

Representing the different selected metrics, all these input variables are quantitative variables that have numeric values measurable. The number of possible values for these variables can be infinite, depending on the considered test system. In order to facilitate the definition of the probabilities of entry nodes, we need to transform these variables into discrete variables having a limited number of values which are:"Small" and "big".

For example in Table 2, the probabilities are used to define the Node Probability Table (NPT) of AMMIC node, a probability of 35% to be "Small" and 65% to be "big".

Table 2. The NPT of AMMIC entry node

Small	0.35
Big	0.65

Intermediate Nodes

The two nodes DesignMetrics and ImplementationMetrics are not directly measurable. They are defined or influenced by their parent nodes. The intermediate nodes have an associated probability table. For example, DesignMetrics variable is defined by both parents AMMIC and OMMIC.

This is to find the conditional probability of the DesignMetrics node: P (Design Metrics / AMMIC & OMMIC). But as the relation between parent nodes AMMIC and OMMIC and their children node DesignMetrics is definitional, the strong presence of these metrics defines also the strong presence of design metrics (Design Metrics). A possible scenario for the Node Probability Table (NPT) of Design Metrics node is shown in the following table 3:

Table 3. The NPT of the intermediate node DesignMetrics

AMMIC	Small		Big	
OMMIC	Small	Big	Small	Big
Yes	0.2	0.4	0.4	0.8
No	0.8	0.6	0.6	0.2

6 Bayesian Editor « *Impact Bayes Net* »

Bayesian networks have been used as a fundamental tool for the representation and manipulation of belief in artificial intelligence. There have been implementations of Bayesian networks in a variety of formats and languages.

IBN « Impact Bayes Net» is a system that manages Bayesian networks: it calculates the marginal probabilities and expectations, and allows the user to import, create, edit and export networks. *IBN* is a complete implementation of Bayesian networks in Java. A Java implementation has several advantages. First, Java is still the best language truly portable, a package written in Java can be exported and execute under Unix, Macintosh Windows.

Second, Java has been adopted by Internet navigators. A program or an application written in Java can be closely linked with web pages. This is a good OO language, and has a multi-threaded functionality which can be very useful for parallelization.

The IBN system is a set of tools for the creation and manipulation of Bayesian networks. The system is composed of a graphical editor which allows you to create and modify Bayesian networks in a user-friendly interface. It also allows importing Bayesian networks in a variety of formats. IBN manipulates data structures that represent Bayesian networks. It can produce:

- The marginal probability of a variable in a Bayesian network.
- Configurations with a maximum of posterior probability.

6.1 Inference Algorithm under IBN

We used the inference algorithm of the exact approach, which exploit conditional independence contained in the network and give each inference posterior probabilities accurate. The algorithm of the junction tree is detailed as follows:

6.2 Junction Tree

Inference is used to calculate the probability of a hypothesis based on the observation of the evidences [13].

The method of the junction tree was introduced by Lauritzen & Spiegelhalter [10] Jensen, Lauritzen & Olesen [11]. It is also called JLO method (for Jensen, Lauritzen, Olessen). It is applicable to every structure of DAG as opposed to the local messages method.

Junction trees have traditionally been constructed by the following method: [12]

- Moralize the belief network by adding arcs connecting every pair of parents of any variable in the network, and dropping the directions on all arcs.
- Triangulate the (now undirected) network by adding fill arcs until no chord-less cycles of length greater than three remain.
- Create a tree whose vertices are the cliques of the triangulated graph and which are connected such that the junction tree property holds: if any two cliques contain a particular variable V, then every clique on the path between those two cliques must a particular variable V, then every clique on the path between those two cliques must also contain that variable.

7 Experiments and Results

Once the structure of the graph and all Node Probabilities Tables (NPT) defined, we can proceed to Bayesian inference. The result is an update of all the conditional probabilities of all nodes.

After implementation of the impact Network by using the probability tables obtained by the study of Abdi et al. [6], we applied the inference algorithm of exact approach junction-tree in the Bayesian editor *IBN* "Bayes Net Impact":

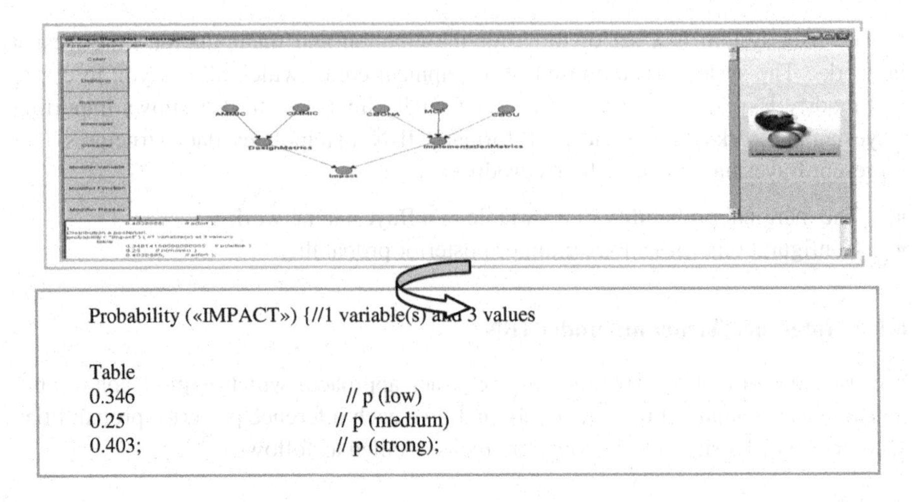

Fig. 2. The change impact Network after inference (Scenario 1)

Affecting three states "Low", "Medium" and "Strong" to Impact node, and the input data used in the four scenarios shown in Table (Tab 4):

Table 4. The NPT of input nodes

Scenarios	Scenario 1		Scenario 2		Scenario 3		Scenario 4	
Input nodes	Small	Big	Small	Big	Small	Big	Small	Big
AMMIC	0.35	0.65	0.5	0.5	0.05	0.95	0.05	0.95
OMMIC	0.6	0.4	0.5	0.5	0.5	0.5	0.5	0.5
CBONA	0.75	0.25	0.9	0.1	0.05	0.95	0.9	0.1
CBOU	0.45	0.55	0.9	0.1	0.05	0.95	0.9	0.1
MPC	0.2	0.8	0.5	0.5	0.5	0.5	0.5	0.5

The execution of this network has generated these impact estimates (probability values of impact). We can conclude that for example the change impact has 40.3% chance of being "strong". Figure 2 shows that decreasing the values of the metrics CBONA and CBOU, the change impact is weakening more (its probability of being "Low" pass from 34.6% to 37.8%). In scenario 2 (Tab.4) by increasing this time the metric values CBONA and CBOU, the change impact becomes increasingly strong. The probability of the state "Strong" pass from 37.1% to 44.4% with a gain of more than 7 points (Tab.5).

Finally, the last scenario executed shows that by reducing the metric values CBONA, CBOU and increasing the value of AMMIC, the change impact becomes a bit stronger. The probability of state "Strong" pass from 37.8% to 41.5%. We used the environment BNJ (Bayesian Network tools in Java) to confirm our results obtained by IBN. The reasons for using BNJ to compare the results of change impact with IBN are:

BNJ is a set of Java tools of research and development of Bayesian networks. This project was developed in the laboratory KDD University of Kensas.

- This tool uses the junction tree algorithm for Bayesian inference.

Table 5. Results of impact after inference "The four scenarios"

SCENARIOS	IMPACT		
	LOW	MEDIUM	STRONG
SCENARIO1 BNJ	34.6 %	25.0 %	40.4%
SCENARIO1 IBN	0.346	0.25	0.403
SCENARIO 2 BNJ	37.8 %	25.0%	37.2%
SCENARIO 2 IBN	0.378	0.25	0.371
SCENARIO 3 BNJ	30.6 %	25.0%	44.4%
SCENARIO 3 IBN	0.3058	0.25	0.444
SCENARIO 4 BNJ	34.0 %	25.0%	41.0%
SCENARIO 4 IBN	0.339	0.25	0.410

8 Conclusion

We proposed in this paper a probabilistic approach using Bayesian networks in order to deal with the problem of change impact analysis. We used a model proposed by Abdi et al. [06]. This model is a Bayesian network whose the entry nodes are design and implementation metrics. Then, we compared the results of inference under BNJ vs IBN to validate our developed tool for Bayesian Networks. Through the present study and our future work, we try to exploit the advantages of BNs in the context of change impact analysis in OO systems. We seek to provide further explanation of the factors responsible for the actual impact of the change and its evolution. Finally, we believe that the use of automatic learning capabilities offered by BNs allow us to have a short-term perspective, a better prediction accuracy and so more convincing results.

References

1. Wuillemin, P., Leray, P., Pourret, O., Becker, A.: Réseaux bayésiens, Eyrolles edn. (2004)
2. Lee (LiLi), M.L.: Change Impact Analysis for Object-Oriented Software. PhD thesis. George Mason University, Virginia, USA (1998)
3. Lee, M.L., Offutt, A.J.: Algorithmic Analysis of the Impact of Changes to Object-Oriented Software. In: ICSM 1996, pp. 171–184 (1996)
4. Kung, D., Gao, J., Hsia, P., Wen, F., Toyoshima, Y., Chen, C.: Change Impact Identification in Object Oriented Software Maintenance. In: ICSM 1994, Victoria, B.C., Canada, pp. 202–211 (September 1994)
5. Tang, A., Nicholson, A., Jin, Y., Han, J.: Using bayesian belief networks for change impact analysis in architecture design. J. Syst. Softw. 80, 127–148 (2007)
6. Abdi, M.K., Lounis, H., Sahraoui, H.A.: Predicting change impact in object-oriented applications with bayesian networks. In: COMPSAC (1), pp. 234–239 (2009)
7. Mirarab, S., Hassouna, A., Tahvildari, L.: Using bayesian belief networks to predict change propagation in software systems. In: Proceedings of the 15th IEEE International Conference on Program Comprehension, pp. 177–188 (2007)

8. Zhou, Y., Würsch, M., Giger, E., Gall, H.C., Lü, J.: A bayesian network based approach for change coupling prediction. In: Proceedings of the 2008 15th Working Conference on Reverse Engineering, pp. 27–36. IEEE Computer Society, Washington, DC (2008)
9. Alikacem, E.H., Snoussi, H.: BOAP 1.1.0. Manuel d'utilisation. In: CRIM (2002)
10. Local computations with probabilities on graphical structures and their application to expert Systems. Journal of the Royal Statistical Society B 50(2), 157–224 (1988)
11. Bayesian updating in causal probabilistic networks by local computations. Computational Statistics Quaterly 4, 269–282 (1990)
12. Draper, D.L.: Clustering Without (Thinking About) Triangulation. In: Conference on Uncertainty in Artificial Intelligence (UAI), University of Washington Seattle (1995)
13. Zoghlami, A.: Approche probabiliste pour l'analyse de l'impact des changements dans les programmes orientés objet, Master of Science (MSc), University of Montreal (2011)

Author Index

Printed in the United States
By Bookmasters

Printed in the United States
By Bookmasters